Vice President of Editorial — Melissa Wagner
Director of Acquisitions — Mikaela Felty
Director of Book Development — Jocelynn Hessler
Production Manager — Mike Brehm
Cover Design — Jose Rosado

Published in the United States of America by
IGI Global Scientific Publishing
701 East Chocolate Avenue
Hershey, PA, 17033, USA
Tel: 717-533-8845 | Fax: 717-533-7115
Website: https://www.igi-global.com E-mail: cust@igi-global.com

Library of Congress Cataloging-in-Publication Data

Names: Vardopoulos, Ioannis editor | Passas, Ioannis editor
Title: Futureproofing sustainability amidst climate change polycrisis / edited by Ioannis Vardopoulos, Ioannis Passas.
Description: Hershey, PA : IGI Global Scientific Publishing, [2026] | Includes bibliographical references and index. | Summary: "This book seeks to explore and present innovative approaches to transforming urban spaces while also analyzing the governance and financial mechanisms that underpin sustainable development"-- Provided by publisher.
Identifiers: LCCN 2025008045 (print) | LCCN 2025008046 (ebook) | ISBN 9798337323275 hardcover | ISBN 9798337323282 paperback | ISBN 9798337323299 ebook
Subjects: LCSH: Sustainable development--Case studies
Classification: LCC HC79.E5 F89 2026 (print) | LCC HC79.E5 (ebook)
LC record available at https://lccn.loc.gov/2025008045
LC ebook record available at https://lccn.loc.gov/2025008046

British Cataloguing in Publication Data
A Cataloguing in Publication record for this book is available from the British Library.

Futureproofed Sustainability Amidst Climate Change Polycrisis:

Advancing SDGs and ESG for the Cities of Tomorrow

Ioannis Vardopoulos
https://orcid.org/0000-0002-7593-4188
Open University of Cyprus, Cyprus

Ioannis Passas
https://orcid.org/0000-0002-1261-2878
Hellenic Mediterranean University, Greece

IGI Global
Scientific Publishing
Publishing Tomorrow's Research Today

Table of Contents

Detailed Table of Contents

Smart cities utilize IoT, AI, and data analytics to enhance urban living through improved resource management, public services, and quality of life. However, challenges like device interoperability, high costs, and vast data management hinder progress. Socio-economic disparities pose issues, as underserved communities often lack access to the necessary technology, increasing inequality. Governance challenges, including political resistance, regulatory inconsistencies, and insufficient frameworks, further slow smart city projects. Privacy and security concerns arise due to extensive data collection, raising risks of cyberattacks and unauthorized access. Environmental sustainability is also questioned, as connected systems require significant energy and produce electronic waste. Examples like South Korea's Songdo and India's Smart Cities Mission illustrate both opportunities and obstacles. To ensure sustainability, equity, and security, addressing technological, social, and governance challenges is essential for realizing the full potential of smart cities.

The rapidly increasing and continuous progress of innovative technologies, such as artificial intelligence (AI), Blockchain, and Fintech (Financial Technologies), which are examined in this chapter, have changed how businesses and governments manage compliance with Environmental, Social, and Governance (ESG) factors. Compliance with ESG criteria is crucial for all organizations, prioritizing operating sustainably and creating added value for all stakeholders. While businesses prioritize ESG compliance, the need to improve corporate governance has become imperative, and organizations worldwide are struggling to respond to challenges such as climate change, social inequalities, and corruption. To effectively comply with ESG,

environment, social and governance (ESG) sustainability of green materials in circular economies during 2015-2023 utilizing bibliometric investigation. This review adopted bibliometric analysis method in direction to illustrate the latest things as well as the new areas of interest in the domain of ESG sustainability of green materials domain in circular economies. In this regard, VOSviewer bibliometric tool are utilized to analyze the sustainability of green materials in circular economies. A sum of 306 research studies were recognized to identify and map the research structure of ESG sustainability of green materials in circular economies during 2015-2023.

As urbanization accelerates and environmental concerns grow, green buildings have emerged a crucial component of global sustainability efforts. This chapter explores green architecture's environmental, economic, and social dimensions, emphasizing its role in reducing carbon emissions, enhancing energy efficiency, and promoting occupant well-being. A bibliometric analysis was conducted using VOSviewer and Biblioshiny to map key research trends, co-citation networks, and thematic developments in green building studies from 2001 to 2025. Findings reveal a growing emphasis on ESG frameworks, smart building technologies, and circular economy principles. The chapter also highlights the interdisciplinary nature of green architecture, intersecting with urban governance, renewable energy, and sustainable materials. Ultimately, the study underscores the need for collaborative policy interventions and technological innovations to accelerate the transition towards sustainable built environments.

This chapter explores the implementation of Blue-Green Infrastructure (BGI) in Mediterranean cities through a comparative analysis of Madrid Río (Madrid, Spain) and the Appia Antica Regional Park (Rome, Italy), applying key insights to the Antonis Tritsis Metropolitan Park (ATMP) in Athens, Greece. Using a multidisciplinary approach – combining literature review, spatial analysis, stakeholder interviews and policy evaluation – it identifies major institutional, environmental and financial

of GIS and AI-driven climate risk assessments to enhance urban planning. Major barriers are analyzed with policy recommendations for participatory governance. This research provides a multidimensional framework that harmonizes global ESG standards with local socio-environmental contexts, offering actionable strategies for policymakers, urban planners, and financial stakeholders.

Despite the global emphasis on Environmental, Social, and Governance (ESG) principles, their adoption in Tunisia remains limited. Investor awareness is low, and corporate ESG disclosure is almost nonexistent. While regulatory measures encourage sustainable investing, ESG integration is still in its early stages. Emerging technologies like AI, blockchain, and big data offer potential solutions by enhancing transparency, sustainability assessments, and ESG reporting. However, little is known about how Tunisian investors perceive these technologies in ESG integration and investment decisions. This study examines investor perspectives within the Corporate Digital Responsibility (CDR) framework. Findings from investor interviews indicate optimism about technology's role in ESG but concerns over data accuracy, regulation, and technological expertise. Investors also stress the need for ethical technology commitments from companies.

The subject of this paper is the contribution of the rural landscape to the sustainable development of the island area. The research objective of this study is the protection, improvement and promotion of the rural island area with appropriate proposals to enhance environmental, economic and social resilience. The methodology chosen includes primary research of autopsy and free interviews and secondary research of literature study. International, European and national guidelines and the corresponding international and national institutional framework for landscape protection and management with a focus on rural areas are analyzed.

Heritage sites are facing growing threats every day due to the ever-increasing, abrupt changes brought on by climate change, like extreme weather conditions. It is important to take urgent measures to protect these valuable sites from the damage caused by these environmental shifts. This discussion explores how climatic solutions can be integrated into heritage conservation by different measures. Important parameters related to climate change and heritage conservation have been identified as a basis for understanding their impacts. It also gives significance to risk assessment and prioritization as brought about by multi-hazard frameworks through systematic approaches to the identification and mitigation of cultural assets that are most vulnerable. Community involvement in the construction of means of adaptation is important because it suggests that conservation initiatives are sustainable and culturally relevant. These results open pathways to future research to focus on the development of adaptive strategies and further refinement of technological applications.

Preface

This book, Futureproofed Sustainability Amidst Climate Change Polycrisis: Advancing SDGs and ESG for the Cities of Tomorrow, emerges at a pivotal historical juncture. The convergence of climate instability, systemic inequality, technological acceleration, and urban overexposure to risk defines a new and complex reality. Cities—both as physical environments and as socio-political constructs—are simultaneously the sites of greatest vulnerability and the most promising laboratories for transformative change. As such, the built environment is no longer a passive backdrop to global challenges; it is a frontline domain where the future of sustainability, equity, and resilience will be negotiated.

Equally pressing is the need to examine and strengthen the systemic frameworks that govern the behavior of institutions, corporations, and public authorities in this time of crisis. ESG (Environmental, Social, and Governance) standards, sustainable finance mechanisms, and emerging regulatory norms are not merely adjuncts to sustainability—they are essential levers for implementation. Yet these mechanisms remain fraught with tensions: between profit and accountability, global frameworks and local contexts, technological capacity and ethical responsibility.

This volume proposes a response that is both integrative and grounded. It seeks to bridge the material transformations of cities—manifest in architecture, infrastructure, and ecological design—with the institutional architectures that support them—ranging from green finance and auditing to policy ethics and participatory governance. The chapters herein reflect a multi-disciplinary and multi-scalar perspective, embracing complexity without collapsing it into simplified prescriptions. They are organized thematically across four parts: foundational frameworks, financial mechanisms, infrastructural innovations, and localized applications.

The intended audience of this book is diverse. For scholars in sustainability science, urban studies, environmental economics, public administration, and finance, the book offers conceptual frameworks and empirical insights. For practitioners—municipal officials, ESG consultants, architects, digital engineers, and financial auditors—it provides applied tools, case studies, and critiques of implementation.

For students and policymakers, the volume serves as both a primer on urban sustainability in the Anthropocene and a platform for envisioning ethically grounded, data-driven solutions.

Each chapter offers a distinct contribution to the collective inquiry:

1. **The Role of Digital Technologies, Governance, and Sustainability in Unlocking the Smart Cities Challenges** initiates the conversation by examining the intersection of digital infrastructures, participatory governance, and urban sustainability. It identifies the enabling conditions necessary for smart cities to serve as platforms for resilience rather than vectors of inequality.
2. **Leveraging Artificial Intelligence, Blockchain, and FinTech to Enhance ESG Compliance and Urban Governance** focuses on the deployment of emerging technologies in automating ESG reporting and enhancing regulatory compliance. It illustrates how AI and blockchain are being integrated into urban governance systems to foster accountability.
3. **Decoding Greenwashing - Consumer Skepticism and Trust in Decathlon's Sustainability Claims** deepens the discussion on corporate sustainability narratives through a case study on consumer perceptions in the retail sector. It underscores the critical role of transparency and corporate accountability in shaping the credibility of ESG commitments.
4. **Sustainability of Green Materials in Circular Economies: A Bibliometric Analysis on ESG** adopts a bibliometric methodology to map the scientific discourse on green materials and ESG. It provides insights into thematic concentrations, research trends, and policy linkages in the field of sustainable materials management.
5. **Greener Cities, Brighter Future: The Critical Role of Green Buildings** examines the environmental and economic performance of green building practices. It situates architectural innovation within broader urban climate strategies, identifying buildings as essential components of sustainable urban transitions.
6. **Implementation Challenges for Blue-Green Infrastructure to Enhance Resilience of Mediterranean Cities** focuses on the barriers to nature-based solutions in Mediterranean urban settings. It addresses institutional fragmentation, financial limitations, and socio-cultural dynamics that constrain large-scale ecological infrastructure deployment.
7. **From Trash to Treasure: Innovations in Waste Management towards Sustainable Energy Solutions** examines the intersection of urban waste and energy through the lens of circular economy. It explores how technological innovation and municipal reform can transform waste into a source of energy resilience.

8. **A Structured Approach to Zero-Emission Urban Regeneration: Integrating Planning, Innovation and Governance for Scalable Urban Transformation** offers a framework for reconfiguring urban systems around decarbonization. It weaves together governance, participatory planning, and technological foresight to deliver zero-emission urban futures.
9. **Climate Crisis and Tourism Industry: The Roles of Coastal and Shipping Quality Service** interrogates the climate vulnerabilities of the tourism sector, with a specific focus on maritime transport and coastal infrastructure. It identifies sustainability gaps and proposes strategies for resilience in tourism-dependent regions.
10. **Sustainable Urban Development in Northeast India: Integrating Traditional Ecological Knowledge with Modern Urban Resilience Frameworks Amidst Climate** bridges indigenous ecological knowledge and contemporary planning methodologies. It offers a contextual model of urban resilience grounded in cultural heritage and socio-ecological sensitivity.
11. **Digital Sustainability and Corporate Digital Responsibility in ESG Adoption. Investor Perceptions in Tunisia** investigates the digital ethics of ESG in emerging markets. The chapter captures how investors perceive corporate digital responsibility, especially in the context of ESG disclosures and compliance in Tunisia.
12. **The Insular Rural Landscape as a Factor of Sustainable Development** challenges the dominant urban bias in sustainability discourse by highlighting the development potential of island and rural areas. It explores land-use strategies that balance conservation with socio-economic growth.
13. **Building Heritage Resilience in the Age of Climate Change: A Discussion on Cultural Preservation and the Future of Conservation** reframes urban resilience to include cultural continuity and heritage protection. It argues that historical preservation must be integrated into climate adaptation and urban policy frameworks.

Together, these chapters construct a comprehensive and multidimensional narrative of sustainability in the urban age. They advance both theoretical and practical insights, offering interdisciplinary lenses through which to understand and act upon the climate polycrisis. In situating the built environment within broader financial, regulatory, and ethical systems, this volume offers a novel and necessary contribution to scholarship and practice. It invites urbanists, scientists, policymakers, and civic actors to adopt integrative, anticipatory, and accountable approaches to shaping cities that can withstand the complexities of the 21st century.

We are indebted to the authors whose intellectual generosity, diverse perspectives, and rigorous scholarship have made this volume possible. We thank the reviewers for their critical insights and thoughtful engagement. As editors, we offer this book as both a reference and a provocation—a resource for those committed to shaping resilient cities not only through technology and planning, but also through justice, care, and imagination.

Ioannis Vardopoulos
Open University of Cyprus, Cyprus

Ioannis Passas
Hellenic Mediterranean University, Greece

Chapter 1

The Role of Digital Technologies, Governance, and Sustainability in Unlocking the Smart Cities Challenges

Kuldipsinh Mori
https://orcid.org/0009-0008-3093-7095
Gujarat University, India

Kiranbhai R. Dodiya
https://orcid.org/0009-0001-9409-7303
Gujarat University, India

ABSTRACT

Smart cities utilize IoT, AI, and data analytics to enhance urban living through improved resource management, public services, and quality of life. However, challenges like device interoperability, high costs, and vast data management hinder progress. Socio-economic disparities pose issues, as underserved communities often lack access to the necessary technology, increasing inequality. Governance challenges, including political resistance, regulatory inconsistencies, and insufficient frameworks, further slow smart city projects. Privacy and security concerns arise due to extensive data collection, raising risks of cyberattacks and unauthorized access. Environmental sustainability is also questioned, as connected systems require significant energy and produce electronic waste. Examples like South Korea's Songdo and India's Smart

DOI: 10.4018/979-8-3373-2327-5.ch001

Cities Mission illustrate both opportunities and obstacles. To ensure sustainability, equity, and security, addressing technological, social, and governance challenges is essential for realizing the full potential of smart cities.

INTRODUCTION

The imaginative and prescient of smart cities revolves around growing advanced city environments wherein digital technologies, governance, and sustainability converge to raise high-quality lifestyles, enhance financial boom, and deal with environmentally demanding situations. By leveraging technology such as the Internet of Things (IoT), artificial intelligence (AI), and big facts analytics, smart towns promise a future of optimized aid management and streamlined public offerings. However, despite their capability, enforcing clever city processes faces vast challenges. These include technological limitations, including interoperability and information privacy; governance boundaries regarding coverage integration and collaborative frameworks; and socio-financial elements, mainly the digital divide and aid disparities among regions. Collectively, those demanding situations create a complex landscape that cities should navigate to gain smart city targets effectively (Paola and Rosenthal-Sabroux 2014).

Research analyzing cities advanced without the integral additives of virtual technologies, governance frameworks, and sustainability practices presents a perception of these demanding situations (Ghisellini, Cialani, and Ulgiati 2016). For instance, many urban regions perform on legacy infrastructures that need more technological foundations for smart metropolis applications, resulting in inefficiencies in carrier shipping and aid usage. Similarly, cities with underdeveloped governance systems frequently need more transparent regulations and restrained public-private collaboration, which restricts the execution of sustainable and inclusive city initiatives. Furthermore, without sustainable practices, cities can also revel in increased environmental degradation and social inequalities, as traditional infrastructure struggles to guide a developing population's needs. These observations highly (Smart Cities: Big Data, Civic Hackers, and the Quest for a New Utopia by Anthony M. Townsend | SMART CITIES: Big Data, Civic Hackers, and the Quest for a New Utopia n.d.).

This book chapter addresses those important additives by inspecting the technological, governance, and sustainability-demanding situations and solutions necessary for smart city development (Kitchin 2014). The first segment delves into technological constraints, specializing in interoperability troubles, legacy infrastructure integration, and records privacy. The second segment explores governance-demanding situations, consisting of coverage consistency, citizen engagement, and the position of public-private partnerships in fostering an adaptable governance model (Cui et al. 2018).

Finally, the chapter analyzes sustainability necessities, addressing environmental effects, useful resource management, and socio-economic inclusivity. Through this shape, the bankruptcy gives a roadmap for knowledge and implementing the middle pillars—digital technologies, governance, and sustainability—which can be critical for developing resilient and inclusive clever cities. Smart cities are designed as cutting-edge urban ecosystems. Where generation, management, and sustainability meet to beautify the notable of existence. Improving operations, promoting a financial boom, and resolving environmental problems are essential. It holds promise for a future of better aid management and public services. But this imaginative and prescient comes with complex, annoying conditions that could sluggish or derail progress. One of the most critical obstacles lies in an era: cities regularly battle with things like interoperability. The excessive initial value of the infrastructure and the sheer volume of records generated are a few different undertakings. Beyond technology and socio-financial inequality, the virtual divide between wealthy and poor populations may exacerbate inequality. This makes it more difficult for marginalized agencies to benefit from smart cities. Governance troubles also play a vital role because the cities should navigate a maze of political conflicts: inconsistent rules and the complexity of collaboration among the public and private sectors. Smart Citi's extended quantity of records increases privacy and security issues (Cagáňová, Cehlár, and Horňáková 2024). Ensuring the safety and integrity of these records while protecting residents' privacy is a first-rate obstacle. Ultimately, environmental sustainability is a supply of competition. The purpose of a smart city is to reduce environmental impact while maximizing the strength of use. Rapid technological progress and the accumulation of e-waste complicate these goals. The promise of smart cities is undeniable(Addas, 2023).

Table 1. An overview of the introduction

Challenge	Description	Proposed Solution
Interoperability Challenges	Difficulty in integrating diverse systems and standards across devices and platforms	Adoption of open standards and protocols
Legacy Infrastructure	High costs and complexity in upgrading outdated city infrastructure	Modular, phased upgrades
Data Management and Privacy	Large data volumes and privacy concerns require advanced analytics and secure storage.	Implementation of AI-based analytics and GDPR-compliant privacy policies
High Initial Costs	Significant upfront investments for building smart infrastructure	Public-Private Partnerships (PPPs) and scalable solutions
Cybersecurity Concerns	Increased risk of cyberattacks on interconnected systems	Robust cybersecurity framework and regular audits

1. TECHNOLOGY LIMITATIONS

1.1 Literature Survey

A) Interoperability Challenges

Interoperability the ability of diverse systems to work together is a key challenge. Smart cities rely on various devices and platforms that often use different communication standards, making integration difficult. Studies like Zanella et al. (2014) highlight the problem of incompatible protocols, while Neirotti et al. (2014) discuss the lack of universal standards, which leads to data silos and inefficient systems. Without seamless data exchange, cities cannot fully leverage smart technologies.

B) Legacy Infrastructure Integration

Many cities rely on outdated infrastructure that needs to be designed to interact with modern technologies. Komninos (2013) points out that integrating these legacy systems with new smart technologies is resource-intensive and costly. A gradual, modular approach to upgrading these infrastructures may be a more feasible solution.

C) Data Management and Privacy

The vast amounts of data generated by smart city technologies create storage, processing, and security challenges. Davenport and Harris (2017) emphasize the need for advanced analytics to manage big data, while Binns (2018) raises concerns about data privacy and security. Ensuring that data is used responsibly while adhering to regulations like GDPR is crucial for the success of smart cities.

1.2. Solutions to Technological Barriers

A) Open Standards

To address interoperability, adopting open standards is key. Open standards enable systems from different vendors to work together, reducing fragmentation and integration costs, as Zanella et al. (2014) suggested.

B) Modular Upgrades

Rather than overhauling entire legacy systems, cities can focus on modular upgrades that allow for gradual integration with modern technologies, as Komninos (2013) advocates.

Technological infrastructure is the backbone of innovative city initiatives. Still, with IoT (Internet of Things), AI, large-scale records, and cloud computing Situations, many demands arise in the application of today's technology.

Technological infrastructure is the cornerstone of any brilliant city initiative (Vardopoulos et al., 2024). It allows for the integration of various systems and services. However, improving city life using contemporary technologies such as the Internet of Things (IoT), artificial intelligence (AI), big data analytics, and cloud computing poses several significant challenges. These technological limitations can hinder the development and scalability of intelligent cities. Below, we dive into the main technology barriers: the latest data, examples, and insights from reputable sources back them (Syed et al. 2021).

1.3 Collaboration Problems

The main challenge in developing smart cities is achieving interoperability between systems, software, and devices. Interoperability refers to the capabilities of various systems and organizations from different vendors. To work together smoothly, exchange information, and use it efficiently.

Figure 1. Interoperability Challenges in Smart Cities

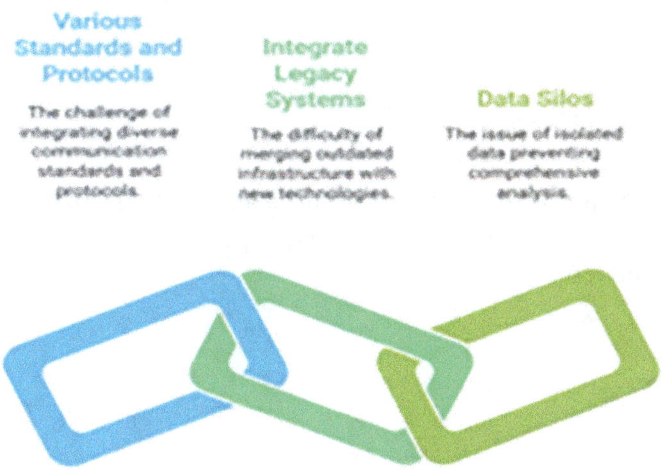

A) **Various standards and protocols:** Smart cities have many devices and systems, each of which may follow different communication standards and protocols. This diversity can lead to compatibility problems, making integrating systems such as traffic management, Energy grid, and public safety difficult.
B) **Integrate legacy systems:** Many cities rely on infrastructure that interacts with legacy systems rather than modern, innovative technology. Re-engineering these systems to work together often requires significant time and resources.
C) **Data silos:** If there is no standard. Data is often stuck within silos. This limits the ability to conduct comprehensive analysis and gain actionable insights.

Consider an intelligent city that aims to integrate traffic management systems with public transport and emergency services. If each system uses different data formats and communication protocols, coordinating real-time responses during peak hours or emergencies is challenging and may affect efficiency and safety.

A study by McKinsey & Company (2022) revealed that interoperability issues can increase the overall cost of a smart city project by up to 30% due to custom integration requirements and development time frames. Use for a long time.(Antonios, Konstantinos, and Christos 2023).

Table 2. Common Interoperability Standards in Smart Cities

Standard/Protocol	Application Area	Description
MQTT	IoT Device Communication	Lightweight messaging protocol for small sensors and mobile devices
HTTP/HTTPS	Web Services	Standard protocols for web-based applications
OPC UA	Industrial Automation	Platform-independent communication standard for industrial automation

1.4 High Initial Costs

Smart cities require significant upfront investment to build a robust infrastructure. High initial costs can be a major deterrent, especially for developing cities with limited financial resources.

Figure 2. Challenges of High Initial Costs in Smart City Development

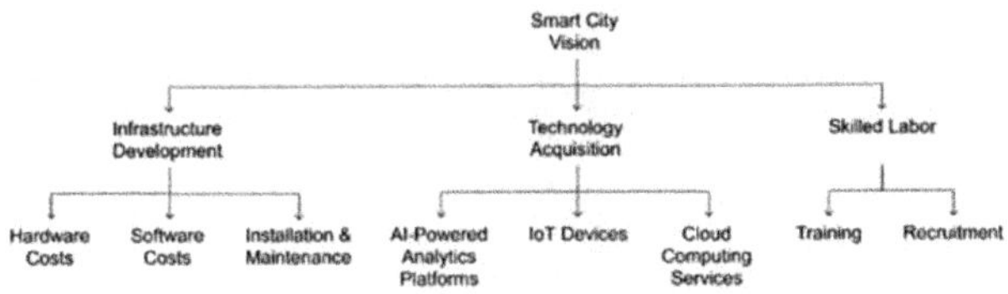

A) **Infrastructure Development:** Building a Smart Grid, Advanced transportation systems, and a comprehensive sensor network requires significant investment, including hardware, software, installation, and maintenance costs.
B) **Technology Acquisition:** Acquiring cutting-edge technology such as AI-powered analytics platforms, IoT devices, and cloud computing services has significant costs.
C) **Skilled labor:** Using and maintaining smart city technology requires skilled labor. This may require additional investment in training and recruitment.

Deploying a citywide IoT network for real-time utility monitoring involves purchasing sensors, installing communication networks, and integrating data management systems. The preliminary investment can quickly run into thousands of bucks for a city with a population of a million.

According to a record by the International Data Corporation (IDC, 2023), the global smart city market is projected to attain $820 billion by 2025, pushed through the want for advanced infrastructure. However, the high initial costs still need to be addressed, mainly for cities in rising economies.(Pandiyan et al. 2023).

Figure 3. Breakdown of Initial Costs for Smart City Infrastructure

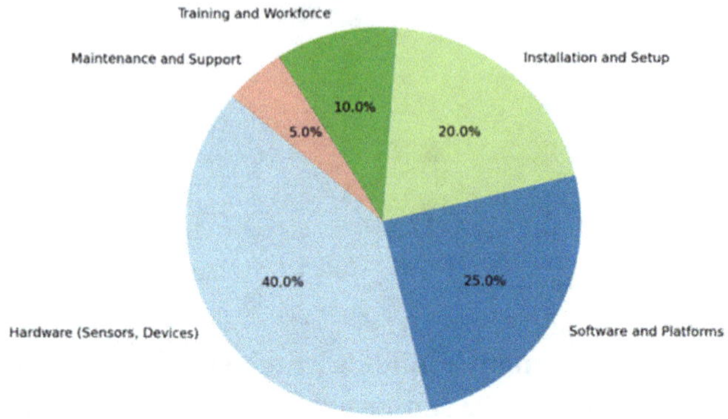

1.5 Data Handling Issues

Smart cities generate sizeable records via diverse connected devices and systems. Managing, storing, and analyzing this data poses widespread infrastructure, safety, and scalability challenges.

Figure 4. Data Handling Challenges in Smart Cities

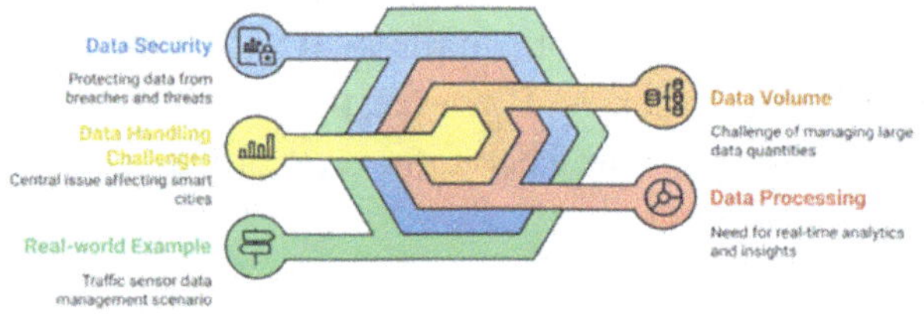

A) **Data Volume:** The sheer quantity of extensive data generated via IoT gadgets, sensors, and other intelligent technologies may be overwhelming. Efficient storage solutions and scalable infrastructure are critical to dealing with this data influx.
B) **Data Processing and Analysis:** Extracting meaningful insights from vast data requires superior analytics tools and understanding. Real-time information processing is vital for programs like traffic control and emergency response.

C) **Data Security and Privacy:** Protecting vast amounts of information gathered is paramount. Ensuring information privacy and securing against cyber threats are ongoing challenges that require robust security features.

In a smart city, giant traffic sensors gather real-time data on car motion, pedestrian drift, and congestion stages. Processing this information in real-time to modify alerts and inform drivers through bright signage requires handling the information and dealing with abilities to ensure well-timed and accurate responses.

As per IDC (2021), IoT devices had been projected to generate approximately 79. 4 zettabytes of data collection by 2025. By 2023, estimates propose this collection has multiplied, underscoring the escalating need for robust records infrastructure. Additionally, a document by Gartner (2023) suggests that information managing inefficiencies can lead to up to 20% losses in the capability of intelligent cities advantages because of delayed or inaccurate statistics processing.

Figure 5. Projected Data Growth from IoT Devices (2020-2025)

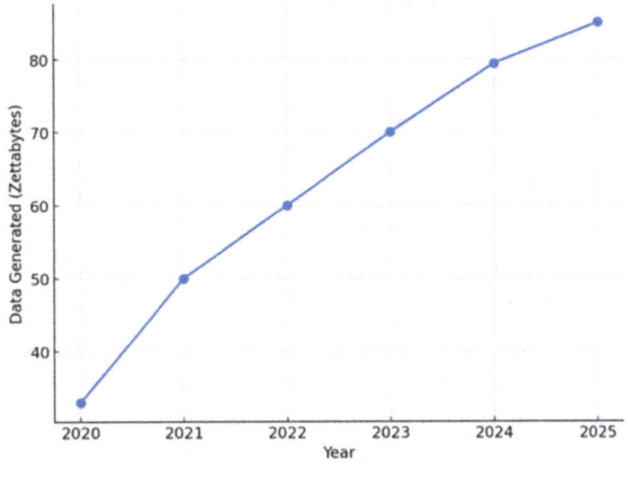

Recent improvements in part computing and AI-pushed data analytics are helping mitigate some information handling challenges by allowing quicker information processing and decreasing the reliance on centralized information facilities. However, those solutions require further refinement and extensive adoption to be effective. (Pandiyan et al. 2023).

1.6 Addressing Technological Limitations

Figure 6. Strategies for Smart Urban Development

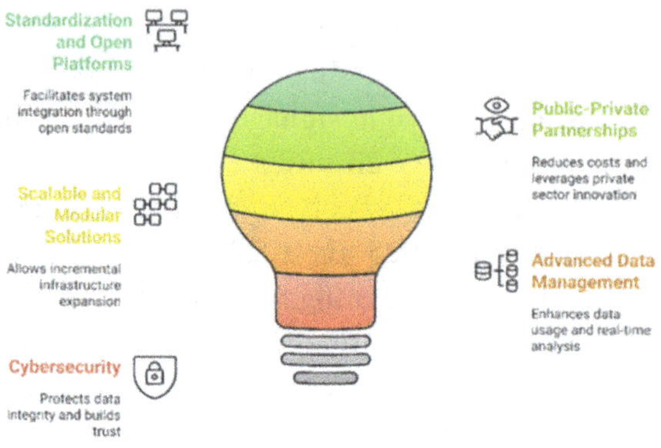

To triumph over these technological limitations, cities can undertake several techniques:

A) Standardization and Open Platforms:

Encouraging the adoption of open standards and interoperable systems can facilitate smoother integration of various systems.

Example: The use of the Open Geospatial Consortium (OGC) requirements for spatial records interoperability.

B) Public-Private Partnerships (PPPs):

Collaborating with good private sector partners can help reduce the monetary burden of preliminary investments and leverage information.

Partnerships among cities, governments, and companies to install innovative lighting fixtures systems.

C) Scalable and Modular Solutions:

Implementing scalable and modular technology lets cities extend their smart infrastructure incrementally, managing charges more successfully.

Example: Start with smart parking in a previously heavily trafficked area that grew up in many cities.

D) Advanced Data Management Technologies:

AI-based computing and machine learning can improve data usage capabilities and enable real-time analysis.

Using AI algorithms to predict high traffic patterns and optimize traffic signal time dynamically.

E) Focus on Cybersecurity:

Implementing a robust cybersecurity framework is essential to protecting data integrity and security.

Adopt the need for encryption and conduct regular security audits to protect intelligent cities' information(Embracing the Future: The Significance of Open Standards in Technology - ETHARTM I The XR Experience Platform n.d.).

2. SOCIO-ECONOMIC DISPARITIES

Implementing the new urban strategy reinforces the structure of smart city living (Vardopoulos, 2024). Without carefulness of efficiencies, these urban developments increase social and financial imbalance. Differences regardless of inclusivity Socio-monetary disparities correlate with the unequal distribution of assets, services, and opportunities across one-of-a-kind different societal organizations. In the context of smart cities, the disparities take the area in several ways, from unequal access to technology to the discrimination of inclined populations (Bibri et al. 2024).

Figure 7. Bridging Socio-Economic Gaps in Smart Cities

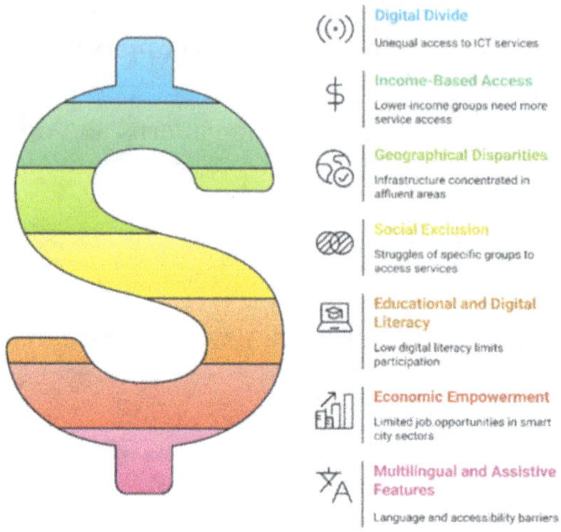

2.1 Digital Divide

The virtual divide refers to the gap between individuals who get modernized information and communication technology (ICT) and those who do not. Smart cities rely more on online platforms for transportation, healthcare, training, and public protection. However, getting entry to these technologies is not even across socio-economic groups, limiting the capability of disadvantaged populations to gain from intelligent metropolis innovations fully.(Bibri et al. 2024).

A) **Income-Based Access:** According to a 2020 Pew Research Center, 29% of low-earning adults in the U.S. lack broadband at home. Similarly, internet penetration and telephone ownership frequently decrease in developing international locations in rural and low-income regions compared to urban, prosperous areas (Salvati & Vardopoulos, 2024). This disparity prevents economically disadvantaged individuals from accessing critical, intelligent city services, including online education, telehealth, and online banking[7].

B) **Geographical Disparities:** In many cities, prosperous neighborhoods have higher access to high-paced internet and intelligent infrastructure, while low-income neighborhoods have substandard connectivity. This imbalance hinders inclusive participation in innovative smart city services, especially in telecommunication, remote work, and e-governance regions(Insights Into Cities in The Digital Era: A People-Centered Approach to Smart Cities n.d.).

In New York and San Francisco, high-earnings residents revel in seamless access to intelligent services like innovative traffic structures and high-pace net. Meanwhile, many low-earnings neighborhoods need more infrastructure, creating inequality and decreasing citizens' participation in digital activities.

A 2020 survey via Pew Research Center discovered that 44% of U.S. Adults in low-earning families depend on their smartphones for net access, compared to just 13% of adults from high-profit households. This "cellphone-only" gets less access to their potential to interact with more complicated responsibilities that require a stable broadband connection, completing activity programs, or attending virtual classes (RKMA Market Research Reports n.d.).

Table 4. Broadband Access by Income Group in the U.S. (2020)

Income Group	Percentage with Broadband Access
High-income (> $75k)	94%
Middle-income ($30k–$74k)	79%
Low-income (< $30k)	71%

2.2 Social Exclusion

Social exclusion occurs when some people are excluded and unable to participate entirely in societal activities, frequently because of socio-economic elements. In the context of smart cities, the focus on technology can unintentionally benefit wealthier or more privileged individuals, individuals who are skilled and knowledgeable about technology, while overlooking the needs of excluded populations, including low-income groups, elderly individuals, and people with disabilities.

A) **Tech-Driven Solutions:** Many intelligent city innovations, such as smart houses, connected transportation, and online services, cater to those already talented with generation and have the financial way to fund smart devices. This can isolate populations who are much less skilled and knowledgeable about technology or who can't manage to pay for superior technology, deepening the gap between people with assets and people without.

B) **Language and Accessibility Barriers:** For many excluded groups, language barriers or bodily disabilities may additionally exclude their right to access digital services. Intelligent cities often design structures and packages that require some level of digital literacy and physical ability, which could exclude those who do not meet these assumptions(Samant Raja and Group n.d.).

In cities like Singapore, where the smart city era is closely integrated into daily life, older adults have reported problems navigating digital devices used for accessing healthcare, transportation, and government services (Kempton et al., 2022). This digital divide between the more youthful, tech-savvy populace and older individuals can cause an experience of isolation and exclusion for the latter organization.

A report via the International Telecommunication Union (ITU) in 2023 indicated that in many developing countries, the most straightforward 30% of the population of elderly 65 and above use the net, as compared to 78% of individuals aged 18-34. This exclusion from digital participation influences their capacity to interact with innovative city technologies that offer good services.

Figure 8. Internet Usage by Age Group in Developing Countries (2023)

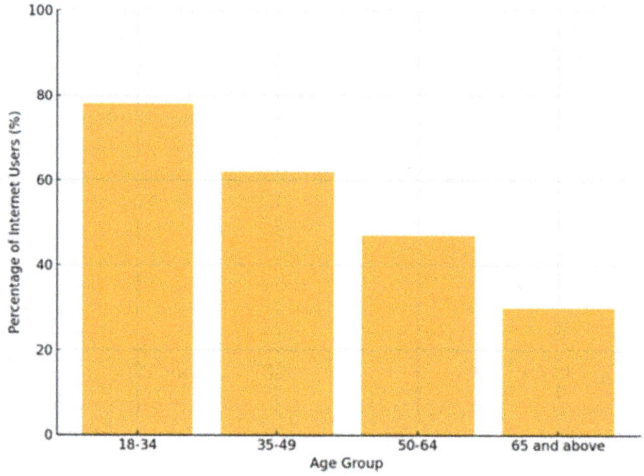

2.3 Address the Socio-Monetary Disparities

Addressing Socio-Economic Disparities Creating inclusive smart cities is crucial. That may be exacerbated with the aid of era-targeted improvement. Here are a few strategies to make sure that smart cities benefit all residents(Lockie et al. 2024).

2.3.1 Bridging the Digital Divide

A) Public Wi-Fi Networks:

Establishing free or low-cost public Wi-Fi networks in low-income and excluded neighborhoods can help expand internet availability.

Example: Barcelona's "Wifi in Public Spaces" initiative provides low internet entry to public areas, reaping the rewards for excluded groups lacking home broadband connections.

2.3.2 Inclusive Design and Accessibility

A) **User-Friendly Platforms:** Smart metropolis services must be designed with customer-satisfactory interfaces, ensuring they're reachable to humans of all ages, technical abilities, and bodily competencies.

B) **Multilingual and Assistive Features:** Offering services in multiple languages and integrating assistive technologies for people with disabilities, such as screen readers or voice commands, can improve access for excluded groups.

Example: The metropolis of London's accessibility suggestions for smart technology emphasize growing virtual services that can be utilized by elderly and disabled individuals, with capabilities like larger text sizes and voice navigation(Lockie et al. 2024).

2.3.3 Education and Digital Literacy Programs

A) **Community-Based Training:** Delivering digital learning programs within centers of communities, libraries, and colleges can empower citizens, especially those from excluded groups

B) **Targeted Programs for Seniors:** Technology training specifically for seniors can help bridge the virtual generational divide, for example, with the "Silver Infocomm Initiative." In Singapore, the event includes workshops to help seniors increase their digital skills and participate in the presentation of smart technology.

2.3.4 Economic Empowerment Through Technology

A) **Smart City Jobs and Training:** Smart city projects can create new ways of working in Statistical data analysis, infrastructure development, and cyber security. Learning about technology and the functionality skills available to low-income workers in these occupations can stimulate economic growth.

B) **Access to Online Services:** Ensure that excluded people have access to essential services. Tools like digital health care learning and job opportunities portals can empower them economically and socially(Digital skills: succeeding in a digital world: Week 2: 2 | OpenLearn - Open University n.d.).

The "LinkNYC" service in New York City offers free Wi-Fi and online services to less developed regions, encouraging people to enter the labor force, urban workers, and educational resources.

3. GOVERNANCE AND POLICY ISSUES

Table 5. Governance & Policy Issues

issue	Description	Example/Insight	Impact
Lack of Standardized Policies	Different cities have varied regulations, complicating multi-city initiatives.	Amsterdam and Barcelona follow different standards for data governance.	Only 20% of smart cities adhere to recognized standards.
Political Resistance	Political disagreements and concerns over funding can delay projects.	India's Smart Cities Mission faced resistance due to resource allocation debates.	35% of city leaders reported political resistance as a major barrier.
Building Capacity	Governments need to educate officials and build expertise in managing smart city technologies.	Singapore's Smart Nation initiative focuses on capacity building for smart governance.	N/A

The governance and initiatives that underpin smart cities are critical to their success. Smart cities rely on fact-based decision-making, Which seeks to create a sustainable governance structure, Clear guidelines, and good policy compliance. However, developing and establishing such a framework requires addressing various issues, from inconsistent policies to political resistance and difficulties in public-private partnerships. Below, we discover the important governance issues and coverage troubles faced through smart cities, supported by insights, examples, and records.

3.1 Lack of Standardized Policies

One of the most critical obstacles to innovative city governance is the need for standardized rules. Brilliant city work covers cities. Each city updates its rules and guidelines based on local priorities, prison design, and future resources. This problem creates conditions that demand the implementation of uniform best practices, especially when or if cities Collaborate on large-scale initiatives. Competent city generation operators operate by jurisdictions.

A) **Inconsistent Regulations:** Different cities have precise guidelines regarding records privacy, era deployment, and infrastructure standards. This lack of uniformity complicates the implementation of technologies, especially when dealing with multi-metropolis solutions, local transportation systems, or inter-city strength grids.
B) **Data Sharing and Security Standards:** A lack of standardized statistics-sharing frameworks throughout cities can restrict cooperation between municipalities. Data privacy legal guidelines can also differ extensively, making it challenging to establish secure facts exchange structures among cities that comply with numerous felony requirements.

In the European Union, cities like Amsterdam and Barcelona have superior clever Citi projects. However, due to the lack of a pan-European regulatory framework for the smart cities approach, they each observe specific standards for statistics governance, privacy, and security. This creates hurdles in sharing best practices and leveraging nearby answers throughout borders.

Figure 9. Adoption of Smart City Standards by Region (2022)

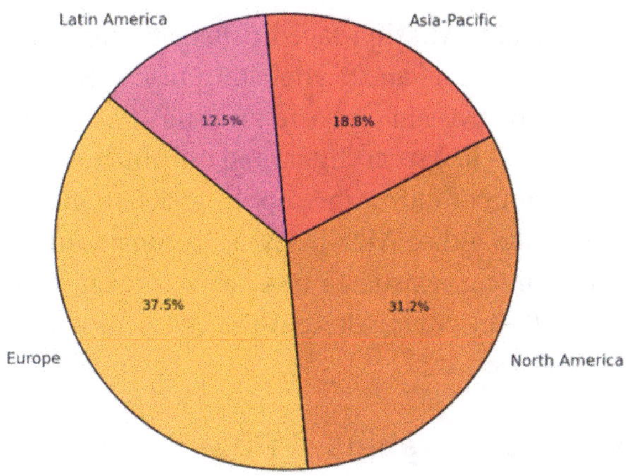

In 2022, a file from the International Organization for Standardization (ISO) highlighted that the most effective 20% of clever cities globally adhere to internationally identified requirements for records control, infrastructure, and provider transport. The absence of standardized policies makes it tough for technology vendors to provide scalable solutions across multiple cities.

3.2 Political Resistance

Political resistance can be a good barrier to successfully implementing smart city tasks. Political assistance is needed at diverse stages—local, local, and countrywide-to expand smart cities. However, political leaders have frequent disagreements concerning priorities, resource allocation, and the fee of smart city technology. Political ideologies and beliefs can also lead to pushback in opposition to positive tasks, delaying or halting progress.

A) **Financial Constraints:** Political leaders may also resist clever city projects because they worry about high expenses. Large-scale investments are required to construct vital infrastructure. Sometimes, politicians may view those prices as unjustifiable, especially when other urgent issues exist, such as housing or healthcare.
B) **Ideological Opposition:** Some political leaders may ideologically oppose the improved reliance on generation and data-pushed governance. Privacy problems and lack of distribution of smart city vendors may lead to skepticism and resistance.

In India, the Smart Cities Mission faced resistance from nearby politicians in numerous cities due to concerns about allocating funds. Some politicians argued that the assets could be better spent on basic offerings like water and sanitation instead of high-tech initiatives like smart lights and information centers. This political pushback slowed the rollout of clever metropolis tasks in favorable areas.

A 2021 survey with the aid of McKinsey & Company found that 35% of city leaders globally said political resistance is a key barrier to clever metropolis implementation, specifically in cities wherein the political panorama is noticeably fragmented. Public Pushback 15%

3.3 Public-Private Partnerships (PPPs)

Public-private partnerships (PPR) are often used in smart city projects to harness the knowledge and resources of private quarters. However, these partnerships can create situations that cause management stress, especially when the goals of private companies compete with public pastimes. Sole proprietorship often places a premium on financial prudence. This might prevail in situations where public desire is overshadowed by expediency. Resources for entrepreneurial dreams that undermine the social benefits of smart city initiatives(Public-Private Partnerships (PPPs): Definition, How They Work, and Examples n.d.).

A) **Profit vs. Public Interest:** Private companies involved in smart initiatives may prioritize profit over public interest. For instance, they might invest heavily in lucrative sectors like smart retail or services for the wealthy, potentially neglecting essential services for marginalized groups. It is necessary to strike a balance that ensures all sections of society benefit from smart city initiatives.
B) **Accountability Issues:** When private companies manage critical city services, such as water, electricity, or transportation, there is a threat that their obligations to the public may be compromised. Transparency and oversight mechanisms must be robust to ensure private location involvement does not undermine public attention.

In Rio de Janeiro, a smart city initiative involving private businesses focused on constructing high-tech infrastructure in wealthier neighborhoods, while low-income areas were left with minimal enhancements. This created a public belief that smart city technologies only benefit the prosperous, leading to dissatisfaction and protests.

A report using the World Economic Forum (WEF) 2023 found that forty percent of clever city initiatives funded through PPPs skilled conflicts among public dreams and personal area hobbies, especially in regions associated with less costly housing and public transportation.

Table 6. Key Issues in Public-Private Partnerships (2023 WEF Report)

Issue	Percentage of Projects Affected
Profit vs. Public Interest	45%
Accountability and Oversight	30%
Unequal Resource Allocation	25%

3.4 Addressing Governance and Policy Challenges

To triumph over governance and policy troubles in smart cities, it's essential to put into effect the following strategies:

A) Standardized Policies and Regulations:

Governments at regional and national levels need to collaborate to create standardized frameworks for improving smart metropolises, particularly in areas such as information governance, cybersecurity, and infrastructure requirements.

Example: The European Union's "Digital Europe Programme" targets to create a unified regulatory framework for smart city technologies throughout member states, ensuring consistency in implementation and data safety.

B) Political Alignment and Stakeholder Engagement:

Political leaders should actively engage in planning and decision-making techniques to build consensus and ensure alignment on smart cities' priorities.

Example: In Copenhagen, everyday consultations with political leaders and community stakeholders have helped ensure that smart metropolis initiatives align with local priorities and gain widespread support.

C) Balancing Public and Private Interests:

Clear guidelines and contracts must be installed in public-personal partnerships to ensure that the general public's wishes remain a priority. Governments must implement strict accountability measures to prevent other private companies from prioritizing profits over public welfare.

Example: In Vienna, smart city tasks funded through PPPs are concerned with rigorous public oversight to ensure the private corporation's role is aligned with the smart city's long-term social and environmental needs.

D) Building Capacity for Smart Governance:

City governments must develop the internal ability to manage smart metropolis initiatives correctly. This includes educating government officials, hiring specialists in data statistics governance and digital technology, and fostering a culture of innovation (Vardopoulos et al., 2023).

Example: Singapore's Smart Nation initiative focuses on capacity building in government. Carry out deployment services to civil servants in smart technology and data management (The United Indian | Government Sector | Strategies for Enhancing Digital Literacy in India n.d.)t.

4. PRIVACY AND SECURITY CONCERNS

Table 7. Privacy and Security Concerns

Privacy and Security Concerns	Description	Example
Data Privacy	Concerns over how personal data is collected, stored, and used in smart cities.	Toronto's Quayside Survey faced backlash over data usage concerns.
Cybersecurity Threats	Increased risk of cyberattacks targeting interconnected smart city infrastructure.	2020 Baltimore ransomware attack paralyzed public services.
Data Misuse	Risk of personal data being misused for profiling or commercial exploitation.	Concerns over personal data being sold without consent.
IoT Vulnerabilities	Cybercriminals exploiting IoT device weaknesses to gain access to critical infrastructure.	Ukraine's 2015 power grid cyberattack.

One of the most pressing problems in developing smart cities is ensuring the privacy and security of the large amounts of data generated by interconnected devices and digital structures. When cities are more connected through IoT devices, sensors, and digital infrastructure, the potential for privacy breaches and cybersecurity threats will also increase exponentially. Understanding these challenges is essential for building trust with innovative city technologies and protecting digital infrastructure.(Kaiser 2024).

4.1 Data Privacy

Smart cities generate large amounts of information sharing from various resources, such as cameras, smartphones, sensors, and other related things. Although this position is important for optimizing services such as handling traffic, public protection, and health services, it raises serious concerns about how this data is collected, stored, and used. Ensuring the security of citizens' records and respecting citizen's privacy rights is a complex and demanding challenge.(Rejeb et al. 2022).

A) **Data Collection:** Smart cities collect information on residents' movement, behavior, and interactions. This information can range from area-wise traffic patterns and power usage to additional personal information and location recordings from smartphones or wearable devices. The large number of records makes it difficult to track how and where public information is being used.

B) **Regulatory Compliance:** Legal frameworks, including the EU General Data Protection Regulation (GDPR), are designed to protect individuals by imposing strict data processing and consent requirements. However, many cities are struggling to comply. Follow these rules. Most must manage statistics from multiple sources and from who provides them.

C) **Data Misuse:** There is a risk that the data accumulated in smart cities can be used for other purposes, including monitoring, profiling, or commercial exploitation. Without the knowledge or consent of the resident, records accumulated for public service may be sold to third parties, which is causing great concern to the public.

Toronto's Quayside Survey, conducted through Sidewalk Labs (an Alphabet subsidiary), raised concerns about the number of personal records being collected and how this information is used intentionally for recognition or commercial purposes. After public backlash and personnel concerns, the project was rejected, which demonstrated the importance of raising personnel concerns early.(Ismagilova et al. 2020).

According to a 2022 McKinsey report, 60% of global city managers agree that people's records are a key concern for residents, and many are calling for strict data protection laws to regulate how personal data is handled.

Table 8. Key Privacy Concerns in Smart Cities (2022 McKinsey Report)

Privacy Concern	Percentage of Cities Reporting It
Data Collection without Consent	40%
Inadequate Data Protection	35%
Commercial Use of Personal Data	25%

4.2 Cybersecurity Threats

Smart cities' interconnected nature makes them the primary targets for cyberattacks. As more critical infrastructure becomes digital, cities face increasing threats from hackers aiming to disrupt important services or steal records. Successful cyber network attacks in smart cities can lead to dangerous results, from shutting down the network to shutting down public transport infrastructure.

A) **Ransomware and Malware Attacks:** Smart cities are vulnerable to ransomware attacks. In 2020, Baltimore faced a major ransomware attack that paralyzed the smart subway system, forcing the city to revert to traditional systems for several weeks. The attack affected services like water billing and assets tax bills, underscoring the urgent need for more robust cybersecurity measures.

B) **IoT Vulnerabilities:** The extensive deployment of IoT gadgets in smart cities introduces many factors of access for cybercriminals. Many devices, including digital cameras, traffic sensors, and linked home equipment, won't have sturdy security features, making them clean targets for hackers. Once an IoT device is compromised through cyber attacks, it could function as a gateway to access different elements of the smart city's server.

C) **Critical Infrastructure Attacks:** Hackers targeting key infrastructure, power supplies, water supplies, or public transportation pose an extreme threat to city protection. In a smart metropolis surroundings, such attacks can cause chaos and disrupt the daily lives of hundreds of thousands of people. For instance, in 2015, Ukraine's energy grid was compromised via a cyberattack, leading to tremendous outages(What Are the Cybersecurity Risks for Smart Cities? - Institute for Defense and Business n.d.).

In 2021, the Colonial Pipeline, a major gas delivery line inside the United States, was hit by a ransomware attack that compelled its operators to close the entire pipeline. Although now not a smart city-precise incident, this attack is a stark reminder of the vulnerabilities in critical infrastructure, which are increasingly interconnected in smart cities.

According to a report using the Cybersecurity and Infrastructure Security Agency (CISA), the wide variety of cyberattacks concentrated on smart cities infrastructure multiplied with the aid by 25% between 2019 and 2021, with transportation systems, power grids, and public utilities being the most common objectives.

Figure 10. Common Cybersecurity Threats to Smart Cities (2021 CISA Report)

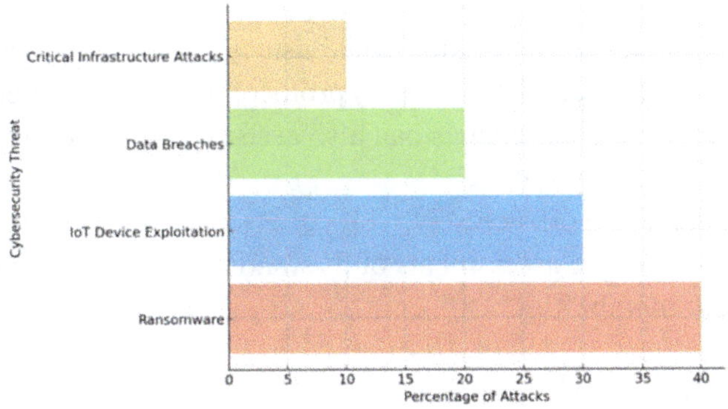

4.3 Addressing Privacy and Security Concerns

To shield privacy and shield against cybersecurity threats, smart cities must adopt a multi-faceted technique that mixes data governance, advanced cybersecurity measures, and continuous tracking. Here are some key methods:

A) Strengthening Data Governance and Privacy Protections:

Cities need to increase clean data or records governance frameworks that define how records are accumulated, saved, and used. Citizens need to know how their information is being used and be able to offer informed consent.

The city of Barcelona has implemented a "Data Commons" version, in which citizens manipulate their information and may determine how the city utilizes its information.

B) Enhancing Cybersecurity Infrastructure:

Cities must spend money on sturdy cybersecurity infrastructure to protect against high attacks. This includes deploying encryption, multi-factor authentication, and continuous hazard-tracking structures.

Singapore's Cybersecurity Strategy includes creating a Cybersecurity Agency tasked with tracking and defending crucial infrastructure, including transportation and electricity structures, from cyberattacks.

C) Collaborating with Technology Providers and Experts:

Cities should collaborate with high-tech businesses, cybersecurity experts, and regulatory bodies to ensure updated security protocols are accompanied. Establishing partnerships with private companies can also assist in implementing cutting-edge cybersecurity-related solutions.

In Helsinki, the city works closely with cybersecurity corporations to conduct regular audits and pressure assessments of its smart infrastructure to become aware of and clear the vulnerabilities.

D) Public Awareness and Education:

It is critical to educate the general public about information privacy rights and how to guard themselves from cyber threats. Cities can launch consciousness campaigns and provide resources to help citizens understand their digital system in a smart city environment.

London's "Cyber Aware" campaign offers residents practical recommendations on shielding their devices and data from cyber threats, helping to create a more stable and secure city environment(What Are the Cybersecurity Risks for Smart Cities? - Institute for Defense and Business n.d.).

5. ENVIRONMENTAL SUSTAINABILITY

Table 9. Environmental Sustainability

Challenge	Example
Energy Consumption	Data centers may consume 8% of global energy by 2030.
E-Waste	Asia produced 24.9 million metric tons of e-waste in 2019.
Toxic E-Waste Components	Harmful materials like lead and mercury pollute the environment
Short Lifespan of Devices	IoT and smartphones discarded without proper recycling.

Although smart cities aim to promote green infrastructure and sustainability by utilizing resources efficiently and reducing pollution, these cities face many challenges arising from environmental and technological factors. Surprisingly, several technologies developed to promote sustainability also contribute to environmental problems, such as increasing energy use and electronic waste.

5.1 Energy Consumption

Managing the data center, maintaining an IoT network, and powering linked gadgets in smart cities require high and distributed energy. As each town expands as an intelligent city, Energy demand for Technological infrastructure is proliferating. This ever-increasing electricity demand contradicts the sustainability dreams of decreasing carbon footprints and increasing electricity efficiency.

A) **Data Centers' Energy Demand:** Data centers are critical to smart cities, as they save and process vast amounts of data generated by IoT gadgets and other systems. However, these facilities consume vast quantities of energy. By 2030, data centers may consume as much as 8% of the world's energy supply, which offers a significant challenge for intelligent cities striving to balance energy efficiency with digital infrastructure development.

B) **Smart Devices' Energy Requirements:** The operation of smart devices, from street lights to sensors and connected transportation structures, adds to a city's total electricity intake. While those structures are regularly designed to optimize performance, their power demands can stress current power grids, especially in cities that depend heavily on non-renewable energy resources(Toli and Murtagh 2020a, 2020b).

According to a 2020 file in Nature, innovative city projects in places like Singapore and Dubai have significantly increased electricity use, especially in data facilities assisting city-extensive IoT infrastructure. The mission is to transition to greener strength sources while providing the developing technological needs of those cities. A 2020 Nature article highlights that energy consumption by information centers of data is expected to rise by about 7% per year, probably achieving 8% of world electricity usage by 2030, a major challenge for cities aiming for carbon neutrality.

Figure 11. Projected Energy Consumption of Data Centers (2020-2030)

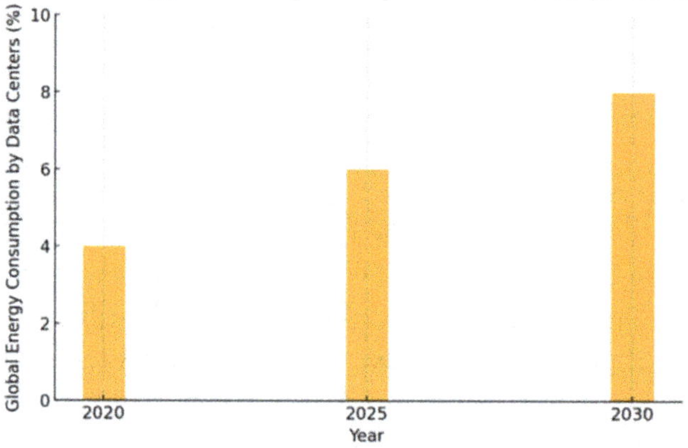

5.2 E-Waste

The rapid pace of technological advancements in smart cities means that devices and structures become outdated due to a widespread accumulation of digital waste (e-waste). Managing e-waste is a critical environmental challenge, as tons of this waste aren't always well recycled, and toxic substances in electronics can harm the environment if not treated well.

Figure 12. Global E-Waste Accumulation (2019)

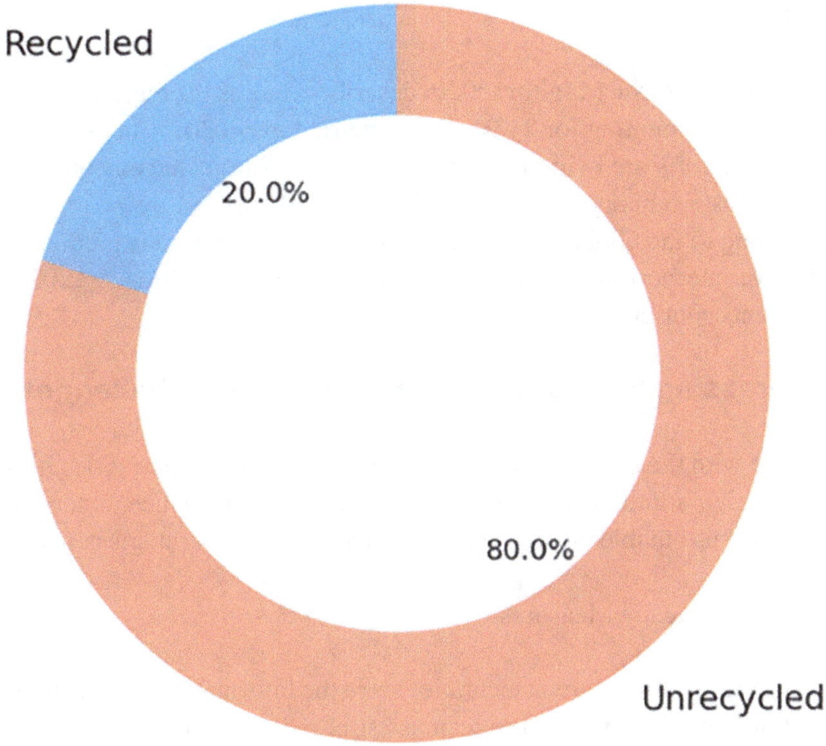

A) **Unrecycled E-Waste:** Global e-waste reached an amazing 53.6 million metric lots in 2019, but the handiest 20% of that waste changed into recycled nicely, according to the Global E-Waste Monitor (2020). The final 80% turned either incinerated, dumped in landfills, or improperly controlled, contributing to environmental degradation and animals' health dangers.

B) **Toxic Components:** Many digital devices include dangerous materials like lead, mercury, and cadmium, which could leach into the soil and water when improperly disposed of. This makes e-waste one of the most risky sorts of waste in terms of longer environmental impact.

C) **Short Lifespan of Devices:** The fast improvement cycles of IoT gadgets, smartphones, and other smart city devices shorten their lifespans, creating e-waste hassles. As new technologies are implemented, older devices become useless and are often discarded without proper recycling((PDF) Impact Factor: 5.2 IJAR 2020 n.d.).

The Global E-Waste Monitor (2020) describes Asia as the region that created the most e-waste, accounting for 24.9 million metric tons in 2019. In cities like Tokyo and Beijing, efforts are being made to improve e-waste management, but the high volume of waste generated creates significant challenges.

According to the Global E-Waste Monitor (2020), e-waste is predicted to grow by 2 million metric tons annually and reach nearly 75 million by 2030 if these developments continue.

5.3 Addressing Environmental Sustainability Challenges

To cope with the energy intake and e-waste problems, smart cities must undertake progressive solutions concentrating on energy performance, renewable power sources, and sustainable e-waste control practices. Some strategies encompass:

A) Energy Efficiency Measures:

Implementing power-green technologies in the information centers of data facilities and metropolis infrastructure can assist in reducing the general power intake of smart cities.

Google has pioneered using AI to optimize the electricity usage of its statistics data centers, decreasing cooling energy intake by 40%. Smart cities can undertake similar AI-pushed strategies to optimize energy supply.

B) Renewable Energy Integration:

Encouraging the use of renewable energy sources, such as solar, wind, and hydropower, to meet the electricity needs of smart city infrastructure can greatly decrease carbon emissions.

Example: In Reykjavik, Iceland, smart city goals are nearly completely powered by renewable energy from geothermal and hydropower assets, demonstrating how smart cities can align the tech era with sustainability.

C) Sustainable E-Waste Management:

Cities want to increase the number of complete e-waste recycling applications that ensure the secure disposal and recycling of outdated devices. Partnerships with tech businesses can also help create recycling applications for obsolete electronics.

A circular economic system for dealing with e-waste has been implemented in Sweden. Non-useable electronic devices are refurbished or dismantled for elements rather than being despatched to landfills.

D) **Circular Economy for Technology:** Embracing a circular economy in smart city technologies can amplify gadgets' lifecycles, reducing the overall quantity of e-waste. This includes refurbishing outdated devices, using sustainable materials, and ensuring the right recycling mechanisms are in place(Top 10: Smart cities driving sustainability around the world | Sustainability Magazine n.d.).

The city of Amsterdam has followed a circular economic system approach where parts from decommissioned digital devices are reused in new technology, minimizing waste.

While intelligent cities offer innovative solutions for environmental sustainability, they also face challenges associated with accelerated energy consumption and the growing e-waste problem. To achieve the dream of sustainable energy, Smart cities must incorporate renewable energy sources, improve electricity use, and manage electronic waste. Doing so puts them in a position to coordinate technology development with environmental problem management.

6. CASE STUDIES

Innovative city projects around the world are experiencing varying levels of implementation. Two excellent examples Songdo in South Korea, India's Smart Cities Mission, Hudson Yards in the USA, Barcelona Smart City in Spain, and Copenhagen, Denmark- illustrate the complexities, challenges, and opportunities these projects face. By evaluating those instances, we benefit from insights into the different approaches used by cities and the lessons that may utilize the shape of future smart cities of the world.

1. Songdo, South Korea

Songdo, one of the earliest intelligent city projects, was built from scratch on reclaimed land and designed to integrate superior digital technology into everyday city lifestyles. The city is built with IoT-enabled infrastructure, smart energy grids, automated waste disposal systems, and expansive green spaces geared toward developing a futuristic, green metropolis environment.

Figure 13. Songdo: A Model for Smart Cities

A) **Advanced Technology:** Songdo has been at the forefront of technological integration since its inception. The metropolis uses a traffic management system with sensors, Centralized data points, and intelligent building structures, which can serve as a benchmark for urban innovation.
B) **Sustainability:** Songdo focuses on green energy with energy-efficient houses, a strong public transport network, and efforts to reduce carbon dioxide emissions
C) **High Standard of Living:** The metropolis values the convenience of its residents, with green spaces, motorcycle lanes, and smart city planning that integrates technology into daily life.

Despite its innovations, the city must contend with challenges and threats

A) **Low Occupancy Rates:** Despite its existing infrastructure, Songdo has trouble attracting residents and companies. As a result, many apartments and corporate spaces are not utilized.
B) **Lack of Community:** Songdo's fast development is usually for commercial operations professionals. This results in a noticeable lack of a connected community. This makes it sound more like a company's area than a residence.
C) **High Costs:** The enormous financing required to build Songdo makes it unaffordable for the middle and lower middle class. Limited variety and stop prolonged growth

The case study provides valuable Key Takeaways, such as

A) **Community Building:** Songdo's key lesson is the importance of fostering a sense of connectedness along with technological advancements. Smart cities must focus on creating inclusive environments where residents can connect.
B) **Balance between technology and reasonableness:** Although advanced technology is preferred, builders must maintain reasonable prices to attract and retain large populations.

2. India's Smart Cities Mission

Table 11. India's Smart Cities Mission

Element	Details
Improved Infrastructure	Water management, sewage, public transport infrastructure.
Economic Growth	Stimulating growth, attracting investment, creating jobs.
Local Development	Tailored improvements based on specific regional needs.
Challenges	Inconsistent policy implementation, slow project timelines, digital divide.
Lessons Learned	Inclusive development, effective governance and coordination

India's Smart City Mission, Launched in 2015, aims to expand to 100 smart cities across India and improve urban infrastructure, good governance, and sustainability using technology. Each town has to deal with precise, demanding situations due to cultural diversity and geographical differences which deal with waste control, infrastructure development to traffic volume control.

Figure 14. India's Smart City Mission

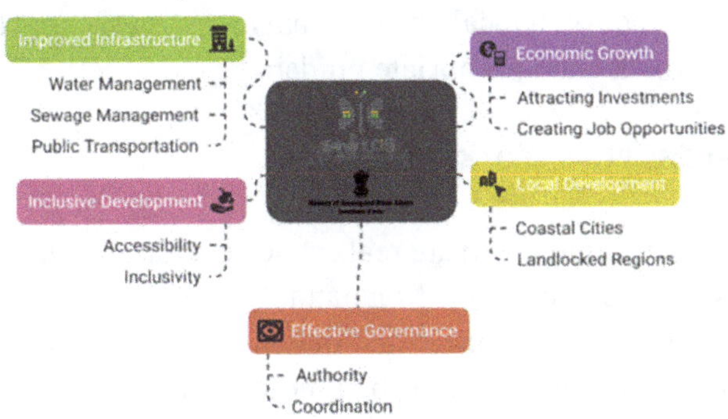

A) **Improved Infrastructure:** The challenge is to improve urban infrastructure, water management, sewage management, and infrastructure for public transportation
B) **Economic Growth:** The mission aims to stimulate economic growth, attract investment, and create jobs, especially in the technology and service sectors, by upgrading city infrastructure.
C) **Local Development:** Each city focuses on specific needs, such as water conservation in coastal areas or waste management in landlocked regions, promoting targeted urban improvements.

Despite its innovations, the city must contend with challenges and threats

A) **Inconsistent Policy Implementation:** Progress varies widely between smart cities due to inconsistent coordination among local and central government policymakers.
B) **Slow Project Timelines:** Many tasks lag behind schedule because of bureaucratic delays and investment problems, making it tough to fulfill within the timeline.
C) **Digital Divide:** The mission faces traumatic conditions in bridging the space between wealthy residents who can access smart city services without problems and poorer populations who may have issues due to a lack of resources or digital awareness to access all facilities.

The case study provides valuable Key Takeaways, such as

A) **Inclusive Development:** A key lesson from India's smart city mission is prioritizing integration and ensuring that smart technologies and services are accessible to all population segments.
B) **Effective Governance:** Establishing lines of authority, coordination, and governance framework between local and national levels is essential to ensure timely project management and appropriate guidance.

3. Hudson Yards, New York, USA

Hudson Yards is the biggest private real estate development in the United States of America's History, positioned on Manhattan's West Side. Completed in 2019, it encompasses residential, enterprise, and public regions and integrates advanced technology for energy power control, urban mobility, and sustainability.

A) **Smart Infrastructure:** The improvement capabilities of smart, effective constructing technologies, efficient energy systems, and actual-time data analytics to screen and manage useful resource utilization.

B) **Public Spaces:** The city has parks, cultural venues, and public artwork installations to enhance community engagement and accessibility.
C) **Economic Growth**: Hudson Yards pursued the goal of attracting new businesses, creating jobs, and generating economic activity, contributing to New York City's economy.

Despite its innovations, the city must contend with challenges and threats

A) **High Costs:** The project has faced scrutiny for its high prices, which prohibit affordability for lower-earning citizens, and for elevating issues about gentrification.
B) **Limited Inclusivity:** The attention to luxurious living has skewed the demographics, making it difficult to create diverse population groups.
C) **Privacy Concerns:** With a high statistical data collection system utilizing smart technologies, There are problems controlling and protecting confidential information.

The case study provides valuable Key Takeaways, such as

A) **Balancing Luxury and Accessibility:** Future smart city initiatives must balance high-yield development and integration to promote diverse communities of population.
B) **Data Privacy and Transparency:** Creating clean data control practices is essential to maintaining human privacy while using technology for urban development.

4. Barcelona Smart City, Spain

Barcelona is committed to a progressive approach to urban growth and innovation. It focuses on integrating technology to improve sustainability, productivity, and resident participation. The Smart City Initiative, introduced in 2009, concealed many elements of the city's way of life.

A) **Sustainability Initiatives:** The city uses smart lighting, energy-saving buildings, and complete public transportation, greatly reducing carbon dioxide emissions.
B) **Citizen Engagement:** Barcelona encourages public participation through digital systems that allow residents to express their opinions and participate in the city's proposals.

C)**Smart Infrastructure:** Integrating IoT Creation in Waste Control, dealing with traffic and energy distribution. It helps enhance the efficiency of the city's services.

Despite its innovations, the city must contend with challenges and threats

A) **Balancing innovation and heritage:** while maintaining the cultural and historical identity of the metropolis, putting it together with modern technology is challenging.
B) **Digital Divide:** Differences in access to technology between different classes of society can hinder the effectiveness of smart city accountability.
C) **Data Privacy Issues:** The collection and use of information raise questions about human privacy and safety, requiring a strong management framework.

The case study provides valuable Key Takeaways, such as

A) **Inclusivity is Key:** Successful smart city projects must prioritize integration to ensure all residents benefit from technological upgrades.
B) **Integration of Technology and Culture:** Urban innovation must be sensitive to the cultural context and combine modern solutions with historical presence(Case Study on Smart City Projects in India: An analysis of Nagpur, Allahabad and Dehradun n.d.).

5. Copenhagen, Denmark

Copenhagen is known to be a leader in smart city development. Aiming for carbon neutrality by 2025, the city is integrating smart solutions with concrete energy regulation and transport and planning for a smart city, making it a sustainable version.

A) **Sustainable Urban Solutions:** the city uses a smart grid, green and efficient infrastructure, and beautiful circular infrastructure that promotes sustainable housing.
B) **Community Involvement:** Copenhagen promotes public participation through participatory planning strategies and ensures citizens have a voice in improvements.
C) **Data-pushed choice-making:** Real-time data allow metropolises to optimize resources and improve city facilities.

Despite its innovations, the city must contend with challenges and threats

A) **High Initial Investments:** Although the long-term benefits are enormous, the initial fee for achieving smart technologies may be challenging.
B) **Scalability Issues:** Ensuring that effective solutions can scale and adapt to the needs of urban areas remains challenging.
C) **Equity in Access:** As technology evolves, Long-term success is critical to ensuring that all population engagement has equal access to smart solutions.

The case study provides valuable Key Takeaways, such as

A) **Sustainability Requires Collaboration:** Effective cooperation between government, society, and the business class is essential for sustainable development.
B) **Focus on Long-Term Planning:** Smart city concerns must focus on long-term impacts rather than short-term benefits and ensure lasting benefits for future generations

Table 12. Comparative Analysis of Smart City Initiatives

Aspect	Songdo, South Korea	India's Smart Cities Mission	Hudson Yards, New York, USA	Barcelona Smart City, Spain	Copenhagen, Denmark
Approach	Top-down, master-planned development	Decentralized, city-specific solutions	Private real estate development	Progressive urban innovation	Integrated sustainability approach
Governance	Heavy governmental involvement	Varying levels of local governance	Influenced by private investors	Strong municipal governance	Collaborative governance
Technological Integration	Advanced tech-driven from inception	Gradual tech integration	Smart infrastructure from the outset	Integration of IoT and citizen engagement	Smart grids and extensive data analytics
Inclusivity	Focus on high-income residents	Emphasis on inclusivity but uneven progress	Limited affordable options	Digital divide among socio-economic groups	Strong community involvement but scalability issues
Economic Viability	High costs leading to gentrification	Economic growth potential but inconsistent	High initial investment	Focus on local economic development	Long-term sustainability with public support
Challenges	Low occupancy rates, lack of community	Inconsistent policy implementation, slow timelines	High costs, limited inclusivity	Balancing innovation with cultural heritage	High initial investment, equity in access

continued on following page

Table 12. Continued

Aspect	Songdo, South Korea	India's Smart Cities Mission	Hudson Yards, New York, USA	Barcelona Smart City, Spain	Copenhagen, Denmark
Privacy and Data Security	Concerns over data privacy and transparency	Data management challenges	Privacy concerns with extensive data collection	Privacy issues with data usage	Strong data policies but continual need for improvement
Key Lessons Learned	Need for community engagement	Importance of local adaptability	Balance luxury development with affordability	Prioritize inclusivity and cultural context	Collaboration is essential for success

7. RESULT AND DISCUSSION

Smart city initiatives also demonstrate varied governance arrangements, with top-down governance models evident in Songdo and Hudson Yards in stark contrast with bottom up or municipal models reflected in the India Smart Cities Mission, Barcelona and Copenhagen. This side-by-side comparison provides the rationale for robust governance frameworks that are dynamic and adaptable to community-specific priorities, to ensure appropriate implementation responsive to the local context and public interest. Third, inclusion and community become important differentiators. Although Songdo and Hudson Yards embody a paradigmatic ultra-exclusive model, other initiatives in India, Barcelona and Copenhagen emphasize participatory planning and integration into the existing urban fabric, as inclusivity is necessary for truly sustainable cities.

Smart city initiatives also need to consider not only technologies involved but also economic viability. The large initial investments and affordability challenges seen in Songdo and Hudson Yards starkly contrasts the strategies of affordability and the economic spectrum of peoples that are addressed in cities like India, Barcelona and Copenhagen; demonstrating a clear need for financially sound and economically inclusive smart cities in the near future. Advanced technology deployment: the building block of Songdo and Hudson Yards are deploying advanced technologies up front while technology integrations are becoming a more gradual process in India, Barcelona and Copenhagen which calls for adaptive solutions that balance needed innovation with data privacy and security. Together, these comparative lessons reinforce that, to one degree or another, inclusive smart city implementation and long-term evolution rely upon equitable community involvement, calculation in economic strategies, and progressive technological adoption across a variety of global contexts.

Energy usage must be a first-priority consideration in all smart city technology implementations. That means employing energy-efficient hardware (LED lighting, low-power sensors), intelligent control systems (AI and data analytics to eliminate wasteful usage) and waste energy (for instance, reusing data center heat, or using regenerative braking in smart transit). Conservancy reduces the minimum requirement to be fulfilled, enabling other solutions e.g. renewables to be more valuable.

Comments Challenges smart cities can accelerate renewable energy deployment ahead of normal sector reform Smart cities can perform catalyst role for accelerating adoption of renewable energy Cities create demand and lead by example for the private sector by pledging to power their digital infrastructure by solar, wind, geothermal, or other renewables. Enabling policies from fast-tracking solar-for-rooftops permits, to ensuring investment in solar microgrids, to partnering with utilities for green energy supply ensure that increased demand for IoT and data services does not get locked-in to fossil fuel dependence. When you combine a smart city with carbon-free power, as Reykjavik and Tokyo have set out to do, you have the foundation for sustainable over the long-term carbon neutrality.

8. CONCLUSIONS POLICY IMPLICATIONS AND FUTURE PERSPECTIVES

Smart cities represent an ambitious jump into destiny, combining technological innovation with sustainable urban planning to create environments that are more efficient, related, and conscious of the wishes of their citizens, considering all classes. However, as with any transformative vision, the road to carrying out fully purposeful, smart cities comes with demanding challenges. Technological infrastructure that enables smart cities, From IoT networks to AI and massive data analytics, provides enormous capacity. Still, it also comes with data security and management, high cost, and interoperability. With the right infrastructure and continuous integration, these cities will likely be more efficient than smart and modern cities. Socioeconomic disparities highlight the risk of increasing gaps between different communities. Maintaining the smart city facilities reach every resident, regardless of financial conditions or ability to use digital technologies, is an ethical code of practice and a real necessity for long-term urban durability. Governance issues are compounded by inconsistent policies and political opposition, complicating the race toward smarter city development. Committed governance, cooperation between the public and private sectors, and a good regulatory framework are paramount in overcoming these obstacles. Privacy of data and safety issues are also important.

The huge amount of data generated by smart cities must include strong cyber security measures and transparent data management to ensure public trust. In addition, renewable energy and environmental sustainability are two guarantees for a smart city facing conflict. High energy consumption and the ever-increasing amount of e-waste challenge the ideals smart cities aim to uphold. In the end, the smart city vision has begun. But success depends on addressing these diverse challenges. By promoting a comprehensive approach to Guaranteed data security, Balancing technology with social needs, and maintaining a commitment to sustainability, Smart cities can evolve into urban landscapes that truly benefit all residents and improve quality of life and resiliency for generations to come.

REFERENCES

Addas, A. (2023). The Concept of Smart Cities: A Sustainability Aspect for Future Urban Development Based on Different Cities. *Frontiers in Environmental Science, 11*, 1241593. DOI: 10.3389/fenvs.2023.1241593

Antonios, P., Konstantinos, K., & Christos, G. (2023). A Systematic Review on Semantic Interoperability in the IoE-Enabled Smart Cities. *Internet of Things : Engineering Cyber Physical Human Systems, 22*, 100754. DOI: 10.1016/j.iot.2023.100754

Bibri, S. E., Huang, J., Jagatheesaperumal, S. K., & Krogstie, J. (2024). The Synergistic Interplay of Artificial Intelligence and Digital Twin in Environmentally Planning Sustainable Smart Cities: A Comprehensive Systematic Review. *Environmental Science and Ecotechnology, 20*, 100433. DOI: 10.1016/j.ese.2024.100433 PMID: 38831974

Cagáňová, D., Cehlár, M., & Horňáková, N. (Eds.). 2024. "*Smart Cities: Importance of Management and Innovations for Sustainable Development.*" DOI: 10.1007/978-3-031-56533-5

"Case Study on Smart City Projects in India: An Analysis of Nagpur, Allahabad and Dehradun." https://dl.acm.org/doi/fullHtml/10.1145/3184558.3191522 (October 18, 2024).

Cui, L., Xie, G., Qu, Y., Gao, L., & Yang, Y. (2018). Security and Privacy in Smart Cities: Challenges and Opportunities. *IEEE Access : Practical Innovations, Open Solutions, 6*, 46134–46145. DOI: 10.1109/ACCESS.2018.2853985

"Digital Skills: Succeeding in a Digital World: Week 2: 2 | OpenLearn - Open University." https://www.open.edu/openlearn/mod/oucontent/view.php?id=105154&section=2 (October 18, 2024).

"Embracing the Future: The Significance of Open Standards in Technology - ETHAR™ | The XR Experience Platform." https://www.ethar.com/2024/01/01/embracing-the-future-the-significance-of-open-standards-in-technology/ (October 18, 2024).

Ghisellini, P., Cialani, C., & Ulgiati, S. (2016). A Review on Circular Economy: The Expected Transition to a Balanced Interplay of Environmental and Economic Systems. *Journal of Cleaner Production, 114*, 11–32. DOI: 10.1016/j.jclepro.2015.09.007

"Insights Into Cities in The Digital Era: A People-Centered Approach to Smart Cities." https://www.cdpp.co.in/articles/insights-into-cities-in-the-digital-era-a-people-centered-approach-to-smart-cities (October 18, 2024).

Ismagilova, Elvira, Laurie Hughes, Nripendra P. Rana, and Yogesh K. Dwivedi. 2020. "Security, Privacy, and Risks Within Smart Cities: Literature Review and Development of a Smart City Interaction Framework." *Information Systems Frontiers 2020 24:2* 24(2): 393–414. .DOI: 10.1007/s10796-020-10044-1

Kaiser, Z. R. M. A. (2024). Smart Governance for Smart Cities and Nations. *Journal of Economy and Technology*, *2*, 216–234. DOI: 10.1016/j.ject.2024.07.003

Kempton, Y., Salvati, L., & Vardopoulos, I. (2022). Long-term planning and development for urban and regional inclusion, safety, resilience, and sustainability. Insights from Singapore. *Region & Periphery*, *14*(10), 58–78. DOI: 10.12681/rp.32607

Kitchin, R. (2014). The Real-Time City? Big Data and Smart Urbanism. *GeoJournal*, *79*(1), 1–14. DOI: 10.1007/s10708-013-9516-8

Lockie, Sean, Pasqual Capizzi, Tim Hawley, Luke Thompson, Max Russels, Ralf-Martin Soe, Semira Osmanovic Design, et al. 2024. "Acknowledgements Smarter and Inclusive Cities Module 1 Inclusivity for Smarter City Development Smarter and Inclusive Cities Smarter Governance Planning for Smarter and Inclusive Cities Smarter and Inclusive Cities Projects and Pilots Introduction to the Course."

Pandiyan, P., Saravanan, S., Usha, K., Kannadasan, R., Alsharif, M. H., & Kim, M. K. (2023). Technological Advancements toward Smart Energy Management in Smart Cities. *Energy Reports*, *10*, 648–677. DOI: 10.1016/j.egyr.2023.07.021

Paola, D. R., & Rosenthal-Sabroux, C. 2014. "Smart City How to Create Public and Economic Value with High Technology in Urban Space" eds. Renata Paola Dameri and Camille Rosenthal-Sabroux. *Smart City How to Create Public and Economic Value with High Technology in Urban Space*: VIII, 238. https://www.springer.com/series/10440 (November 14, 2024).

"Public-Private Partnerships (PPPs): Definition, How They Work, and Examples." https://www.investopedia.com/terms/p/public-private-partnerships.asp (October 18, 2024).

Rejeb, A., Rejeb, K., Simske, S., Treiblmaier, H., & Zailani, S. (2022). The Big Picture on the Internet of Things and the Smart City: A Review of What We Know and What We Need to Know. *Internet of Things : Engineering Cyber Physical Human Systems*, *19*, 100565. DOI: 10.1016/j.iot.2022.100565

"RKMA Market Research Reports." https://www.rkma.com/SFPL/cuim24/ (October 18, 2024).

Salvati, L., & Vardopoulos, I. (2024). Intricacies of Mediterranean urban and regional dynamics. *Euro-Mediterranean Journal for Environmental Integration*, *9*(1), 277–292. DOI: 10.1007/s41207-023-00434-7

Samant Raja, Deepti, and World Bank Group. "Digital Dividends Bridging the Disability Divide through Digital Technologies Bridging the Disability Divide through Digital Technologies." https://www.worldbank.org/en/publication/wdr2016. (October 18, 2024).

"Smart Cities: Big Data, Civic Hackers, and the Quest for a New Utopia by Anthony M. Townsend | SMART CITIES: Big Data, Civic Hackers, and the Quest for a New Utopia." https://ssir.org/books/excerpts/entry/smart_cities_big_data_civic_hackers_and_the_quest_for_a_new_utopia (November 14, 2024).

Syed, Abbas Shah, Daniel Sierra-Sosa, Anup Kumar, and Adel Elmaghraby. 2021. "IoT in Smart Cities: A Survey of Technologies, Practices and Challenges." *Smart Cities 2021, Vol. 4, Pages 429-475* 4(2): 429–75. .DOI: 10.3390/smartcities4020024

"The United Indian | Government Sector | Strategies for Enhancing Digital Literacy in India." https://theunitedindian.com/news/blog?Digital-literacy-in-India&b=193&c=3 (October 18, 2024).

Toli, A. M., & Murtagh, N. (2020a). The Concept of Sustainability in Smart City Definitions. *Frontiers in Built Environment*, *6*, 496662. DOI: 10.3389/fbuil.2020.00077

"Top 10: Smart Cities Driving Sustainability around the World | Sustainability Magazine." https://sustainabilitymag.com/top10/top-10-smart-cities-around-the-world (October 18, 2024).

Vardopoulos, I. (2024). Urban planning policies and tools for sustainable development: insights into adaptive reuse, smart cities, and urban metabolism. In *Socio-enviromental change: The path toward sustainability* (pp. 121–140). Nova Science Publishers.

Vardopoulos, I., Giannopoulos, K., Kalli, S., & Salvati, L. (2024). From tech-centric to human-centric smart tourist destinations: innovations in real estate, urban planning, and sustainability. In *Socio-enviromental change: the path toward sustainability* (pp. 65–96). Nova Science Publishers.

Vardopoulos, I., Papoui-Evangelou, M., Nosovo, B., & Salvati, L. (2023). Smart 'tourist cities' revisited: Culture-led urban sustainability and the global real estate market. *Sustainability (Basel)*, *15*(5), 4313. DOI: 10.3390/su15054313

“What Are the Cybersecurity Risks for Smart Cities? - Institute for Defense and Business.” https://www.idb.org/what-are-the-cybersecurity-risks-for-smart-cities/ (October 18, 2024).

Chapter 2
Leveraging Artificial Intelligence, Blockchain, and FinTech to Enhance ESG Compliance and Urban Governance

Aikaterini Petropoulou
https://orcid.org/0009-0005-7528-4057
Hellenic Mediterranean University, Greece

Alexandros Garefalakis
https://orcid.org/0000-0001-7980-3743
Hellenic Mediterranean University, Greece

Ioannis Passas
https://orcid.org/0000-0002-1261-2878
Hellenic Mediterranean University, Greece

ABSTRACT

The rapidly increasing and continuous progress of innovative technologies, such as artificial intelligence (AI), Blockchain, and Fintech (Financial Technologies), which are examined in this chapter, have changed how businesses and governments manage compliance with Environmental, Social, and Governance (ESG) factors. Compliance with ESG criteria is crucial for all organizations, prioritizing operating sustainably and creating added value for all stakeholders. While businesses prioritize ESG compliance, the need to improve corporate governance has become imperative, and organizations worldwide are struggling to respond to challenges

DOI: 10.4018/979-8-3373-2327-5.ch002

such as climate change, social inequalities, and corruption. To effectively comply with ESG, integrating advanced technologies such as AI, Blockchain, and Fintech is crucial, as they can ensure greater transparency, accountability, and efficiency. This chapter examines how AI, Blockchain, and FinTech technologies can enhance compliance with ESG factors and the criteria they contain and improve urban governance.

1. THE CONTRIBUTION OF ARTIFICIAL INTELLIGENCE (AI) TO COMPLIANCE WITH ESG STANDARDS

AI is revolutionizing the way organizations manage compliance against ESG standards. As more people demand responsible business actions from companies, Artificial Intelligence has a crucial role in confronting ESG performance. From environmental monitoring to social accountability and governance structures, AI software makes organizations conform to ESG criteria and enhance world sustainability initiatives (Zopounidis et al., 2020).

The contribution of AI to environmental compliance is mainly focused on effectively utilizing data to track, manage, and minimize environmental effects. One of the biggest things AI does is process huge amounts of data, like satellite pictures and sensor information. For example, AI models can help identify environmental degradation like deforestation or pollution, enabling organizations to monitor their production operations and supply chains more efficiently (Rane, Choudhary & Rane, 2024). Technology provides organizations with channels by which they can comply better with environmental regulations, mitigate risks, and attain sustainability goals. Besides, AI-powered predictive analytics can forecast environmental risks, allowing companies to predict and avoid potential negative impacts (Riyadh Jariya, 2024). Not only does this allow companies to make sure everyone complies with the rules, but it also keeps them up to date with tough global environmental standards. Better compliance boosts everyone's playing fair and keeps everyone aligned with high goals.

The application of AI to environmental data also includes energy management, where AI systems can optimize energy use and reduce waste (Vardopoulos et al., 2023 and Vardopoulos, 2025) Nowadays, in the world of manufacturing and energy generation, a great opportunity exists with this technology called AI. AI can find where we are being wasteful, like cleaning up operations and reducing carbon (Chen & Zhang, 2024). Also, these new technologies help companies write more accurate sustainability reports that align their performance with ESG standards and guidelines. It is about impressing worldwide stockholders and investors who care that their investments benefit the environment.

Social ESG metrics involve ensuring that firms do business in a way that supports human rights, labor, and community well-being. So, one useful thing about AI is that it helps companies keep track of and improve their regard for labor practices and community involvement. Oversight of daily work practices, including fair compensation, compared to having to rely on word of mouth and observation, and research related to the well-being of people in communities and relationships, really makes it easier. It is crucial because companies can use AI to keep everything ordered and consistent accountability over time, projects, and people. AI tools can analyze data from social media, surveys, and employee feedback to identify social risks and issues that could potentially damage an organization's reputation (Chen, 2024). Using sentiment analysis, AI systems can determine public mood sense up-and-coming hot topics and help a company react quickly and deal with stakeholder worries.

AI's social impact goes beyond risk detection. It can also drive positive social outcomes. By integrating AI into employee training programs, organizations can build more inclusive and diverse workplaces and implement compliance with labor laws and ethical standards (Sulkowski, 2024). Furthermore, AI systems can identify diversity and inclusion gaps and point out areas for improvement, which can cause organizations to move towards more balanced practices (Rozkrut, Tarczynska-Luniewska, Singh, Piwowarski, 2024). Also, AI helps companies improve their social responsibility initiatives, contributing positively to employees, communities, and society.

Governance refers to the structures, processes, and practices by which organizations are governed and controlled to provide accountability and transparency. Due to AI, government operations are much smoother and more efficient. It automates processes like monitoring compliance rules and catching suspicious activities or fraud in real time. At the same time, it helps make decisions faster and more consistent than humans alone. For example, AI solutions can scan companies' communications, transactions, and reports for irregularities to ascertain if they follow legal and ethical standards (Lee et al., 2024). Having AI infused into systems that handle administration and governance helps companies better report sustainability information. That helps stakeholders like investors or activists evaluate that information much better.

1.1 The Role of Big Data Analytics and Machine Learning

The role of Big Data analytics and machine learning in strengthening ESG practices has become more significant in recent years (Ragazou et al., 2023). With all the technological improvements each day, organizations can look at huge data sets and line up their work to use resources wisely. Integrating ESG criteria into business operations requires data-driven insights primarily achieved through big data and machine learning techniques. They can figure out how they affect the world and bigger social problems. At the same time, they are also figuring out how to run their own business. According to Riyath and Jariya (2024), it helps people make better decisions and simultaneously helps compliance with sustainable criteria.

Machine learning tools are very helpful when assessing and forecasting factors that impact the environment, society, and governance. The tools of AI help pick up patterns from big data sets, like grades from school or weather predictions. Using advanced algorithms turns into conventional ways of looking at big piles of data. These technologies help find patterns right out of huge stores of information that are normally hard to recognize. Having the ability to look at this complex data helps us to find out the key risks and opportunities very accurately and efficiently. For example, machine learning can discern connections between sustainability performance and financial outcomes. It shows companies clearly and directly how compliance with the environmental criteria of ESG has a positive economic impact (Chen, 2024). These technologies make all processes more transparent, which is important for handling rules and important for investors who care about social responsibility (Sulkowski, 2024).

Big data analytics help manage huge amounts of different data immediately, which is important for checking ESG performance. Real-time data processing continuously evaluates an organization's ESG performance, allowing businesses to adapt swiftly to changing environmental, social, and regulatory conditions. This approach helps get things moving at companies, and it helps them keep monitors on the changing landscape and be prepared for any shifts that might come up, meaning they can adapt quickly to new trends and avoid risks. For example, machine learning can facilitate monitoring carbon emissions, resource consumption, and labor practices across global supply chains, offering companies actionable insights that improve their sustainability efforts (Li et al., 2024).

Big databases and artificial intelligence are important and help blend sustainability data from various sources and formats. This helps to see the big picture and understand how the company values sustainability. This is important for industries with complicated supply chains and getting data from different businesses and people. The information that comes in handy includes environmental impact reports, social impact analysis, and governance audits (Cheliatsidou et al., 2023; Kagias et al., 2023; Yiannoulis et al., 2025). Simultaneously searching all that data, businesspeople can see parts of their work that could be reworked or improved. They handle potential dangers and risks better if they can predict them properly (Rozkrut et al., 2024). This approach helps the organization untangle its work to match global sustainability goals and fulfill its ESG commitments.

Big Data and artificial intelligence systems allow organizations to transcend and faithfully adhere to reporting guidelines related to sustainability. They can now move towards strategic and proactive ways to look at sustainability that follow the rules, think ahead, and do a huge amount of strategic and cutting-edge work on sustainability. Using predictive analytics, these tools can also help spot future trends regarding ESG considerations (Passas et al., 2022). Companies can figure out what kind of trouble might happen and be ready to take advantage of the good opportunities when they show up.

1.2 Forecasts through Predictive Models and Risk Management

Predictive models have become crucial tools for companies wanting to predict the future, find out where trouble's brewing ahead, and make better decisions. Having a good grip on what the future might bring means companies are much better prepared and make better judgments and decisions. By looking at historical datasets and spotting trends, these prediction systems help businesspeople make forecasts, plan better, and deal with possible risks more efficiently. AI plays a more central role, especially regarding metrics important to ESG elements, according to Rane, Choudhary, and Rane (2024). AI models handle important jobs and make a big difference in their work. Companies keep getting better at predicting things because of this smart intelligence. Planning looks beyond money-related risks and considers all the social stuff, too, like how management is at the table. This integration means companies can see potential societal and environmental disruptions early enough to be nimble and quick to respond. Preemptive action can get them ahead of such setbacks.

Integrating AI into predictive models opens a new level of risk management that allows us to analyze large data sets much faster and accurately. Chen and Zhang (2024) argue that AI applications, particularly those driven by machine learning, can assess historical data and real-time information, making it possible to predict potential risks more accurately. This skill is very important for companies that manage the ESG criteria seriously. By using AI, businesses can improve their ability to forecast better. They can develop strong strategies for situation reduction that consider both near dangers and longer-term uncertainties, which can help comply with ESG requirements.

Models that use AI also supercharge the quality of reporting Sustainability, ESG or otherwise. Integrating digital tech into monitoring for ESG criteria has much better assessments of risks. This allows us to process huge amounts of information quickly. According to Chen (2024), using AI improved the quality of data analysis, such as updates on the environment and sustainability reporting. We have gotten much better at using metrics and gauging performance now with powerful data analyses because of AI. This enhanced forecasting capability is crucial in risk management, as it empowers companies to identify and address emerging ESG-related challenges before they escalate, fostering a culture of proactive risk mitigation rather than reactive adjustments.

In the world of power and energy (Vardopoulos et al., 2024), advanced AI can also be key to playing the risk champ. For example, it is a great way to figure out risks that energy systems confront (Li et al., 2024). This year, studies note that systems built with AI can forecast increases in energy demand. They can also anticipate all sorts of risks related to sustainability because of smart systems with AI. Using ESG metrics, these models also can say something about the environmental impacts and rank the long-term sustainability of different methods of energy production. Forecasting powers let energy companies eliminate the less useful waste and ensure sustainability. It lets them do everything with the book properly and efficiently.

As the importance of ESG factors in decision-making grows, so does the need for advanced predictive models that integrate these criteria into risk assessments. Sulkowski (2024) highlights the potential of AI in improving sustainability reporting through enhanced data collection and analysis. Predictive models that consider ESG risk factors are good for helping companies meet all sorts of challenges like climate change, government regulations, and what the public thinks (Ribeiro Hoffmann et al., 2024). Knowing which risks lie ahead allows companies to make strategic moves where they consider compliance with ESG criteria in the process. It means making smart financial calls and doing things according to the directions of ESG criteria.

In conclusion, Machine learning, a subset of Artificial Intelligence AI and part of computer science refers to the creation of algorithms in which computer systems "*learn*" from the data they have received and improve their effectiveness without the need for many contributions of human effort. Therefore, machine learning can contribute to processing data collected by organizations. The data mentioned above are related to the three factors of the ESG criteria. More specifically, data such as CO2 emissions and energy consumption are collected for the environmental factor. For the social factor, data such as working conditions and other social issues, such as gender equality, are collected (Stavropoulou et al., 2023, Vardopoulos & Konstantinou 2017, Georgiou et al., 2025).

The governance factor includes data on transparency in decision-making and risk management systems, which do not depend directly but indirectly on businesses and organizations. AI algorithms contribute to processing data fed to them by the company or organization and through the analysis of this data. They can predict risks that will lead to non-compliance with ESG criteria and offer possible solutions that suit each company or organization. For example, suppose a company does not meet the conditions set by the ESG criteria regarding the environmental factor, such as producing more CO2 than acceptable. In that case, it can, based on other data about the company, such as its financial data, suggest how it can solve this problem and achieve the desired compliance (Sulkowski, 2024).

The financial factor of AI has contributed to the development of tools that help investors identify companies that perform well in compliance with ESG criteria. To support sustainable finance and avoid investing in companies that contribute to the deterioration of environmental and societal risks. AI technologies make it possible to predict the impacts on compliance with ESG criteria through predictive models. The prediction concerns both the long-term plan, which is the strategy plan of the company or organization, and the short-term plan, which is the tactics based on which it deals with changes in the environment to achieve the goal of the long-term plan or strategy. For example, through predictive models, companies can determine how their future actions will affect CO2 emissions or social responsibility with a possible increase in production and overtime (Rane, Choudhary, and Rane, 2024).

2. BLOCKCHAIN'S CONTRIBUTION TO TRANSPARENCY AND GOVERNANCE

Blockchain is taking a new step; this technology helps more people find trust, which is important, especially for finance and corporate reporting. One of the things about blockchain is its decentralized and tamper-resistant nature. That important quality will help start dealing with transparency, which has been a sore subject for traditional governance models of old. By keeping all sorts of financial and documentation notes on a shared digital log, the blocks ensure that information is open, verified, and just plain bulletproof. This feature significantly reduces the risk of fraud, manipulation, and errors in financial transactions and reporting systems.

2.1 The Role of Blockchain in Securely Recording ESG Data

Blockchain has taken off and has some big implications for keeping track of the planet and how companies can be responsible. It allows us to be secure about following up and provides a high level of security and verification to lend trust to environmental and business accountability. This is especially true for industries that must be serious about going green and being responsible. These days, everyone, from rich people to those in charge of money rules and from people who use money to people who use it to consume sustainability data (Yu, 2021).

This data comes into play with a lot related to the environment, how the business relates to people in its local community, and important leadership stuff. We are talking about important stuff like that. By looking at this kind of sustainable data, people want reassurance that companies are protecting and sticking to good sustainability plans. Using traditional ways to check and gather ESG data rarely results in clean, smooth-sailing figures that can be all over the map, streaks of inconsistency are common, and it is hard to follow through. Blockchain addresses these challenges by providing a secure and immutable platform for recording ESG data (Yu, 2021).

Using blockchain helps prevent those schemes where people try to tamper or mess with data once transactions are entered. Everything gets recorded reliably and in an ironclad way, so there is no room for fraud. This immutability is particularly important in maintaining the integrity of ESG disclosures, as it guarantees that the reported information is authentic and verifiable. Yu (2021) emphasized that blockchain technology significantly enhances the transparency and reliability of ESG information by creating an immutable ledger that stakeholders can independently verify.

This matters for many people who need a lot of faith and trust from the public. Lots of folks count on being able to trust what the government and important public figures are doing and saying, and this feature is essential for keeping that trust. Companies that work in big sectors like energy and construction must show they are serious about going green (Pavlides et al., 2023; Yu, 2021).

2.2 Transparency in Governance Systems

Regarding transparency from governments, the win comes with blockchain, which is a secure and easy journal for different business transactions and important papers. Recently, governments worldwide have started to zoom in on how to make things a lot simpler through this new technology. As Rerung et al. (2023) noted, blockchain innovations are increasingly integrated into financial systems to ensure that financial transactions are accurate, transparent, and accessible to all stakeholders. Blockchain is all about limiting middlemen and going directly, which can be helpful because often those middlemen interfere and cause all sorts of partiality and sometimes unfairness.

Open access to data speeds up accountability and is not just for private companies. However, it is also for governments and any team that wants transparency to be real and things to run fairly. More and more people understand that when information is readily available and easily accessible, things like government performance can be scrutinized and audited more closely. Transparency ensures that it is always clear to see when people in charge are doing something. They cannot just waltz around waving their white Flags and pretend things are going fine when they are not. They cannot stop cutting corners because they have power and cannot dodge proper supervision and control. Keeping things transparent helps ensure everyone is held accountable and actions are taken fairly and well (Rerung et al., 2023).

Blockchain offers transparency when managing supply chains. Now, companies can nail the authenticity of what they sell and provide smoothly. By using blockchain technology, companies can track every step of everything they make, getting to watch data that people can check. People who use it and regulators also see it through this script. Liu et al. (2019) discuss how a blockchain-enabled framework for ESG reporting can improve transparency across supply chains, ensuring that companies disclose accurate data related to their sustainability efforts. With this openness, both people and investors can make smart choices, so companies use good leadership and good rules inside them.

Blockchain goes beyond finance and supply chain management and helps improve governments' and institutions' operations. In the context of green finance and sustainable investing, we have been focusing lately. Udeh et al. (2023) highlight the importance of blockchain in enhancing transparency in green finance markets. Blockchain helps secure the straightforward recording of environmental efforts and activities so that investors and people who stake their interests know that information about the environmental impact of the money they put in is right and verifiable. Building this kind of transparency builds trust on a whole new level in green finance and encourages more investment into sustainable initiatives.

In addition to letting everything be more open and clearer, blockchain makes it efficient and much more secure for managing data and improving governance. Using smart contracts can help automate deal details and enforce things so that the whole process no longer needs human mediators. That way, everyone is less likely to mess things up or take longer because of it. There is less human error that way, too. Gong et al. (2023) noted that blockchain technology allows for secure data management in various sectors, including construction, where the technology ensures the transparency of ESG data related to construction projects. This automated process improves efficiency and helps maintain high governance standards by providing real-time access to accurate data.

In conclusion, Blockchain, the technology that allows for a ledger in which all transactions and data are linked together, has the potential to provide a comprehensive picture of how activities and their compliance with ESG standards are recorded and verified. By the nature of this technology, Blockchain is immutable, transparent, and secure, verifying compliance with ESG criteria and preventing data manipulation that contributes to a false picture (Rerung et al., 2023).

Through Blockchain, businesses and organizations can ensure the integrity of all ESG-related activities, creating a reporting system that cannot be tampered with. Applying this technology to financial services makes it possible to track the progress of green bonds and finance tools that help identify companies that operate based on ESG criteria and support sustainable investments. Blockchain technology and platforms enhance transparency in Governance processes as they allow stakeholders (investors, organizations, governments) to monitor companies' activities and ensure the integrity of ESG data. Blockchain technology enhances corporate governance, as all stakeholders can verify compliance with ESG standards in real-time, i.e., what is happening now and not with past data, which increases accountability and responsibility (Gong et al., 2023, Vortelinos, Passas, et al., 2024).

3. THE CONTRIBUTION OF FINTECH INDUSTRIES TO FINANCIAL SERVICES

FinTech has emerged as one of the most significant factors shaping the future of financial services. Since its inception, the FinTech industry has greatly influenced how consumers and businesses interact with financial systems. Financial technology has revolutionized payments, lending, insurance, asset management, and financial advisory services. In this chapter, we will explore the contribution of the FinTech industry to financial services, focusing on innovation, improved transparency and accountability, and supporting sustainability principles as defined by ESG practices. FinTech has been recognized for radically reshaping the financial landscape. Due to innovative solutions, banking transactions are quicker, safer, and easier to access now. FinTech solutions enable users to perform transactions and payments in real-time, access funding instantly, and manage investments in new ways. Technology integration has improved customer experience and reduced costs for financial institutions (Rerung, Paranita, Agung, Budiandru, Tandililing, 2023).

Innovations like those for digital payments and peer lending platforms shake things up and change the whole sector in a big way. These techy systems are getting people excited and taking things to a new level. FinTech platforms provide users with the ability to exchange money, invest, or obtain loans without the mediation of traditional banks, thus increasing efficiency and reducing the need for high fees (Rerung, Paranita, Agung, Budiandru, Tandililing, 2023). One of the most positive outcomes of introducing FinTech technologies into financial services is enhancing transparency and accountability. FinTech innovations improve the ability of financial organizations to monitor, record, and manage data in real-time, enabling greater transparency in financial markets. Using technologies such as blockchain can ensure transparency in financial transactions, reducing the need for intermediaries and strengthening consumer trust in financial services. Integrating blockchain technology can ensure financial institutions have reliable and secure data and account records because everything that goes down is rock solid (Duran & Tierney, 2023).

Using fintech in finance offers tools that work continuously to report sustainability matters, and that matters because it helps to build stronger accountability among banks and financial companies. Think of it like setting up tools that work around the clock to look at people's finances and track indicators that people care about, like sustainability, water conservation, or recycling, and knowing who does what well help customers feel secure in knowing their banks are doing their part. Putting that data together lets companies better follow all the rules and show they care about following through with green practices. The growing attention to sustainability and social responsibility has enhanced the connection between FinTech technology and ESG practices. The FinTech world has become important as it helps people manage investments that are green and good for the environment and society simultaneously (Giakoumelou et al., 2024; Passas I., 2024).

FinTech companies can integrate technologies that enhance the sustainability of financial services. Through tools for ESG reporting, businesses can monitor their performance on sustainability issues and meet legal regulations. Blockchain technology helps track ESG performance, allowing us to have records that are traceable and impossible to change (Giakoumelou et al., 2024; Passas I., 2024).

3.1 The Role of Green Bonds and Sustainable Investments

Pairing financial technology with sustainable finance opens new chances for bond development that is green or has a social benefit. One big upside for green bonds is new tech that adds more transparency and ensures investors know where funds go towards clean projects and sustainable stuff. Investors love using fintech apps nowadays. They are swiping through their portfolios daily, tracking how green investments perform immediately (Garrido & Alberto, 2023). Everything is there for a quick and current look at effectiveness and performance.

We have seen some real wins in the fintech world in Europe, particularly in new financial technologies. One of the big benefits of this new wave has been a shift from strategies that focus on just making money to strategies that also champion sustainability and social responsibility. Projects that work to support the environment and those who are not so well off are getting much more attention now. Fintech has helped make these important things a big deal, too. FinTech companies have become important players in helping move funding flows centered on sustainable goals. They are also key to strengthening financial sectors and helping businesses make better decisions (Galeone et al., 2024). Innovative FinTech technologies provide financial institutions and investors with the necessary tools to achieve sustainability goals, creating new opportunities for funding projects that promote social justice and environmental responsibility. The FinTech industry is not stagnant; its continuous evolution is expected to lead to new applications and capabilities. As financial institutions continue to recognize the importance of ESG criteria, FinTech will play a central role in enhancing compliance.

The emergence of green bonds and sustainable investments is like introducing a new instrument into a musical ensemble. Like music qualities like harmony and dynamics, investment decisions are becoming richer and more significant, with environmental elements putting performance and political factors in the background.

As climate concerns intensify and people realize sustainability is huge, green bonds have become awesome tools for financing cool green projects. Like other sustainable investments, green bonds allow investors to channel their capital into initiatives that promote a positive environmental impact, such as renewable energy, clean water, and low-carbon technologies.

The role of green bonds in the financial ecosystem is multifaceted. These bonds are specifically designed to finance projects that contribute to environmental sustainability, allowing governments, corporations, and other organizations to raise funds for such initiatives. The key characteristic of green bonds is the use of proceeds, which must be directed toward projects that address environmental challenges. This clear focus on sustainability distinguishes green bonds from traditional bonds, where proceeds may be used for various purposes unrelated to environmental goals.

The growth of green bonds has been driven by the increasing demand from investors seeking to align their portfolios with their values, particularly those interested in promoting sustainability and combating climate change. As investors become more conscious of the long-term impacts of their investments, green bonds offer an attractive option for those looking to support environmentally responsible projects. Furthermore, because corporations and governments feel increasing regulatory pressure to adopt greener practices, there is a rise in the number of people likely to want to buy bonds that invest in green stuff. These pressures and growing concerns about climate change have come together nicely to make green bonds attractive and productive (Baldi & Pandimiglio, 2022).

The contribution of green bonds to sustainable finance is financial and strategic. By setting up a separate fund for very green projects, "green bonds" are a true nudge to companies and countries who want to score big and score environmentally. This means they are more likely to think hard about how to be greener now to make their investment pay off ten years from now. This financing bridges the big gap for projects where other usual financial instruments might see them as high risk or expensive. It ensures that if something is cool but hard to fund, there will be a way to bridge that pocket. Green bonds set high standards because issuers must report very transparently, like giving a lecture open to the class. They need to report on how green their projects are. That means keeping nature clean and clear (Maltais Nyqvist, 2020).

More and more people are now looking for sustainable investments to handle longer-term risks related to pollution and global warming. The importance of these investments is shown by the growing popularity of financial instruments focusing on ESG criteria. As noted by Garrido and Alberto (2023), ESG-driven innovation in the FinTech industry plays a crucial role in facilitating the development and implementation of green bonds. These innovations, such as advanced data analytics and blockchain technology, enhance the ability to track and verify the environmental outcomes of projects funded by green bonds, making it easier for investors to monitor their impact and ensure that funds are being used effectively.

Moreover, integration into the broader investment world is moving along, thanks to the growing awareness we are seeing now of the social plus super financial perks that come with practicing greener and more sustainable stuff. Green bonds contribute to reducing carbon emissions and promoting renewable energy, which are central to achieving global sustainability goals. As demand for green bonds continues to rise, the financial sector will likely see more innovation in the types of sustainable investments available to investors. The continued development of green bonds and other sustainable investment products is a testament to the growing importance of environmental and social responsibility in modern finance.

3.2 The Role of ESG Data in Corporate Finance

Integrating ESG data into corporate finance has become increasingly important in recent years. Companies are now looked at in terms of how much money they make and how they affect the world, from society to the planet (Passas I., 2023). The role of ESG data in corporate finance is crucial for improving transparency, enabling better decision-making, and ensuring long-term sustainability. ESG data gives insights for people casting a financial eye, policy folks, and other major players who study a company for potential risks and chances regarding things like the environment, social factors, and governance standards.

Using ESG data in corporate finance allows companies to understand better and address the risks associated with their operations. For example, environmental risks such as climate change and resource scarcity can significantly impact a company's financial performance. Companies that ignore these risks may face reputational damage, regulatory penalties, or higher operational costs. By weaving ESG data into financial strategies, businesses improve their lowdown sustainability and lower risks in the long haul, as noted by Duran and Tierney (2023).

In addition to managing risks, ESG data also helps businesses identify new opportunities. Social factors, such as changing consumer preferences for sustainable products or services, can give companies a competitive edge. Companies incorporating ESG considerations into their product development, marketing strategies, and supply chains can attract socially conscious consumers and investors. Lee, Serafin, and Courteau (2023) write that there has been a surge in demand for environmentally friendly and sustainable things. Because of this big jump in greenery, companies are using ESG practices—themes related to Environment, Society, and Governance as part of their business plans. Using those practices often leads companies to achieve better financial outcomes and makes them lovelier to their customers over time. So, they have seen an uptick in companies wanting to use green practices to be happier with finances and to make their customers love them even more.

From a governance perspective, ESG data improves corporate accountability and transparency. Investors and other people who invest in and care about a company in one way or another can investigate how the company runs internally and at the top. It is all about looking at what kinds of people sit on board (that important board of supervisors that watches over the company), who gets paid high piles of cash, and how shareholders (basically people who have invested in the company) are all treated. In this regard, ESG data lets investors make very smart choices about the companies they decide to back with their money by allowing them to understand important environmental, social, and governance factors. For instance, companies with strong governance structures will likely be better equipped to navigate regulatory changes and mitigate legal risks, contributing to more stable financial returns (Gherghina, 2023).

Financial institutions and people investing money increasingly find that using data related to ESG matters is essential. ESG goes way beyond merely ensuring there's not too much pollution or careless waste disposal. It also touches on social matters like working conditions for employees and more, as well as governance by accounting for fairness and accountability settings between that firm's workforce and others. Using such data is key for any institutional investor or financial institution seeking to add value in the market going forward. Major investors ramp up environmental, social, and governance factors as part of their ultimate investment picks process. As ESG data becomes more standardized and available, investors can compare companies based on environmental, social, and governance performance. According to Wang, Jiang, Dittrich, and Osterlund (2023), the growing integration of ESG criteria in the investment process is prompting companies to improve their ESG disclosures as investors seek to allocate capital toward companies that are aligned with sustainable development goals (Passas I., 2024).

ESG data also plays a critical role in corporate finance by influencing the cost of capital. Companies with strong ESG practices may be able to access capital at a lower cost due to their perceived lower risk profile. More and more investors are eager to get their money into companies that show a real dedication to sustainability because these businesses get dressed down a lot less for fines, regulatory criticism, and a hit to their reputation. Alqudah, García, and García-Benau (2024) noted that companies that adopt sustainable finance technologies are better positioned to attract investors and secure funding for future growth.

Technological advancements further amplify the role of ESG data in corporate finance. With more and more tools for data analytics and fintech solutions flying off the shelves, many companies can scoop up, analyze, and report information about sustainability. This has led to developing more sophisticated ESG metrics and reporting standards, allowing companies to assess their ESG performance more accurately and efficiently. The growing use of technology in ESG reporting and data management is helping to build trust and credibility with investors and other stakeholders, thereby enhancing a company's financial reputation (Garrido Alberto, 2023).

The role of ESG data in corporate finance has become a fundamental aspect of modern financial strategies. ESG data is valuable because it helps companies manage risks while finding new chances to move forward. Working with better governance improves a firm's standing and helps an organization draw in more investment. For companies, dealing skillfully with these issues shows their dedication and care towards social and environmental matters, and it can make them attractive. Due to the ongoing developments when it comes to analyzing and reporting on ESG data, companies are much better at weaving sustainability into key matters and their contributions.

In conclusion the Fintech industry includes all technologies that automate and improve financial services, such as online banking, which has grown rapidly in recent years. The Fintech sector offers digital financing tools that facilitate compliance with ESG criteria and the sustainable development of investments in businesses that enhance it. Investors can benefit from Fintech platforms, as they can identify companies that comply with ESG criteria and invest money in them to strengthen sustainable investments. Green bonds are bonds issued to finance environmentally friendly projects. The Fintech industry helps facilitate the issuance of tools that monitor companies' compliance with ESG indicators, enhancing responsibility in financial markets. The Fintech industry uses ESG data to help businesses and organizations evaluate their investments to increase investment efficiency. The combination of Fintech and Blockchain technologies allows the development of integrated ecosystems that enhance transparency and accountability in the creation and financing of sustainable projects (Passas I., 2024).

4. THE CONTRIBUTION OF AI, BLOCKCHAIN, AND FINTECH TECHNOLOGIES TO STRENGTHENING CORPORATE GOVERNANCE

AI, blockchain, and FinTech have become significant drivers for improving corporate governance. Including these emerging technologies helps firms generate ideas that make things simpler and fairer and improve how the whole company runs. Several contemporary research studies highlight their impact on business governance (Weston Nnadi, 2023). AI will play a key role in promoting corporate governance with its capacity to read large amounts of data to ascertain anomalies and risks. With very advanced machine learning, companies can ascertain indicators that there are massive problems with complying with the rules or that books are not kept upright. The application of AI in data analysis gives speed, accuracy, and greater dependability to reports generated by an organization. As stated by Adeyelu, Ugochukwu, and Shonibare (2023), the use of artificial intelligence is not only able to improve productivity but also make decisions far more accurate.

Blockchain technology is also highly beneficial to corporate governance due to its transparency and security features. Blockchain. It makes it possible to store decentralized records of all transactions, and each transaction is traceable and undatable. That extra accountability minimizes fraud risk and accelerates trust from shareholders and others who rely on the business. In a recent detailed examination of blockchain technology by Basdekidou and Papapanagos (2023), they highlight the value of its use for enhancing transparency and efficiency immensely in governmental processes. FinTech plays an enormous role in bolstering how business is governed. It aids access to any money service as well as brings newer ideas of finances, like the idea of green bonds. Green bonds are new; they raise funding for projects meant for the environment's safety. So, with their advances and the further means of acquiring monies, corporations and other financial institutions that have a role to play within society are yet further facilitated in being run efficiently and forward-looking. Such instruments, founded upon technologies like blockchain, promote clarity and further increase corporate social responsibility by linking funding with green and sustainable projects (Ragazou et. al., 2023). Ojukwu et al. (2023) identified that financial innovation through FinTech technology leads to more effective and clear financial activity.

However, combining blockchain and AI in finance has ethical and regulatory consequences. AI-aided decision-making must follow strict ethical guidelines to maintain the process of dealing with all the stakeholders open and equitable. Data privacy and issues of not making mistakes just by generating misinformation are also significant. Devey and Ferraris (2023) highlight the need for better data governance and algorithmic accountability to minimize ethical concerns and offenses due to AI usage. Artificial intelligence uses problems, such as ethical creation and its application in business governance and must be provided with adequate policies and control measures. Datta (2023) emphasizes that the ethical consideration of AI is imperative for integrating AI into organizational planning, and AI governance must be regulated strictly to ensure transparency and effectiveness.

While there are immense advantages, there are also genuine risks involved with this technology, including issues of data security and the possibility of doing nasty things with this technology. Hence, their deployment must be carried out cautiously so that corporate governance is stronger and not weaker. According to Trotta, Rania, and Strano, for 2023, businesses should be capable of seeing the ethical problems brought about by using these new technologies and developing ways to ensure everybody is playing by the rules. There is transparency about what is going on.

The continuous improvement of these technologies also requires adjustment in regulatory legislation. Policymakers must consider the danger and opportunity of adopting these technologies and utilize technology to support the economy without generating new dangers and inequalities. Wang et al. (2023) suggest that there is a need for establishing and enforcing legislation that would make it easier to adopt these technologies in the sphere of corporate governance in the interests of increased transparency and accountability. Thus, AI, blockchain, and FinTech technologies hold tremendous potential for strengthening corporate governance through the availability of tools to improve transparency, efficiency, and accountability in organizations. However, their application must be done responsibly, with the continued development of policies and regulations that will enable these technologies' ethical and secure use in business.

4.1 AI's Contribution to Governance and Enhancing Ethical Decision-Making

AI has come into its own for government work, making fair decisions and just these days. As highlighted by Winfield and Jirotka (2019), ethical governance is crucial for building trust in robotics and AI systems. Without a solid ethical framework, applying AI algorithms in government can lead to poor decisions that may violate fundamental human rights. Researchers are concerned about how important it is to put ethical values right into AI systems - that way, they do not make unfair decisions, and things stay neat (Winfield & Jirotka, 2019).

These technologies bolster transparency and accountability in the context of government agencies' use of AI. Devey and Ferraris (2023) argue that AI can enhance accountability in public administration, ensuring that government institutions remain answerable to the public. Automating various processes helps eliminate human mistakes and makes decisions work much better. Additionally, it fosters transparency, allowing citizens to track government actions better and increasing their trust in the government (Devey Ferraris, 2023).

The ethical dimension of AI is vital for ensuring justice and fairness in governmental decisions. Adeyelu, Ugochukwu, and Shonibare (2023) also point out that it is important to have strong controls in place to ensure that AI algorithms do not have any prejudiced tendencies that might affect decisions. Sometimes, vulnerable populations may not receive adequate protection, or decisions may not align with societal moral standards. The researchers advocate for collaboration between technologists and public policymakers to create ethical algorithms grounded in shared values, promoting justice and equality (Adeyelu, Ugochukwu, Shonibare, 2023).

To effectively leverage AI in governance, ongoing monitoring, and evaluation of systems are essential. Macchiavello and Siri (2020) emphasize the need for regular updates to policies and procedures governing AI to ensure that these systems function effectively and efficiently. Blockchain is one of those shiny new technologies that blew up in the last ten years. It has taken off big time in industries ranging from finance to supply chains and general business processes. One of the most interesting and critical applications of Blockchain is its contribution to governance and enhancing accountability, two fundamental elements for the effective operation of organizations and governmental institutions.

4.2 Blockchain's Contribution to Governance and Strengthening Accountability

One of blockchain's big strengths is its trustworthiness because it is transparent and unchangeable once data is written down. This super helpful transparency helps ensure that everyone is doing what they say they do. Accountability is going to be improved with this level of transparency and clarity. Through its decentralized nature, Blockchain ensures that everyone involved in a process has access to the same, completely unchanged data and does not need any middleman. That openness means others can see what we are doing and can watch to make sure things follow the rules and rules are not being gamed for personal advantage. Hajiyev (2023) states that implementing Blockchain technology in EU sustainability reporting has provided clear examples of how technology can improve monitoring compliance with environmental and social goals.

The immutable nature of Blockchain is also critical for strengthening accountability within organizations. As data is recorded in each block of the chain and cannot be modified without the agreement of the parties involved in the network, it creates an irrefutable record of transactions and decisions. This feature is important for building clear systems that eliminate bad management and shady business, especially in high places like governance. Trotta, Rania, and Strano (2023) highlighted how Blockchain enhances the accountability of business decisions, especially in integrating technology with ESG criteria.

Blockchain technology can surely make government staff run better through smart contracts. Smart contracts are like having an online contract that can execute itself automatically when certain conditions are met. They can help to make government processes more efficient by cutting out many middlemen and reducing the risk of mistakes and backroom dealings. It is like having a detailed playbook of who does what on a team and how things work inside a digital safe where only the right people can open it, assuring that the right actions are taken at the right time. These self-executing contracts record the terms on the Blockchain, ensuring that all agreements are executed without human intervention or arbitration. As stated by Basdekidou and Papapanagos (2023), the application of smart contracts has the potential to reshape the way companies and governments collaborate, increasing the efficiency of governance processes by automating and ensuring compliance.

Furthermore, Blockchain can enhance accountability in the financial sector by integrating it into FinTech. Using Blockchain to handle cash transactions ensures everything is spot-on and completely clear for all parties involved. This cuts down blunders and safeguards us against fraudsters. According to Cui, Li, and Bai (2023), developments in FinTech in China have demonstrated how Blockchain technology enhances corporate ESG performance, promoting accountability and responsibility in business practices by facilitating more reliable reporting and decision-making (Ragazou et al., 2024).

While Blockchain has enormous potential to strengthen governance and accountability it is important to also deal with some challenges that come along. Limitations in technical expertise, regulatory gaps, and concerns about data security remain obstacles to fully realizing the potential of this technology. So, this neat progress is happening with Blockchain technology, and more and more government and business folks are cooperating on it. If those changes ever stick, the possibilities for improving the government are enormous and super important. These issues are explored further by Sanni (2023), who emphasizes the challenges of integrating smart contracts into supply chains and their role in enhancing operational transparency. By weaving all these elements together, Blockchain technology can make a big difference in making systems for managing stuff like politics and other companies stronger and smarter.

The combination of AI, Blockchain, and Fintech technologies offers organizations the tools to strengthen corporate governance by ensuring accountability, transparency, and ethical decision-making. The technologies mentioned enable businesses to comply with ESG standards, as they can monitor corporate practices in real-time, ensuring that they operate according to the necessary ethical guidelines. Artificial intelligence has enhanced corporate governance through the decision-support systems it offers and the processing of complex data it can perform, providing useful information. This constitutes better national decision-making, ensuring that businesses prioritize environmental protection, societal well-being, and effective governance with all this entails (Hajiyev, 2023).

An example to reinforce the above development of the theory is how governance platforms that have integrated artificial intelligence can analyze proposals received from shareholders, monitor the diversity of the composition of the board of directors, and monitor the impact that corporate actions have on society. These technologies and the tools they provide ensure that corporate boards make decisions consistent with the long-term sustainability of all ESG factors, i.e., environment, society, and governance, and do not aim exclusively at ensuring short-term profits at the expense of general well-being Basdekidou and Papapanagos (2023).

Blockchain technology is crucial in enhancing accountability and providing an immutable record of decisions and actions taken by businesses and organizations. By using Blockchain for corporate governance, companies can guarantee that their decisions are transparent, reducing the chances of corruption and/or unethical behavior. Additionally, Blockchain provides security, protecting sensitive personal data to assess compliance with ESG criteria. Therefore, it is an ideal solution for organizations that want to have integrity in data and processes related to compliance with ESG standards (Trotta, Rania, and Strano, 2023).

5. CHALLENGES AND OPPORTUNITIES FROM THE USE OF AI, BLOCKCHAIN, AND FINTECH TECHNOLOGIES IN ESG COMPLIANCE AND URBAN GOVERNANCE

Over the past few years, requests have grown for a way to do business that is more friendly to the environment and better city governance. So big tech hits like AI, blockchain, and fintech have come on to help us with that. These technologies are pivotal in improving ESG compliance and urban governance. Integrating AI, blockchain, and fintech in these sectors offers a range of challenges and opportunities, shaping the future of sustainable development.

Artificial intelligence is gaining significant attention for its potential to improve ESG compliance. One of the key advantages AI offers is the ability to process and analyze huge amounts of data in real time, facilitating more accurate and efficient ESG assessments. AI can enable financial institutions to evaluate a company's ESG performance by automating data collection from various sources, such as social media, company reports, and market data (Xu, 2024). This capability can make reports on ESG scores transparent. Transparency is important for organizations that try to stay compliant with all the rules and standards they must follow.

However, the use of AI also brings several challenges. One of the primary concerns is the potential for biases in AI algorithms, which could undermine the fairness of ESG assessments. For instance, AI systems may favor certain companies or sectors based on historical data, perpetuating existing inequalities or overlooking emerging sustainable practices (Nwogugu, 2023). The thing is AI systems depend critically on the information quality they are given and using bad or insufficient data can result in incorrect evaluations of corporate sustainability. Bad data sends companies the message that they are not doing a good job of going green, and such mistakes can also cause them to deny they have any sustainability issues or claim they are doing quite well at it falsely. Furthermore, there are concerns regarding the accountability of AI decisions in ESG compliance. As AI systems become more autonomous, it becomes increasingly difficult to trace the rationale behind certain outcomes, which could challenge regulatory frameworks designed to ensure transparency and fairness.

Despite these challenges, AI also presents opportunities to enhance ESG compliance. Automating ESG reporting can significantly reduce the administrative burden on companies, enabling them to allocate resources more efficiently. Moreover, AI can smoothly talk up some forecasting, too, letting companies look forward at what kind of green and ethical troubles could be coming and then proactively taking steps to sidestep them. That is very useful advice from Shahzad, Ghaemi Asl, and Tedeschi (2023), and it can be incredibly powerful for companies. Foreseeing risks before they happen lets companies avoid fines or untold damage to their reputation when they miss rules.

Blockchain technology, known for its transparency, immutability, and decentralization, has emerged as another tool to support ESG compliance. Blockchain can create transparent and tamper-proof records of a company's ESG activities, ensuring that reported information is verifiable and trustworthy. This is particularly relevant in industries with complex supply chains, where ensuring compliance with ESG standards can be difficult. Blockchain's distributed ledger system can track the movement of goods and resources across supply chains, providing verifiable proof that suppliers adhere to environmental and social standards (Schulz, Gstrein, Zwitter, 2020).

However, implementing blockchain in compliance with Environment, Social, and Governance (ESG) standards is challenging. One of the major hurdles is the scalability of blockchain networks. While blockchain offers significant advantages in terms of transparency, the energy consumption required to maintain these networks is a growing concern, particularly in environmental sustainability (Tanchangya et al., 2023). Some blockchain protocols are really energy intensive, using proof of work systems. Using vast energy means there is an issue with ESG compliance; ESG is about aligning how things get done with social good. So, blockchain sustainability is blemished when it uses such a huge amount of power. This results in some questions about whether this technology is truly sustainable.

Furthermore, while things are still fuzzy regarding regulation and how blockchain can help ensure better standards in sustainable business practices such as ESG Compliance, we are still figuring that out. One big challenge for companies that want to use this cutting-edge blockchain technology on important things like ESG reports might be that there are no unified rules or standards. When a company does not have clear rules and guidelines, it can get lost in whether they are doing right by law when using blockchain to keep tabs on sustainability. Additionally, implementing blockchain solutions can be prohibitive, especially for smaller companies lacking the resources to invest in such technologies (Hasan et al., 2023).

On the flip side, blockchain opens huge opportunities for boosting ESG compliance. Stuff like ensuring clean energy use and honest transactions is all greatly improved thanks to this technology. By creating immutable records of ESG-related activities, blockchain can help mitigate the risks of greenwashing, where companies falsely claim to meet environmental standards (Kazachenok, Stankevich, Chubaeva, Tyurina, 2023, Vortelinos et al., 2024). One great aspect of transparency that blockchain offers is accountability, which lets people involved in a company have access to clear information about that company's efforts around sustainability. Customers and other stakeholders can count on records generated using blockchain to be solid and factual.

Fintech technologies, encompassing digital payment systems, crowdfunding platforms, and online lending, play an increasingly important role in urban governance. The rise of green finance and sustainable investment has been closely linked to fintech innovations. Fintech platforms enable the efficient flow of capital into sustainable projects, making it easier for cities to finance green infrastructure, renewable energy projects, and social initiatives (Mhlanga, 2023, Vardopoulos et all, 2024 and Vardopoulos, 2024). Fintech has the potential to make finance accessible to many more people, especially those who have been left out before. They can also ensure their business practices are sustainable by offering them tools.

Sure, Fintech has much potential, but it also poses some big challenges for urban management. One of the big worries is financial inclusion. While fintech can potentially increase access to capital, it may also exacerbate inequalities if the technologies are not designed to be inclusive. For instance, underserved communities in developing regions may lack the digital literacy or infrastructure to participate in fintech-enabled green finance projects (Addy et al., 2023). One of the tricky things about fintech is that it can be a magnet for fraudsters and people who use computer tricks to get into things that should not be. This makes people worry about whether things are safe when money moves from one place to another. In this case, that safety is how governments run cities and allocate resources.

Another challenge is the regulatory environment surrounding the fintech. Governments and regulators face difficulties in keeping up with the rapid pace of fintech innovation, and the absence of clear regulations can hinder the widespread adoption of fintech for ESG purposes. The lack of standardized ESG metrics and reporting frameworks for fintech solutions further complicates ensuring compliance with ESG standards in urban governance (Rane, Choudhary, Rane, 2023).

Nevertheless, fintech offers considerable opportunities for improving urban governance. Through digital platforms, cities can facilitate more efficient and transparent financial transactions for sustainable projects. By using some deep tech like AI and machine learning, Fintech companies are like detectives, studying lots and lots of data, looking for patterns, and making predictions that relate to helping cities be greener, and that helps policymakers make smart decisions (Elias et al., 2023). Fintech can fuel innovation by helping to fund green projects that would otherwise struggle to get traditional financing—it is like giving green-light projects the boost they need through financial backing (Garrido & Alberto, 2023).

In conclusion, the Use of AI Blockchain and Fintech Technologies in ESG Compliance and Urban Governance Challenges and Opportunities Undoubtedly, the infusion of technologies such as AI, blockchain, and fintech into civic governance and ESG compliance has many pros. However, the cons exist in equal measures. The primary challenge is that innovation must be standardized across businesses and organizations for ESG data to become homogeneous and comparable across enterprises. Secondly, governments and/or small firms with low resources may find it daunting to adopt these technologies because of their complexity. Even challenges cannot be victories. Such tools will be effective by joining hands with tech givers, market players, and more within the field. Also, the growth of Fintech Blockchain and AI tools brings chances for wide use across many areas (Elias et al., 2023).

6. CONCLUSIONS: THE CONTRIBUTION OF TECHNOLOGY TO ENHANCING ESG COMPLIANCE AND URBAN GOVERNANCE

Integrating AI, Blockchain, and Fintech technologies offers a new path to enhance ESG compliance. Businesses and organizations can now have useful tools to increase transparency, efficiency, and accountability in recording and monitoring the data required for ESG compliance. From monitoring the decision-making process and financing sustainable investments, the technologies discussed in this chapter offer solutions for ESG compliance and enhance the trust and accountability key elements for successful urban governance. AI, Blockchain, and Fintech technologies provide a solution for organizations to evolve into responsible and transparent entities that act in their financial interests while ensuring the benefit of the environment and society to achieve sustainable and ethical development (Elias et al., 2023).

REFERENCES

Addy, W. A., Ofodile, O. C., Adeoye, O. B., Oyewole, A. T., Okoye, C. C., Odeyemi, O., & Ololade, Y. J. (2023). *Data-driven sustainability: How fintech innovations are supporting green finance*. ResearchGate.

Alqudah, M. Z., García, L. S., & Garcia-Benau, M. A. (2024). *Influence of ESG among Jordanian fintech businesses: The moderating role of sustainable finance technology*. Springer Link., DOI: 10.1007/s43621-024-00712-6

Baldi, F., & Pandimiglio, A. (2022). The role of ESG scoring and greenwashing risk in explaining the yields of green bonds: A conceptual framework and an econometric analysis. *Global Finance Journal, 52*, 1044028322000138. DOI: 10.1016/j.gfj.2022.100711

Basdekidou, V., & Papapanagos, H. (2023). Blockchain technology adoption for disrupting FinTech functionalities: A systematic literature review for corporate management, supply chain, banking industry, and stock markets. *ResearchGate*. https://www.researchgate.net/publication/383969110_Blockchain_Technology_Adoption_for_Disrupting_FinTech_Functionalities_A_Systematic_Literature_Review_for_Corporate_Management_Supply_Chain_Banking_Industry_and_Stock_Markets

Cheliatsidou, A., Sariannidis, N., Garefalakis, A., Passas, I., & Spinthiropoulos, K. (2023). Exploring Attitudes towards Whistleblowing in Relation to Sustainable Municipalities. Administrative Sciences 2023, Vol. 13, Page 199, 13(9), 199. DOI: 10.3390/admsci13090199

Chen, R., & Zhang, T. (2024). Artificial intelligence applications implication for ESG performance: Can digital transformation of enterprises promote sustainable development? *Chinese Management Studies*, *18*(1). Advance online publication. DOI: 10.1108/CMS-11-2023-0653

Chen, S. (2024). The influence of artificial intelligence and digital technology on ESG reporting quality. *International Journal of Global Economics and Management*, *3*(1), 301–310. Advance online publication. DOI: 10.62051/IJGEM.v3n1.36

Cui, Z., Li, Q., & Bai, M. (2023). Does fintech improve corporate ESG performance? Evidence from China. *SSRN*. https://papers.ssrn.com/sol3/papers.cfm?abstract_id=5103526

Datta, K. (2023). AI-driven public administration: Opportunities, challenges, and ethical considerations. *ResearchGate*. https://www.researchgate.net/publication/387297421_ai-driven_public_administration_opportunities_challenges_and_ethical_considerations

Devey, J., & Ferraris, C. (2023). Ensuring ethical AI: Strengthening data governance and algorithmic accountability in the digital era. *ResearchGate*. https://www.researchgate.net/publication/384628034_Ensuring_Ethical_AI_Strengthening_Data_Governance_and_Algorithmic_Accountability_in_the_Digital_Era

Duran, R. E., & Tierney, P. (2023). Fintech data infrastructure for ESG disclosure compliance. *Journal of Risk and Financial Management*, *16*(8), 378. DOI: 10.3390/jrfm16080378

Elias, O., Awotunde, O. J., Oladepo, O. I., Azuikpe, P. F., Adeleye, O. S., Oladele, O. R., & Ogunruku, O. O. (2023). The evolution of green fintech: Leveraging AI and IoT for sustainable financial services and smart contract implementation. *ResearchGate*. https://www.researchgate.net/publication/382785523_The_evolution_of_green_fintech_Leveraging_AI_and_IoT_for_sustainable_financial_services_and_smart_contract_implementation

Galeone, G., Ranaldo, S., & Fusco, A. (2024). ESG and FinTech: Are they connected? *Research in International Business and Finance*, *69*, 0275531924000175. DOI: 10.1016/j.ribaf.2024.102225

Garrido, V., & Alberto, M. (2023). ESG-driven innovation in the fintech industry. *ResearchGate*. https://www.researchgate.net/publication/373238616_ESG-driven_innovation_in_the_fintech_industry

Georgiou, C., Maialetti, M., Zorpas, A. A., Salvati, L., & Vardopoulos, I. (2025). Policy and planning for "zero-net energy" real estate development in rural and exurban districts. In *Environmental Sustainability and Global Change* (pp. 291–303). Elsevier., DOI: 10.1016/B978-0-443-31596-1.00020-9

Giakoumelou, A., Salvi, A., Bekiros, S., & Onorato, G. (2024). ESG and fintech funding in the EU. *Research in International Business and Finance*, *69*, 0275531924000254. DOI: 10.1016/j.ribaf.2024.102233

Gong, X., Tao, X., Zhang, M., Xu, Y., Kwok, H. H. L., Dai, J., & Cheng, J. C. P. (2023). Secure environmental, social, and governance (ESG) data management for construction projects using blockchain. *OUCI*. https://ouci.dntb.gov.ua/en/works/loegEKz7/

Hajiyev, S. (2023). Blockchain technology in EU sustainability reporting: Legal analysis. *Lund University Publications*. https://lup.lub.lu.se/student-papers/search/publication/9157333

Hasan, M. H., Hossain, M. Z., Hasan, L., & Dewan, M. A. (2023). FinTech and sustainable finance: How is FinTech shaping the future of sustainable finance? *ResearchGate*. https://www.researchgate.net/publication/386548986_FinTech_and_Sustainable_Finance_How_is_FinTech_Shaping_the_Future_of_Sustainable_Finance

Hossain, M. Z., & Ali, S. (2023). Exploring the role of artificial intelligence in ESG disclosure. *Journal of Business Ethics*, *168*(3), 1–17. DOI: 10.1007/s10551-023-04924-3

Kagias, P., Sariannidis, N., Garefalakis, A., Passas, I., & Kyriakogkonas, P. (2023). Validating the Whistleblowing Maturity Model Using the Delphi Method. Administrative Sciences 2023, Vol. 13, Page 120, 13(5), 120. https://doi.org/DOI: 10.3390/admsci13050120

Kazachenok, O. P., Stankevich, G. V., Chubaeva, N. N., & Tyurina, Y. G. (2023). Economic and legal approaches to the humanization of FinTech in the economy of artificial intelligence through the integration of blockchain into ESG finance. *ResearchGate*. https://www.researchgate.net/publication/370027001_Economic_and_legal_approaches_to_the_humanization_of_FinTech_in_the_economy_of_artificial_intelligence_through_the_integration_of_blockchain_into_ESG_Finance

Lee, J., Serafin, A. M., & Courteau, C. (2023). Corporate disclosure, ESG and green fintech in the energy industry. *SSRN*. https://papers.ssrn.com/sol3/papers.cfm?abstract_id=4090792 DOI: 10.1093/jwelb/jwad018

Lee, S. U., Perera, H., Liu, Y., Xia, B., Lu, Q., Zhu, L., Cairns, J., & Nottage, M. (2024). Integrating ESG and AI: A comprehensive responsible AI assessment framework. *arXiv*. https://doi.org//arXiv.2408.00965DOI: 10.48550

Li, Q., Zou, G., Zeng, W., Gao, J., He, F., & Zhang, Y. (2024). ESG guidance and artificial intelligence support for power systems analytics in the energy industry. *Scientific Reports*, *14*(1), 11347. DOI: 10.1038/s41598-024-61491-8 PMID: 38762582

Liu, X., Wu, H., Wu, W., Fu, Y., & Huang, G. Q. (2019). Blockchain-enabled ESG reporting framework for sustainable supply chain. *ResearchGate*. https://www.researchgate.net/publication/344274654_Blockchain-Enabled_ESG_Reporting_Framework_for_Sustainable_Supply_Chain

Macchiavello, E., & Siri, M. (2020). Sustainable finance and fintech: Can technology contribute to achieving environmental goals? A preliminary assessment of 'green FinTech'. *SSRN*. https://papers.ssrn.com/sol3/papers.cfm?abstract_id=3672989 DOI: 10.2139/ssrn.3672989

Mhlanga, D. (2023). FinTech, financial inclusion, and sustainable development: Disruption, innovation, and growth. *Routledge*. https://www.routledge.com/FinTech-Financial-Inclusion-and-Sustainable-Development-Disruption-Innovation-and-Growth/Mhlanga/p/book/9781032657912?srsltid=AfmBOoo5NDLStJnH0bUpvjOzreX0FwibhTUcJ3boSiBRJBLMO1Td0zqc

Nwogugu, M. C. (2023). ESG, inequality and international political economy problems inherent in regulation of DeFi, Web5/Web3, cryptocurrencies and fractional-NFTs. *SSRN*. https://papers.ssrn.com/sol3/papers.cfm?abstract_id=4164776

Passas, I. Accounting for Integrity: ESG and Financial Disclosures: The Challenge of Internal Fraud in Management Decision-Making. Ph.D. Thesis, Hellenic Mediterranean University, Heraklion, Greece, 2023

Passas, I. (2024). The Evolution of ESG: From CSR to ESG 2.0. *Encyclopedia*, *4*(4), 1711–1720. DOI: 10.3390/encyclopedia4040112

Passas, I., Ragazou, K., Zafeiriou, E., Garefalakis, A., & Zopounidis, C. (2022). ESG Controversies: A Quantitative and Qualitative Analysis for the Sociopolitical Determinants in EU Firms. Sustainability 2022, Vol. 14, Page 12879, 14(19), 12879. https://doi.org/DOI: 10.3390/su141912879

Pavlides, T., Vardopoulos, I., Papamichael, I., Voukkali, I., Stylianos, M., & Zorpas, A. A. (2023). Environmental sustainability assessment of excavation, construction, and demolition waste conditions and practices across Greece and Cyprus. *IOP Conference Series. Earth and Environmental Science*, *1196*(1), 012037. DOI: 10.1088/1755-1315/1196/1/012037

Ragazou, K., Lemonakis, C., Passas, I., Zopounidis, C., & Garefalakis, A. (2024), "ESG-driven ecopreneur selection in European financial institutions: entropy and TOPSIS analysis", *Management Decision*, Vol. ahead-of-print No. ahead-of-print. https://doi.org/DOI: 10.1108/MD-12-2023-2425

Ragazou, K., Passas, I., Garefalakis, A., Galariotis, E., & Zopounidis, C. (2023). Big Data Analytics Applications in Information Management Driving Operational Efficiencies and Decision-Making: Mapping the Field of Knowledge with Bibliometric Analysis Using R. *Big Data and Cognitive Computing*, *7*(1), 13. DOI: 10.3390/bdcc7010013

Ragazou, K., Passas, I., Garefalakis, A., Zafeiriou, E., & Kyriakopoulos, G. (2022). The Determinants of the Environmental Performance of EU Financial Institutions: An Empirical Study with a GLM Model. *Energies*, *15*(15), 5325. DOI: 10.3390/en15155325

Rane, N., Choudhary, S., & Rane, J. (2024). Artificial intelligence driven approaches to strengthening Environmental, Social, and Governance (ESG) criteria in sustainable business practices: A review. *SSRN*. https://doi.org/DOI: 10.2139/ssrn.4843215

Rane, N. L., Choudhary, S. P., & Rane, J. (2023). Blockchain and artificial intelligence (AI) integration for revolutionizing security and transparency in finance.

Ribeiro Hoffmann, A., Sandrin, P., & Doukas, Y. E. (Eds.). (2024). Climate change in regional perspective. European Union and Latin American initiatives, challenges, and solutions. Cham, Switzerland: United Nations University Series on Regionalism, Vol. 27, Springer Nature. https://doi.org/DOI: 10.1007/978-3-031-49329-4

Riyath, M. I. M., & Jariya, A. M. I. (2024). The role of ESG reporting, artificial intelligence, stakeholders and innovation performance in fostering sustainability culture and climate resilience. *Journal of Financial Reporting and Accounting*. https://doi.org/DOI: 10.1108/JFRA-10-2023-0621

Rozkrut, D., Tarczynska-Luniewska, M., Singh, G. A., & Piwowarski, M. (2024). The use of big data in ESG monitoring and evaluation. In Grima, S., Maditinos, D., Noja, G. G., Stankevičienė, J., Tarczynska-Luniewska, M., & Thalassinos, E. (Eds.), *Exploring ESG challenges and opportunities: Navigating towards a better future* (Vol. 116, pp. 25–40). Emerald Publishing Limited., DOI: 10.1108/S1569-375920240000116003

Schulz, K. A., Gstrein, O. J., & Zwitter, A. J. (2020). Exploring the governance and implementation of sustainable development initiatives through blockchain technology. *Futures*. https://papers.ssrn.com/sol3/papers.cfm?abstract_id=3663149

Shahzad, U., Ghaemi Asl, M., & Tedeschi, M. (2023). Is there any market state-dependent contribution from blockchain-enabled solutions to ESG investments? Evidence from conventional and Islamic ESG stocks. *Research in Economics*, *86*, 139–154. https://ideas.repec.org/a/eee/reveco/v86y2023icp139-154.html

Stavropoulou, E., Spinthiropoulos, K., Ragazou, K., Papademetriou, C., & Passas, I. (2023). Green Balanced Scorecard: A Tool of Sustainable Information Systems for an Energy Efficient Business. *Energies*, *16*(18), 6432. DOI: 10.3390/en16186432

Sulkowski, A. J. (2024). AI, ESG, and law: Potential, limitations, and strategies concerning artificial intelligence in sustainability reporting. *SSRN*. https://doi.org/DOI: 10.2139/ssrn.4972787

Tanchangya, T., Raihan, A., Mia, M. R., Tafsirun, U., Naher, K., Islam, N., Rashid, F., & Sarker, S. R. (2023). Greenwashing within the context of financial technology and sustainable development: Conceptual frameworks and theoretical perspectives.

Trotta, A., Rania, F., & Strano, E. (2023). Exploring the linkages between FinTech and ESG: A bibliometric perspective.

Vardopoulos, I. (2024). Urban planning policies and tools for sustainable development: insights into adaptive reuse, smart cities, and urban metabolism. In *Socio-enviromental change: The path toward sustainability* (pp. 121–140). Nova Science Publishers.

Vardopoulos, I. (2025). Cities in flux: Waste, the shadow of circularity, and the (elusive) promise of sustainability. *Sustainability (Basel)*, *17*(2), 705. DOI: 10.3390/su17020705

Vardopoulos, I., Giannopoulos, K., Kalli, S., & Salvati, L. (2024). From tech-centric to human-centric smart tourist destinations: innovations in real estate, urban planning, and sustainability. In *Socio-enviromental change: the path toward sustainability* (pp. 65–96). Nova Science Publishers.

Vardopoulos, I., & Konstantinou, Z. (2017). Study of the possible links between CO2 emissions and employment status. Sustainable Development, Culture, Traditions Journal, 1(B), 100–112. https://doi.org/DOI: 10.26341/issn.2241-4002-2017-1b-8

Vardopoulos, I., Papoui-Evangelou, M., Nosovo, B., & Salvati, L. (2023). Smart 'tourist cities' revisited: Culture-led urban sustainability and the global real estate market. *Sustainability (Basel)*, *15*(5), 4313. DOI: 10.3390/su15054313

Vardopoulos, I., Santamouris, M., Zorpas, A. A., Barone, G., Italos, C., & Vassiliades, C. (2024). A comparative study on discrepancies in residential building energy performance certification in a Mediterranean context. *Buildings (Basel, Switzerland)*, *14*(4), 1009. DOI: 10.3390/buildings14041009

Vortelinos, D., Menegaki, A., Passas, I., Garefalakis, A., & Viskadouros, G. (2024). Heterogeneous Responses of Energy and Non-Energy Assets to Crises in Commodity Markets. Energies 2024, Vol. 17, Page 5438, 17(21), 5438. DOI: 10.3390/en17215438

Vortelinos, D., Passas, I., Floros, C., & Garefalakis, A. (2024). Corporate Social Responsibility and Country Governance: An International Comparative Study Amid the COVID-19 Pandemic. International Journal of Financial Studies 2024, Vol. 12, Page 110, 12(4), 110. DOI: 10.3390/ijfs12040110

Wang, C., Jiang, Q., Dittrich, Y., & Østerlund, C. (2023). ESG in fintech: An overview. *IT University of Copenhagen*. https://pure.itu.dk/ws/portalfiles/portal/101319385/TR-2023-209.pdf

Wang, D., Peng, K., Tang, K., & Wu, Y. (2023). Does fintech development enhance corporate ESG performance? Evidence from an emerging market. *SSRN*.

Weston, P., & Nnadi, M. (2023). Evaluation of strategic and financial variables of corporate sustainability and ESG policies on corporate finance performance. *SSRN*. DOI: 10.1080/20430795.2021.1883984

Winfield, A. F. T., & Jirotka, M. (2019). *Ethical governance is essential to building trust in robotics and artificial intelligence systems*. ResearchGate.

Xu, J. (2024). AI in ESG for financial institutions: An industrial survey. *arXiv*. https://ideas.repec.org/p/arx/papers/2403.05541.html

Yiannoulis, Y., Vortelinos, D., & Passas, I. (2025). Exploring Audit Opinions: A Deep Dive into Ratios and Fraud Variables in the Athens Exchange. Accounting and Auditing 2025, Vol. 1, Page 3, 1(1), 3. DOI: 10.3390/accountaudit1010003

Zopounidis, C., Garefalakis, A., Lemonakis, C., & Passas, I. (2020). Environmental, social and corporate governance framework for corporate disclosure: A multicriteria dimension analysis approach. *Management Decision*, *58*(11), 2473–2496. DOI: 10.1108/MD-10-2019-1341

Chapter 3
Decoding Greenwashing: Consumer Skepticism and Trust in Decathlon's Sustainability Claims

Rahul Makker
School of Commerce, Manav Rachna International Institute of Research and Studies, Faridabad, India

Ashwarya Kapoor
https://orcid.org/0000-0003-3858-3945
School of Commerce, Manav Rachna International Institute of Research and Studies, Faridabad, India

ABSTRACT

Sustainability has become a key factor in consumer decision-making with growing awareness of environmental responsibility influencing brand choices. Decathlon, a global leader in sporting goods retailing, promotes itself as a green brand through eco-designed products, recycling programs and a responsible supply chain. However, greenwashing—where brands exaggerate or mislead consumers about environmental efforts—has led to skepticism and trust deficits. This study adopts a mixed-method approach combining surveys, interviews and literature research to explore factors contributing to consumer skepticism. Key areas of analysis include transparency in communication, consistency in environmental claims and the role of third-party certifications. Findings reveal a dual consumer perception: while many recognize Decathlon's sustainability efforts, others remain skeptical due to unverifiable or inconsistent information. This lack of trust negatively impacts both purchasing decisions and long-term brand loyalty.

DOI: 10.4018/979-8-3373-2327-5.ch003

1. INTRODUCTION

Sustainability has emerged as a very critical issue for businesses across all industries. The increasing consumer awareness and demand for environmentally responsible practices have pushed businesses to adapt (Szabo and Webster, 2021; Adamkiewicz et al., 2022, Vortelinos et al., 2024). In the retail sector, companies like Decathlon have been very proactive in positioning themselves as champions of sustainable development through initiatives such as eco-designed products, recycling programs and responsible supply chain management (Pimonenko et al., 2020; Free et al., 2024). These initiatives resonate with more global objectives to minimize environmental effects and encourage environmentally friendly consumption. However, the challenge of greenwashing when businesses exaggerate or even lie about their contributions to the environment has led to rampant consumer skepticism (Rahman and Nguyen-Viet, 2023, Passas et al., 2022). False claims have made customers highly critical of the validity and workability of corporate sustainability initiatives. Greenwashing is not a recent phenomenon (Qayyum et al., 2023). As according to Delmas and Burbano (2011), it is observed that brands utilize greenwashing as a marketing technique to attract eco-conscious consumers but later reputation damage is inflicted once the claims prove to be baseless. Similarly, Lyon and Maxwell (2011) state that whereas short-term benefits may be enhanced by greenwashing, long-term trust and credibility might be lost. For instance, with Decathlon being a very sustainable brand which values access to sports, it is what the consumer puts his or her trust in these initiatives. Recent development, misleading claims by retailers have further complicated this relationship between consumer expectation and brand communication. Thus, aspects that influence the trust and disbelief of consumers with regards to sustainability communications require due attention. These issues include role of transparency, appropriate communication strategy and third-party certification validation.

The purpose of this research paper is to explore consumer insights into sustainability practices by focusing on Decathlon to evaluate the gap between the company's sustainability claims and consumer perceptions. The study attempts to explore how transparency, communication and trust influence consumer attitudes towards Decathlon's environmental initiatives. Thus, this study strives to achieve the following objectives:

- To analyze consumer perceptions of Decathlon's sustainability claims and their credibility.
- To identify key factors contributing to consumer skepticism or trust in Decathlon's environmental initiatives.

- To examine the impact of greenwashing on consumer purchasing behaviour and brand loyalty in the retail sector.

2. LITERATURE REVIEW

2.1 Sustainability and Greenwashing from marketing perspective

Sustainability marketing involves promoting practices that align with environmental and social responsibility, addressing consumer needs while minimizing ecological impact (Peattie & Crane, 2005; Yang et al., 2020). Companies that authentically incorporate sustainability into their operations benefit from increased trust, customer loyalty, and a willingness among consumers to pay premium prices.

Beyond eco-friendly products, sustainability marketing often emphasizes ethical labor practices, supply chain transparency, and community initiatives. Storytelling plays a vital role, as brands use narratives to connect emotionally with consumers by showcasing their sustainability efforts (Furlow, 2010; Yildirim,2023). However, the effectiveness of such campaigns hinges on credibility and clarity, as skepticism toward green claims remains high.

By making exaggerated or inaccurate statements about a company's environmental policies, greenwashing sabotages sustainability initiatives. Terra Choice (2009) "Seven Sins of Greenwashing" framework classifies these strategies, which include unsubstantiated information, ambiguous claims, and hidden trade-offs. In addition to harming the reputations of specific brands, this practice undermines consumer confidence in sustainability marketing in general (Lopes et al., 2023).

2.2 Sustainability and Brand Loyalty

Sustainability has shifted from being a niche problem to a universal expectation, therefore, it becomes a vital part of the brand. Studies show that when choosing brands, consumers—particularly the Millennials and Generation Z—are prioritizing ethical and responsible practices (Cone Communications, 2017). Eco-friendly materials, a reduction in carbon footprints and community development, among others, thus become essential practices to stay competitive in competitive markets.

Brand loyalty is the repeated preference and loyalty of a customer toward repurchasing of a particular brand (Kapoor et al., 2022a, 2024a, b). Sustainability has been proven to be a determinant in this loyalty by fostering trust and emotional attachment.

2.3 Consumer Skepticism and Decision-Making

Consumer skepticism in sustainability is the growing distrust towards the environmental claims of brands, especially because of greenwashing. As companies are increasingly advertising their eco-friendly practices, consumers have become cautious and question whether these claims are genuine (de Freitas Netto et al., 2020). This skepticism leads them to carefully evaluate green statements seeking clear evidence through certifications, transparency reports and reviews before making purchasing decisions. Vagueness and exaggeration may push consumers to delay or even avoid purchases and they opt for brands that have history-of-sustainability. This demand for authenticity and accountability has forced brands to be more transparent as consumers are unwilling to back up the companies that can't validate their claims (Badhwar et al., 2024).

2.3.1 Creating Mistrust

Mistrust in sustainability arises when consumers doubt the authenticity of a brand's environmental or ethical claims. This mistrust is frequently tied to practices such as greenwashing, where companies exaggerate or misrepresent their sustainability efforts to appear environmentally conscious (Nemes et al., 2022). Delmas and Burbano (2011) define greenwashing as a deliberate attempt to promote deceptive environmental practices, which undermines consumer trust and damages brand reputation. Over time such practices create a broader skepticism toward corporate sustainability claims, even when they are genuine.

2.3.2 Impact of Vague or Misleading Sustainability Labels on Consumer Trust

Sustainability labels are designed to guide consumers in making environmentally conscious purchasing decisions. However, vague or misleading labels can significantly undermine consumer trust. Terms like "eco-friendly" "green" or "natural" often lack standardized definitions or measurable criteria making it difficult for consumers to verify the authenticity of a product's claims (Duchin et al., 2024). This lack of clarity fosters confusion and skepticism as consumers struggle to distinguish between genuinely sustainable products and those using greenwashing tactics.

2.3.3 Misalignment Between Brand Claims and Actual Practices

The misalignment between a brand's sustainability claims and its actual practices is a significant driver of consumer mistrust and skepticism. While brands often position themselves as environmentally or socially responsible, discrepancies between their marketed image and real-world actions can lead to accusations of greenwashing (Butt et al., 2021). This misalignment is characterized by overstated or unsubstantiated claims that fail to align with measurable outcomes or transparent practices (Delmas & Burbano, 2011).

For instance, brands may promote limited sustainable initiatives while continuing harmful practices in other areas, a tactic often referred to as selective disclosure. Such inconsistencies undermine consumer confidence, with studies showing that consumers are quick to penalize brands that fail to deliver on their promises (Leonidou et al., 2013; Nygaard and Silkoset, 2023). Cognitive dissonance plays a key role, as consumers who perceive this misalignment feel deceived leading to reduced brand loyalty and trust. Moreover, when third-party watchdogs or media expose these gaps, it amplifies reputational damage and fosters industry-wide skepticism (Lyon & Montgomery, 2015, Cheliatsidou et al., 2023; Kagias et al., 2023, 2024; Passas, 2023; Ragazou et al., 2022; Yiannoulis et al., 2025; Zafeiriou et al., 2023).

2.4 Consumer Perceptions of Greenwashing

Consumer insights into greenwashing reveal a strong correlation between perceived deception and reduced brand loyalty. Studies show that consumers increasingly scrutinize sustainability claims, relying on transparency, third-party certifications and measurable metrics to differentiate authentic efforts from greenwashed initiatives (Ritch, 2023). Perceptions of greenwashing are particularly pronounced in retail sectors where environmental claims are prominently advertised such as fashion and sportswear (Leonidou et al., 2011).

Capgemini (2022) reported that 79% of consumers prefer products from brands they perceive as sustainable and 41% of consumers actively research a brand's sustainability practices before making purchases.

2.5 Role of Transparency in Building Consumer Trust

Transparency in sustainability practices is essential for fostering consumer trust and mitigating greenwashing accusations. Research highlights the importance of clear communication, third-party validation, and accessible sustainability reporting in strengthening consumer confidence. Brands that disclose detailed information about their eco-friendly initiatives such as supply chain data, product life cycle impacts and carbon neutrality goals—are more likely to gain consumer trust.

2.6 Consumer Behaviour Towards Eco-Friendly Products

2.6.1 Willingness to Pay a Premium for Sustainable Products

Consumer willingness to pay a premium for sustainable products has become a key area of interest in sustainability research (Williams, 2024). Studies suggest that many consumers are inclined to spend more on products perceived as environmentally friendly or ethically produced driven by personal values, environmental awareness and social responsibility. However, this willingness varies across demographics and depends on factors such as income, education and cultural context.

2.6.2 Influence of Product Certifications and Eco-Labels on Buying Behaviour

Product certifications and eco-labels significantly influence consumer buying behaviour by serving as indicators of trust and credibility for sustainable products (Dempere et al., 2024). Research shows that certifications from reputable third-party organizations reduce consumer skepticism and increase perceived product quality and environmental impact (Testa et al., 2015). Eco-labels such as Fair Trade, USDA Organic or Energy Star provide consumers with a sense of assurance that the product meets specific environmental or ethical standards. This perceived credibility plays a crucial role in shaping purchase decisions, especially among environmentally conscious consumers (Thøgersen et al., 2010).

2.7 Social Media's Role in Consumer Insights

Social media plays a pivotal role in shaping consumer insights by serving as a platform for interaction, feedback, and brand-consumer engagement. Through social media, companies gain access to real-time data on consumer opinions, preferences and behaviors which can inform marketing strategies and product development (Kapoor et al., 2022b, 2023a). Platforms like Instagram, Facebook and Twitter allow brands to monitor trends, analyze customer sentiment and gauge public perception through tools such as social listening and sentiment analysis (Kaplan & Haenlein, 2010).

2.7.1 How consumers Use Social Media to Evaluate and Share Opinions on Sustainability Claims

Consumers increasingly rely on social media to evaluate and share opinions on sustainability claims, making it a critical platform for shaping public perception of a brand's environmental and ethical practices. Social media provides consumers with access to diverse information sources, including official brand accounts, third-party reviews and peer-generated content enabling them to assess the credibility of sustainability claims. According to a study by Chen and Chang (2013), consumers use social media to verify claims by cross-referencing brand statements with external evidence, such as news articles, certifications and activist reports.

2.7.2 Impact of Viral Campaigns and Online Reviews on Consumer Trust

Viral campaigns and online reviews have a profound impact on consumer trust, shaping perceptions of brands and influencing purchasing decisions (Kapoor et al., 2025). Viral campaigns which often involve compelling, shareable content that spreads rapidly across social media can significantly enhance brand visibility and credibility when they highlight a brand's positive attributes or sustainability efforts (Kapoor et al., 2024c; 2023b, 2021a, b). According to studies by Kietzmann et al. (2011), viral marketing strategies capitalize on consumers' social networks creating a sense of trust through peer endorsement and organic engagement. When consumers encounter messages from their social circles or influencers they trust, they are more likely to perceive these brands as credible and reliable.

On the other hand, viral campaigns that expose negative aspects of a brand such as accusations of greenwashing or unethical practices can have devastating effects on consumer trust. Negative viral content spreads quickly and has a lasting impact on public perception as consumers are highly influenced by social proof and word-of-mouth (Tuten & Solomon, 2017). For instance, scandals or misleading claims often gain traction on social media where users actively share their concerns and amplify the issue, prompting widespread scrutiny. Such campaigns can damage a brand's reputation leading to consumer boycotts or a significant decline in brand loyalty.

2.7.3 Role of Influencers in Shaping Perceptions of Sustainability

Influencers play a significant role in shaping consumer perceptions of sustainability, as they have the ability to amplify brand messages, influence attitudes, and drive consumer behavior. With their established credibility and large, engaged followings, influencers serve as trusted sources of information, especially in the realm of social and environmental issues. Studies indicate that consumers are more likely to trust sustainability claims endorsed by influencers, particularly those who align with ethical or eco-conscious values (De Veirman et al., 2017). Influencers, especially those in the green or ethical lifestyle space, bridge the gap between brands and consumers by providing relatable and personalized endorsements of sustainable products or practices.

3 RESEARCH METHODOLOGY

The study had adopted an exploratory and descriptive nature to understand customer behaviour and preferences concerning the sustainability initiatives of Decathlon. The data were collected through personal interviews mainly with an objective to gain a more profound understanding. The sample size selected for the study consisted of 46 respondents in a way that it represented various diverse ranges of views. It was balanced because it involved male and female participants as a sample unit to obtain consumer attitudes. The study selected customers who visited the Decathlon store at Saket to be the sample. This location was chosen specifically to gain a localized view of consumer behaviour. Respondents were those who came to shop at Decathlon, therefore providing firsthand insights into their views and experiences with the brand's efforts toward sustainability.

4. ANALYSIS AND FINDINGS

To fully understand consumer perception of Decathlon's sustainability initiatives, Authors have conducted an in-depth case study on the company's sustainability policies and practices. Consumer feedback is very valuable to evaluate the effectiveness of corporate sustainability initiatives. Thus, authors carried out qualitative interviews with 43 Decathlon customers. These will be used to elicit their knowledge, opinions and experience regarding Decathlon's environmental commitments and practices.

Interviews were conducted on several aspects of Decathlon's sustainability strategy such as use of environment-friendly materials, product lifecycle management and efforts to lower the carbon footprint and transparency in communicating sustainability goals. Participants were questioned on their awareness of the company's initiatives, how they thought the company's authenticity in dealing with environmental issues and if the same actions influenced their purchase decisions. This method enabled a closer look at whether Decathlon's attitudes toward profitability were being balanced out by some kind of conservation.

In addition to gathering insights into sustainability practices, the study sought to investigate the possibility of greenwashing when businesses present a false image of environmental responsibility. Customers were encouraged to share whether they believe Decathlon's claims are consistent with its actions and how such perceptions affect their trust in the brand. The interviews also sought to identify any knowledge gaps among consumers about Decathlon's initiatives as well as the role of communication in shaping these perceptions.

This study seeks to close the gap between Decathlon's stated sustainability commitments and how these are perceived by its customers. By analyzing customer feedback, the study identifies key trends, areas for improvement and the efficacy of Decathlon's communication about sustainability policies.

4.1 "How familiar are you with Decathlon's sustainability initiatives and environmental commitments?"

Table 1. Responses

Response	Number of Customers
Somewhat familiar: "I've heard about their sustainability efforts but don't know details."	16
Not very familiar: "I know they do something but can't specify what."	11
Not at all familiar: "I'm unaware of Decathlon's sustainability initiatives."	19

The results indicate that the levels of awareness by customers over Decathlon's sustainability efforts were not uniform. 16 of the respondents expressed that they are somewhat familiar about the efforts put by the brand. Meaning, they know about its efforts but don't understand the details about them. Hence, it reflects that although the brand's sustainability messages have hit a section of its public, it doesn't inform them of the details of its efforts.

On the other hand, 11 respondents reported that they were not very familiar. They knew sustainability efforts exist but did not know what it was all about. This group indicates a possible lack of communication in which the brand's efforts are known but not clearly articulated or understood by its customers.

The most worrying insight is that 19 of the respondents, the largest group were not at all familiar with Decathlon's sustainability initiatives. This shows a huge lack of awareness among a large part of the target group and thus Decathlon needs to have more powerful and effective communication and outreach strategies to educate its customers.

4.2. What comes to mind when you hear that Decathlon is working toward being a more sustainable brand?

Table 2. responses

Response	Number of Customers
Recycling initiatives: "They must be focusing on recycling programs."	16
Greenwashing concerns: "It might just be marketing without real impact."	20
No clear thoughts: "I'm not sure what sustainability means for them."	10

16 respondents associated Decathlon's sustainability efforts with recycling programs. This means that recycling is a recognizable and relatable concept for customers meaning that Decathlon's communication around these programs may not be very effective. This, however, is a relatively small number of respondents suggesting that the messaging needs to be stronger to make recycling initiatives a central feature of Decathlon's sustainability narrative.

Of those 20, a large majority thought the brand would be accused of greenwashing. This translates to a lack of confidence in Decathlon's message as truthful. This mistrust might stem from a lack of transparency about the brand's sustainable practices, third-party verification lacking, or generic marketing language that consumers do not feel convinces them of a brand's commitment to sustainability. Addressing these issues is what will build and then maintain trust with customers.

10 admitted that they did not know what sustainability means for Decathlon. This points to a lack of consumer education and awareness, which indicates that the brand needs to provide clearer and more accessible information about its sustainability goals and initiatives to foster greater engagement and understanding.

The above areas will assist Decathlon in developing strength in all manners and therefore, better aligned sustainability initiatives with expectations of consumers.

4.3 "How transparent do you feel Decathlon is about their sustainability goals and achievements?"

Table 3. responses

Response	Number of Customers
Somewhat transparent: "They share some details but not enough."	12
Uncertain: "I'm not sure how transparent they are."	34

Somewhat Transparent: 12 respondents reported that Decathlon shares some details but not in sufficient detail. This again indicates that while the brand is actually trying to communicate its sustainability goals, the efforts are half-baked or lack substance to be fulfilled by customers. Probably, the customers desire more precise, detailed and measurable information so that they can actually understand and become confident in the sustainability commitments of Decathlon.

Uncertain: A much larger group, 34 respondents, were uncertain and reported that they do not know how transparent Decathlon is about its sustainability efforts. This points to a huge gap in customer awareness and understanding which may be as a result of inconsistent or poor communication strategies. Uncertainty may deter trust since the consumers will perceive the unclear information as a potential indicator of greenwashing or superficial commitment.

4.4. "What areas do you think Decathlon could improve in to make their sustainability efforts more impactful?"

Table 4. responses

Response	Number of Customers
Increase transparency: "Provide more detailed information about efforts."	18
Improve product sustainability: "Use more eco-friendly materials."	13
Enhance communication: "Promote their initiatives more effectively."	15

Increase Transparency: Since 18 respondents said that there is a need for more information about the efforts, it can be derived that transparency is a major requirement for many customers. Hence, the present communication may not be that deep or precise to satisfy consumer expectations entirely. Customers now look for open brands that discuss their processes and goals along with challenges of implementing sustainability practices. To do this, Decathlon can report more on tangible metrics, such as reduced carbon emissions, percentage of green product sales or recyclables' impact. Thematic reporting through sustainability reports, engaging web content or customer mailers could greatly improve consumers' confidence and trust in the brand's environmental commitment.

Make Products Greener: 13 respondents opined that Decathlon should use more green materials in its products. This feedback underscores a growing consumer demand for sustainable products that minimize environmental impact. Decathlon has an opportunity to innovate in material sourcing and product design, ensuring that eco-design principles are applied across more product categories. By offering detailed information about the materials used and the lifecycle of its products, Decathlon can position itself as a leader in sustainable retail. Working together with material science experts having certifications of sustainable materials and promoting product innovation in-store and online may further solidify the commitment of the company towards sustainability.

Enhance Communication: 15 respondents claimed that the company should communicate its sustainability initiatives better. Decathlon is failing in relation to how it communicates its efforts now. Even though the brand is taking immense sustainability actions, poor communication might result in low awareness or skepticism on the part of customers. In response, Decathlon might implement some innovative and varied forms of communication: engaging storytelling, outreach via social media, engaging in-store experiences or teaming up with influencers and environmental groups. The campaign may educate the consumer on initiatives the brand is taking to understand how purchasing decisions will create broader change.

4.5 "Does Decathlon's sustainability policy affect your decision to shop with them? If so, in what way?

Table 5. responses

Response	Number of Customers
Neutral: "It doesn't influence my decision; I shop for quality and price."	21
Negative skepticism: "I'm unsure if their policies are genuine, so it doesn't matter."	13
Unaware: "I didn't know they had a sustainability policy, so it hasn't influenced me."	12

Neutral (21 Respondents): This group is the largest. The respondents believe that sustainability policies are not relevant to their decisions and rely on quality and price when purchasing. This indicates that even though sustainability may be a desirable added value for some, it is not yet an important purchase driver for many. Decathlon could seek ways to include sustainability in quality and price stories making environmentally friendly practices fit or improve these core attributes.

Negative Skepticism (13 Respondents): Some respondents questioned the legitimacy of Decathlon's policy, indicating that lack of trust can work to nullify the sustainability initiatives put forward. This group represents a deficit of trust, likely a result of insufficient transparency or even greenwashing. The answer to this will require building credibility through verifiable claims, third-party certifications and clear communication regarding the impact of the effort.

Unaware (12 Respondents): A smaller number admitted to being unaware of Decathlon's sustainability policy. This finding underlines a lack of effective communication and outreach. To increase awareness, Decathlon could implement educational campaigns, in-store signage and digital content that highlights its sustainability initiatives and how they contribute to a better shopping experience for environmentally conscious customers.

4.6 "What more could Decathlon do to make you feel confident in their commitment to sustainability?"

Table 6. responses

Response	Number of Customers
Share more data: "Provide detailed reports and measurable progress."	22
Transparent communication: "Clearly explain their goals and achievements."	24

Share More Data (22 Respondents): Customers demanded detailed reports and measurable progress. This means that many consumers want to see concrete evidence of Decathlon's sustainability efforts, such as specific metrics, achievements and challenges. Providing data-driven insights through sustainability reports, infographics or dashboards will help Decathlon respond to this demand, demonstrate accountability, and gain the trust of its audience.

Transparent Communication- 24: More prominent, a bigger group spoke of the need for greater transparency in how goals are being achieved and communicated. Again, this feedback points to a gap that needs to be addressed by the sustainability narrative that Decathlon is offering its customers. Customers ask for simplicity, clarity and consistency in messaging knowing what the brand is doing and what difference that makes. This gap can easily be bridged with video updates by the leadership team, third-party certifications and case studies on successful initiatives.

5. DISCUSSION

In addition to understanding from the literature review, conclusions drawn from the study enable a comprehensive understanding of how customers think about Decathlon's sustainability practices. There is a gap between the level of trust which consumers have in companies' claims on sustainability and the data collected from the literature and that empirically gathered from consumer interviews, even though sustainability is increasingly emerging as a factor in the decision-making process of consumers.

Table 7.

Parameter	Existing Literature Review	Findings and Analysis
Awareness of Greenwashing	Literature indicates that while environmental claims are increasingly prevalent, many customers struggle to identify greenwashing.	Most respondents showed limited understanding of greenwashing practices emphasizing the need for consumer education.
Impact of Transparency	Studies show that transparent environmental initiatives enhance brand trust and loyalty among consumers.	Customers preferred detailed information about Decathlon's sustainability efforts, citing transparency as a trust factor.
Perception of Initiatives	Research highlights that vague or unsupported claims often lead to skepticism towards a brand's sustainability initiatives.	Respondents expressed doubt about Decathlon's claims when specific examples or certifications were not presented.
Influence of Certifications	Verified certifications and third-party endorsements are shown to increase consumer trust in sustainable practices.	Respondents demonstrated higher trust levels when Decathlon's initiatives were linked to recognized certifications.
Consumer Behavior	Previous studies indicate that customers are more likely to support brands whose sustainability aligns with their personal values.	Findings showed that while many customers valued sustainability, only a fraction considered it a deciding purchase factor.
Effectiveness of Communication	Research stresses the importance of clear communication in conveying environmental efforts to avoid misconceptions.	Customers felt Decathlon's communication was insufficiently engaging suggesting the need for more visual and interactive approaches.
Price vs. Sustainability	Literature often points out that cost can deter consumers from opting for sustainable products.	Respondents indicated price sensitivity, with sustainability being secondary unless costs were comparable to alternatives.

The results of the study and the literature review suggest a strong correlation between consumer education, transparency and perceived legitimacy of sustainability initiatives. These elements together determine whether consumers believe that a brand is actually committed to sustainable practices or is merely greenwashing. According to the literature, Greenwashing can severely damage a brand's reputation and customer trust (Walker & Wan, 2012). While consumers are likely aware of the potential ability of Decathlon to support sustainability, the results seem to suggest that skepticism will prevail unless the company strengthens its efforts at transparency and communication.

6. RECOMMENDATIONS FOR DECATHLON

A few critical suggestions regarding the further improvement of sustainability efforts for Decathlon as well as how consumer perception might be increased about Decathlon's concern for the environment may be extracted from these findings and analyses.

6.1 Increase Transparency and Communication

One of the main issues that consumers were raising was that Decathlon did not clearly disclose its sustainability activities. To address this, Decathlon should give more clear information about its sustainability activities both online and in-store. This could be in terms of publishing annual reports with measurable data on carbon emissions, sustainable materials, and other key metrics. In addition, Decathlon shall upgrade communication by incorporating sustainability messages to marketing campaigns, products labelling, and introducing display in the stores and through which the customers should understand the environmental impact their purchased products have.

6.2 Third-Party Certification and Partnerships

Consumers expressed a strong desire for third-party certification to validate Decathlon's sustainability claims. The brand could partner with trusted environmental organizations or obtain certifications such as Fair Trade, Global Organic Textile Standard (GOTS), or Carbon Trust to provide credible validation of its sustainability practices. These certifications would not only build trust but also demonstrate Decathlon's commitment to maintaining high standards in environmental and social responsibility.

6.3 Educational Content and Consumer Engagement

Consumers often asked for more educational content on the brand's sustainability efforts. Decathlon could produce online content in the form of videos, articles, and infographics explaining the company's sustainability goals and progress. In-store, this content could be shown through digital signage or educational pamphlets that explain the benefits of sustainable products and the company's green initiatives. Engaging customers on the aspect of sustainability through storytelling, whether from product packaging or social media campaigns, would also increase consumer awareness and create a deeper emotional bond with the brand.

6.4 Implement Recycling and Product Lifecycle Programs

Some of the consumers were interested in product life cycle extension programs, such as recycling or repair services. Decathlon can institute a take-back program where the customers return used or worn-out products for recycling, upcycling, or repurposing. Repair services for damaged products or information on extending product life can also serve to demonstrate Decathlon's commitment to reducing waste and supporting sustainability beyond the point of sale.

6.5 Improve Visibility of Green Initiatives

Most customers said that Decathlon's green initiatives are not easily visible and communicated in-store or on the website. To make sustainability part of the brand experience, Decathlon should embed sustainability messages into product displays through clear signage to show consumers which products are environmentally friendly. Sustainability incorporated in product marketing campaigns and separate sections on the website dedicated to eco-friendly products could be helpful in raising awareness for customers.

6.6 Build Trust through Measurable Impact

Decathlon should focus on highlighting the measurable outputs of its sustainability efforts to eliminate the concerns of greenwashing. The customers would have tangible evidence of the brand's commitment to sustainability if it would provide specific data regarding the quantity of waste diverted through recycling programs, the number of products produced from recycled materials, or the quantity of carbon emissions reduced. Customers would be updated and engaged with Decathlon's progress through frequent updates on these initiatives.

7. CONCLUSION

Overall, this paper presents an understanding of customers' perceptions of Decathlon's sustainability initiatives and how such perceptions subsequently shape consumers' purchasing decisions. Customers are aware of the fact that Decathlon can be a driver of sustainability but a huge knowledge gap emerges as a result of the extremely insufficient communication around the initiatives and transparency of the same. Besides the visibility of Decathlon's sustainability efforts through marketing and labelling on products as well as in-store displays, customers also wanted more specific and measurable information about the footprints left by the company.

The study epitomized also the urgent clarion call to all brands to substantiate their sustainability claims through verifiable proofs-such as third-party certification and transparent reporting-and with greenwash skepticism. By addressing all those very counts, Decathlon will be likely to enrich customer loyalty to the brand and build trust, alongside making sustainability an important part in brand experience.

Finally, the above research brings to light that broad-mindedness, education and communication need to form part of the solutions that will result in high consumer trust toward sustainability claims. It is clear that to balance this increasing demand of responsible consumption, organizations like Decathlon need to always upgrade their whole activities along with considering the fact that sustainability should be easily visible and accessible to consumers.

8. LIMITATION OF THIS STUDY

This research has limited its geographical scope because it focused on only one group of consumers from the Saket Decathlon store in North India. The scope would have been much more widespread had consumers from South India also been included thereby encompassing regional differences in the perception and behavior of consumers. Differences in cultural and environmental priorities across regions can offer valuable insights into the perceptions of sustainability efforts by the country's different parts. Moreover, this study spans a relatively short period, hence long-term consumer trends or the effect of a season-related factor such as holiday shopping or promotions may not be considered. An extended study period may yield a better perspective on how all these affect consumer attitudes. Finally, although this study engaged male and female respondents, the study does not determine how demographics like age, income and education influence one's opinion regarding sustainability. These elements tend to heavily influence consumers' views so their presence might provide deeper insights into how diverse these opinions really are toward environmental initiatives. Future studies should consider including these dimensions for a more nuanced and comprehensive analysis.

9. SCOPE OF FURTHER RESEARCH

Future studies can expand their scope by covering customers from different regions in India, such as South, East and West India, to understand regional differences in consumer perceptions and behaviors toward sustainability initiatives. Such studies could reveal how cultural, economic and environmental factors influence expectations and trust in sustainability claims (Panagiotopoulos et al., 2024). For example, customers in water-scarce regions might prioritize water conservation initiatives, while urban customers may value waste management efforts more (Vardopoulos et al., 2021). Comparative research analysis may also be conducted among different brands to compare against Decathlon's sustainability-related initiatives. Benchmarking, thus allows Decathlon to identify areas of strength while determining improvement areas and potential best practices from other companies' strategies. This would open avenues for insights into how consumers perceive the differential commitments of various brands toward improving their environmental positions and also improve strategies to enhance a sustainability leader like Decathlon within its market.

Data Availability Statement

Data will be made available on request.

REFERENCES

Adamkiewicz, J., Kochańska, E., Adamkiewicz, I., & Łukasik, R. M. (2022). Greenwashing and sustainable fashion industry. *Current Opinion in Green and Sustainable Chemistry, 38*, 100710. DOI: 10.1016/j.cogsc.2022.100710

Badhwar, A., Islam, S., Tan, C. S. L., Panwar, T., Wigley, S., & Nayak, R. (2024). Unraveling green marketing and greenwashing: A systematic review in the context of the fashion and textiles industry. *Sustainability (Basel), 16*(7), 2738. DOI: 10.3390/su16072738

Butt, A., Khan, S. N., Zakir, M. M., Rana, W., & Laila, A. (2021). Firms greenwashing practices and consumers' perception: Role of marketing and non-marketing external stake holders in firms greenwashing practices. *Indian Journal of Economics and Business, 20*(3), 689–706.

Capgemini (2022) 'Consumer perspectives on sustainability: Trends and insights', *Capgemini Research Institute*.

Cheliatsidou, A., Sariannidis, N., Garefalakis, A., Passas, I., & Spinthiropoulos, K. (2023). Exploring Attitudes towards Whistleblowing in Relation to Sustainable Municipalities. *Administrative Sciences 2023, Vol. 13, Page 199, 13*(9), 199. DOI: 10.3390/admsci13090199

Cone Communications. (2017). *2017 Cone Communications CSR Study*. Cone Communications.

de Freitas Netto, S. V., Sobral, M. F. F., Ribeiro, A. R. B., & Soares, G. R. D. L. (2020). Concepts and forms of greenwashing: A systematic review. *Environmental Sciences Europe, 32*(1), 1–12. DOI: 10.1186/s12302-020-0300-3

De Veirman, M., Cauberghe, V., & Hudders, L. (2017). Marketing through Instagram influencers: The impact of number of followers and product divergence on brand attitude. *International Journal of Advertising, 36*(5), 798–828. DOI: 10.1080/02650487.2017.1348035

Delmas, M. A., & Burbano, V. C. (2011). The drivers of greenwashing. *California Management Review, 54*(1), 64–87.

Dempere, J., Alamash, E., & Mattos, P. (2024). Unveiling the truth: Greenwashing in sustainable finance. *Frontiers in Sustainability, 5*, 1362051. DOI: 10.3389/frsus.2024.1362051

Duchin, R., Gao, J., & Xu, Q. (2024). Sustainability or greenwashing: Evidence from the asset market for industrial pollution. *The Journal of Finance*.

Free, C., Jones, S., & Tremblay, M. S. (2024). Greenwashing and sustainability assurance: A review and call for future research. *Journal of Accounting Literature*. Advance online publication. DOI: 10.1108/JAL-11-2023-0201

Furlow, N. E. (2010). 'Greenwashing in the new millennium', *Journal of Applied Business and Economics*, 10(6), pp. 22-25.

Kagias, P., Garefalakis, A., Passas, I., Kyriakogkonas, P., & Sariannidis, N. (2024). Whistleblowing Based on the Three Lines Model. *Administrative Sciences 2024, Vol. 14, Page 83, 14*(5), 83. DOI: 10.3390/admsci14050083

Kagias, P., Sariannidis, N., Garefalakis, A., Passas, I., & Kyriakogkonas, P. (2023). Validating the Whistleblowing Maturity Model Using the Delphi Method. *Administrative Sciences 2023, Vol. 13, Page 120, 13*(5), 120. DOI: 10.3390/admsci13050120

Kaplan, A. M., & Haenlein, M. (2010). Users of the world, unite! The challenges and opportunities of Social Media. *Business Horizons, 53*(1), 59–68. Retrieved December 13, 2024, from. DOI: 10.1016/j.bushor.2009.09.003

Kapoor, A., Sindwani, R., & Goel, M. (2021a). Ranking mobile wallet service providers using fuzzy multi-criteria decision-making approach. [IJEBR]. *International Journal of E-Business Research, 17*(4), 1–21. DOI: 10.4018/IJEBR.2021100102

Kapoor, A., Sindwani, R., & Goel, M. (2021b). Prioritising the key factors influencing the adoption of mobile wallets: An Indian perspective in covid-19 era. *Journal of Information Technology Management, 13*(4), 161–182.

Kapoor, A., Sindwani, R., & Goel, M. (2022a). Assessing mobile banking service quality dimensions using multi-criteria decision making. In *Future of Work and Business in Covid-19 Era: Proceedings of IMC-2021* (pp. 131-147). Singapore: Springer Nature Singapore. DOI: 10.1007/978-981-19-0357-1_12

Kapoor, A., Sindwani, R., & Goel, M. (2023a). Exploring quick commerce service experience: A moderated mediated investigation using SEM and fsQCA. *Total Quality Management & Business Excellence, 34*(13-14), 1896–1919. DOI: 10.1080/14783363.2023.2213653

Kapoor, A., Sindwani, R., & Goel, M. (2023b). Evaluating mobile wallet acceptance factors using best worst method. *International Journal of Process Management and Benchmarking, 13*(4), 449–469. DOI: 10.1504/IJPMB.2023.129822

Kapoor, A., Sindwani, R., & Goel, M. (2024a). A novel framework for understanding the interplay between the mobile wallet service quality dimensions and loyalty intention. *The TQM Journal.*

Kapoor, A., Sindwani, R., & Goel, M. (2024b). Exploring the symmetric and asymmetric effects of mobile wallet service experience on brand loyalty in India: Evidence from SEM and fsQCA. *Journal of Financial Services Marketing*, *29*(4), 1–18. DOI: 10.1057/s41264-024-00280-w

Kapoor, A., Sindwani, R., & Goel, M. (2024c). Mobile wallets: Theoretical and empirical analysis. *Global Business Review*, *25*(2, suppl), S211–S228. DOI: 10.1177/0972150920961254

Kapoor, A., Sindwani, R., & Goel, M. (2025). Assessing mobile wallet service quality using analytic hierarchy process and fuzzy evaluation method. *International Journal of Services and Operations Management*, *50*(2), 246–266. DOI: 10.1504/IJSOM.2025.144481

Kapoor, A., Sindwani, R., Goel, M., & Shankar, A. (2022b). Mobile wallet adoption intention amid COVID-19 pandemic outbreak: A novel conceptual framework. *Computers & Industrial Engineering*, *172*, 108646. DOI: 10.1016/j.cie.2022.108646 PMID: 36118531

Leonidou, L. C., Katsikeas, C. S., & Samiee, S. (2013). Marketing strategy determinants of export performance: A meta-analysis. *Journal of International Business Studies*, *44*(5), 493–510. Retrieved December 13, 2024, from. DOI: 10.1057/jibs.2013.13

Lopes, J. M., Gomes, S., & Trancoso, T. (2023). The dark side of green marketing: How greenwashing affects circular consumption? *Sustainability (Basel)*, *15*(15), 11649. DOI: 10.3390/su151511649

Lyon, T. P., & Maxwell, J. W. (2011). Greenwash: Corporate environmental disclosure under threat of audit. *Journal of Economics & Management Strategy*, *20*(1), 3–41. Retrieved December 13, 2024, from. DOI: 10.1111/j.1530-9134.2010.00282.x

Lyon, T. P., & Montgomery, A. W. (2015). 'The means and ends of greenwashing', *Organization & Environment*, 28(2), pp. 223-249. Retrieved December 13, 2024, from. DOI: 10.1108/SAMPJ-06-2017-0181

Nemes, N., Scanlan, S. J., Smith, P., Smith, T., Aronczyk, M., Hill, S., Lewis, S. L., Montgomery, A. W., Tubiello, F. N., & Stabinsky, D. (2022). An integrated framework to assess greenwashing. *Sustainability (Basel)*, *14*(8), 4431. DOI: 10.3390/su14084431

Nygaard, A., & Silkoset, R. (2023). Sustainable development and greenwashing: How blockchain technology information can empower green consumers. *Business Strategy and the Environment, 32*(6), 3801–3813. DOI: 10.1002/bse.3338

Panagiotopoulos, P., Vardopoulos, I., Maialetti, M., Ciaschini, C., Koundouri, P., & Salvati, L. (2024). Reimagining sustainable development and economic performance indicators: a human-centric Maslow–Bossel blueprint. *Economies, 12*(12), 338. DOI: 10.3390/economies12120338

Passas, I. (2023). *Accounting for integrity: ESG and financial disclosures: the challenge of internal fraud in management decision-making* [Hellenic Mediterranean University]. DOI: 10.12681/eadd/55359

Passas, I., Ragazou, K., Zafeiriou, E., Garefalakis, A., & Zopounidis, C. (2022). ESG Controversies: A Quantitative and Qualitative Analysis for the Sociopolitical Determinants in EU Firms. *Sustainability 2022, Vol. 14, Page 12879, 14*(19), 12879. DOI: 10.3390/su141912879

Peattie, K., & Crane, A. (2005). Green marketing: Legend, myth, farce or prophesy? *Qualitative Market Research, 8*(4), 357–370. Retrieved December 13, 2024, from. DOI: 10.1108/13522750510619733

Pimonenko, T., Bilan, Y., Horák, J., Starchenko, L., & Gajda, W. (2020). Green brand of companies and greenwashing under sustainable development goals. *Sustainability (Basel), 12*(4), 1679. DOI: 10.3390/su12041679

Qayyum, A., Jamil, R. A., & Sehar, A. (2023). Impact of green marketing, greenwashing and green confusion on green brand equity. *Spanish Journal of Marketing - ESIC, 27*(3), 286–305. DOI: 10.1108/SJME-03-2022-0032

Ragazou, K., Passas, I., & Garefalakis, A. (2022). It Is Time for Anti-Bribery: Financial Institutions Set the New Strategic "Roadmap" to Mitigate Illicit Practices and Corruption in the Market. *Administrative Sciences 2022, Vol. 12, Page 166, 12*(4), 166. DOI: 10.3390/admsci12040166

Rahman, S. U., & Nguyen-Viet, B. (2023). Towards sustainable development: Coupling green marketing strategies and consumer perceptions in addressing greenwashing. *Business Strategy and the Environment, 32*(4), 2420–2433. DOI: 10.1002/bse.3256

Ritch, E. L. (2023). Sustainable fashion marketing: green or greenwash? In *Marketing fashion: Critical perspectives on the power of fashion in contemporary culture*. Routledge. DOI: 10.4324/9781003263753-16

Szabo, S., & Webster, J. (2021). Perceived greenwashing: The effects of green marketing on environmental and product perceptions. *Journal of Business Ethics, 171*(4), 719–739. DOI: 10.1007/s10551-020-04461-0

TerraChoice. (2009) 'The Seven Sins of Greenwashing: Environmental Claims in Consumer Markets', *TerraChoice Group*. Available at: https://sinsofgreenwashing.com/ (Accessed: 13 December 2024).

Vardopoulos, I., Konstantopoulos, I., Zorpas, A. A., Limousi, L., Bennici, S., Inglezakis, V., & Voukkali, I. (2021). Sustainable metropolitan areas perspectives through assessment of the existing waste management strategies. *Environmental Science and Pollution Research International, 28*(19), 24305–24320. DOI: 10.1007/s11356-020-07930-1 PMID: 32072414

Vortelinos, D., Passas, I., Floros, C., & Garefalakis, A. (2024). Corporate Social Responsibility and Country Governance: An International Comparative Study Amid the COVID-19 Pandemic. *International Journal of Financial Studies 2024, Vol. 12, Page 110, 12*(4), 110. DOI: 10.3390/ijfs12040110

Williams, J. (2024). Greenwashing: Appearance, illusion and the future of 'green' capitalism. *Geography Compass, 18*(1), e12736. DOI: 10.1111/gec3.12736

Yang, Z., Nguyen, T. T. H., Nguyen, H. N., Nguyen, T. T. N., & Cao, T. T. (2020). Greenwashing behaviours: Causes, taxonomy and consequences based on a systematic literature review. *Journal of Business Economics and Management, 21*(5), 1486–1507. DOI: 10.3846/jbem.2020.13225

Yiannoulis, Y., Vortelinos, D., & Passas, I. (2025). Exploring Audit Opinions: A Deep Dive into Ratios and Fraud Variables in the Athens Exchange. Accounting and Auditing 2025, Vol. 1, Page 3, 1(1), 3. DOI: 10.3390/accountaudit1010003

Yildirim, S. (2023). Greenwashing: A rapid escape from sustainability or a slow transition? *LBS Journal of Management & Research, 21*(1), 53–63. DOI: 10.1108/LBSJMR-11-2022-0077

Zafeiriou, E., Garefalakis, A., Passas, I., & Ragazou, K. (2023). Illicit and Corruption Mitigation Strategy in the Financial Sector: A Study with a Hybrid Methodological Approach. Sustainability 2023, Vol. 15, Page 1366, 15(2), 1366. DOI: 10.3390/su15021366

Chapter 4
Sustainability of Green Materials in Circular Economies: A Bibliometric Analysis on ESG

Pankaj Singh
https://orcid.org/0000-0002-7559-9762
School of Business, Galgotias University, India

Jyoti Kushwaha
https://orcid.org/0009-0005-9467-8852
Government Museum, Mathura, India

Ruchi Kushwaha
https://orcid.org/0009-0000-0953-6985
School of Studies in Economics, Jiwaji University, Gwalior, India

Sneha Rajput
https://orcid.org/0000-0001-8130-7941
Prestige Institute of Management, Gwalior, India

Pooja Jain
Amity Business School, Amity University Madhya Pradesh, Gwalior, India

ABSTRACT

Environmental, social, and governance (ESG) sustainability of green materials refers to the assessment of the environmental impact, social responsibility, and corporate governance practices associated with the production, use, and disposal of materials that are marketed as environmentally friendly or sustainable. The main impetus of

DOI: 10.4018/979-8-3373-2327-5.ch004

the current paper is to provide a review of standard research contributions on environment, social and governance (ESG) sustainability of green materials in circular economies during 2015-2023 utilizing bibliometric investigation. This review adopted bibliometric analysis method in direction to illustrate the latest things as well as the new areas of interest in the domain of ESG sustainability of green materials domain in circular economies. In this regard, VOSviewer bibliometric tool are utilized to analyze the sustainability of green materials in circular economies. A sum of 306 research studies were recognized to identify and map the research structure of ESG sustainability of green materials in circular economies during 2015-2023.

1. INTRODUCTION

Environmental, social, and governance (ESG) sustainability of green materials refers to the assessment of the environmental impact, social responsibility, and corporate governance practices associated with the production, use, and disposal of materials that are marketed as environmentally friendly or sustainable (Senadheera et al. 2022, Zopounidis et al., 2020). Subsequently, environmental sustainability of green materials involves evaluating the impact of the materials on the natural environment, including factors such as resource depletion, greenhouse gas emissions, pollution, and waste generation throughout their life cycle (Patil et al. 2021). This includes assessing the sustainability of the sourcing, production, transportation, and disposal of the materials (Ditlev-Simonsen, 2022). However, social sustainability of green materials focuses on the impact of the materials on people and communities, including workers' rights, labor practices, health and safety, and social justice (Popescu et al. 2022). This includes evaluating the social responsibility of the companies producing and using these materials, as well as the social benefits they provide to local communities (Munoz-Torres et al. 2019). Consequently, governance sustainability of green materials involves assessing the corporate governance practices of companies involved in the production and use of these materials (Sadiq et al. 2023). This includes evaluating the transparency of the supply chain, ethical business practices, and board diversity, among other factors (Markopoulos et al. 2020). In general, assessing the ESG sustainability of green materials in a circular economy involves taking a holistic approach to evaluate the environmental, social, and governance impacts associated with these materials to ensure that they are truly sustainable and not just marketed as such within the context of a circular economy framework (Opferkuch et al. 2022). This is important for promoting sustainable development and reducing the environmental and social impact of the materials we use while creating economic value through circular economy practices (Velenturf & Purnell, 2021). There have been developing calls from different partners all over the planet for straightforward estimation and exposure of data about monetarily material environmental, social, and governance (ESG) concerns (Corona et al. 2019).

Subsequently, environmental sustainability of green materials involves evaluating the impact of the materials on the natural environment, including factors such as resource depletion, greenhouse gas emissions, pollution, and waste generation throughout their life cycle. This includes assessing the sustainability of the sourcing, production, transportation, and disposal of the materials. However, social sustainability of green materials focuses on the impact of the materials on people and communities, including workers' rights, labor practices, health and safety, and social justice. This includes evaluating the social responsibility of the companies producing and using these materials, as well as the social benefits they provide to local communities. Consequently, governance sustainability of green materials involves assessing the corporate governance practices of companies involved in the production and use of these materials. This includes evaluating the transparency of the supply chain, ethical business practices, and board diversity, among other factors. In general, assessing the ESG sustainability of green materials in a circular economy involves taking a holistic approach to evaluate the environmental, social, and governance impacts associated with these materials to ensure that they are truly sustainable and not just marketed as such within the context of a circular economy framework. This is important for promoting sustainable development and reducing the environmental and social impact of the materials we use while creating economic value through circular economy practices (Papamichael et al., 2023). There have been developing calls from different partners all over the planet for straightforward estimation and exposure of data about monetarily material environmental, social, and governance (ESG) concerns.

Many scholars have conveyed analysis regarding the significance of green materials in circular economies in different sectors but a comprehensive review on this topic is still missing in existing literature specifically in context of ESG sustainability of green materials in circular economy. In this regard a bibliometric analysis on ESG sustainability of green materials is crucial as the demand of green material is increasing day by day due to global sustainability concerns.

1.1 Comparative Evaluation of Green Material Sustainability across Industries

Sustainability in material selection is a growing priority across industries. However, the criteria, standards, and priorities for evaluating the sustainability of "green materials" can differ significantly between sectors. This chapter explores these differences by comparing two prominent industries: construction and packaging. While both industries strive toward sustainability, their evaluation methods reflect their distinct operational demands. A truly sustainable material must meet industry-specific performance metrics while minimizing environmental harm across its life cycle. Future convergence through shared frameworks (e.g., expanded LCA databases, cross-industry certifications) could streamline evaluations and foster innovation.

1.1.1 Evaluation Criteria in the Construction Industry

In the construction sector, sustainability evaluations often emphasize long-term performance, life cycle assessment (LCA), and building certification standards (Vardopoulos et al., 2024). Key evaluation factors include:

- **Durability and Lifespan**: Materials must last decades with minimal maintenance.
- **Embodied Energy and Carbon**: Measurement of emissions from raw material extraction to installation.
- **Indoor Environmental Quality**: Use of non-toxic, low-VOC (volatile organic compound) materials.
- **Renewable or Recycled Content**: Use of rapidly renewable materials (e.g., bamboo) or recycled aggregates.
- **Certifications**: LEED, BREEAM, WELL, and Living Building Challenge influence material choices.

1.1.2 Evaluation Criteria in the Packaging Industry

In packaging, the focus shifts toward resource efficiency, end-of-life disposal, **and** supply chain impact due to high material turnover and short usage cycles. Evaluation factors include:

- **Material Source**: Preference for bio-based or recycled feedstock (e.g., bioplastics, recycled paper).
- **Recyclability and Compostability**: Emphasis on designing for recyclability or industrial/home composting.
- **Lightweighting**: Reducing material usage while maintaining functionality.
- **Carbon Footprint**: Often measured per package unit or ton of packaging.
- **Certifications and Labels**: FSC (Forest Stewardship Council), OK compost, Cradle to Cradle, etc.

1.1.3 Industry-Specific Challenges and Innovations

- **Construction**: Adoption is slow due to code requirements and the inertia of traditional practices (Pavlides et al., 2023). However, modular building and digital tools (e.g., BIM for LCA integration) are accelerating green material uptake.
- **Packaging**: While innovation is rapid, challenges include greenwashing, infrastructure limitations (e.g., lack of composting facilities), and balancing sustainability with product protection.

1.2 Problem Formulation

Reviewing the state of art of related literature environmental, social and governance (ESG) sustainability of green materials in circular economies exposes that very slight specific qualitative analysis has ever been done by past scholars. It offers enough of an analysis gap to carry out a qualitative analysis on the selected topic. This analysis is also important to strategist and planners in global perspective on climate change impact on multinationals corporations. This chapter's analysis problem is therefore well-defined by the authors as "A bibliometric analysis on environmental, social and governance (ESG) sustainability of green materials in circular economies during 2015-2023"

1.3 Chapter's Objectives

The following purposes are expressed as follows in order to address the aforementioned queries on sustainability of green material in circular economies following objectives are formulated as follows:

- To visualize the research trends of ESG sustainability of green materials in circular economy
- To analyze the most influential articles and academic affiliations in ESG sustainability of green materials in circular economy
- To detect the authorship pattern, prolific authors and keyword analysis over time in ESG sustainability of green materials in circular economy

1.4 Chapter's Methodology

In the present study, we have employed a bibliometric analysis methodology to carry out this analysis. The structure literature search on electronic bibliometric database has been utilized to extract the standard research contributions on environment, social and governance (ESG) sustainability of green materials in circular economies utilizing bibliometric investigation. In this regard Vosviewer tool is deployed to create network and cluster diagram.

1.5 Chapter Organization

The rest of the chapter is organized as follows. In the first phase, we start with a brief overview of the ESG sustainability of green materials in the circular economy along with a brief overview of related work. Second, we provide a material and method section. Third, a bibliometric analysis is discussed. Fourth, the findings and discussion is deliberated. At last, the conclusion is portrayed and further extensions are determined.

2. OVERVIEW OF RELATED WORK

There is a growing body of literature that focuses on the environmental, social, and governance (ESG) sustainability of green materials within a circular economy framework. In summary, the literature suggests that a holistic approach is needed to ensure the ESG sustainability of green materials within a circular economy (Ferreira Gregorio et al. 2018). This requires consideration of the interrelationships between environmental, social, and governance factors, as well as the need for collaboration and engagement across different sectors and stakeholders (Ahmad et al. 2024). Below are some main themes and outcomes from recent research on ESG sustainability of green materials in circular economy:

a) Environmental sustainability

In perspective of environmental dimensions. Researchers have highlighted the significance of measuring the life cycle impacts of green materials in a circular economy, including their sourcing, production, use and end-of-life management (Walker et al. 2018). This includes evaluating the potential environmental benefits of using green materials, such as reduced greenhouse gas emissions and resource depletion, as well as potential negative impacts, such as increased waste generation and pollution (Pajula et al. 2017).

b) Social sustainability

Studies have highlighted the potential social benefits of green materials within a circular economy, such as job creation, improved working conditions, and increased access to affordable and sustainable products (Padilla-Rivera et al, 2020). However, researchers also note that there are challenges associated with ensuring that the benefits of circular economy practices are shared equitably across different communities and groups (Marchesi & Tweed, 2021).

c) Governance sustainability

Scholars have stressed the importance of transparent and accountable governance structures for green materials within a circular economy, including the need for strong regulatory frameworks, effective stakeholder engagement, and responsible supply chain management (Kulakovskaya et al, 2023). This includes addressing issues such as worker rights, labor practices, and human rights, as well as promoting ethical and sustainable business practices (Torres et al, 2023).

2.1 Types of Green Materials

Green materials are materials that have a lower environmental impact than conventional materials, due to factors such as their composition, sourcing, and end-of-life disposal. In a circular economy, green materials are those that can be reused, recycled, or repurposed at the end of their life, rather than being discarded as waste (Chatziparaskeva et al. 2022). Green materials in a circular economy are those that are environmentally sustainable and can be reused, recycled, or repurposed at the end of their life, rather than being discarded as waste. The use of green materials in a circular economy can help to reduce the environmental impact of material production and consumption, and promote greater resource efficiency and sustainability (Purchase et al. 2022). Some examples of green materials in a circular economy include:

a) Biodegradable materials

These are materials that can decompose naturally, without causing harm to the environment. Examples include bio plastics, which are made from renewable sources such as corn or sugarcane, and can be composted at the end of their life.

b) Recycled materials

These are materials that are made from recycled waste, such as recycled plastics, metals, and paper. By using recycled materials, less virgin material is needed, reducing the environmental impact of material production.

c) Natural materials

These are materials that are sourced from renewable natural resources, such as wood, bamboo, and hemp. These materials are often biodegradable and have a lower environmental impact than materials sourced from non-renewable sources.

d) Sustainable textiles

These are textiles that are produced using sustainable methods and materials, such as organic cotton, hemp, and bamboo. Sustainable textiles are often biodegradable and have a lower environmental impact than conventional textiles.

e) Green building materials

These are materials that are used in the construction of buildings and infrastructure, such as recycled concrete, sustainably sourced wood, and insulation made from recycled materials.

2.2 Integrating ESG in Circular Economies

To fully realize the potential of green materials in circular economies, ESG principles must be integrated at every stage as mentioned below:

- *Design*: Products should be designed for longevity, reparability, and recyclability.
- *Production*: Manufacturing processes should prioritize renewable energy, minimal waste generation, and sustainable resource use.
- *Usage*: Encouraging responsible consumption and maintenance of products to extend their lifespan.
- *End-of-Life*: Efficient systems for recycling and repurposing materials to ensure they remain in the production cycle.

2.3 ESG Metrics in Practice: GRI and SASB Standards

Environmental, Social, and Governance (ESG) metrics have become central to evaluating corporate sustainability performance. As stakeholder expectations evolve, companies increasingly rely on standardized reporting frameworks such as the Global Reporting Initiative (GRI) and the Sustainability Accounting Standards Board (SASB) to disclose ESG information. This literature review explores scholarly and professional works that examine the application, strengths, and limitations of these frameworks in practice. GRI and SASB represent two dominant frameworks with differing philosophies. GRI emphasizes **stakeholder-oriented** reporting, encouraging organizations to disclose impacts on the economy, environment, and society, while SASB focuses on **financial materiality**, prioritizing ESG issues most relevant to investors and specific industries. The literature reveals that both GRI and SASB play critical roles in shaping ESG disclosures, with their distinct approaches catering to different audiences and strategic priorities. An integrated approach that aligns GRI's stakeholder inclusiveness with SASB's investor relevance appears to be the emerging best practice. Continued academic research and regulatory harmonization will be essential in refining ESG metrics for practical and strategic use.

2.4 Adoption and Prevalence of ESG Standards: GRI vs. SASB

Numerous studies have evaluated the uptake of ESG reporting standards across industries and regions. According to the KPMG Survey of Sustainability Reporting, over 78% of top companies globally use GRI guidelines, indicating its dominance in corporate sustainability reporting. Conversely, SASB is gaining traction particularly in North America, where investor pressure has encouraged adoption. Multiple scholars have compared GRI and SASB on materiality, scope, and audience (Passas, 2023). GRI adopts a double materiality perspective, incorporating both financial and non-financial impacts, while SASB focuses solely on single materiality, i.e., financial relevance to investors. Both GRI and SASB also provides a joint guide for companies to integrate both frameworks, encouraging a hybrid reporting approach to address the needs of both stakeholders and investors (Table 1).

Table 1. Comparative analysis GRI vs. SASB

Framework	Materiality Approach	Audience	Scope	Integration
GRI	Double Materiality	Stakeholders	Broad ESG	UN SDGs, CSRD
SASB	Financial Materiality	Investors	Industry-specific	TCFD, IFRS/ISSB

2.5 Region-Specific Trends in Green Materials Adoption

1. European Union
    - **Policy Drivers:** European Green Deal, Circular Economy Action Plan, reach regulation.
    - **Trends:** High adoption of **bioplastics**, **recycled aluminum**, and **bio-based composites**. Strong growth in **green construction materials** (e.g., hempcrete, cross-laminated timber). Emphasis on **eco-design** and extended producer responsibility (EPR) for packaging.
2. North America

- **Policy Drivers:** Inflation Reduction Act (U.S.), Green Building Standards (LEED), EPA Safer Choice.
- **Trends:** Packaging sector prioritizes **compostable polymers** and **PCR (post-consumer recycled)** content. Automotive and aerospace sectors focus on **lightweight bio-composites**. Construction materials are increasingly **certified low-VOC** and **carbon-neutral.**

3. Asia-Pacific
   - **Policy Drivers:** China's Green Product Certification, Japan's Circular Economy Vision, India's CE Roadmap.
   - **Trends:** China leads in **recycled plastics infrastructure** and **eco-labeling programs**. India promotes **natural fiber-based packaging and construction materials** (e.g., jute, coir). Southeast Asia adopting **green materials in electronics and textiles.**
4. Latin America
   - **Policy Drivers:** Brazil's National Solid Waste Policy, Chile's REP Law.
   - **Trends:** Use of **agricultural waste** (e.g., bagasse, coffee husks) in packaging and construction. Growth in **recycled PET applications** in the beverage and textile sectors. Emerging emphasis on **social inclusion** in material recovery (e.g., waste picker cooperatives).

Table 2. Sector-Specific Trends in Green Materials Adoption

Sector	Key Green Materials	Adoption Drivers	Challenges
Construction	Hempcrete, bamboo, recycled steel, CLT	Carbon reduction goals, green certifications (LEED, BREEAM)	Cost, regulatory hurdles
Packaging	Bioplastics, paper pulp, mushroom-based foams	Single-use plastic bans, EPR mandates	Scale, barrier properties of materials
Automotive	Recycled aluminum, flax fiber composites, natural rubber	Lightweighting for fuel efficiency, consumer preference	Performance testing, consistency
Textiles	Organic cotton, recycled polyester, Tencel	Sustainable fashion movement, eco-labels (OEKO-TEX)	Price, durability
Electronics	Recycled rare earths, biodegradable casings, bio-resins	E-waste regulation, demand for ethical sourcing	Material availability, recyclability

2.6 Comparative Life Cycle Assessment (LCA) Framework Table

The global transition to green materials is driven by regulatory, economic, and social pressures. Adoption patterns vary by region and sector, with construction and packaging leading innovation. Life Cycle Assessment (LCA) provides a robust comparative tool to assess sustainability, though standardization and sector-specific customization remain ongoing challenges. Table 3 compares how green materials are evaluated across sectors using LCA approaches.

Table 3. Life Cycle Assessment (LCA) of green materials in different sector

LCA Stage	Construction Sector	Packaging Sector	Automotive Sector
Goal & Scope	Building lifespan (50–100 years)	Single-use or short-term usage	Lifecycle of vehicle (10–15 years)
Material Inputs	Embodied carbon of concrete, steel, wood	Renewable sources (PLA, paper), additives	Lightweight materials (aluminum, composites)
Production Phase	Energy intensity of material extraction	Emissions from polymerization or pulping	VOCs and toxicity in part fabrication
Use Phase	Thermal insulation, durability	None or limited	Impact on fuel economy or battery range
End-of-Life	Recyclability, reuse potential (e.g., deconstruction)	Compostability, recyclability, incineration	Dismantling, recyclability of metals and polymers
Impact Categories	GWP, acidification, resource depletion	GWP, eutrophication, marine toxicity	GWP, smog formation, human toxicity
Assessment Tools	One Click LCA, SimaPro, GaBi	OpenLCA, COMPASS, Ecoinvent	GREET, GaBi, EcoInvent

3. MATERIAL AND METHODS

This section devoted to tool and software's that are utilized in the current analysis.

3.1 Overview of Bibliometric Approach

The review analysis of any prior studies provide a comprehensive insight on previously research accomplished by others (Singh & Agrawal, 2019). Subsequently, the review analysis provide a pathway to future researchers to conduct their studies in prospective areas (Singh, 2024). Bibliometrics was first introduced in 1969 by Allan Pritchard in his study "Statistical Bibliography or Bibliometrics" (Roy & Basak, 2013, Passas, 2024). Bibliometric analysis is significant in two ways. First one is descriptive analysis of searched articles in which we count academic productivity, i.e. the number of articles. This analysis involves institutions, countries, journals, authorship pattern and focus areas. The second perspective is evaluation analysis, within which we count literature usage, network and citations analysis.

Both analysis enables evaluation and observation of research trends and scholars' preferences for publication output.

3.2 Source of Literature and Collection Strategy

The 306 research studies were recognized to identify and map the research structure of ESG sustainability of green materials in circular economies. These 306 articles are extracted from Scopus databases which publishes articles from 2015 to 2023 through advance search by using keyword sustainability of green materials in circular economies. Biblioshiny interface (Aria & Cuccurullo, 2017, Petropoulou et al., 2024; Ragazou et al., 2022, 2023; Zafeiriou et al., 2023) and VOSviewer software (Van Eck & Waltman, 2010) along with Voyant tools (Sinclair & Rockwell, 2020)are utilized to picture the latest things as well as the new areas of interest in the domain of ESG sustainability of green materials domain in circular economies for identified articles which are relevant and appropriate to search keywords. Subsequently, bibliometric parameters are increasingly used to summarize the academic output researches and journals articles which summarize the current research interest and provide direction for future research.

4. ANALYSIS AND RESULTS

A bibliometric analysis of ESG sustainability of green materials in circular economy can provide insights into the publication trends, authorship patterns, affiliation network and collaborations in this field.

4.1 Analysis Based on Research Publication Contribution

The 306 publications are acknowledged and are arranged under five primary types and their recurrence is illustrated in Figure 1. The biggest class of documents is articles (i.e., journal papers), contributing to 52.0%.

Figure 1. Contribution of Publication Documents

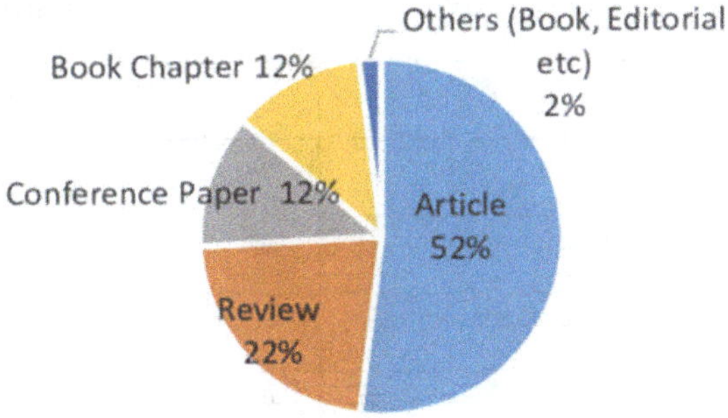

(Source: Own elaboration)

4.2 Authors Publication Contribution Analysis

The authors' information recover from the Scopus data set, a sum of 162 authors took part in performing research on the ESG sustainability of green materials in circular economies and contributed a total of 306 published articles. The data was attained by providing a full count of entire authors for a particular article. Out of these 106 authors (65.43%) published an individual article and just eight authors (4.93%) had more than three publications. The distribution of the number of authors' participation and article publications, as mentioned in Table X.1.

Table 4. Number of authors' publication contribution

No. of authors publication counts	Authors contribution	Percentage
Authors having seven publications	1	0.61%
Authors having three publications	8	4.93%
Authors having two publications	47	29.01%
Authors having one publications	106	65.43%
Total	162	

(Source: Own elaboration)

4.3 Citation Analysis of Most Influential Articles

The citation analysis of most cited publications recognizes the ranking of top cited articles along with their authors, affiliation, publication year and number of citations as presented in Table X.2.

Table 5. Analysis of most cited publications

Rank	Author	Affiliation	Year	Journal	No. of Citations
1	Andrea et al.	University of Sheffield, United Kingdom	2017	Omega	698
2	Yuliya et al.	Chalmers University of Technology, Sweden	2018	*Resources, Conservation and Recycling*	638
3	Mohan et al.	Indian Institute of Chemical Technology	2016	*Bioresource Technology*	524
4	Roger	University of the Witwatersrand, Johannesburg, South Africa	2018	*ACS Sustainable Chemistry and Engineering*	474
5	Hong and Chen	Shanghai Institute of Organic Chemistry, China	2017	Green Chemistry	395

(Source: Own elaboration)

4.4 Analysis of Published Articles Over Time

The initial study on the sustainability of green materials in circular economies was published in 2015. From 2015 to 2022 there is an increasing trend in publication was observed that is continuous in year 2023 too as depicted in Figure 2.

Figure 2. Yearly trends of published articles

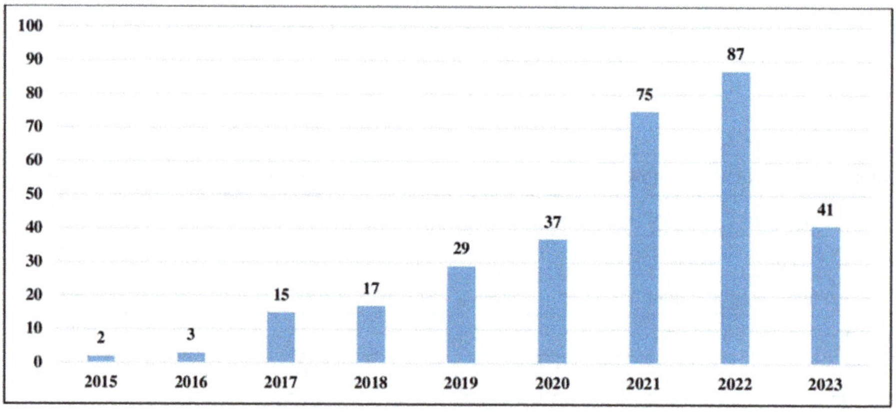

Source: Own elaboration; 306 papers (2015-2023)

4.5 Most Influential Journal Analysis

The influential journal analysis are performed in direction to identify the most relevant journals in the area of ESG sustainability of green materials in circular economies. The retrieved 306 articles are published in different 160 journals. The utmost publications are published in the Sustainability journal (17 articles) followed by Journal of Cleaner Production 9 articles as portrayed in Figure3.

Figure 3. Most relevant authors

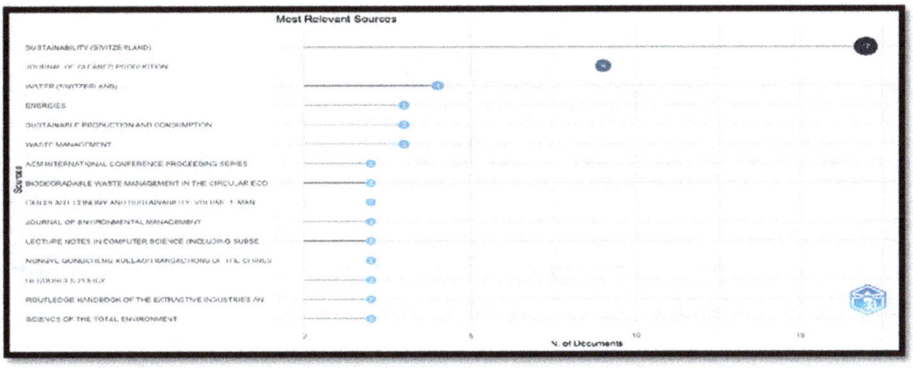

(Source: Own elaboration)

4.6 Authors Country Analysis

The author country analysis are used to identify the collaborative patterns among institutions of different countries. In the present analysis, Italy is the most influential country (72 articles) followed by Spain (31 articles) and United Kingdom with 30 articles as illustrated in Figure 4.

Figure 4. Analysis of article publication based on authors' country

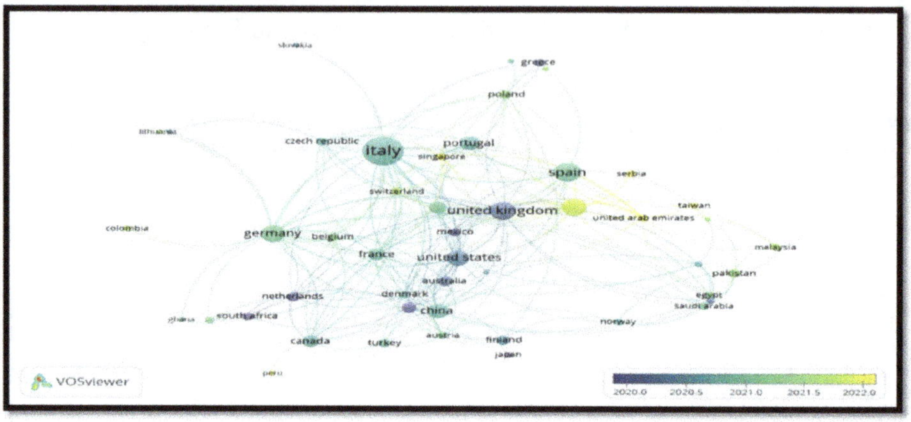

(Source: Own elaboration)

4.7 Keyword Analysis

The keyword analysis are used to analyze the most commonly used keywords in the field of ESG sustainability of green materials in circular economy. It is found in keyword analysis that circular economy is the frequently used keyword as illustrated in brown cluster followed by environmental sustainability in red cluster as portrayed in Figure 5. It is clearly shown that sustainable development, circular economy and economic, social, governance (ESG) approach is the most frequently occurring keyword in studies on ESG sustainability of green materials in circular economy.

Figure 5. Keyword Analysis

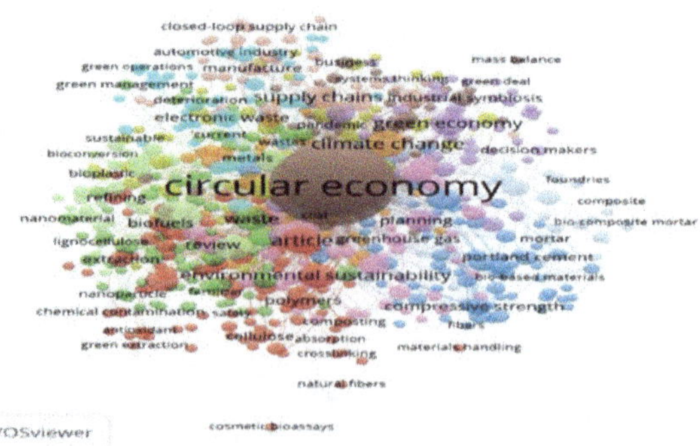

(Source: Own elaboration)

4.8 Co-Authorship Analysis

The co-author analysis are conducted in order to identify the most influential network pattern among co-authors. It was found in analysis that Saeli, M. is the most influential co-authorship network with seven articles as illustrated in Figure 6.

Figure 6. Co-authorship network analysis

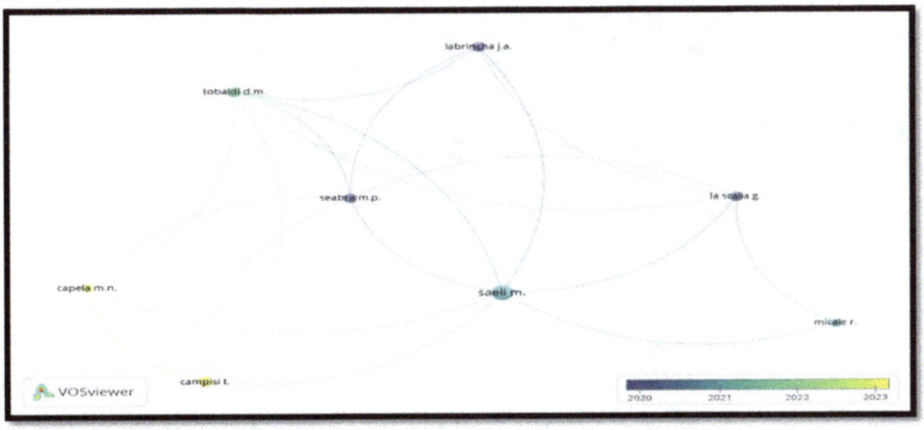

4.9 Word Tree Map Analysis

The word tree map analysis are conducted in order to detect the network of significant words through tree diagram. In Figure 7, word tree diagram determine the mapping of hierarchical pattern of significant words in the area of ESG sustainability of green materials in circular economies. As per illustration, sustainable development (9 percent) and circular economy (5 percent) is the most frequently words in tree diagram followed by waste management.

Figure 7. Analysis of Word Tree Map

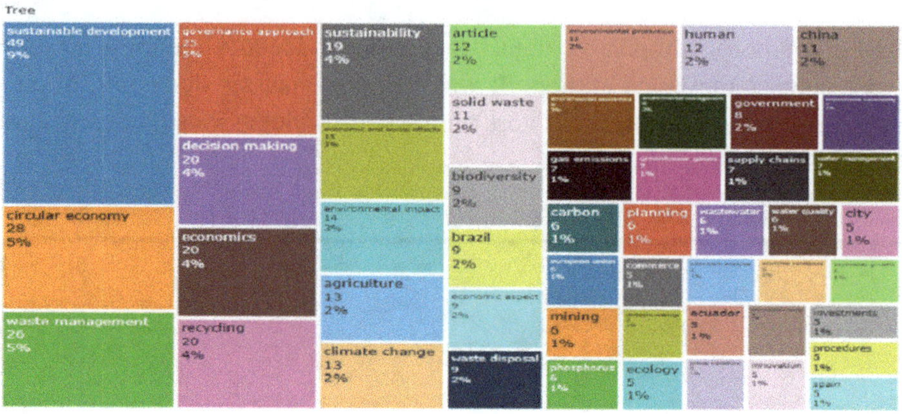

(Source: Own elaboration)

4.10 Word Cloud Analysis

The word cloud analysis is performed in order to identify text mining of frequently occur keywords in source data. In present analysis word cloud illustration shows that sustainable term is used majority of times followed by environmental and development as depicted in Figure 8.

Figure 8. Word Cloud Analysis

(Source: Own elaboration)

4.11 Analysis of CO-OCCURRENCE LINKS in TEXT DATA

A text mining and word cloud investigation has been performed for 306 articles separated from Scopus by utilizing the web text insightful programming "Voyant Instruments" (Sinclair & Rockwell, 2020). Voyant devices are utilized for automated text investigation and create co-occurrence networks and correlations of the particular terms that show up across the examination of text information. It is taken note from the text mining analysis that the term "green materials" is straightforwardly connected with the term sustainable and circular economy in the text information of 306 articles was extricated from the Scopus data set as shown in Figure X.9.

Figure 9. Analysis of co-occurrence links based on text mining

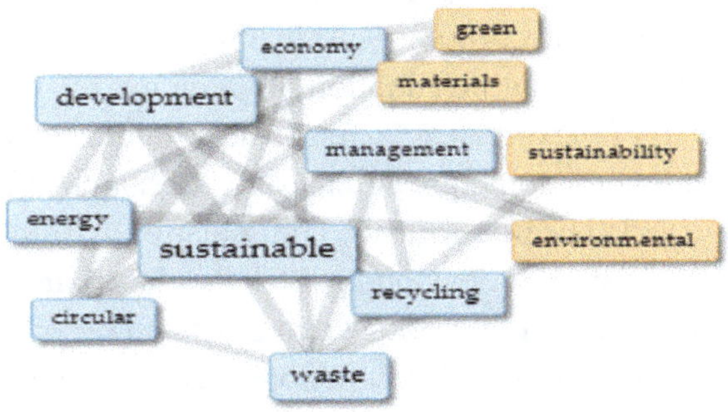

(Source: Own elaboration)

4.12 Thematic Synthesis

Apart from bibliometric analysis, a qualitative analysis on Sustainability of Green Materials has been also conducted. In this regard, an expert interview with ten experts of green materials were organized in order to attain valuable insights from qualitative data. The reason behind incorporating the qualitative data analysis into this study, particularly through expert interviews, provides rich, contextual insights that complement quantitative analyses. The findings attained from expert interview are mentioned below:

- *Stakeholder Perceptions and Attitudes*: Experts generally view green materials favorably, associating them with environmental benefits and sustainable practices. However, there is often a need for greater awareness and education about specific green materials and their advantages. Some stakeholders express skepticism about the performance and cost-effectiveness of green materials. Common barriers include perceived higher costs, lack of performance data, and limited availability.

- *Policy and Regulatory Insights*: Stakeholders highlight the importance of supportive policies and incentives in promoting the adoption of green materials. Examples include tax breaks, subsidies, and stringent environmental regulations. Inconsistent and fragmented regulatory environments across different regions can hinder the adoption of green materials. Stakeholders call for harmonized standards and clearer guidelines.
- *Economic Considerations*: Businesses emphasize the need for green materials to be cost-competitive with traditional materials. Initial high costs can be a significant barrier, although long-term savings and benefits are recognized. There is a growing market demand for sustainable products, driven by consumer awareness and corporate sustainability commitments. Stakeholders see this as a key driver for the adoption of green materials.
- *Technological and Logistical Challenges*: Stakeholders identify a gap in technological readiness for some green materials. Research and development are needed to improve their performance and scalability. Efficient supply chains for green materials are still developing. Stakeholders point out logistical challenges, including the sourcing of raw materials and recycling infrastructure.
- *Cultural and Social Factors*: Consumer preferences and behavior significantly impact the adoption of green materials. Stakeholders note the importance of consumer education and engagement to drive demand. Cultural factors play a role in the acceptance of green materials. In some regions, traditional materials and practices may be preferred, requiring tailored communication strategies.
- *Environmental, Social, and Governance (ESG) Considerations*: Stakeholders emphasize the importance of integrating ESG criteria into the evaluation of green materials. This includes assessing environmental impacts, social benefits, and governance practices. There is a call for greater transparency and standardized reporting on the sustainability performance of green materials. Stakeholders advocate for clear and consistent ESG metrics.

Above qualitative findings highlight the importance of understanding stakeholder perspectives, addressing barriers, and leveraging opportunities to promote the sustainability of green materials in circular economies. Future research should continue to integrate qualitative and quantitative approaches to develop comprehensive and actionable strategies for advancing green materials and achieving sustainable development goals.

5. DISCUSSION AND RESULTS

The integration of Environmental, Social, and Governance (ESG) sustainability with the use of green materials in circular economies presents a promising avenue for fostering sustainable development. This discussion section explores the multifaceted impact of ESG criteria on green materials within circular economies, highlighting key benefits, challenges, and future directions.

- *Environmental Sustainability*: Green materials, by design, aim to minimize environmental impact through reduced resource consumption and lower emissions. In a circular economy, the emphasis on reusing and recycling materials further amplifies these environmental benefits. The closed-loop system characteristic of circular economies ensures that materials are continually repurposed, significantly reducing the need for virgin resource extraction. This not only conserves natural resources but also mitigates environmental degradation associated with mining and deforestation. Additionally, the use of green materials often results in lower carbon footprints, as recycled materials typically require less energy to process compared to new materials. However, the challenge lies in ensuring the scalability and economic viability of green materials, which necessitates ongoing innovation and investment in sustainable technologies.
- *Social Sustainability*: From a social perspective, the use of green materials in circular economies has the potential to improve community well-being and safety. Sustainable sourcing practices ensure that materials are obtained without exploiting labor or causing harm to local communities. Furthermore, the non-toxic nature of many green materials enhances occupational health and safety for workers and reduces health risks for consumers. The shift towards circular economies also promises significant social benefits through job creation in sectors such as recycling, remanufacturing, and sustainable design. However, achieving social sustainability requires stringent adherence to fair labor practices and the creation of inclusive opportunities that support marginalized communities. Policymakers and businesses must work together to establish frameworks that protect workers' rights and promote social equity.
- *Governance Sustainability*: Effective governance is crucial for the successful integration of ESG principles in circular economies. Transparency and accountability in sourcing and production processes are

Green materials, also known as sustainable or eco-friendly materials, have gained significant importance in the contemporary scenario due to growing concerns about the impact of human activities on the environment. Green materials are those that are produced using processes that have a minimal impact on the environment and human health. In contemporary scenario, there has been a significant increase in the use of green materials in various sectors, including construction, automotive, packaging, and consumer goods. Green materials offer several benefits such as reduced carbon footprint, improved energy efficiency, and reduced waste generation. In the construction sector, green materials such as bamboo, recycled steel, and natural insulation materials are being increasingly used to construct energy-efficient buildings that have a minimal impact on the environment.

In the automotive sector, green materials such as recycled plastic, bio-based materials, and natural fibers are being used to make lighter and more fuel-efficient vehicles (Koronis et al. 2013). The use of green materials in packaging is also gaining popularity as it helps reduce the use of non-biodegradable materials such as plastics. Biodegradable packaging materials such as compostable plastics and biodegradable paper are increasingly being used by companies to reduce their environmental impact (Moshood et al. 2022). In the consumer goods industry, green materials are being used to make products such as clothing, shoes, and accessories (Singh et al. 2024a). Materials such as organic cotton, hemp, and bamboo are being used to make eco-friendly clothing that has a reduced impact on the environment (Singh et al. 2024b). By and large, the scope for green materials in the contemporary scenario is significant, as they offer an effective solution to reduce the impact of human activities on the environment. As more and more companies and consumers become aware of the benefits of green materials, the demand for these materials is likely to increase in the future (Bradu et al. 2022).

5.1 Outcome Matrix

To assess the real-world impact of research on the sustainability of green materials in circular economies, it is essential to define specific and measurable outcome metrics (Singh et al. 2024c). These metrics should cover various aspects, including environmental, economic, social, and governance factors. Table 4 depicted the detailed outcome metrics that can be used to evaluate the effectiveness and impact of the research:

continued on following page

Table 6. Continued

Table 6. Outcome Matrix

Matrix	Outcome	Metric	Measurement
Environment matrix	Carbon Footprint Reduction	Tons of CO2 equivalent (CO2e) reduced per unit of green material compared to traditional materials.	Lifecycle assessment (LCA) from production to end-of-life, comparing emissions of green vs. traditional materials.
	Resource Efficiency	Percentage reduction in the use of virgin raw materials.	Amount of recycled or renewable materials used in the production of green materials.
	Waste Reduction	Tons of waste diverted from landfills through recycling and reuse of green materials.	Quantitative tracking of waste streams before and after the adoption of green materials.
	Water Usage	Reduction in water usage per unit of green material produced.	Water consumption data from production processes.
Economic Metrics	Cost Savings	Cost savings achieved through the use of green materials over their lifecycle.	Comparative analysis of lifecycle costs of green materials versus traditional materials, including production, maintenance, and disposal costs.
	Market Penetration	Market share of green materials in relevant industries.	Sales data and market analysis reports indicating the percentage of green materials used compared to traditional materials.
	Job Creation	Number of new jobs created in the green materials sector.	Employment statistics from companies and industries adopting green materials.
Social Metrics	Consumer Adoption and Satisfaction	Percentage of consumers using and satisfied with products made from green materials.	Surveys and market research on consumer preferences and satisfaction levels.
	Community Impact	Number of community initiatives and programs supporting the use of green materials.	Reports and documentation from community organizations and local governments.
	Health and Safety Improvements	Reduction in health risks and improvements in safety due to non-toxic and safer green materials.	Health and safety incident reports and studies comparing traditional and green materials.

Matrix	Outcome	Metric	Measurement
Governance Metrics	Compliance with Environmental Regulations	Percentage of green material producers compliant with environmental regulations.	Audits and compliance reports from regulatory bodies
	ESG Reporting and Transparency	Number of companies publishing ESG reports that include data on green materials.	Analysis of ESG reports and sustainability disclosures from relevant companies.
	Stakeholder Engagement	Frequency and quality of stakeholder engagement activities related to green materials.	Documentation and evaluation of stakeholder meetings, surveys, and collaborative initiatives.

The above detailed outcome metrics provide a comprehensive framework for assessing the real-world impact of research on the sustainability of green materials in circular economies. By regularly measuring and analyzing these metrics, researchers, policymakers, and industry stakeholders can track progress, identify areas for improvement, and make informed decisions to enhance the sustainability and circularity of green materials.

5.2 Significance of Sustainability of Green Materials in Circular Economies

The ESG sustainability of green materials in a circular economy offers several advantages and merits, which are significant in promote ng sustainable development. These benefits make a strong case for the adoption of sustainable practices in material production, use, and disposal. Some of these advantages include:

- *Reduced environmental impact*
- *Increased resource efficiency*
- *Economic benefits*
- *Improved social outcomes*
- *Enhanced corporate reputation*
- *Contribution to sustainable development goals*

5.3 Challenges to Sustainability of Green Materials in Circular Economies

While the circular economy presents a promising framework for promoting the sustainability of green materials, there are also several challenges that need to be addressed to ensure their continued sustainability. Largely, addressing these challenges will be critical to ensuring the continued sustainability of green materials in a circular economy. This will require a concerted effort by multiple actors, including governments, businesses, and consumers, to promote sustainable practices and infrastructure. Some of these challenges include:

- *Complexity of supply chains*
- *Limited infrastructure*
- *Economic viability*
- *Consumer awareness*
- *Policy and regulatory challenges*

5.4 Future Prospects of Sustainability of Green Materials in Circular Economies

The future prospects of ESG sustainability of green materials in circular economy are promising. The increasing recognition of the importance of sustainability, coupled with the growing awareness of the need for a more circular economy, is driving demand for more sustainable materials and circular economy practices. Generally, the future prospects for ESG sustainability of green materials in a circular economy are positive, but will require a concerted effort from multiple stakeholders to promote sustainable practices and infrastructure. The continued adoption and refinement of circular economy practices, coupled with advances in technology and increased collaboration among stakeholders, will be critical to promoting the sustainability of green materials in a circular economy. Some of the key future prospects of ESG sustainability of green materials in a circular economy include:

- *Technological innovations:* Advances in technology are likely to play an important role in improving the sustainability of green materials in a circular economy. For example, new technologies are being developed that can enable the more efficient and effective recycling and reuse of materials, reducing waste and promoting a more circular economy.
- *Collaboration and partnerships:* Collaboration and partnerships among stakeholders across different sectors will be critical to driving the adoption of circular economy practices and promoting the sustainability of green materials. This includes collaborations among businesses, governments, NGOs, and consumers, to develop sustainable supply chains and infrastructure, promote sustainable consumption patterns, and address policy and regulatory challenges.
- *Government policies and regulations:* Governments can play an important role in promoting the sustainability of green materials in a circular economy through the development of supportive policies and regulations. This

includes promoting sustainable production practices, developing incentives for the adoption of circular economy practices, and implementing regulations to reduce waste and promote sustainable consumption patterns.

- *Consumer demand:* Increasing consumer demand for sustainable products and practices is likely to drive the adoption of green materials in a circular economy. This includes consumers being willing to pay a premium for sustainable products and engaging in sustainable consumption practices, such as recycling and reducing waste.
- *Increased awareness and education:* Increasing awareness and education about the benefits of green materials and circular economy practices can help to drive their adoption and promote sustainability. This includes educating consumers about the environmental, social, and economic benefits of green materials, as well as promoting sustainable consumption patterns and waste reduction practices.

5.5 Social Equity in the Circular Economy

The circular economy (CE) is often viewed through environmental and economic lenses—recycling, resource efficiency, and product life extension. However, social equity remains a critical, often under-addressed pillar in making CE accessible and inclusive.

Social equity in the CE means ensuring that all communities—especially marginalized, rural, or low-income groups—can access the benefits, participate in decision-making, and gain employment or entrepreneurial opportunities created through circular systems. It also involves recognizing the informal sector's role in waste management and resource recovery.

6. CONCLUSION

The concept of Environmental, Social, and Governance (ESG) sustainability, especially in the context of green materials within circular economies, is multifaceted and critical for fostering sustainable development. ESG sustainability of green materials in circular economy is a rapidly growing field of research that has gained significant attention in recent years. A review of the literature on this topic reveals several key findings and trends. First, there is a growing recognition of the importance of ESG sustainability practices in material production, use, and disposal. This includes the adoption of green materials and circular economy practices, which can significantly reduce the environmental impact of material production, promote greater resource efficiency, and offer economic and social benefits (Mohanty et al. 2002). Second, there is a growing body of research on the ESG sustainability of green materials in circular economy, with a focus on topics such as material selection, recycling and reuse, life cycle assessment, and stakeholder engagement. This research has led to the development of new tools, methods, and frameworks for assessing and improving the sustainability of material production and consumption (Shchipunov, 2012). Third, there is a growing emphasis on the role of collaboration and partnerships in promoting ESG sustainability of green materials in circular economy. This includes collaborations between companies, government agencies, academic institutions, and civil society organizations to promote sustainable practices and policies (Irimia-Vladu, 2014). Fourth, there are several challenges and barriers to the ESG sustainability of green materials in circular economy, including technical, economic, and social factors. The findings of current analysis is found compliance and generalization with results of other related work in terms of ESG sustainability in circular economies (Petrica et al. 2024; Khursid and Islam 2024; Senadheera 2022).

The result of the integration of ESG sustainability with green materials in circular economies represents a holistic approach to sustainable development (Singh et al. 2023b). By addressing environmental, social, and governance aspects, stakeholders can ensure that economic growth aligns with the principles of sustainability and equity (Kushwaha et al. 2023a). Moving forward, concerted efforts are required to overcome economic, technological, and educational barriers, fostering a sustainable and inclusive future (Jain & Singh, 2023). This approach not only enhances environmental protection and social well-being but also strengthens governance frameworks, paving the way for resilient and sustainable economic systems. Incorporating ESG criteria into the use of green materials within circular economies presents a robust pathway towards sustainable development. By addressing environmental, social, and governance aspects, businesses and policymakers can ensure that economic growth does not come at the expense of the planet or people, fostering a more sustainable and equitable future (Jain et al. 2024).

These challenges require innovative solutions and approaches to promote sustainable development (Kushwaha et al. 2023b). In general, the literature on the ESG sustainability of green materials in circular economy highlights the importance of sustainable practices in material production and consumption, and the need for collaboration and innovation to address the challenges and barriers to sustainable development (Singh & Kushwaha, 2024). This research has significant implications for companies, policymakers, and other stakeholders seeking to promote sustainable development and advance the transition to a circular economy. Future research is needed to continue to develop and refine strategies for promoting sustainable green materials within a circular economy framework (Kushwaha et al. 2024).

7. FURTHER DIRECTION AND ESG LIMITATION OF GREEN MATERIALS IN CIRCULAR ECONOMIES

While the integration of ESG principles in the use of green materials within circular economies holds great promise, several limitations must be acknowledged (Singh, 2014). These limitations span environmental, social, and governance dimensions and pose significant challenges to the widespread adoption and effectiveness of sustainable practices (Singh & Agrawal, 2020).

7.1 Environmental Limitations

- *Resource Availability and Quality*: The availability of high-quality recycled materials can be inconsistent, leading to reliance on virgin resources. This is particularly challenging for industries requiring specific material standards.
- *Technological barriers*: Current recycling technologies are not always capable of efficiently processing all types of materials. Advanced materials and composites used in modern products can be difficult to separate and recycle effectively.
- *Environmental Trade-offs*: Some green materials, while reducing carbon footprints, may have other environmental impacts. For example, bioplastics derived from agricultural products can contribute to deforestation and biodiversity loss if not sourced sustainably.
- *End-of-Life Management*: Effective end-of-life management systems for green materials are often underdeveloped, leading to challenges in ensuring that materials are properly recycled or repurposed.

7.2 Social Limitations

- *Equity and Access*: The benefits of green materials and circular economies may not be evenly distributed. Marginalized communities often lack access to the benefits of sustainable practices and may bear disproportionate environmental and economic burdens.
- *Job Displacement*: Transitioning to circular economies may lead to job displacement in traditional manufacturing and resource extraction industries. Without adequate retraining and support programs, workers in these sectors could face significant hardships.
- *Consumer Acceptance*: There is often a gap between consumer awareness and behavior. Despite growing awareness of sustainability issues, many consumers are reluctant to change purchasing habits or pay premiums for sustainable products.

- *Health and Safety Concerns*: While green materials are generally safer, there is still a need for rigorous testing and standards to ensure that they do not pose unforeseen health risks to consumers and workers.

7.3 Governance Limitations

- *Regulatory Challenges*: Regulatory frameworks for circular economies and green materials are still evolving. Inconsistent regulations across regions can create barriers to the adoption of sustainable practices.
- *Lack of Standardization*: There is a lack of standardized metrics and reporting for ESG criteria, making it difficult to compare and assess the sustainability performance of different companies and materials.
- *Transparency Issues*: Ensuring transparency in the supply chain for green materials can be challenging. Without robust verification mechanisms, claims of sustainability can be misleading or exaggerated.
- *Corporate Resistance*: Some companies may resist adopting ESG practices due to perceived costs, complexity, or a lack of understanding of the long-term benefits. Overcoming this resistance requires strong leadership and commitment at all levels.

7.4 Addressing the ESG Limitations

By addressing these limitations, the potential of ESG sustainability in green materials and circular economies can be more fully realized, contributing to a more sustainable and equitable future (Singh & Agrawal, 2022). To overcome these limitations, a multi-faceted approach is required (Singh et al. 2021b).

- *Innovation and Research*: Continued investment in research and development can address technological barriers and improve the efficiency and effectiveness of recycling processes for a wider range of materials.
- *Policy and Regulation*: Developing robust, harmonized regulatory frameworks that support sustainable practices and provide incentives for compliance can drive broader adoption.
- *Education and Awareness*: Educating consumers and businesses about the benefits of green materials and circular economies can help bridge the gap between awareness and action.
- *Equity and Inclusion*: Ensuring that the transition to circular economies is inclusive and equitable, with targeted support for marginalized communities and workers affected by the transition, is crucial.

- *Transparency and Accountability*: Enhancing transparency through standardized reporting and robust verification mechanisms can build trust and ensure that ESG claims are credible and meaningful.

7.5 Future Direction

Despite the promising benefits, several challenges must be addressed to fully realize the potential of ESG sustainability in green materials and circular economies (Singh et al. 2021a). One major challenge is the economic feasibility of producing and scaling green materials (Singh et al. 2023a). While environmentally beneficial, these materials can be cost-prohibitive, limiting their widespread adoption. Public and private sector collaboration is needed to provide financial incentives and subsidies that support sustainable production practices.

Moreover, consumer awareness and education are critical to driving demand for green materials. Businesses and governments must engage in robust awareness campaigns that inform consumers about the environmental and social benefits of green materials and circular economies (Saraswat & Singh, 2025). Another challenge is the need for continuous innovation in material science to develop new green materials that meet performance standards without compromising sustainability (Kumar et al. 2025).

7.6 Geographical Scope of the Study

To expand the geographical scope and broaden the analysis of the sustainability of green materials in circular economies, it is essential to consider regional differences in resources, policies, technological capabilities, and cultural practices among circular economies worldwide (Singh & Kushwaha, 2025). By expanding the geographical scope and incorporating regional diversity into the analysis, research on the sustainability of green materials in circular economies can become more comprehensive and applicable globally, leading to more effective and inclusive solutions (Anand & Singh, 2025).

- *Global Research Networks*: Establish global research networks and consortia dedicated to the study of green materials and circular economies, ensuring diverse regional representation.
- *Funding and Grants*: Secure funding and grants for region-specific research projects, encouraging local researchers and institutions to contribute to the global body of knowledge.

- *Publications and Dissemination*: Encourage the publication and dissemination of research findings in open-access journals and international conferences to reach a broader audience.
- *Pilot Projects*: Implement pilot projects in diverse regions to test the feasibility and performance of green materials in real-world conditions, providing valuable data for further research.

7.7 Actionable Recommendations

Sustainability of green materials in circular economies is a dynamic field (Singh et al. 2025). Some actionable and detailed recommendations for future research several emerging trends and identified gaps are following. By addressing follwing research recommendations, the field can advance towards creating sustainable, economically viable, and widely accepted green materials that support the principles of circular economies. This holistic approach will help bridge the existing gaps and leverage emerging trends for a more sustainable future

- *Focus on Comprehensive life cycle assessments (LCAs)*: Current LCAs often lack depth in certain stages of a product's life. Research should aim to develop more comprehensive LCAs that include detailed analyses of raw material extraction, production, use, and end-of-life phases. Develop standardized LCA methodologies specific to green materials to ensure consistency and comparability of results across different studies and regions.
- *Bio-based and Renewable Materials Innovation and developments*: Invest in the development and optimization of bio-based materials that are renewable and biodegradable. Research should focus on enhancing their performance to match or exceed conventional materials. Explore the use of nanotechnology to improve the properties of green materials, such as strength, durability, and resistance to environmental factors.
- *End-of-life strategies*: Advance research in recycling technologies that are specific to green materials, ensuring that they can be efficiently and economically recycled. Promote research in product design that facilitates easy disassembly and separation of materials at the end of their life cycle, making recycling and reuse more feasible.
- *Environmental Impact and Performance*: Perform comparative studies between green materials and traditional materials to quantify the environmental benefits and identify areas for improvement. Research the long-term performance and durability of green materials under various environmental conditions to ensure they meet industry standards.

REFERENCES

Ahmad, H., Yaqub, M., & Lee, S. H. (2024). Environmental-, social-, and governance-related factors for business investment and sustainability: A scientometric review of global trends. *Environment, Development and Sustainability*, *26*(2), 2965–2987. DOI: 10.1007/s10668-023-02921-x PMID: 36714213

Anand, A., & Singh, P. (2025). Interdisciplinary Studies: Issues, Perspectives, and Challenges From National Education Program 2020. *Integrating Personalized Learning Methods Into STEAM Education*, 233-260.

Aria, M., & Cuccurullo, C. (2017). Bibliometrix: An R-tool for comprehensive science mapping analysis. *Journal of Informetrics*, *11*(4), 959–975. DOI: 10.1016/j.joi.2017.08.007

Bradu, P., Biswas, A., Nair, C., Sreevalsakumar, S., Patil, M., Kannampuzha, S., Mukherjee, A. G., Wanjari, U. R., Renu, K., Vellingiri, B., & Gopalakrishnan, A. V. (2022). AV. Recent advances in green technology and Industrial Revolution 4.0 for a sustainable future. *Environmental Science and Pollution Research International*, *30*(60), 1–32. DOI: 10.1007/s11356-022-20024-4 PMID: 35397034

Chatziparaskeva, G., Papamichael, I., Voukkali, I., Loizia, P., Sourkouni, G., Argirusis, C., & Zorpas, A. A. (2022). End-of-life of composite materials in the framework of the circular economy. *Microplastics.*, *1*(3), 377–392. DOI: 10.3390/microplastics1030028

Corona, B., Shen, L., Reike, D., Carreón, J. R., & Worrell, E. (2019). Towards sustainable development through the circular economy—A review and critical assessment on current circularity metrics. *Resources, Conservation and Recycling*, *151*, 104498. DOI: 10.1016/j.resconrec.2019.104498

Ditlev-Simonsen, C. D. (2022). Sustainability and finance: environment, social, and governance (ESG). *A Guide to Sustainable Corporate Responsibility: From Theory to Action*, 189-206. Cham. DOI: 10.1007/978-3-030-88203-7_9

Ferreira Gregorio, V., Pié, L., & Terceño, A. (2018). A systematic literature review of bio, green and circular economy trends in publications in the field of economics and business management. *Sustainability (Basel)*, *10*(11), 4232. DOI: 10.3390/su10114232

Irimia-Vladu, M. (2014). "Green" electronics: Biodegradable and biocompatible materials and devices for sustainable future. *Chemical Society Reviews*, *43*(2), 588–610. DOI: 10.1039/C3CS60235D PMID: 24121237

Jain, P., & Singh, P. (2023). Lethargic approach in expediting financial reporting system: A case of construction firm. *International Journal of Teaching and Case Studies*, *14*(1), 1–10. DOI: 10.1504/IJTCS.2023.131663

Jain, P., Singh, P., & Kushwaha, R. (2024). Investigating change management as a novel digital wallet service innovation in banks: A case of Indian bank. *International Journal of Teaching and Case Studies*, *15*(1), 67–81. DOI: 10.1504/IJTCS.2025.143552

Koronis, G., Silva, A., & Fontul, M. (2013). Green composites: A review of adequate materials for automotive applications. *Composites. Part B, Engineering*, *44*(1), 120–127. DOI: 10.1016/j.compositesb.2012.07.004

Kulakovskaya, A., Knoeri, C., Radke, F., & Blum, N. U. (2023). Measuring the economic impacts of a circular economy: An evaluation of indicators. *Circular Economy and Sustainability*, *3*(2), 657–692. DOI: 10.1007/s43615-022-00190-w PMID: 37408852

Kumar, A., Singh, P., Parihar, A.S. & Saraswat, S. (2025), "Analyzing the impact of psychological capital on the entrepreneurial intentions of students through mediating role of individual expectancy using PLS-SEM", *International Journal of Social Economics*, Vol. ahead-of-print No. ahead-of-print. DOI: 10.1108/IJSE-03-2024-0252

Kushwaha, J., Sharma, A., & Singh, P. (2023). b. Exploration and prioritization of enablers to organization work–family balance planning for working sole Indian mothers integrating fuzzy Delphi and AHP. *International Journal of Social Economics*, *50*(3), 398–418. DOI: 10.1108/IJSE-05-2022-0348

Kushwaha, J., Singh, P., & Kushwaha, R. (2024). Predicting working sole Indian mothers' satisfaction towards work–family balance integrating the Kano model and weighted average method. *International Journal of Social Economics*, *51*(6), 725–740. DOI: 10.1108/IJSE-02-2023-0074

Kushwaha, J., Singh, P., & Sharma, A. (2023). a, "Modelling the enablers of work-family balance among working single mothers integrating expert-mining and TISM-MICMAC". *Journal of Organizational Effectiveness: People and Performance*, *10*(4), 501–526. DOI: 10.1108/JOEPP-05-2022-0106

Marchesi, M., & Tweed, C. (2021). Social innovation for a circular economy in social housing. *Sustainable Cities and Society*, *71*, 102925. DOI: 10.1016/j.scs.2021.102925

Markopoulos, E., Kirane, I. S., Gann, E. L., & Vanharanta, H. (2020). A democratic, green ocean management framework for environmental, social and governance (ESG) compliance. In *Human Interaction, Emerging Technologies and Future Applications II: Proceedings of the 2nd International Conference on Human Interaction and Emerging Technologies: Future Applications (IHIET–AI 2020), April 23-25, 2020, Lausanne, Switzerland* (pp. 21-33). Springer International Publishing. DOI: DOI: 10.1007/978-3-030-44267-5_4

Mohanty, A. K., Misra, M., & Drzal, L. T. (2002). Sustainable bio-composites from renewable resources: Opportunities and challenges in the green materials world. *Journal of Polymers and the Environment*, *10*(1-2), 19–26. DOI: 10.1023/A:1021013921916

Moshood, T. D., Nawanir, G., Mahmud, F., Mohamad, F., Ahmad, M. H., & Abdul Ghani, A. (2022). Biodegradable plastic applications towards sustainability: A recent innovations in the green product. *Cleaner Engineering and Technology*, *100404*, 100404. Advance online publication. DOI: 10.1016/j.clet.2022.100404

Munoz-Torres, M. J., Fernández-Izquierdo, M. Á., Rivera-Lirio, J. M., & Escrig-Olmedo, E. (2019). Can environmental, social, and governance rating agencies favor business models that promote a more sustainable development? *Corporate Social Responsibility and Environmental Management*, *26*(2), 439–452. DOI: 10.1002/csr.1695

Opferkuch, K., Caeiro, S., Salomone, R., & Ramos, T. B. (2022). Circular economy disclosure in corporate sustainability reports: The case of European companies in sustainability rankings. *Sustainable Production and Consumption*, *32*, 436–456. DOI: 10.1016/j.spc.2022.05.003

Padilla-Rivera, A., Russo-Garrido, S., & Merveille, N. (2020). Addressing the social aspects of a circular economy: A systematic literature review. *Sustainability (Basel)*, *12*(19), 7912. DOI: 10.3390/su12197912

Pajula, T., Behm, K., Vatanen, S., & Saarivuori, E. (2017). Managing the life cycle to reduce environmental impacts. *Dynamics of long-life assets: From technology adaptation to upgrading the business model*, 93-113. https://doi.org/.DOI: 10.1007/978-3-319-45438-2_6

Papamichael, I., Voukkali, I., Loizia, P., Stylianou, M., Economou, F., Vardopoulos, I., Klontza, E. E., Lekkas, D. F., & Zorpas, A. A. (2023). Measuring Circularity: Tools for monitoring a smooth transition to Circular Economy. *Sustainable Chemistry and Pharmacy*, *36*, 101330. DOI: 10.1016/j.scp.2023.101330

Passas, I. (2023). Accounting for integrity: ESG and financial disclosures: the challenge of internal fraud in management decision-making [Hellenic Mediterranean University]. https://doi.org/DOI: 10.12681/eadd/55359

Passas, I. (2024). Bibliometric Analysis: The Main Steps. Encyclopedia 2024, Vol. 4, Pages 1014-1025, 4(2), 1014–1025. https://doi.org/DOI: 10.3390/encyclopedia4020065

Patil, R. A., Ghisellini, P., & Ramakrishna, S. (2021). Towards sustainable business strategies for a circular economy: environmental, social and governance (ESG) performance and evaluation. *An introduction to circular economy*, 527-554. DOI: DOI: 10.1007/978-981-15-8510-4_26

Pavlides, T., Vardopoulos, I., Papamichael, I., Voukkali, I., Stylianos, M., & Zorpas, A. A. (2023). Environmental sustainability assessment of excavation, construction, and demolition waste conditions and practices across Greece and Cyprus. *IOP Conference Series. Earth and Environmental Science*, *1196*(1), 012037. DOI: 10.1088/1755-1315/1196/1/012037

Petrica, I. M., Caraiania, C., Lungua, C. I., & Anica-Popaa, L. E. (2024). The interconnectivity of ESG research within the realm of sustainability: A bibliometric analysis. *Journal of Accounting and Management Information Systems*, *23*(1), 261–290. DOI: 10.1007/978-981-15-8510-4_26

Petropoulou, A., Angelaki, E., Rompogiannakis, I., Passas, I., Garefalakis, A., & Thanasas, G. (2024). Digital Transformation in SMEs: Pre- and Post-COVID-19 Era: A Comparative Bibliometric Analysis. Sustainability 2024, Vol. 16, Page 10536, 16(23), 10536. DOI: 10.3390/su162310536

Popescu, C., Hysa, E., Kruja, A., & Mansi, E. (2022). Social innovation, circularity and energy transition for environmental, social and governance (ESG) practices—A comprehensive review. *Energies*, *15*(23), 9028. DOI: 10.3390/en15239028

Purchase, C. K., Al Zulayq, D. M., O'Brien, B. T., Kowalewski, M. J., Berenjian, A., Tarighaleslami, A. H., & Seifan, M. (2022). Circular economy of construction and demolition waste: A literature review on lessons, challenges, and benefits. *Materials (Basel)*, *15*(1), 76. DOI: 10.3390/ma15010076 PMID: 35009222

Ragazou, K., Passas, I., Garefalakis, A., & Dimou, I. (2022). Investigating the Research Trends on Strategic Ambidexterity, Agility, and Open Innovation in SMEs: Perceptions from Bibliometric Analysis. *Journal of Open Innovation*, *8*(3), 118. DOI: 10.3390/joitmc8030118

Ragazou, K., Passas, I., Garefalakis, A., Galariotis, E., & Zopounidis, C. (2023). Big Data Analytics Applications in Information Management Driving Operational Efficiencies and Decision-Making: Mapping the Field of Knowledge with Bibliometric Analysis Using R. Big Data and Cognitive Computing 2023, Vol. 7, Page 13, 7(1), 13. DOI: 10.3390/bdcc7010013

Roy, S. B., & Basak, M. (2013) Journal of Documentation: a bibliometric study. Library Philosophy and Practice (e-Journal). Paper.;945.: https://digitalcommons .unl.edu/libphilpracPart of the Library and Information Science Commons.

Sadiq, M., Ngo, T. Q., Pantamee, A. A., Khudoykulov, K., Ngan, T. T., & Tan, L. P. (2023). The role of environmental social and governance in achieving sustainable development goals: Evidence from ASEAN countries. *Economic research-. Ekonomska Istrazivanja, 36*(1), 170–190. DOI: 10.1080/1331677X.2022.2072357

Saraswat, S. & Singh, P. (2025), "The moderating role of perceived trust in predicting the adoption intention of consumers toward online pharmacies using the application of UTAUT model", *International Journal of Pharmaceutical and Healthcare Marketing*, Vol. ahead-of-print No. ahead-of-print. DOI: 10.1108/IJPHM-05-2024-0042

Senadheera, S. S., Gregory, R., Rinklebe, J., Farrukh, M., Rhee, J. H., & Ok, Y. S. (2022). The development of research on environmental, social, and governance (ESG): A bibliometric analysis. *Sustainable Environment, 8*(1), 2125869. DOI: 10.1080/27658511.2022.2125869

Shchipunov, Y. (2012). Bionanocomposites: Green sustainable materials for the near future. *Pure and Applied Chemistry, 84*(12), 2579–2607. DOI: 10.1351/PAC-CON-12-05-04

Sinclair, S., & Rockwell, G. (2020). "Voyant-tools". available at: http://voyant-tools .org (accessed 2 June 2022). DOI: DOI: 10.1002/9781118680605.ch19

Singh, A. K., Meshram, S. A., Khandelwal, V., Tiwari, P., & Singh, P. (2024)c. Implementation of Industry-Based Perspective in Technical Education: A Qualitative and Quantitative Analysis. *Industrial and Manufacturing Designs: Quantitative and Qualitative Analysis*, 321-346. https://doi.org/.DOI: 10.1002/9781394212668.ch11

Singh, P. (2014). Unorganized sector and India's informal economy: challenges and prospects. *New Man International Journal of Multidisciplinary Studies, 1*(2), 209-218. https://www.newmanpublication.com/dash/issueworkfiles/71.pdf?1721465584

Singh, P. (2024). Analyzing Responsible Leadership Effectiveness for Attainment of Sustainable Development Goals in Developing Countries Using Environmental Scanning Techniques. In Bhattacharya, S., Venkatesh, V. G., & Chatterjee, S. R. (Eds.), *Responsible Corporate Leadership Towards Attainment of Sustainable Development Goals. Responsible Leadership and Sustainable Management*. Springer., DOI: 10.1007/978-981-97-1386-8_13

Singh, P. (2024). Weather index insurance viability in mitigation of climate change impact risk: A systematic review and future agenda. *Journal of Science and Technology Policy Management, 15*(1), 142–163. DOI: 10.1108/JSTPM-07-2021-0102

Singh, P., & Agrawal, G. (2019). Efficacy of weather index insurance for mitigation of weather risks in agriculture: An integrative review. *International Journal of Ethics and Systems, 35*(4), 584–616. DOI: 10.1108/IJOES-09-2018-0132

Singh, P., & Agrawal, G. (2020). Development, present status and performance analysis of agriculture insurance schemes in India: Review of evidence. *International Journal of Social Economics, 47*(4), 461–481. DOI: 10.1108/IJSE-02-2019-0119

Singh, P., & Agrawal, G. (2022). Mapping the customer centric weather index insurance service design using quality function deployment. *The TQM Journal, 34*(6), 1800–1822. DOI: 10.1108/TQM-08-2021-0236

Singh, P., Agrawal, G., & Kushwaha, R. (2021, November) a. Evaluating Customer Satisfaction of Weather Index Insurance Service Quality Using Kano Model. In *2021 IEEE Bombay Section Signature Conference (IBSSC)* (pp. 1-6). IEEE. **DOI:** DOI: 10.1109/IBSSC53889.2021.9673370

Singh, P. & Kushwaha, R. (2024), "Envisaging Indian farmers' desires from agricultural index insurance integrating rank sum weighting method and MoSCoW technique: an approach to requirements prioritization", *International Journal of Social Economics*, Vol. ahead-of-print No. ahead-of-print. DOI: 10.1108/IJSE-03-2024-0233

Singh, P., & Kushwaha, R. (2025), "Conceptualizing the corporate-social accountability and sustainability assessment of weather-risk management policy towards stakeholders using qualitative insights: a case of weather index insurance", *Qualitative Research in Financial Markets*, Vol. ahead-of-print No. ahead-of-print. DOI: 10.1108/QRFM-05-2024-0113

Singh, P., Kushwaha, R., & Kushwaha, J. (2021) b. Assessing visitors' satisfaction about museum service quality using Kano model. In *2021 IEEE 9th Region 10 Humanitarian Technology Conference (R10-HTC)* (pp. 1-6). IEEE. **DOI:** DOI: 10.1109/R10-HTC53172.2021.9641707

Singh, P., Kushwaha, R., & Kushwaha, J. (2023)a. Analyzing Accountability of Weather Index Insurance Service in Attainment of Sustainable Development Goals: A Sustainable Accounting Perspective Review. In Pratik Arte et al. (Eds): *Sustainable International Business, 2023*. Springer Nature, 2023. doi: , 531340. DOI: 978-3-031-43784-7

Singh, P., Kushwaha, R., & Kushwaha, J. (2023)b. Evaluating Farmer Participation in Weather Index Insurance in a Socio-Demographic Perspective: Evidence From the Crop Insurance Data of India. In A. Popoola, H. Magidimisha-Chipungu, & L. Chipungu (Eds.), *Handbook of Research on Managing the Urban-Rural Divide Through an Inclusive Framework* (pp. 69-88). IGI Global. DOI: 10.4018/978-1-6684-6258-4.ch005

Singh, P., Kushwaha, R., & Kushwaha, J. (2024)a. Foresight the future of Industry 5.0 in perspective of sustainable services in developing countries. *Modern Technologies and Tools Supporting the Development of Industry 5.0*, 381. DOI: .DOI: 10.1201/9781003489269-11

Singh, P., Kushwaha, R., & Kushwaha, J. (2025), "A Qualitative Study on Analysing the Role of Mimicking Technology in Electronic Customer Relationship Management: A Case of Service Sector", Kumar, A., Ciddikie, M.D., Kashyap, A.K. and Akram, H.W. (Ed.) *Marketing 5.0*, Emerald Publishing Limited, Leeds, pp. 103-119. DOI: 10.1108/978-1-83797-815-120251008

Singh, P., Kushwaha, R., Kushwaha, J., & Singh, A. K. (2024)b. A Qualitative Study to Rank Non-Conventional Energy Sources for Industrial Sustainability and Energy Management Decisions Using MoSCoW Prioritization Method. *Industrial and Manufacturing Designs: Quantitative and Qualitative Analysis*, 105-137. https://doi.org/.DOI: 10.1002/9781394212668.ch4

Torres, L., Ripa, D., Jain, A., Herrero, J., & Leka, S. (2023). The potential of responsible business to promote sustainable work–An analysis of CSR/ESG instruments. *Safety Science*, *164*, 106151. DOI: 10.1016/j.ssci.2023.106151

Van Eck N & Waltman L. (2010). Software survey: VOSviewer, a computer program for bibliometric mapping. scientometrics. 84 (2):523-38. .DOI: 10.1007/s11192-009-0146-3

Vardopoulos, I., Santamouris, M., Zorpas, A. A., Barone, G., Italos, C., & Vassiliades, C. (2024). A comparative study on discrepancies in residential building energy performance certification in a Mediterranean context. *Buildings (Basel, Switzerland)*, *14*(4), 1009. DOI: 10.3390/buildings14041009

Velenturf, A. P., & Purnell, P. (2021). Principles for a sustainable circular economy. *Sustainable Production and Consumption*, *27*, 1437–1457. DOI: 10.1016/j.spc.2021.02.018

Walker, S., Coleman, N., Hodgson, P., Collins, N., & Brimacombe, L. (2018). Evaluating the environmental dimension of material efficiency strategies relating to the circular economy. *Sustainability (Basel)*, *10*(3), 666. DOI: 10.3390/su10030666

Zafeiriou, E., Garefalakis, A., Passas, I., & Ragazou, K. (2023). Illicit and Corruption Mitigation Strategy in the Financial Sector: A Study with a Hybrid Methodological Approach. Sustainability 2023, Vol. 15, Page 1366, 15(2), 1366. DOI: 10.3390/su15021366

Zopounidis, C., Garefalakis, A., Lemonakis, C., & Passas, I. (2020). Environmental, social and corporate governance framework for corporate disclosure: A multicriteria dimension analysis approach. *Management Decision*, *58*(11), 2473–2496. DOI: 10.1108/MD-10-2019-1341

Chapter 5
Greener Cities, Brighter Future:
The Critical Role of Green Buildings

Erasmia Angelaki
https://orcid.org/0009-0006-1336-4315
Hellenic Mediterranean University, Greece

Alexandros Garefalakis
https://orcid.org/0000-0001-7980-3743
Hellenic Mediterranean University, Greece

Ioannis Sitzimis
Hellenic Mediterranean University, Greece

Ioannis Rompogiannakis
Hellenic Mediterranean University, Greece

Ioannis Passas
https://orcid.org/0000-0002-1261-2878
Hellenic Mediterranean University, Greece

ABSTRACT

As urbanization accelerates and environmental concerns grow, green buildings have emerged a crucial component of global sustainability efforts. This chapter explores green architecture's environmental, economic, and social dimensions, emphasizing its role in reducing carbon emissions, enhancing energy efficiency, and promoting occupant well-being. A bibliometric analysis was conducted using VOSviewer and Biblioshiny to map key research trends, co-citation networks, and

DOI: 10.4018/979-8-3373-2327-5.ch005

thematic developments in green building studies from 2001 to 2025. Findings reveal a growing emphasis on ESG frameworks, smart building technologies, and circular economy principles. The chapter also highlights the interdisciplinary nature of green architecture, intersecting with urban governance, renewable energy, and sustainable materials. Ultimately, the study underscores the need for collaborative policy interventions and technological innovations to accelerate the transition towards sustainable built environments.

INTRODUCTION

The urgency of climate change (Ribeiro Hoffmann et al., 2024), rapid urbanization (Vardopoulos et al., 2024), and resource depletion (Salvati, 2025), has placed unprecedented pressure on the built environment to adopt sustainable solutions (Vardopoulos et al., 2024). Buildings account for nearly 40% of global energy consumption and 30% of greenhouse gas emissions, making them a crucial sector for climate action . The concept of green architecture has emerged as a vital strategy to mitigate environmental degradation while improving economic and social well-being.

Green buildings are designed to minimize their ecological footprint through energy efficiency, carbon neutrality, and resource conservation (Vardopoulos et al., 2023). The United Nations Environment Programme (UNEP) emphasizes that sustainable buildings can cut global carbon emissions by 50% by 2050 if widely implemented (UNEP, 2022). These buildings integrate innovative technologies, sustainable materials, and efficient water management systems to reduce environmental impact while providing healthier indoor environments for occupants.

Economic benefits also drive the green building movement (Vassiliades et al., 2023). Studies indicate that green – certified buildings have lower operational costs due to reduced energy and water consumption (Vassiliades et al., 2024). A 2020 report by the World Green Building Council found that green buildings yield a 7% increase in asset value and a 14% reduction in operational expenses over conventional buildings (World Green Building Council, 2013). Moreover, governments and financial institutions are providing incentives, tax rebates, and sustainability-linked financing accelerate the adoption of green construction practices (Karytsas et al., 2019).

Socially, green buildings contribute to occupant health and well-being by improving indoor air quality, maximizing natural light, and incorporating biophilic design elements (Cortesi et al., 2022). Research from Harvard University's T.H. Chan School of Public Health revealed that individuals working in green-certified buildings experience 26% higher cognitive function scores compared to those in conventional buildings (Allen et al., 2016). These findings underscore the broader benefits of sustainable architecture beyond environmental considerations.

In recent years, the green building movement has evolved from a niche environmental initiative into a mainstream, multidisciplinary solution for addressing the interviewed crises of climate, health, and equity (Liu et al., 2022). The global construction sector, long criticized for its carbon-intensive practices, is undergoing a profound transformation. Sustainability is no longer an optional feature but a foundational requirement driven by demand, regulatory pressure, and the financial sector's shift towards Environmental, Social, and Governance (ESG) criteria (Christensen et al., 2021).

This transformation is particularly evident in cities, where more than half of the world's population now resides (Escrivà Saneugenio et al., 2024). Urban centers are both the cause and the victim of climate stress (Vardopoulos, 2024). They generate most of the world's CO_2 emissions, yet are also the most vulnerable to its impacts, such as heatwaves, flooding, and energy instability (Glaeser & Kahn, 2008). Green buildings are a direct response to this dilemma-structures that not only use fewer resources but are designed to perform adaptively and support community resilience.

The evolution of the green building concept reflects broader shifts in how we define progress and prosperity. Where once economic growth was measured by GDP and expansion, the modern definition of development includes social well-being, environmental stewardship and intergenerational equity. Green architecture represents this new paradigm in action: it measures success not only in square meters built but in carbon saved, water reused, energy produced, and lives improved (Hariram et al., 2023).

The increasing adoption of certification systems like LEED, BREEAM, Green Star, and WELL Building Standard is evidence of this cultural shift (Da Silva & Ramirez, 2021). These frameworks provide both the language and benchmarks needed to evaluate performance holistically. But green building is not just about ticking of ''sustainability boxes''. It requires a rethinking of the deign process itself-from the sourcing of materials and building orientation to how spaces support physical and mental health. Architects, urban planners, and engineers are now expected to collaborate with climatologists, public health experts, and economists.

In parallel, technological advancements have unlocked new possibilities. Smart urban and building technologies (Vardopoulos et al., 2023; 2024), such as sensors and automated systems, now allow real-time optimization of energy and water use (Okedara & Odubiyi, 2024). Innovations in material science have introduced low-carbon concrete, recycled composites, and living materials that can self-repair or sequester carbon. Digital tools like Building Information Modeling (BIM) and digital twins provide simulations that anticipate a building's long-term environmental footprint even before a single brick is laid (Apanaviciene et al., 2020).

However, the diffusion of green practices is uneven. While some regions lead with bold sustainability mandates and investment incentives, others lag due to economic constraints, lack of awareness, or institutional inertia. For example, Europe and North America dominate global green building certifications, whereas developing nations-despite being more vulnerable to climate risks – face challenges in adopting green technologies at scale (Maier & Gluszak, 2024; Varma & Palaniappan, 2019). This disparity raises critical questions about global equity and how knowledge, financing, and policy frameworks can be shared more effectively.

The impact of Covid-19 pandemic (Alaimo et al., 2022), has further amplified the urgency of sustainable building design. The crisis exposed weakness in indoor environmental quality, particularly ventilation and air purification (Kaklauskas et al., 2021; Passas, 2023). As a result, post-pandemic building design is increasingly centered on health resilience, ensuring that indoor environments support immune health, reduce pathogen transmission, and provide psychological comfort. This health-driven shift has expanded the definition of sustainability to include not only energy and emissions but also well-being and safety (Navaratnam et al., 2022).

Aimed this context, the role of sustainable buildings in achieving broader sustainability goals has become a central focus of academic research, government policy, and corporate strategy (Vassiliades et al., 2025). Internationally, green buildings are explicitly referenced in several Sustainable Development Goals (SDGs), including SDG 11 (Sustainable Cities), SDG 12 (Responsible Consumption), and SDG 13 (Climate Action) (Olabi et al., 2025; Ragazou, Passas, et al., 2022; Ragazou et al., 2023; Sinha et al., 2013). The construction industry is now recognized as both contributor to global warming and a potential driver of the low carbon transition (Pavlides et al., 2023).

To contribute to this evolving discourse, this chapter undertakes a two – fold mission. First, to provide a comprehensive overview of the environmental, economic, and social benefits of green buildings; and second, to investigate the emerging research themes and knowledge gaps through a bibliometric analysis spanning the last two decades. By combining theoretical insights and empirical data and drawing from interdisciplinary perspectives, this approach offers both breadth and depth.

The bibliometric analysis is central to deepening our understanding of how buildings contribute to sustainability. It seeks to answer two key research questions:

RQ1: How do green buildings shape sustainability?
RQ2: What is the role of green buildings in achieving sustainability targets?

This chapter delves into the role of green buildings in achieving global sustainability targets. It explores how energy-efficient designs, smart technologies, and sustainable materials contribute to long-term resilience and ecological balance.

Furthermore, real-world case studies will highlight the successes and challenges of green building implementation worldwide. Ultimately, as the global community strives for net-zero emissions and sustainable urban development, the role of green buildings will remain indispensable in shaping a healthier, more resilient future.

KEY FEATURES OF GREEN BUILDINGS

Green buildings incorporate several key features that enhance environmental sustainability, energy efficiency, and occupant well-being. These features reflect a shift away from conventional construction methods toward a holistic, performance-oriented approach. While their specific design may vary by context or certification standard, most green buildings share a foundational commitment to reducing environmental impact, optimizing resource use, and improving quality of life for their occupants (Pavate et al., 2024; Zopounidis et al., 2020).

A fundamental element of green buildings is their emphasis on energy efficiency. This begins at the design phase, where building orientation, passive solar gain, insulation, and natural ventilation are carefully considered. Beyond these passive strategies, green buildings integrate advanced technologies such as Internet of Things (IoT) devices and artificial intelligence (AI)-driven energy management systems. These systems enable real-time monitoring and optimization of energy consumption, leading to significant reductions in energy use. For instance, automated lighting systems and smart thermostats adjust to occupancy patterns, ensuring energy is used only when necessary (Poyyamozhi et al., 2024a).

Additionally, another cornerstone is the integration of renewable energy systems. Building-integrated photovoltaic (BIPV), solar thermal systems, and even small-scale wind turbines are now commonly deployed in green buildings to reduce reliance on fossil fuels. Some buildings function as net zero energy structures-producing as much energy as they consume over a year. In cutting – edge examples, net-positive energy buildings are becoming a reality, contributing excess energy back to the grid (Şirin et al., 2023).

The selection of materials plays a crucial role in the sustainability of green buildings. Prioritizing materials with low environmental impact, such as recycled steel, bamboo, and low-carbon concrete, reduces the overall carbon footprint of construction. Embracing the principles of a circular economy, green buildings focus on the reuse and recycling of materials, thereby minimizing construction waste. This approach not only conserves resources but also promotes environmental responsibility throughout the building's lifecycle (Chaudhari et al., 2024).

Equally vital is the choice of construction materials. Green buildings prioritize the materials that are nontoxic, locally sourced, durable, and have low embodied carbon. The use of recycled and rapidly renewable materials—such as reclaimed wood, bamboo, and recycled steel-reduces the environmental burden associated with resource extraction and manufacturing. Furthermore, green buildings increasingly align with the principle of circular economy, aiming to minimize waste by designing for material reuse and recycling across the building's lifecycle (Chaudhari et al., 2024). This involves cradle-to-cradle thinking, where materials are selected not only for performance but for their potential to be safely reintroduced into the production cycle at the end of the building's use phase.

Water efficiency is another critical feature. Green buildings implement water-saving technologies such as low-flow plumbing fixtures, dual-flush toilets, rainwater harvesting systems, and greywater recycling systems. These reduce demand for potable water while supporting local water conservation efforts. In drought-prone regions, these measures are essential for resilience and regulatory compliance (Al-Qawasmi et al., 2019; Passas et al., 2022; Sheth, n.d.).

Indoor environmental quality (IEQ) also plays a significant role in the definition of green architecture. This encompasses thermal comfort, indoor air quality, lighting quality, and acoustics. High-efficiency air filtration systems, low-VOC (volatile organic compound) materials, and well-designed natural ventilation contribute to cleaner indoor air. Access to natural daylight and exterior views has been shown to reduce stress, improve mood, and enhance cognitive performance among occupants. According to research from Harvard's T.H. Chan School of Public Health, workers in green-certified buildings perform significantly better on cognitive tasks compared to those in conventional buildings (Allen et al., 2016).

One of the most distinctive features of modern green buildings in their biophilic design, the international integration of natural elements into agricultural spaces. Biophilic features include living walls, rooftop gardens, water features, and interior green spaces. These elements reconnect people with nature, n especially important function in dense urban environments. Biophilic design is not only aesthetically pleasing; it has been linked to improvements in mood, productivity, and physical health. It also contributes to urban biodiversity by creating habitats for insects and birds (Tabassum & Park, 2024).

The adaptability and resilience of green buildings are increasingly emphasized in response to climate change. Green buildings are designed to withstand extreme temperatures, flooding, and power outages. They often include features like cool roofs, green roofs, operable windows for natural ventilation during power outages, and stormwater management systems to prevent urban flooding. These design elements are essential for protecting not just the building but the people and ecosystems around it (C. Zhang et al., 2021).

Another emerging dimension is the use of digital tools and smart systems to monitor and optimize building performance. Smart meters, occupancy sensors, and predictive analytics allow building managers to anticipate energy demand, detect faults early, and automate adjustments. These tools feed into BMS that create self-regulating, learning environments (Poyyamozhi et al., 2024b). As buildings become increasingly complex and connected, the line between physical infrastructure and digital architecture continues to blur.

Waste management and construction practices also distinguish green buildings from conventional counterparts. During the construction phase, green building projects emphasize efficient logistics, material tracking, and waste diversion strategies. Many certifications require that a high percentage of construction waste – often above 75% be derived from landfills (Ismaeel & Kassim, 2023). Prefabrication and modular construction techniques are also increasingly used to reduce material waste and improve energy performance by allowing for precision manufacturing in controlled environments (Ragazou, Garefalakis, et al., 2022; Stavropoulou et al., 2023; Vortelinos et al., 2024).

The location and connectivity of green buildings contribute to broader sustainability goals. Ideally, green buildings are situated in walkable neighborhoods with access to public transportation, cycling infrastructure, and essential services. This reduces the need for private vehicle use, cutting transportation-related emissions and fostering healthier communities (Ding et al., 2018). In terms of certification and performance measurement, several standards guide the development and evaluation of green buildings. These include:

- **Leadership in Energy and Environmental Design (LEED):** Developed by the U.S. Green Building Council, LEED is one of the most widely recognized certification systems globally (Pushkar, 2024)
- **Building Research Establishment Environmental Assessment Method (BREAM):** Originating in the UK, BREAM assesses buildings based on sustainability performance in categories such as energy, water, and pollution (Crawley & Aho, 1999)
- **WELL Building Standard**: Focused more specifically on health and wellness outcomes within buildings (Kokatnur et al., 2025)
- **Green Star (in Australia), CASBEE (in Japan), DGNB (in Germany):** Regional rating systems that reflect local priorities and regulatory frameworks (Illankoon et al., 2019).

These certification systems offer third-party validation and benchmarks for continuous improvement. While they differ in emphasis, they all promote a lifecycle perspective that encourages sustainable performance from design through operation and maintenance. Effective green design includes clear signage, accessible controls, feedback mechanisms (such as energy dashboards), and even occupant engagement programs that foster sustainable habits (Doan et al., 2017).

The most successful green buildings also embrace a user-centered design philosophy. Designers are increasingly recognizing the importance of behavioral aspects—how people interact with space, systems, and each other. Buildings that are technically efficient but uncomfortable or confusing to use can underperformed in real-world conditions (Cheliatsidou et al., 2023; Sellers & Fiore, 2013). In recent years, social equity and inclusion have become increasingly important within the green building movement. Green design should not only benefit the wealthy or technologically advanced urban centers; it must be also be accessible, affordable, and adaptable to diverse communities. Affordable housing projects that integrate green design, schools with biophilic learning environments, and public buildings designed for universal access demonstrate the broader societal of sustainable architecture (Obinna Iwuanyanwu et al., 2024).

Green buildings are, at their core, not just technical innovations but cultural expressions reflections of how we prioritize health, equity, and the environment in the places we inhabit. Their key features—ranging from smart energy systems and sustainable materials to biophilic spaces and digital intelligence—demonstrate a commitment to designing with both people and the planet in mind (Mba et al., 2024).

In summary, the key features of green buildings go far beyond energy savings. They encompass a holistic suite of strategies aimed at improving environmental outcomes, human health and long-term resilience. From passive design principles to smart technologies, from water efficiency to wellness-centered interiors, green buildings represent a rethinking of what buildings can and should be in the 21st century. As the global construction industry faces mounting pressure to decarbonize and adapt to climate instability, these features are no longer optional, they are foundation of sustainable, future-ready built environments.

CASE STUDIES IN GREEN BUILDING IMPLEMENTATION

While the features and principles of green buildings are well established, their practical application reveals how these ideals are realized—or challenged—in real-world settings. Case studies serve as tangible examples of the benefits, trade-offs, and lessons learned across diverse social, economic, and geographic contexts. This section examines four representative projects from Europe, Asia, Africa, and North America to illustrate different approaches to sustainability in the built environment.

Bosco Verticale – Milan, Italy

Bosco Verticale (or Vertical Forest) is one of the most prominent examples of biophilic and sustainable architecture in Europe. Completed in 2014, the project consists of two residential towers in Milan—one 80 meters and the other 112 meters tall designed by architect Stefano Boeri. The buildings host over 900 trees and thousands of shrubs and plants, all strategically distributed across their facades (Study & Verticale, 2015).

The primary objective was to address urban air pollution and reduce the heat island effect while enhancing biodiversity within the city. The vegetation acts as living facade, absorbing CO_2, producing oxygen, and filtering particulate matter. It also improves the microclimate around the building, lowering temperatures and acting as natural insulation.

From an energy perspective, the vegetation reduces solar gain, minimizing reliance on mechanical cooling systems. Residents benefit from lower utility costs and increased thermal comfort. Economically, the project boosted local property values, despite its initially higher construction costs. The main challenge has been long-term maintained of the extensive plat life, which requires regular pruning and irrigation system management.

Bosco Verticale has inspired similar greening strategies in cities such as Utrecht and Nanjing, proving that integrating green infrastructure into vertical urban growth is both feasible and scalable.

Punggol Digital District- Singapore

Singapore's Punggol Digital District (PDD) exemplifies the integration of smart city technologies with green building principles. Designed as a mixed-use district combining business parks, a university campus, and residential areas, PDD was conceived as a testbed for digital and environmental innovation.

PDD uses a centralized district cooling system that distributes chilled water to multiple buildings, reducing overall energy use. Smart lightning, waste management, and energy monitoring systems operate via an integrated digital platform, allowing real time optimization of consumption. Buildings are designed according to BCA Green Mark standard, with mandatory energy efficiency targets and green roofs to mitigate urban heat (K. Zhang et al., 2021).

What sets PDD apart is its systemic approach: sustainability is not limited to individual buildings but embedded into the entire neighborhood's infrastructure. The district is also designed for walkability and public transport access, reducing carbon emissions from commuting. Economically, the use of shared infrastructure reduces capital costs for individual buildings and creates opportunities for public and private partnerships. Socially, the integration of educational, residential, and business function within a single zone supports a vibrant, inclusive community with access to green space and smart public amenities. Punggol illustrates how urban-scale planning can elevate the impact of green building strategies when aligned with digital innovation and long-term governance.

Eastgate Centre – Harare, Zimbabwe

The Eastgate Centre in Harare, designed by architect Mick Pearce, is a landmark in climate-responsive architecture. Completed in the late 1990s, the building is modeled after termite mounds, which use natural ventilation to maintain internal temperature. Eastgate's design uses thermal mass, night cooling, and stack ventilation to create comfortable indoor conditions without mechanical air conditioning (Turner & Soar, 2008).

The building remains roughly 90% more efficient in energy use than comparable conventional buildings. Its design significantly lowers operational costs and has a smaller carbon footprint due to reduced reliance on electrical cooling systems. Constructions costs were reportedly 10% lower than similar buildings with HVAC systems, thanks to the passive strategies employed.

Beyond environmental performance, Eastgate has had symbolic and practical value. In a region where infrastructure is often under pressure, the building's independence from unreliable power grids demonstrates how local climatic knowledge can shape cutting-edge architecture. Maintenance has been straightforward, and its performance has remained consistent over two decades. Eastgate is often cited as a powerful example of biomimicry and frugal innovation-proof that sustainable design is not limited by location or budget.

Bullitt Centre – Seattle, USA

The Bullitt Centre in Seattle, completed in 2013, is often referred to as the greenest commercial building in the world. It was designed to meet the ambitious Living Building Challenge, which goes beyond LEED or BREEAM standards by requiring net-zero energy, water, and waste (Peña et al., 2017).

The six-story structure features a large solar photovoltaic array, rainwater harvesting and purification systems, composting toilets, and no connection to a conventional sewer system. Materials were selected based on a ''red list'' of banned toxic substances, and construction waste was minimized through prefabrication.

One of the most innovative features of the Bullitt Center is its occupant-centered design. Tenants agree to sustainability clauses in their leases, such as restrictions on personal heaters or plug load limits. The building includes dashboards and visual feedback tools to engage users in reducing consumption.

Despite high upfront costs around $18.5 million, the Bullitt Centre saves substantially on operating expenses. Energy bills are virtually nonexistent, and tenant turnover is low due to the building's performance, comfort, and prestige. It has also become research site and educational tool, visited by architects, policymakers, and students from around the world.

The Ellinikon Park – Athens Greece

The Ellinikon Park in Athens, currently development, is one of the most ambitious green urban regeneration projects in Europe. Located on the former site of Athens' international airport, the project spans over 6 million square meters and includes residential, commercial, cultural, and recreational zones. At the heart of the development is a large metropolitan park, projected to be one of the biggest coastal parks in the Mediterranean.

The Ellinikon project is not only a real estate venture it is envisioned as a sustainable city within a city, incorporating green infrastructure and climate responsive design across all aspect of construction. A key feature is the integration of a sustainable residential community, designed as a model for low-carbon, resource-efficient living. New homes will be built with green roofs, photovoltaic panels, smart energy systems, and water conservation technologies. The entire district will follow stringent energy and building performance standards in line with the EU's green transition goals (Tousi & Mela, 2024).

From an environmental standpoint, the park's design includes extensive reforestation, restoration of local ecosystems, and biodiversity corridors that aim to offset urban heat and support native flora and fauna. The reuse of former airport materials and infrastructure adds a circular dimension to the project, reducing construction waste and embodied carbon. Economically, the project is expected to generated thousands of jobs and attract green investment to the region. It is also projected to become a flagship destination for sustainable tourism, digital entrepreneurship, and innovation hubs.

The Ellinikon represents a forward-looking vision for post-austerity Athens, blending sustainability, innovation, and resilience. Though still under construction, it stands as a significant case of green urban transformation, redefining what modern coastal cities can look like in the context of climate adaptation and inclusive development.

Across these case studies, several important patterns emerged. First, bioclimatic and passive design strategies remain central to effective green building—whether through Harare's natural ventilation systems or Milan's vertical forests. Second, interdisciplinary collaboration is key; the success of projects like the Bullitt Center, Punggol Digital District, and Ellinikon Park relies on coordination among architects, engineers, developers, and policymakers.

The Ellinikon Park in particular introduces a large-scale dimension green urban regeneration that blends residential sustainability with ecosystem restoration. Unlike isolated green buildings, Ellinikon's ambition lies in transforming an entire district into a low-carbon, smart, and livable environment. This reflects a growing trend where green architecture extends beyond the building level to influence entire neighborhoods and cityscapes.

Additionally, these case studies reinforce the importance of contextual adaptation. Eastgates's passive cooling would be impractical in colder climates, just as Punggol's smart infrastructure would be cost-prohibitive in some developing contexts. Each project is rooted in its specific environmental, cultural, and policy environment, emphasizing that there is no one-size-fits-all model for sustainable development.

Lastly, the collective impact of these buildings shows that green architecture is not merely a technological fix, it's a platform for broader systemic change. From improving biodiversity and reducing emissions to fostering well-being and economic resilience, green buildings are catalysts in the shift toward more sustainable, inclusive, and future-ready cities.

BIBLIOMETRIC ANALYSIS OF GREEN BUILDINGS AND SUSTAINABILITY

To deepen the understanding of how green buildings contribute to Sustainablity, this chapter employs a bibliometric analysis – a quantitative method used to assess patterns, trends, and networks in scientific literature. Bibliometric approaches are particularly useful for mapping the evolution of research fields, identifying influential authors and institutions, and detecting emerging themes (Donthu et al., 2021; Passas, 2024a). In this study, tools such the R-based open-source package Bibliometrix, and the visualization software VOSviewer were used in combination to analyze a curated dataset of peer reviewed research. These tools enabled the creatin of keyword co-occurrence maps, author networks, and citation clusters, offering a structured view of the green building knowledge domain over time (Passas, 2024b; Ragazou et al., 2024).

In the first step, the research terms were carefully defined using Boolean logic to capture the breadth of academic work on green buildings and sustainability. The search string combined key terms such as "green buildings," "sustainable buildings," "sustainability," "ESG," and "sustainability targets" to ensure inclusion of interdisciplinary studies and evolving sustainability frameworks. This step ensured that the dataset would be both comprehensive and relevant, targeting core themes that align with the two main research questions of the chapter.

The second and third steps involved selecting the Scopus database, due to its intensive coverage of peer-reviewed academic journals, and applying strict inclusion criteria. Only original research articles published between 2001 and 2025 were selected, filtered further using the ABS (Academic Journal Guide) and Scimago journal ranking lists. This was done to maintain a high standard of scholarly quality and academic rigor. These filters ensured that the data used for analysis came from reputable sources and reflected the most influential research in the field.

In the fourth step focused on the analysis, which included mapping scientific collaboration between countries, identifying co-citation networks, and analyzing keyword clusters to reveal dominant research themes and their evolution. This phase helped uncover not only the most productive authors and institutions but also how the field has shifted over time - from a narrow focus on energy efficiency to a broader emphasis on ESG integration, smart technologies, and urban resilience. Through analyses like author dynamics, keyword co-occurrence, Tree Maps, and MCA (Multiple Correspondence Analysis), the methodology uncovered both the structural and thematic dimensions of the literature.

Finally, the sixth step involved synthesizing the results to draw meaningful insights and identify gaps and emerging directions in the literature. This synthesis forms the basis for the chapter's conclusions and recommendations, particularly around future research paths in areas like green accounting, digital transformation in the built environment, and sustainable business models. The framework also points to the need for deeper international collaboration and the integration of social equity into green building research.

Figure 1. Flowchart of the methodology

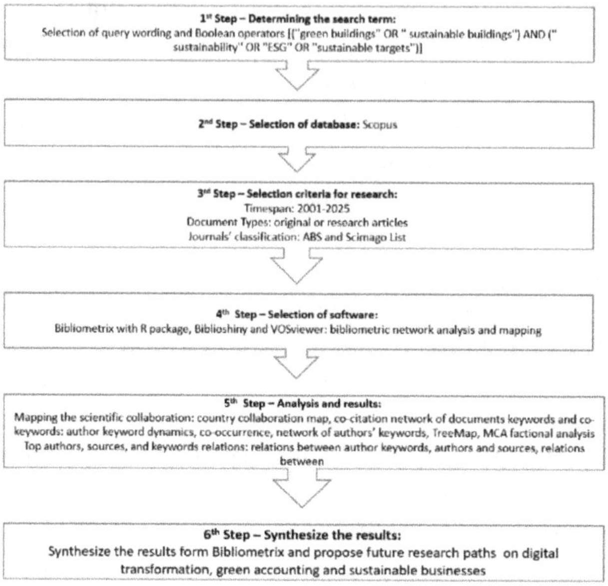

Figure 2, illustrates the evolution of academic research output on green buildings and sustainability from 2001 to 2025. This visualization provides a temporal overview of publication frequency based on the filtered dataset retrieved from Scopus and analyzed using Biblioshiny. The upward trend demonstrates increasing scholarly interest in the intersection of architecture, energy efficiency, and sustainable development.

From 2001 to 2010, research production remained relatively modest, reflecting the early stages of green building discourse in mainstream academic literature. This limited output may be attributed to a lack of standardized terminology, fragmented policy frameworks, and the emerging nature of sustainability as a formalized research area (Doan et al., 2017). However, around 2011, the graph shows a steady upward trajectory, suggesting a growing academic consensus around the importance of sustainable construction. The noticeable spike between 2015 and 2017 aligns with the major international climate initiatives, particularly the adoption of the United Nations Sustainable Development Goals (SDGs) in 2015 and the Paris Agreement the same year. These events catalyzed research focused on decarbonizing the built environment, with green buildings positioned as critical to achieving SDG 11 (Sustainable Cities and Communities) and SDG 13 (Climate Action) (Olabi et al., 2025; UNEP, 2022).

The sharp rise in publications from 2020 to 2023 is particularly significant. This period coincides with the COVID-19 pandemic, which reshaped global conversations around building design, air quality, and resilience. Scholars began to explore how green buildings could address both environmental and public health concerns, leading to an intensified focus on indoor environmental quality, biophilic design, and post-pandemic construction practices (Kaklauskas et al., 2021; Navaratnam et al., 2022)

The peak in 2023, with over 60 articles published, likely reflects the culmination of a decade of accumulated interest and interdisciplinary expansion. By this point, green building research had matured to incorporate adjacent fields such as IoT-based energy management, smart city integration (Apanaviciene et al., 2020),and ESG-aligned investment models. The rise in publications is not just quantitative but also qualitative, with research increasingly focusing on the economic and policy implications of green construction, especially in the context of climate finance and net-zero mandates.

Interestingly, the chart shows a dramatic drop in output in 2025. While this may initially appear to be a decline in interest, it is more likely due to incomplete indexing for the current year. In bibliometric studies, it's common for the latest year's data to appear artificially low, as many articles are still in press or have not yet been fully indexed by databases like Scopus. Thus, this decline should be interpreted with caution and not assumed to reflect a real decrease in academic focus.

Overall, the annual scientific production graph confirms that research on green buildings has not only grown in volume but also responded dynamically to global developments. It validates the relevance of bibliometric methods in capturing the evolutionary trajectory of academic discourse and highlights the increasing urgency with which researchers, policymakers, and professionals are approaching the sustainable building agenda.

Figure 2. Annual Scientific production from Biblioshiny analysis

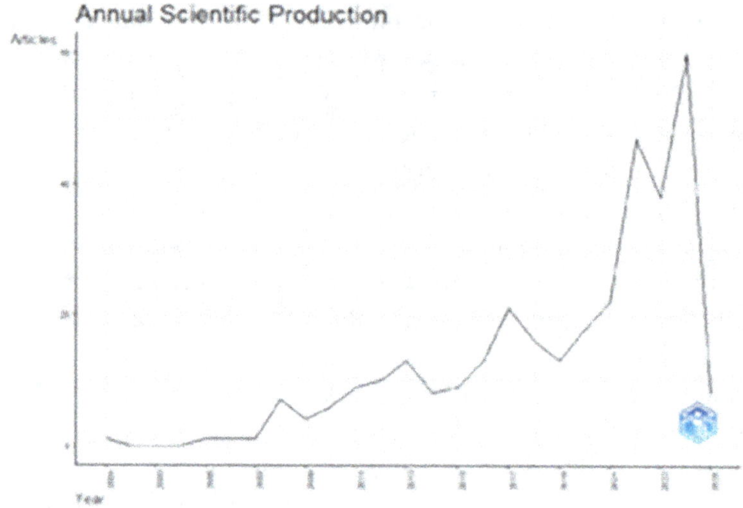

Figure 3, shows the co-citation map analysis, generated from VOSviewer, and visualizes the he intellectual foundations of the field by identifying journals that are frequently cited together in scholarly literature on green buildings, ESG (Environmental, Social, and Governance), and sustainability. In bibliometrics, co-citation analysis is a powerful technique for uncovering the structure of academic knowledge—how researchers cluster around specific journals, and how disciplines intersect.

Each node represents a scientific journal, and the size of the node indicates its overall influence based on the number of times it has been co-cited. The proximity between nodes reveals how closely journals are linked through shared citations, while the colors denote distinct thematic clusters within the field. The largest and most central nodes include “Journal of Cleaner Production,” “Sustainability,” “Energy and Buildings,” “Energy Policy,” and “Journal of Environmental Management.” These publications form the core of the research ecosystem, frequently cited together in literature addressing the Triple Bottom Line (TBL), environmental, economic, and social sustainability. The prominence of the journal of ‘’cleaner production’’ reflects its interdisciplinary focus on life-cycle analysis, energy efficiency, policy, and corporate environmental responsibility critical topics for green building research.

Another notable cluster includes ‘’Renewable and Sustainable Energy Reviews’’, ‘’Applied Energy’’, and ‘’Energies,’’ forming a technical backbone of literature related to energy systems, renewables, and building energy modeling. These journals contribute empirical and engineering-based solutions, anchoring the environmental performance dimension of green construction (Olabi et al., 2025).

The ''Sustainability'' journal, published by MDPI, appears as a major interdisciplinary hub, frequently cited in tandem with both technical and policy journals. This reflects its openness to topics ranging from building design and ESG frameworks to urban resilience and behavioral change. Its role suggest that much of the recent scholarly attention is cross-cutting, integrating economic, environmental, and institutional factors.

The presence of ''Energy and Buildings'', ''Building and Environment," and "Construction and Building Materials" suggests strong connections to the engineering and architectural science communities. These journals focus on building performance metrics, material sustainability, indoor environmental quality, and smart systems areas that have gained visibility in response to the post-COVID built environment agenda (Kaklauskas et al., 2021; Poyyamozhi et al., 2024a).

The overall structure of this map indicates a strong integration between energy studies, environmental science, and sustainability policy literature. This supports the chapter's argument that green buildings are not viewed solely through a technological lens, but rather as part of a complex ecosystem involving regulation, behavior, ESG investment strategies, and climate adaptation.

Importantly, the co-citation patterns reveal which journals act as "knowledge brokers"- publications that connect technical innovation with policy and management discussions. These include Energy Policy, Sustainability, and Journal of Environmental Management, all of which serve as bridges between science and decision-making. This observation is consistent with previous bibliometric studies, which emphasize the importance of interdisciplinary channels in advancing green building adoption

Figure 3. Co-citation Analysis Map of Key publications on ESG, Green Computing, and their Contribution to the TBL Using VOSviewer

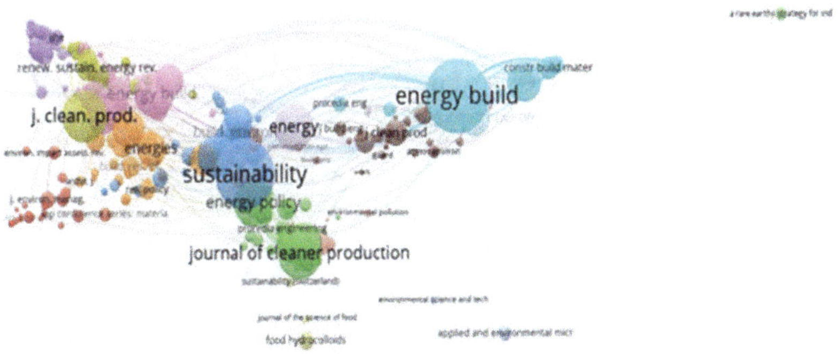

Figure 4 illustrates the geographical distribution of scientific production in the field of green buildings and sustainability, based on bibliometric data extracted from Scopus. Each country is color-coded by the number of documents published, with darker shades indicating higher output. The data shows that Chine and the United States with 118 and 111 publications respectively. This aligns with their economic size, rapid urbanization, and significant investment in both green technologies and sustainability research. China's leadership reflects its aggressive climate policy shifts, investment in eco-cities, and its recent emphasis on achieving carbon neutrality by 2060(International Energy Agency, 2024). Similarly, the US demonstrates strong research output tied to its innovation - driven universities, green building certifications (like LEED), and federal ESG-driven initiatives. Other key contributors include Australia, the UK, India, and Italy, all of which show high research productivity. These countries have well-developed national green building councils, progressive sustainability policies, and active university-led research centers. In Europe, nations like Germany, Sweden, and the Netherlands also show notable activity, often tied to their leadership in energy efficiency standards and climate-resilient infrastructure.

The map also exposes global disparities. Much of Africa, Central Asia, and parts of Latin America remain underrepresented in the green building literature. This imbalance highlights both a research gap and a policy challenge: regions most vulnerable to climate change often contribute the least to academic discourse on sustainability. Encouraging knowledge transfer, international collaboration, and capacity - building in these areas is critical for achieving inclusive progress in sustainable development.

Figure 4. Scientific Output of Leading Countries on Green Buildings and ESG Contributions, by Biblioshiny

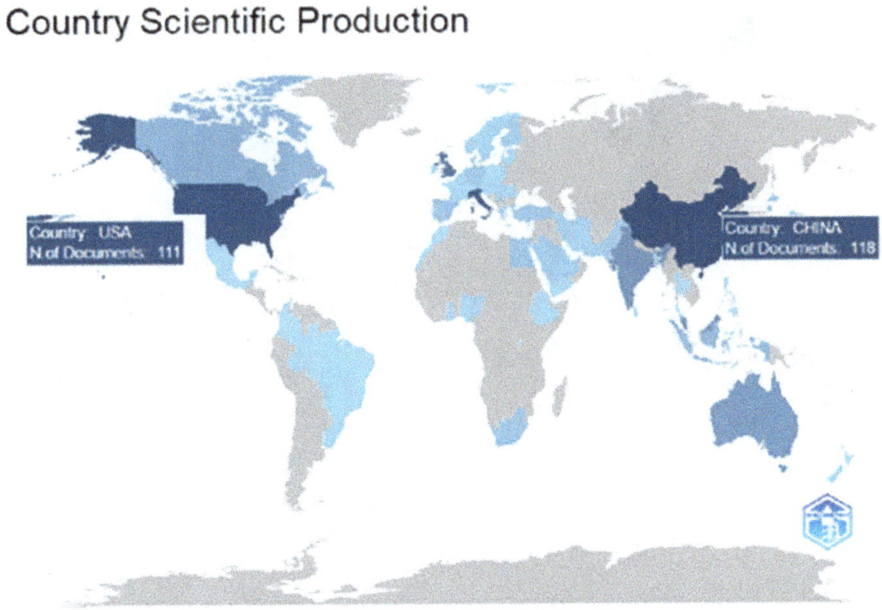

Figure 5, shows the country collaboration map in the field of green buildings and sustainability, generated using Biblioshiny country co-authorship function. It visualizes the geographic spread of cooperative research efforts by showing how often scholars from different countries co-author publications. Lines between countries represent co-authorship links, with line thickness reflecting the frequency of collaboration.

The United Staes, China, the UK, and Australia emerge as the most connected hubs, with multiple lines linking them to other high-output countries such as Germany, Italy, Malaysia, and India. This pattern reflects the global nature of green building research, often requiring multidisciplinary input and international expertise in energy systems, urban planning, ESG frameworks, and policy analysis. The presence of collaboration lines connecting developing economies, such as India and Southeast Asian countries, with more research-intensive nations highlights the growing inclusivity of the field. However, many regions in Africa, Latin America, and Central Asia remain disconnected from this network, reinforcing earlier observations about global research imbalances. Bridging these gaps through North–South and South–South research partnerships is vital to ensure that sustainable building innovations are both globally informed and locally applicable.

Figure 5. Map of Country Collaborations by Biblioshiny

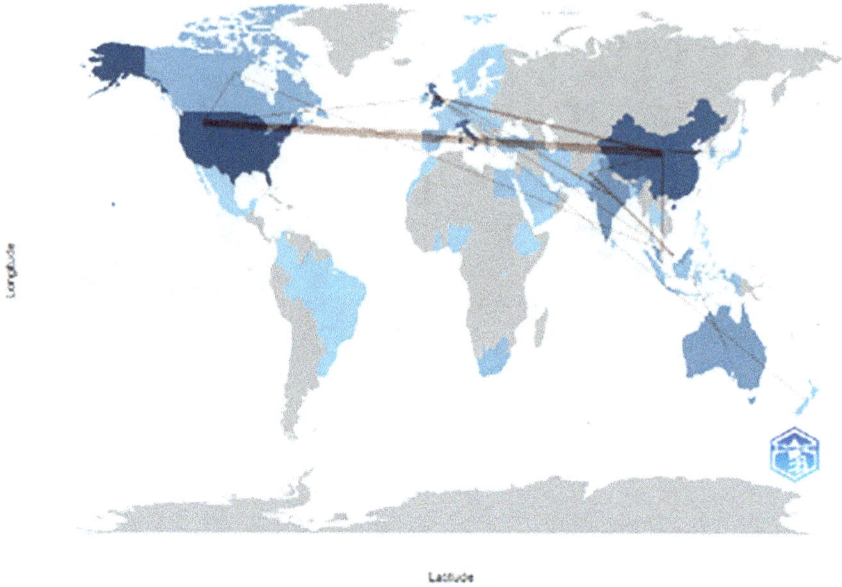

Figure 6, illustrates the keyword co-occurrence map above visualizes the conceptual structure of green building and sustainability research by mapping terms that frequently appear together in article titles, abstracts, and keywords. Generated using VOSviewer, this map uses clustering algorithms to group keywords into distinct thematic areas. The size of each node reflects the frequency of the term's occurrence, while the lines and proximity indicate how closely terms are linked across the literature.

At the core of the map are the most central and frequently co-occurring terms: "sustainability," "green buildings," "energy efficiency," and "sustainable development." These form the conceptual backbone of the field, reflecting the central questions driving academic discourse: How do green buildings contribute to sustainability goals? And what role do energy systems and design strategies play in that process?

Around these anchors, several key thematic clusters emerge:

- The green cluster focuses on construction practices and performance metrics, including terms like "building energy efficiency," "quality of life," and "construction industry." This cluster represents the technical and engineering-focused dimension of sustainable buildings
- The blue cluster includes terms like "sustainable development goals," "urban planning," "circularity," and "green city," highlighting the policy and urban

systems perspective. This aligns closely with the increasing push for city-level sustainability and climate action planning

- The purple clusters capture more conceptual and interdisciplinary terms such as "decision-making," "bioeconomy," "agriculture," and "urban design." This suggests growing interest in cross-sectoral thinking—especially how buildings interact with food systems, land use, and governance.
- The orange and brown clusters feature more niche or emergent topics like "commercial buildings," "phase change materials," and "embodied carbon," indicating an increasing focus on materials innovation and life-cycle assessments.

The map also reveals evolving terminology. Words like "decarbonization," "circularity," "data-driven design," and "digital" suggest that the field is increasingly integrating with the digital transformation agenda, including AI, BIM, and smart infrastructure. This supports the idea that green building research is no longer confined to traditional architecture and engineering, but is expanding into interdisciplinary and data-intensive domains. Notable, isolated terms like "embodied carbon" and "structural equation modeling" appear on the periphery. These terms are underrepresented but potentially emerging areas for future exploration. As policies shift from operational to embodied emissions (especially in Europe), research into low-carbon materials and building lifecycle impacts is expected to grow rapidly

Figure 6. Visualization of co-occurrence of author keywords by VOSviewer software

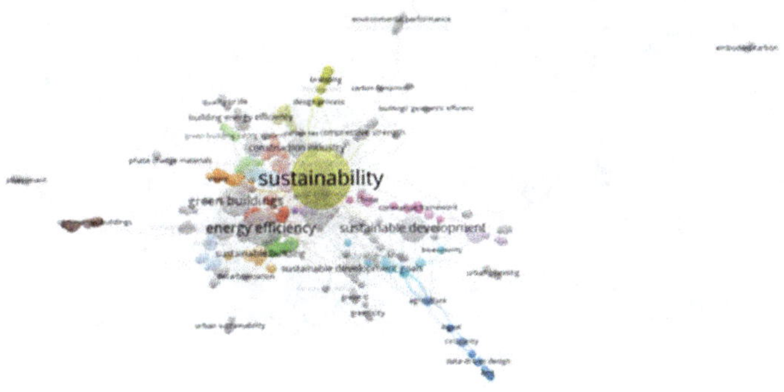

Figure 7, presents the TreeMap which offers a compact, data-rich overview of the most frequently used author keywords in scholarly publications on green buildings and sustainability. Each box represents a keyword, and its size is proportional to the number of occurrences within the dataset. The TreeMap helps identify dominant research themes and shows how the field is structured around core and peripheral concepts.

At the center of the map are the most prominent terms: sustainable development" (139 entries), "energy efficiency" (57), "sustainability" (57), and "green buildings" (51). These four terms clearly anchor the field, aligning with the broader research question explored in this chapter—how green buildings intersect with energy systems and global sustainability targets. Their dominance suggests that sustainability is approached both as a policy and engineering problem, with "energy efficiency" acting as a measurable operational bridge between vision and implementation. Other high – frequency keywords include "climate change," "construction industry," "architectural design," "intelligent buildings," and "environmental impact." These reflect the increasing technical and sectoral granularity of the field. For example, "intelligent buildings" connects to the growing literature on smart energy systems, sensors, and automation in architecture. Similarly, "climate change" reinforces the view of green building not just as a technical strategy, but as a core response to planetary challenges.

Figure 7. Word Tree Map by Biblioshiny

Figure 8 presents, the thematic map which provides a strategic overview of the intellectual structure and evolution of green building and sustainability research. Created using Biblioshiny keyword clustering function, the map classifies themes based on two key dimensions: centrality - How important a theme is to the overall field (i.e., its connectivity to other themes) and density – How developed or mature the theme is (i.e., the strength of its internal keyword cohesion).

The map is divided into four quadrants:

Motor Themes (Top-right quadrant):

This area includes the most central and well-developed themes—those driving the academic conversation forward. The keywords here include "sustainability," "green buildings," "energy efficiency," "sustainable development," and "China." These topics are not only highly connected to other areas of research but also deeply developed, making them foundational to the field. Their centrality confirms their relevance to both academic inquiry and policy agendas like the SDGs and net-zero urban transitions.

Basis Themes (Bottom-right quantrant):

These are themes of high centrality but lower density, meaning they are widely discussed but less conceptually developed. This includes "climate change," "construction industry," and "environmental impact." These terms form the contextual basis for green building research but often lack cohesive sub-frameworks or detailed methodological attention. Their position suggests that while foundational, these areas still require deeper theoretical and practical exploration, particularly around the integration of climate mitigation into construction workflows.

Niche Themes (Top left quantrant):

This quantrant contains specialized but less central topics. Terms like "compressive strength," "fly ash," "green computing," and "renewable energy resources" fall into this category. These are highly focused areas—often technical or material-specific—that may not connect directly with broader sustainability narratives but are still critical for innovation and performance optimization. For example, research on high-performance materials like fly ash concrete plays a vital role in reducing embodied carbon in green buildings.

Emerging or Declining Themes (Bottom-left quantrant):

These themes have both low centrality and density, making them either emerging areas of interest or ones whose relevance is diminishing. Terms like "bio-based," "biomass," and "energy-consumption" appear here. Their presence may reflect shifts in the literature, where once-prominent discussions (e.g., biomass as a renewable fuel) are giving way to more complex debates around circularity, electrification, and embodied energy. Alternatively, they may signal early-stage topics that could grow with further research.

Additionally, keywords like "waste management" and "sustainable construction" near the center suggest cross-cutting themes that connect environmental policy with architectural practice. These themes could potentially transition into more central categories as construction circularity becomes a global priority.

Figure 8. Research themes derived from bibliometric analysis

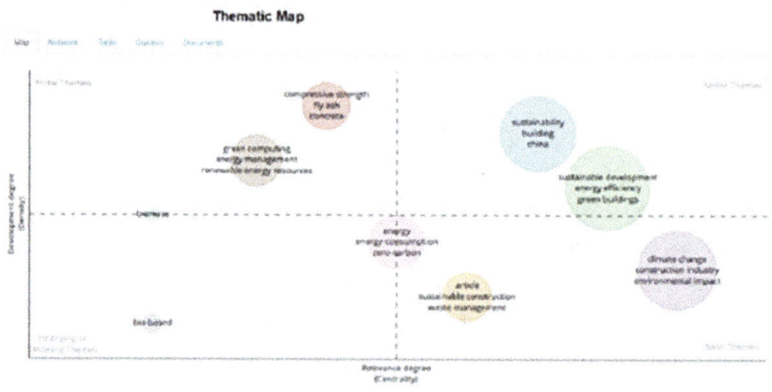

The last figure presents, the Conceptual Structure Map above was generated using Multiple Correspondence Analysis (MCA) to identify semantic clusters in the green building and sustainability literature. It positions keywords based on how often they co-occur, forming a spatial representation of conceptual proximity across the dataset. Terms positioned closer together are more likely to appear in the same documents, suggesting thematic alignment.

The red polygon encloses a wide array of interrelated terms. The upper section is dominated by forward-looking, tech-oriented topics such as "smart city," "Internet of Things," "green computing," and "energy management." These reflect the integration of digital transformation themes into green building research, particularly through smart infrastructure, ICT, and automation.

In the center of the map, core sustainability terms like "energy conservation," "decision-making," "life cycle," "green IT," and "sustainable development" coalesce. This concentration confirms that the central concern across the literature is not only environmental performance, but also strategic governance and systems thinking—emphasizing multi-level decision-making in urban and construction contexts. Toward the lower edge, we find terms like "carbon," "carbon emissions," "greenhouse gases," and "gas emissions," which ground the field in climate-related metrics and impact assessments. These terms suggest a continued research focus on emission reductions and climate mitigation through built environment strategies.

The far right includes more information-oriented themes - "information systems," "behavioral research," "information use"—which likely stem from the ESG and green accounting literature. These ideas reflect the emerging convergence of sustainability with data analytics, corporate disclosure, and decision-support systems. Overall, the map reinforces the interdisciplinary nature of the field. It shows that research on green buildings now spans architecture, computer science, energy systems, behavioral studies, and environmental economics. This broad thematic landscape suggests rich potential for cross-sector innovation and also highlights the need for integrated research frameworks that bridge technology, policy, and environmental outcomes.

Figure 9. Multiple Correspondence Analysis (MCA) using VOSviewer.

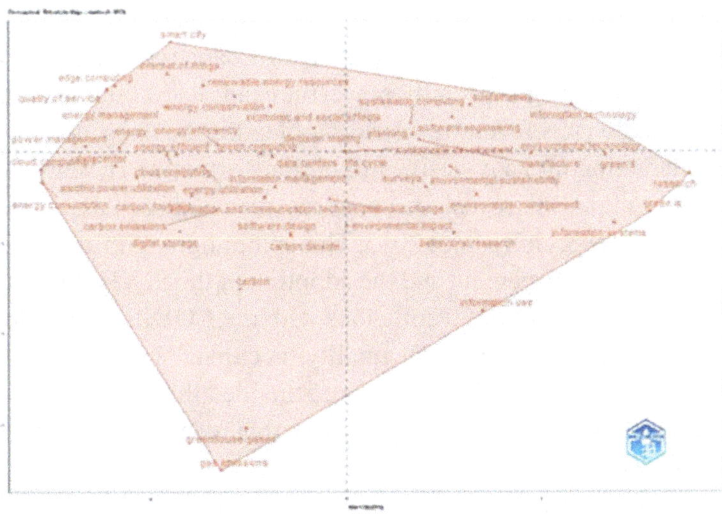

DISCUSSION AND CONCLUSIONS

The findings of this chapter reinforce the centrality of green buildings in global sustainability efforts, especially as urban areas grapple with the intersecting pressures of climate change, population growth, and resource scarcity. Across all sections from the technical features and real-world case studies to the bibliometric analysis several themes consistently emerge: the rising maturity of green building research, the shift toward integrated and digital solutions, and the need for socially inclusive and context-specific strategies.

One of the most compelling insights is the transition of green building discourse from a narrow, energy-efficiency focus toward a more holistic and interdisciplinary paradigm. While energy efficiency remains a cornerstone, the incorporation of biophilic design, smart technologies, and circular economy principles reflects a deeper understanding of sustainability as a systems challenge. The bibliometric analysis supports this evolution, with keywords such as "digital transformation," "urban resilience," "circularity," and "ESG" gaining prominence alongside traditional terms like "energy consumption" and "carbon emissions." This shift reflects broader global narratives around the SDGs, net-zero commitments, and post-pandemic resilience planning.

Crucially, green buildings are not merely technological solutions; they are cultural and policy instruments that signal new values in urban planning. This is evident in the diversity of case studies presented. From the biomimetic Eastgate Centre in Harare to the high-tech, smart Punggol District in Singapore, green buildings demonstrate an ability to embody local climate, economic, and governance conditions. The Ellinikon Park in Athens provides an especially powerful example of how sustainability can scale from the building level to entire districts—blending housing, nature, innovation, and climate adaptation into a cohesive urban transformation strategy.

The bibliometric analysis revealed a consistent revealed a consistent global growth in academic interest, especially since 2015. This spike aligns with international policy shifts such as the Paris Agreement and the adoption of the SDGs. Notably, the sharp increase in publications post 2020 coincides with the COVID-19 pandemic, which accelerated discussions on indoor air quality, occupant well-being, and building health performance. These findings reinforce the idea that green buildings are not static technical solutions but dynamic platforms that respond to evolving social and environmental needs

In terms of geography, the analysis highlighted clear disparities. Countries like China, the U.S., and the UK dominate the literature, driven by robust funding mechanisms, research institutions, and green building mandates. However, parts of Africa, Latin America, and Central Asia remain underrepresented. This lack of inclusion is troubling, given that these regions often face the most severe climate vulnerabilities. Addressing this imbalance will require targeted South-South and North-South research partnerships, capacity-building initiatives, and financing mechanisms that allow these regions to leapfrog into sustainable construction using locally adapted, cost-effective strategies.

The keyword co-occurrence and thematic maps further reinforce the field's complexity. The dominance of terms like "sustainable development," "green buildings," and "energy efficiency" underscores their foundational role, but the emergence of "zero-carbon," "smart city," and "information systems" shows that the future of green architecture lies in its ability to integrate technology, data, and behavioral feedback into building performance. The convergence of green building research with digital infrastructure, ESG reporting, and lifecycle analysis reflects the evolving nature of how buildings are designed, operated, and evaluated.

Additionally, the conceptual map reveals that green buildings are increasingly seen as part of a broader knowledge ecosystem connected not only to engineering and architecture but also to information technology, behavioral science, and governance. Themes such as "decision-making," "carbon footprint," and "data centers" indicate that future research will need to consider not just the physical structure of buildings, but also how information flows, user behaviors, and policy structures influence outcomes.

From a practice standpoint, the case studies illustrate that no one-size-fits-all model exists. The Eastgate Centre's success depends on passive cooling suited to Zimbabwe's climate, while the Bullitt Centre relies on high-tech systems and community engagement in a high-income urban context. The Punggol Digital District represents a future-forward model that integrates green building principles with urban-scale planning and digital innovation, offering a roadmap for smart, sustainable communities. Meanwhile, Ellinikon Park's regenerative ambition illustrates how large-scale urban redevelopment can incorporate biodiversity, heritage, and inclusive economic growth into green design.

These case show that green building is both place-based and globally relevant. It adapts to local constraints while responding to global imperatives, such as reducing emissions, improving health outcomes, and enhancing urban resilience. However, they also expose persistent challenges: the high upfront costs of green technologies, maintenance complexities, unequal access to certifications, and regulatory fragmentation. These barriers underscore the need for stronger policy frameworks, incentives, and public-private partnerships to scale green building adoption.

In terms of research gaps the bibliometric analysis identified several underexplored areas. Topics like embodied carbon, life-cycle thinking, and post-occupancy behavior are still peripheral in the literature but are rapidly gaining relevance, especially in the context of whole-life carbon accounting and circular design. Additionally, social equity and inclusion are not yet central themes, despite being critical to ensuring that green buildings benefit all socioeconomic groups not just affluent urban populations.

Future research should prioritize inclusive design principles, equitable access, and participatory governance in green building strategies. The findings also suggest that interdisciplinary collaboration is no longer optional but essential. Successful green buildings require the convergence of multiple expertise: architects, engineers, data scientists, public health professionals, and environmental economists. Similarly, knowledge sharing between academia, industry, and government needs to be more fluid and outcome-driven. The field would benefit from shared data platforms, living labs, and open innovation ecosystems that allow for experimentation and feedback.

In sight of global climate targets particularly the push for net-zero emissions by mid-century green buildings represent one of the most scalable and impactful tools for decarbonizing urban environments. They reduce operational emissions, enhance resource efficiency, and provide co-benefits in health, well-being, and economic resilience. However, for their full potential to be realized, several conditions must be met:

1. Clear policy alignment across national and local levels:

Successful implementation of green building strategies requires strong regulatory frameworks that bridge national sustainability goals with municipal action plans. Fragmented policies and inconsistent building codes often create uncertainty for developers and investors. Governments should prioritize the integration of green building standards into urban planning legislation, zoning incentives, and procurement guidelines. For example, municipalities could offer expedited permits or tax relief for projects that meet net-zero benchmarks. Aligning regulations with climate targets ensures not just compliance, but accelerates adoption through a stable and supportive policy environment.

2. Investment in training and upskilling:

Green building innovation depends on the capabilities of the workforce behind it. Architects, engineers, contractors, and facility managers must be equipped with the latest knowledge in sustainable materials, energy modeling, and building information systems. This calls for revamped university curricula, continuing education programs, and cross-disciplinary certification schemes. Furthermore, developing countries should prioritize workforce development to avoid dependency on foreign expertise and ensure local capacity to deliver sustainable infrastructure. A skilled, sustainability-literate workforce is foundational for scaling innovation from pilot projects to mainstream practice.

3. Financing mechanisms that support inclusive green development:

The high upfront cost of green buildings remains a major barrier, especially in low- and middle-income economies. Public and private financing schemes must evolve to reflect the long-term value and reduced lifecycle costs of green projects. This includes green bonds, ESG-aligned funds, blended finance models, and subsidized loans for retrofits. Financial tools should also prioritize equity and accessibility, ensuring that affordable housing, schools, and healthcare facilities can meet green standards—not just luxury developments. Democratizing access to green finance is critical to ensure that sustainability is not a privilege, but a right.

4. Integration of digital tools and data:

The future of green buildings lies in intelligent, responsive systems that optimize that go beyond operational metrics to consider embodied emissions and circularity. performance in real time. Smart sensors, building automation, and energy dashboards can provide granular data on occupancy, consumption, and indoor environmental quality. When connected to predictive analytics and machine learning algorithms, these systems enable proactive maintenance, adaptive energy use, and better occupant engagement. Moreover, digital tools play a vital role in monitoring ESG metrics, supporting transparent sustainability reporting. Embedding digital infrastructure in green design enhances not just operational efficiency, but also governance and accountability.

5. Lifecycle-based standards and circularity integration:

Current building codes and certifications often focus on heavily on operational emissions, overlooking he embodied carbon in materials, construction, and end-of-life phases. To meet climate goals, sustainability standards must shift toward whole-life carbon assessments, emphasizing reuse, modularity, deconstruction, and material recovery. Circular construction strategies such as using recycled content, designing for disassembly, and minimizing waste should be embedded from the design stage onward. Lifecycle thinking ensures that buildings deliver long-term value with minimal ecological impact, aligning environmental responsibility with economic performance.

Finally, green buildings should be reframed not just as a response to climate change, but as a proactive investment in social infrastructure. As climate events become more frequent and severe, buildings will increasingly serve as frontline defenses providing shelter, thermal comfort, clean air, and psychological well-being. The post-pandemic shift toward health-centered design only amplifies their relevance in future city planning.

REFERENCES

Al-Qawasmi, J., Asif, M., El Fattah, A. A., & Babsail, M. O. (2019). Water effciency and management in sustainable building rating systems: Examining variation in criteria usage. *Sustainability (Basel), 11*(8), 2416. Advance online publication. DOI: 10.3390/su11082416

Alaimo, L. S., Ciommi, M., Vardopoulos, I., Nosova, B., & Salvati, L. (2022). The medium-term impact of the COVID-19 pandemic on population dynamics: The case of Italy. *Sustainability (Basel), 14*(21), 13995. DOI: 10.3390/su142113995

Allen, J. G., MacNaughton, P., Satish, U., Santanam, S., Vallarino, J., & Spengler, J. D. (2016). Associations of cognitive function scores with carbon dioxide, ventilation, and volatile organic compound exposures in office workers: A controlled exposure study of green and conventional office environments. *Environmental Health Perspectives, 124*(6), 805–812. DOI: 10.1289/ehp.1510037 PMID: 26502459

Apanaviciene, R., Vanagas, A., & Fokaides, P. A. (2020). Smart building integration into a smart city (SBISC): Development of a new evaluation framework. *Energies, 13*(9), 2190. Advance online publication. DOI: 10.3390/en13092190

Chaudhari, P., Xiao, Y., Cheng, M. M. C., & Li, T. (2024). Fundamentals, Algorithms, and Technologies of Occupancy Detection for Smart Buildings Using IoT Sensors. In Sensors (Vol. 24, Issue 7). Multidisciplinary Digital Publishing Institute (MDPI). DOI: 10.3390/s24072123

Cheliatsidou, A., Sariannidis, N., Garefalakis, A., Passas, I., & Spinthiropoulos, K. (2023). Exploring Attitudes towards Whistleblowing in Relation to Sustainable Municipalities. *Administrative Sciences, 2023*(9), 199. Advance online publication. DOI: 10.3390/admsci13090199

Christensen, H. B., Hail, L., & Leuz, C. (2021). Mandatory CSR and sustainability reporting: Economic analysis and literature review. *Review of Accounting Studies, 26*(3), 1176–1248. DOI: 10.1007/s11142-021-09609-5

Cortesi, A., Vardopoulos, I., & Salvati, L. (2022). A partial least squares analysis of the perceived impact of sustainable real estate design upon wellbeing. *Urban Science (Basel, Switzerland), 6*(4), 69. DOI: 10.3390/urbansci6040069

Crawley, D., & Aho, I. (1999). Building environmental assessment methods: Applications and development trends. *Building Research and Information, 27*(4–5), 300–308. DOI: 10.1080/096132199369417

Da Silva, K. P., & Ramirez, K. N. (2021). Green buildings: a worldwide overview on leed, breeam and green star certifications / Green buildings: uma análise do panorama mundial das certificações leed, breeam e green star. *Brazilian Journal of Development*, *7*(8), 82761–82778. DOI: 10.34117/bjdv7n8-471

Ding, Z., Fan, Z., Tam, V. W. Y., Bian, Y., Li, S., Illankoon, I. M. C. S., & Moon, S. (2018). Green building evaluation system implementation. *Building and Environment*, *133*, 32–40. DOI: 10.1016/j.buildenv.2018.02.012

Doan, D. T., Ghaffarianhoseini, A., Naismith, N., Zhang, T., Ghaffarianhoseini, A., & Tookey, J. (2017). A critical comparison of green building rating systems. In *Building and Environment* (Vol. 123, pp. 243–260). Elsevier Ltd., DOI: 10.1016/j.buildenv.2017.07.007

Donthu, N., Kumar, S., Mukherjee, D., Pandey, N., & Lim, W. M. (2021). How to conduct a bibliometric analysis: An overview and guidelines. *Journal of Business Research*, *133*, 285–296. DOI: 10.1016/j.jbusres.2021.04.070

Escrivà Saneugenio, F., Sateriano, A., Salvati, L., & Vardopoulos, I. (2024). The extrinsic manifestation of sprawl: Complex dynamics of urbanization, land imperviousness, and sustainability in Mediterranean landscapes. In *Homage (and criticism) to the Mediterranean city. Regional sustainability and economic resilience* (pp. 65–84). River Publishers. DOI: 10.1201/9781003517306-3

Glaeser, E. L., & Kahn, M. E. (2008). The Greenness of Cities: Carbon Dioxide Emissions and Urban Development. http://www.eia.doe.gov/oiaf/1605/ggrpt/carbon.html

Hariram, N. P., Mekha, K. B., Suganthan, V., & Sudhakar, K. (2023). Sustainalism: An Integrated Socio-Economic-Environmental Model to Address Sustainable Development and Sustainability. *Sustainability (Basel)*, *15*(13), 10682. Advance online publication. DOI: 10.3390/su151310682

Illankoon, I. M. C. S., Tam, V. W. Y., Le, K. N., Tran, C. N. N., & Ma, M. (2019). Review on green building rating tools worldwide: Recommendations for Australia. *Journal of Civil Engineering and Management*, *25*(8), 831–847. DOI: 10.3846/jcem.2019.10928

International Energy Agency. (2024). World Energy Outlook 2024. www.iea.org/terms

Ismaeel, W. S. E., & Kassim, N. (2023). An environmental management plan for construction waste management. *Ain Shams Engineering Journal*, *14*(12), 102244. Advance online publication. DOI: 10.1016/j.asej.2023.102244

Iwuanyanwu, O., Gil-Ozoudeh, I., Okwandu, A. C., & Ike, C. S. (2024). Cultural and social dimensions of green architecture: Designing for sustainability and community well-being. *International Journal of Applied Research in Social Sciences, 6*(8), 1951–1968. DOI: 10.51594/ijarss.v6i8.1477

Kaklauskas, A., Lepkova, N., Raslanas, S., Vetloviene, I., Milevicius, V., & Sepliakov, J. (2021). COVID-19 and green housing: A review of relevant literature. In Energies (Vol. 14, Issue 8). MDPI AG. DOI: 10.3390/en14082072

Karytsas, S., Vardopoulos, I., & Theodoropoulou, E. (2019). Factors affecting sustainable market acceptance of residential microgeneration technologies. A two time period comparative analysis. *Energies, 12*(17), 3298. DOI: 10.3390/en12173298

Kokatnur, T., Faris, F., Gunay, B., O'Brien, W., & Azar, E. (2025). The WELL Building Standard: A literature review and bibliometric analysis of a nascent field. In Journal of Building Engineering (Vol. 103). Elsevier Ltd. DOI: 10.1016/j.jobe.2025.112121

Liu, T., Chen, L., Yang, M., Sandanayake, M., Miao, P., Shi, Y., & Yap, P. S. (2022). Sustainability Considerations of Green Buildings: A Detailed Overview on Current Advancements and Future Considerations. In Sustainability (Switzerland) (Vol. 14, Issue 21). MDPI. DOI: 10.3390/su142114393

Maier, G., & Gluszak, M. (2024). The Diffusion of Green Building Certifications in Europe-A preliminary descriptive analysis. https://www.researchgate.net/publication/386336541

Mba, E. J., Okeke, F. O., Igwe, A. E., Ozigbo, C. A., Oforji, P. I., & Ozigbo, I. W. (2024). Evolving trends and challenges in sustainable architectural design; a practice perspective. *Heliyon, 10*(20), e39400. Advance online publication. DOI: 10.1016/j.heliyon.2024.e39400 PMID: 39502211

Navaratnam, S., Nguyen, K., Selvaranjan, K., Zhang, G., Mendis, P., & Aye, L. (2022). Designing Post COVID-19 Buildings: Approaches for Achieving Healthy Buildings. *Buildings (Basel, Switzerland), 12*(1), 74. Advance online publication. DOI: 10.3390/buildings12010074

Okedara, O., & Odubiyi, M. A. (2024). The Role of Smart Buildings and Automated Energy Management Systems in Enhancing Building Energy Efficiency. https://www.researchgate.net/publication/384107270

Olabi, A. G., Shehata, N., Issa, U. H., Mohamed, O. A., Mahmoud, M., Abdelkareem, M. A., & Abdelzaher, M. A. (2025). The role of green buildings in achieving the sustainable development goals. *International Journal of Thermofluids*, *25*, 101002. Advance online publication. DOI: 10.1016/j.ijft.2024.101002

Passas, I. (2023). Accounting for integrity: ESG and financial disclosures: the challenge of internal fraud in management decision-making [Hellenic Mediterranean University]. DOI: 10.12681/eadd/55359

Passas, I. (2024a). Bibliometric Analysis: The Main Steps. Encyclopedia 2024, Vol. 4, Pages 1014-1025, 4(2), 1014–1025. DOI: 10.3390/encyclopedia4020065

Passas, I. (2024b). Bibliometric Analysis: The Main Steps. *Encyclopedia*, *4*(2), 1014–1025. DOI: 10.3390/encyclopedia4020065

Passas, I., Ragazou, K., Zafeiriou, E., Garefalakis, A., & Zopounidis, C. (2022). ESG Controversies: A Quantitative and Qualitative Analysis for the Sociopolitical Determinants in EU Firms. Sustainability 2022, Vol. 14, Page 12879, 14(19), 12879. DOI: 10.3390/su141912879

Pavate, V., Deshmukh, V. D., Kolekar, A., Mendapara, A., Patil, S., & Amrutatti, S. (2024). Green Building and Energy-Efficient Design. *Journal of Environmental Engineering and Studies*, *9*(2), 33–52. DOI: 10.46610/JoEES.2024.v09i02.003

Pavlides, T., Vardopoulos, I., Papamichael, I., Voukkali, I., Stylianos, M., & Zorpas, A. A. (2023). Environmental sustainability assessment of excavation, construction, and demolition waste conditions and practices across Greece and Cyprus. *IOP Conference Series. Earth and Environmental Science*, *1196*(1), 012037. DOI: 10.1088/1755-1315/1196/1/012037

Peña, R., Meek, C., & Davis, D. (2017). The Bullitt Center: A Comparative Analysis Between Simulated and Operational Performance. *Technology Architecture + Design*, *1*(2), 163–173. DOI: 10.1080/24751448.2017.1354611

Poyyamozhi, M., Murugesan, B., Rajamanickam, N., Shorfuzzaman, M., & Aboelmagd, Y. (2024a). IoT—A Promising Solution to Energy Management in Smart Buildings: A Systematic Review, Applications, Barriers, and Future Scope. In Buildings (Vol. 14, Issue 11). Multidisciplinary Digital Publishing Institute (MDPI). DOI: 10.3390/buildings14113446

Pushkar, S. (2024). Leadership in Energy and Environmental Design Commercial Interior Version 4 (LEED-CI v4) Gold-Certified Office Space Projects: A Pairwise Comparative Analysis between Three Mediterranean Countries. *Buildings (Basel, Switzerland), 14*(3), 815. Advance online publication. DOI: 10.3390/buildings14030815

Ragazou, K., Garefalakis, A., Papademetriou, C., & Passas, I. (2023). Well-being Human Resource Model In The Collaborative Economy: The Keystone of ESG Strategy In *The Tourism Sector. International Conference on Tourism Research*, 6(1), 251–260. DOI: 10.34190/ictr.6.1.1066

Ragazou, K., Garefalakis, A., Zafeiriou, E., & Passas, I. (2022). Agriculture 5.0: A New Strategic Management Mode for a Cut Cost and an Energy Efficient Agriculture Sector. Energies 2022, Vol. 15, Page 3113, 15(9), 3113. DOI: 10.3390/en15093113

Ragazou, K., Lemonakis, C., Passas, I., Zopounidis, C., & Garefalakis, A. (2024). ESG-driven ecopreneur selection in European financial institutions: entropy and TOPSIS analysis. Management Decision, ahead-of-print(ahead-of-print). https://doi.org/DOI: 10.1108/MD-12-2023-2425/FULL/PDF

Ragazou, K., Passas, I., Garefalakis, A., & Dimou, I. (2022). Investigating the Research Trends on Strategic Ambidexterity, Agility, and Open Innovation in SMEs: Perceptions from Bibliometric Analysis. *Journal of Open Innovation, 8*(3), 118. Advance online publication. DOI: 10.3390/joitmc8030118

Ribeiro Hoffmann, A., Sandrin, P., & Doukas, Y. E. (Eds.). (2024). Climate change in regional perspective. European Union and Latin American initiatives, challenges, and solutions. Cham, Switzerland: United Nations University Series on Regionalism, Vol. 27, Springer Nature. DOI: 10.1007/978-3-031-49329-4

Salvati, L. (Ed.). (2025). *Environmental sustainability and global change*. Elsevier., DOI: 10.1016/C2023-0-52570-4

Sellers, B. C., & Fiore, S. M. (2013). Sustainable and user-centered: Applying human factors solutions towards improving the effectiveness of "green" buildings. *Proceedings of the Human Factors and Ergonomics Society*, 550–554. DOI: 10.1177/1541931213571118

Sheth, K. N. (n.d.). WATER EFFICIENT TECHNOLOGIES FOR GREEN BUILDINGS. https://www.researchgate.net/publication/316582554

Sinha, A., Gupta, R., & Kutnar, A. (2013). Održivi razvoj i zelena gradnja. In Drvna Industrija (Vol. 64, Issue 1, pp. 45–53). DOI: 10.5552/drind.2013.1205

Şirin, C., Goggins, J., & Hajdukiewicz, M. (2023). A review on building-integrated photovoltaic/thermal systems for green buildings. In *Applied Thermal Engineering* (Vol. 229). Elsevier Ltd., DOI: 10.1016/j.applthermaleng.2023.120607

Stavropoulou, E., Spinthiropoulos, K., Ragazou, K., Papademetriou, C., & Passas, I. (2023). Green Balanced Scorecard: A Tool of Sustainable Information Systems for an Energy Efficient Business. Energies 2023, Vol. 16, Page 6432, 16(18), 6432. DOI: 10.3390/en16186432

Study, C., & Verticale, B. (2015). Tall buildings: design, construction, and operation | 2015 Issue I. www.iuav.it

Tabassum, R. R., & Park, J. (2024). Development of a Building Evaluation Framework for Biophilic Design in Architecture. *Buildings (Basel, Switzerland)*, *14*(10), 3254. Advance online publication. DOI: 10.3390/buildings14103254

Tousi, E., & Mela, A. (2024). Supralocal Role of Medium to Large Scale Urban Parks, in Greater Athens Region in Greece. Issues of Meso Car Dependence During the Covid-19 Pandemic. *Journal of Sustainable Architecture and Civil Engineering*, *35*(2), 201–215. DOI: 10.5755/j01.sace.35.2.34661

Turner, J. S., & Soar, R. C. (2008). Beyond biomimicry: What termites can tell us about realizing the living building. www.archpaper.com/features/2007_14_imitation.htm

UNEP. (2022). Annual Report 2022.

Vardopoulos, I. (2024). Urban planning policies and tools for sustainable development: insights into adaptive reuse, smart cities, and urban metabolism. In *Socio-enviromental change: The path toward sustainability* (pp. 121–140). Nova Science Publishers.

Vardopoulos, I., D'Agata, A., Escrivà Saneugenio, F., & Salvati, L. (2024). *Sprawl and the city: realizing a sustainable Mediterranean urbanization*. Nova Science Publishers., DOI: 10.52305/GTVV1244

Vardopoulos, I., Giannopoulos, K., Kalli, S., & Salvati, L. (2024). From tech-centric to human-centric smart tourist destinations: innovations in real estate, urban planning, and sustainability. In *Socio-enviromental change: the path toward sustainability* (pp. 65–96). Nova Science Publishers.

Vardopoulos, I., Papoui-Evangelou, M., Nosovo, B., & Salvati, L. (2023). Smart 'tourist cities' revisited: Culture-led urban sustainability and the global real estate market. *Sustainability (Basel)*, *15*(5), 4313. DOI: 10.3390/su15054313

Vardopoulos, I., Santamouris, M., Zorpas, A. A., Barone, G., Italos, C., & Vassiliades, C. (2024). A comparative study on discrepancies in residential building energy performance certification in a Mediterranean context. *Buildings (Basel, Switzerland), 14*(4), 1009. DOI: 10.3390/buildings14041009

Vardopoulos, I., Vannas, I., Xydis, G., & Vassiliades, C. (2023). Homeowners' perceptions of renewable energy and market value of sustainable buildings. *Energies, 16*(10), 4178. DOI: 10.3390/en16104178

Varma, C. R. S., & Palaniappan, S. (2019). Comparision of green building rating schemes used in North America, Europe and Asia. *Habitat International, 89*, 101989. DOI: 10.1016/j.habitatint.2019.05.008

Vassiliades, C., Kyriakides, N., Pissourios, I., & Vardopoulos, I. (2024). Sustainable architectural innovations for energy-efficient multi-storey buildings through a novel taxonomy and evaluation approach. *Strategic Planning for Energy and the Environment, 43*(4), 1195–1224. DOI: 10.13052/spee1048-5236.43415

Vassiliades, C., Minterides, C., Astara, O. E., Barone, G., & Vardopoulos, I. (2023). Socio-economic barriers to adopting energy-saving bioclimatic strategies in a Mediterranean sustainable real estate setting: A quantitative analysis of resident perspectives. *Energies, 16*(24), 7952. DOI: 10.3390/en16247952

Vassiliades, C., Vardopoulos, I., Barone, G., Santamouris, M., & Kalogirou, S. (2025). Exposing the limits in the residential building energy performance certification model. *Energy Reports, 14*, 1008–1019. DOI: 10.1016/j.egyr.2025.07.011

Vortelinos, D., Menegaki, A., Passas, I., Garefalakis, A., & Viskadouros, G. (2024). Heterogeneous Responses of Energy and Non-Energy Assets to Crises in Commodity Markets. Energies 2024, Vol. 17, Page 5438, 17(21), 5438. DOI: 10.3390/en17215438

World Green Building Council. (2013). THE BUSINESS CASE FOR GREEN BUILDING A Review of the Costs and Benefits for Developers, Investors and Occupants Project Partner Principal Sponsors Sponsors. www.worldgbc.org

Zhang, C., Kazanci, O. B., Levinson, R., Heiselberg, P., Olesen, B. W., Chiesa, G., Sodagar, B., Ai, Z., Selkowitz, S., Zinzi, M., Mahdavi, A., Teufl, H., Kolokotroni, M., Salvati, A., Bozonnet, E., Chtioui, F., Salagnac, P., Rahif, R., Attia, S., & Zhang, G. (2021). Resilient cooling strategies – A critical review and qualitative assessment. *Energy and Building, 251*, 111312. Advance online publication. DOI: 10.1016/j.enbuild.2021.111312

Zhang, K., Troitzsch, S., Zhang, S. Y., Teh, E. S. P., Subramanian, L., & Massier, T. (2021). Transactive Energy in an Urban Environment: A Case Study of Local Generation and Flexibility Potentials in a Singaporean Urban District. *Frontiers in Energy Research*, *9*, 632992. Advance online publication. DOI: 10.3389/fenrg.2021.632992

Zopounidis, C., Garefalakis, A., Lemonakis, C., & Passas, I. (2020). Environmental, social and corporate governance framework for corporate disclosure: A multicriteria dimension analysis approach. *Management Decision*, *58*(11), 2473–2496. . DOI: 10.1108/MD-10-2019-1341

Chapter 6
Implementation Challenges for Blue-Green Infrastructure to Enhance Resilience of Mediterranean Cities

Konstantinos Kavouras
https://orcid.org/0009-0007-8800-7571
Harokopio University, Greece

Kalliopi Sapountzaki
https://orcid.org/0000-0002-5408-8955
Harokopio University, Greece

ABSTRACT

This chapter explores the implementation of Blue-Green Infrastructure (BGI) in Mediterranean cities through a comparative analysis of Madrid Río (Madrid, Spain) and the Appia Antica Regional Park (Rome, Italy), applying key insights to the Antonis Tritsis Metropolitan Park (ATMP) in Athens, Greece. Using a multi-disciplinary approach – combining literature review, spatial analysis, stakeholder interviews and policy evaluation – it identifies major institutional, environmental and financial challenges. ATMP, located in the socially and environmentally vulnerable area of western Athens, is presented as a significant yet underutilized BGI asset. The study emphasizes the importance of adaptive governance, participatory planning and long-term strategies for enhancing climate resilience and urban sus-

DOI: 10.4018/979-8-3373-2327-5.ch006

tainability. The chapter concludes with actionable strategies for policymakers and urban planners to strengthen BGI implementation and realize its full potential in Mediterranean urban contexts.

1. INTRODUCTION

In the era of climate crisis, cities face increasing environmental, societal and climatic challenges such as air pollution and urban heat islands, health risks and disaster risks owing to climate extremes and high levels of urban vulnerability. In the Mediterranean cities the continuous and often unregulated expansion of the densely built grey infrastructure coupled with air pollution and the unfavourable microclimatic conditions put immense pressure on natural ecosystems and ecosystem services and undermine urban sustainability and resilience from every point of view. Under the circumstances, local and regional/metropolitan authorities and planners are seeking Blue-Green Infrastructure (BGI) solutions well-fit to the contextual conditions of their cities.

BGI, integrating water (blue) and vegetation (green) into urban areas represents not only an aesthetic upgrade but a multifunctional strategy that offers essential ecosystem services. It helps in flood control, mitigates the urban heat island effect, enhances biodiversity and improves air, urban soil and water quality. These functions ultimately help cities become more resilient in the face of challenges.

However, implementing BGI at a large scale, as in the case of a metropolitan park design and implementation, is not an easy task. There are numerous barriers – technical, geographical, planning, financial, political, social and institutional – making the venture difficult. Fragmented administrative structures and overlapping competences, lack of funds, land use conflicts and public perceptions, social norms and cultural values, land ownership status, maintenance costs and constraints in water availability, all pose significant obstacles. Meeting these challenges presupposes a strong political will and an integrated and collaborative approach involving sponsorship, policymakers, administrative officials, urban planners, environmental scientists and public consultation.

This chapter aims to analyze the key barriers that emerge when integrating BGI in the form of Metropolitan parks in Mediterranean cities and propose practical solutions to avoid or overcome them. Through the specific case study of the Athens' Metropolitan Park "Antonis Tritsis" (ATMP) – the largest park in the Athens urban complex – and learning from the experience and innovative policies that have been followed in other cities facing similar constraints – the present work aims to propose effective implementation processes for Municipal and Regional authorities to fully exploit the potential of BGI and foster long-term sustainability and resilience in urban development.

2. MATERIALS AND METHODS

The main objective of the work is to address in both theoretical and empirical terms the obstacles and constraints involved in the design, implementation and long-term maintenance of Mediterranean city metropolitan greenery parks, i.e., the most important version of large scale BGI. The methodology adopted includes a theoretical and an empirical part, where the second involves a mutual feedback approach between the empirical case of the ATMP under way and the practices in cities where park establishment, integration and operation were crowned with success.

To address this objective, the research adopts a cyclical, comparative and interdisciplinary methodology that links an in-depth case study analysis with Mediterranean city best practices. The study is guided by the following research questions:

1. What are the key challenges in the planning, implementation and long-term sustainability of large-scale BGI parks in Mediterranean urban contexts?
2. What are the specific institutional, planning, financial, environmental and participatory obstacles faced by the ATMP?
3. Which Euro-Mediterranean practices and models of Blue-Green (BG) park management can be transferred and adjusted to the ATMP context?

The research has been structured in the following stages:

- Systematic literature and policy review: An extensive review of academic publications and institutional reports was conducted to identify the main benefits, risks and design and implementation parameters of large-scale BG parks in cities with similar to Athens contextual profile (environmental stress, land use conflicts, institutional complexity, funding limitations).
- Collection and analysis of primary and secondary data on the ATMP and the surrounding urban context: A mixed dataset was compiled, including statistical data, as well as qualitative material from field observation and stakeholder engagement in the areas serviced by the park.
- Semi-structured interviews with eight representatives of stakeholders (two municipal officials, two environmental experts, two community representatives, two planners/academics): These were conducted between January and March 2025 on the basis of a flexible topic guide. A thematic analysis was

employed to identify common patterns and points of convergence and divergence among participants' views.

- Cross-referencing of findings with other Euro-Mediterranean city park cases: The findings from the ATMP were re-evaluated in light of two emblematic Euro-Mediterranean BGI examples – Madrid Río and Appia Antica Park. The comparison focused on governance, funding, participatory processes and spatial integration.

The above-described feedback approach ensures a comprehensive, interdisciplinary and many-sided consideration of the best solutions to contextual challenges that are more or less common in Euro-Mediterranean city park design and implementation processes.

3. LITERATURE REVIEW

3.1 From Urban Green Spaces to Integrated Blue-Green Infrastructure (BGI)

3.1.1 From Green Parks to BGI offering Ecosystem Services

The concentration of the global population in urban areas has been steadily increasing in the last century (Vinci et al., 2023). Mass urbanization after the industrial revolution of the 19th century led to the displacement of nature from the urban landscape (Vardopoulos et al., 2024). Lack of contact with nature and the deterioration of environmental quality in the cities gradually raised awareness about the need to integrate nature into the urban environment (Escrivà Saneugenio et al., 2024). Large urban parks became then an essential component of the plans and structures of (large) cities (Vardopoulos et al., 2024). Additionally, the presence of greenery in cities became gradually a social demand, influencing planning practices over time (Imbrenda et al., 2025).

From the Garden City movement of Ebenezer Howard (Howard, 1965) and the green belts to contain urban sprawl, to the large-scale metropolitan parks of the 20th century, urban green spaces evolved in terms of both their form and purpose. In the mid-19th century, the "Park Movement" emerged in Europe and the United States as a response to the urban population explosion and disconnection from nature. Initiated by Fr. L. Olmsted, the movement's most emblematic creation is Central Park in New York, an early prototype of today's multifunctional metropolitan parks. Similar examples include the Bois de Boulogne in Paris and the Tiergarten in Berlin, both transformed from historic forest areas. In some cases, urban parks emerged from degraded or brownfield sites – such as the Central Park, which was built on a swampy terrain, or Moscow's Gorky Park, developed on top of a former landfill.

These early parks laid the groundwork for understanding the role of nature in cities. Today, as cities expand and face increasing pressures from climate change (CC) and not only (Ribeiro Hoffmann et al., 2024), urban green spaces are recognized not just for their aesthetic or recreational value but also for their functional role in promoting sustainability and resilience (Glantzi & Ramfou, 2012).

With CC the deficit of vegetation and water surfaces in the cities became more evident (Simoniello et al., 2025) as a factor accountable for intensified CC hazards and impacts (Gunawardena et al., 2017). This has led to the emergence of the BGI approach, which expands the traditional notion of Green Infrastructure (GI) to include water-based features. The green and blue elements and their integration into the urban systems become increasingly the focus of spatial planning and particularly land-use planning to enhance urban resilience to climate and environmental stresses (Deely et al., 2020).

According to the European Commission, BGI refers to "*a strategically planned network of natural and semi-natural areas with other environmental features, designed and managed to deliver a wide range of ecosystem services such as water purification, air quality, space for recreation and climate mitigation and adaptation*" (European Commission, 2025). GI (e.g., parks, street trees, green roofs) supports biodiversity, reduces temperatures and regulates air flows. BI (e.g., rivers, canals, ponds, wetlands) replicates natural hydrology, improves water quality and mitigates flood risks. By combining these two types, BGI delivers multiple benefits: connectivity of ecosystems, climate adaptation, disaster risk reduction, public health protection and sustainable urban development (Almaaitah et al., 2021; European Parliament, 2020; European Environment Agency, 2011).

As BGI emphasizes the integration of water-dominated elements (rivers, lakes, wetlands) with land-based green features, it offers an opportunity for better connection between and integration of urban, peri-urban and rural planning.

BGI in spatial terms can be classified into elements/entities of three distinct spatial scales (Figure 1):

- The national/regional scale elements, encompassing large protected areas (crossing or surrounding cities) and river systems,
- The municipal/city scale elements, referring to urban parks or canals,
- The local/neighborhood scale, including pocket parks, green corridors or ponds.

Figure 1. Classification of Blue-Green Infrastructure by Spatial Scale

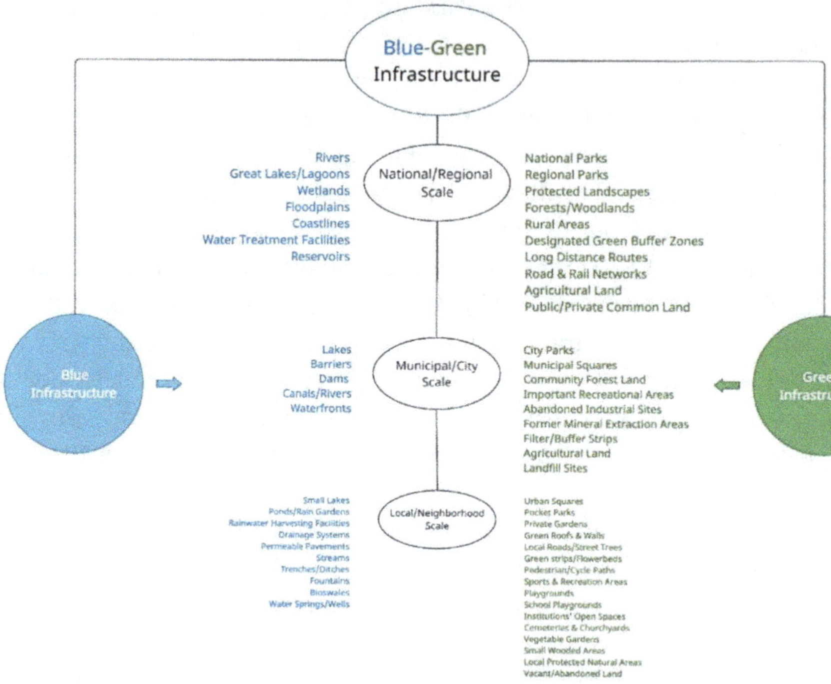

(Designed by the authors)

This classification helps planners and policymakers identify priorities for conservation, enhancement and integration of BGI across scales (Yiannakou & Salata, 2018; Lamond & Everett, 2019; Surrey County Council, 2023). As environmental and climate awareness grows, BGI is increasingly recognized as a fundamental tool for building sustainable and resilient urban communities.

3.1.2 The Contribution of Blue-Green Infrastructure to Urban Resilience

BGI – as the historical evolution of urban greenery – offers critical environmental, social and economic services in urban areas, making it a key lever for enhancing urban resilience (Vardopoulos et al., 2024).

The most obvious and direct environmental benefits of BGI come from vegetation. Through evapotranspiration, plants capture carbon dioxide and release oxygen (Kassios & Perperidou, 2005); vegetated elements improve the urban microclimate by providing shading, influencing air flow (Yiannakou & Salata, 2018) and filtering pollutants such as dust, microparticles, nitrogen and sulfur oxides (Kassios & Perperidou, 2005). Greenery contributes also to water purification (Chiesura, 2004). Additionally, healthy urban trees play a central role in stormwater management by means of three mechanisms: (a) leaves and roots intercept rainwater, reducing surface runoff and delaying overflow, (b) root growth and decomposition enhance soil infiltration capacity and reduce surface flow and (c) tree canopies minimize soil erosion by shielding the ground from direct rainfall (McPherson et al., 2005). Green corridors also serve as biodiversity refuges within dense urban environments (Melissas, 2009), mitigate the Urban Heat Island (UHI) effect through extensive vegetation cover (Akbari & Kolokotsa, 2016) and function as effective noise buffers (Kassios & Perperidou, 2005). Finally, vegetated structures reduce water evaporation from soil and mitigate flooding, erosion and landslides (Kaur & Gupta, 2022).

Water-based infrastructure, or Blue Infrastructure (BI), also plays a crucial role in urban climate regulation and mitigation of CC disaster risks (Escrivà Saneugenio et al., 2024). Water bodies reduce air temperature via evaporation (Oke, 1987) and convection (Spronken-Smith et al., 2000); they are also central to flood mitigation, as they slow down runoff, store rainwater and reduce the burden on urban drainage systems during heavy precipitation events (Chandwani et al., 2021; Ferreira et al., 2024; Wang et al., 2020).

Beyond the environmental dimension, BGI provides essential social benefits. While its traditional value was mainly linked to aesthetics and recreational functions (Wassenhoven, 2018), there is an increasing recognition of its positive effect on physical and mental health (Ulrich, 1981; Loures et al., 2007; World Health Organization, 2016). Green and blue spaces foster relaxation and peace (Kaplan, 1985), contribute to reduced aggression (Kuo et al., 1998) and promote social cohesion. Open and widely accessible green areas transcend social and racial divisions (Hubbart, 1994) and proximity to BGI has been linked to lower crime and delinquency levels, particularly when local communities are engaged in BGI's design and use (Yiannakou & Salata, 2018). Furthermore, BGI can serve emergency gathering operations during crises, e.g. after earthquakes (Melissas, 2009).

In economic terms, BGI elements contribute to increased property values, facilitate job creation and enhance urban areas' attractiveness, supporting tourism and place branding (Loures et al., 2007; Melissas, 2009; Douvas, 2007). These improved urban living conditions translate into broader economic gains, including lower healthcare costs, reduced public spending for pollution mitigation and more efficient resource use (Yiannakou & Salata, 2018). Importantly, BGI supports equitable access to public resources and strengthens community resilience both under normal conditions and in times of crisis.

3.2 Challenges in the Process of Implementation of Blue-Green Infrastructure

Despite its advantages, planning and implementation of BGI is hindered by a variety of factors including governance fragmentation, financial constraints, limited public participation, spatial limitations and technical difficulties.

Table 1 summarizes the key categories of challenges associated with the implementation of BGI, as identified in the literature:

Table 1. Key Categories of Challenges Associated with the Implementation of BGI

Challenge	Description	Key References
Governance and Multi-Level Coordination	- Fragmentation across governance levels (vertical, horizontal, diagonal) - Lack of long-term planning and policy alignment - Weak transdisciplinary collaboration	Deely et al., 2020; Thorne et al., 2015; O'Donnell et al., 2017; Koohestani et al., 2025; Chaurasia et al., 2024; Drosou et al., 2019; Puppim de Oliveira et al., 2022
Financial Limitations and Funding Tools	- High capital and maintenance costs - Lack of clear financial responsibility across actors - Need for locally adaptable funding models - Limited access to innovative instruments	Yiannakou & Salata, 2018; Lamond & Everett, 2019; Surrey County Council, 2023; Koohestani et al., 2025; Deely et al., 2020; Chaurasia et al., 2024
Public Perception, Engagement and Participation	- Low public awareness of BGI benefits - Limited citizen engagement and trust - Need for anticipatory and inclusive planning processes	Lamond & Everett, 2019; O'Donnell et al., 2017, 2021; Koohestani et al., 2025; Puppim de Oliveira et al., 2022; Deely et al., 2020
Environmental, Land Use and Technical Challenges	- Limited available land in dense urban settings - Spatial-functional conflicts in multifunctional spaces - Technical uncertainty and lack of expertise	Gunawardena et al., 2017; Wassenhoven, 2018; Thorne et al., 2015; Mukheibir et al., 2019; Chaurasia et al., 2024; O'Donnell et al., 2017; Koohestani et al., 2025

The following sections analyze the key barriers to the implementation of BGI in contemporary urban contexts.

3.2.1 Governance and Multi-Level Coordination

The successful implementation of BGI requires robust institutional coordination and adaptive, multi-scalar governance mechanisms. Many cities continue to face challenges due to institutional fragmentation, lack of cross-sectoral coordination and limited capacity for integrated planning (Deely et al., 2020; Chaurasia et al., 2024).

This fragmentation is particularly evident in the context of large-scale interventions such as Metropolitan Parks, where governance must operate across vertical, horizontal and diagonal axes.

Vertical coordination is often undermined by policy inconsistency across levels of government. Drosou et al. (2019) highlight how poor coordination between national, regional and municipal authorities results in duplicated efforts and misaligned strategic priorities. Similarly, Thorne et al. (2015) emphasize the lack of confidence among local decision-makers in the continuity of political support for BGI over time, which further discourages long-term investment and institutional commitment.

Lack of horizontal coordination between departments (e.g., planning, environment, transport) is also a persistent barrier. As noted by O'Donnell et al. (2017), institutional resistance to change and a preference for traditional grey infrastructure projects often prevent integrated planning. This is compounded by unclear role allocation and overlapping competences, leading to a lack of accountability and operational confusion (Deely et al., 2020; Koohestani et al., 2025).

Diagonal coordination, involving cross-sectoral collaboration among public bodies, private actors, academia and civil society, is essential but remains underdeveloped. Thorne et al. (2015) and Puppim de Oliveira et al. (2022) point out that transdisciplinary collaboration is frequently hindered by discipline-specific language and siloed institutional cultures, making joint action and knowledge exchange difficult. Chaurasia et al. (2024) describe this condition as "organisational disintegration", where incompatible objectives and the absence of a long-term governance vision undermine planning coherence.

Metropolitan Parks, as flagship BGI interventions, serve not only environmental and hydrological functions but also act as experimental governance platforms (Koohestani et al., 2025). Their successful implementation depends on building institutional trust, establishing participatory mechanisms and facilitating cross-boundary communication. In this sense, they operate as governance laboratories, revealing both the potential and limitations of current administrative architectures to deliver complex, multifunctional urban infrastructure.

3.2.2 Financial Limitations and Funding Tools

Financial constraints represent one of the most significant and recurring obstacles in the implementation of BGI. Urban authorities frequently encounter challenges in securing funding for both the initial capital costs and the ongoing operation and maintenance of such systems. Although the ecosystem service benefits of BGI are increasingly acknowledged in policy frameworks, their value is often not captured in monetary terms, making it difficult to translate them into investable propositions (Deely et al., 2020; Thorne et al., 2015). BGI implementation often requires innovative financing models and fiscal incentives to encourage developers to integrate nature-based solutions into urban development projects (Yiannakou & Salata, 2018).

A persistent issue is the absence of clear financial responsibility across institutional layers. Uncertainty around "who pays" for BGI, especially in multi-actor contexts, contributes to inertia in implementation (Deely et al., 2020). Fragmented and short-term budgetary planning exacerbates this problem, particularly in cases where BGI must be retrofitted into existing grey infrastructure systems (Drosou et al., 2019). Moreover, as highlighted by Koohestani et al. (2025), the complexity of financing transitions in socio-technical systems means that economic instruments must be embedded within broader policy strategies, not treated as stand-alone mechanisms.

The scalability of BGI – from site-specific interventions to metropolitan and regional applications – demands differentiated funding approaches. Research highlights the need to develop locally adaptable financial tools that account for diverse socio-spatial contexts (Yiannakou & Salata (2018); Lamond & Everett (2019); Surrey County Council (2025). Without such flexibility, BGI risks being confined to flagship or pilot projects, unable to achieve systemic impact. In this regard, cities in both developed and developing countries report that insufficient funding streams and institutional capacity for sustaining BGI throughout its lifecycle is the inevitable result of absence of a long-term financial planning (Puppim de Oliveira et al., 2022; Chaurasia et al., 2024).

Beyond public budgets, the literature suggests a need for innovative funding tools – such as green bonds, payment for ecosystem services (PES) and public-private partnerships (PPPs) – to distribute financial risk and encourage investment from non-state actors (Koohestani et al., 2025; Deely et al., 2020). However, many municipalities still lack the institutional frameworks or financial expertise to design and deploy such instruments effectively.

Ultimately, as Chaurasia et al. (2024) note, financial barriers are not simply a matter of lacking resources but reflect broader systemic issues: weak cost-benefit methodologies for BGI, limited access to capital markets and the absence of supportive regulatory incentives.

3.2.3 Public Perception, Engagement and Participation

A core component of a successful BGI implementation is public engagement. BGI elements are most effective when local communities actively support them through stewardship programs and participatory planning processes (Lamond & Everett, 2019; O'Donnell et al., 2021). However, lack of public awareness or interest often hampers these efforts. An essential prerequisite for increasing community involvement is open and continuous public consultation, also communication and learning regarding the multiple benefits of BGI.

Several studies emphasize that the lack of awareness about the functions and long-term benefits of BGI remains a critical challenge. O'Donnell et al. (2017) found that citizens often lack even basic understanding of terms like "blue-green infrastructure", which impedes local uptake and limits community ownership. Similarly, Koohestani et al. (2025) stress that uncertainty and skepticism from users about the viability of BGI systems – especially in contexts requiring behavioral change – can undermine support for systemic transitions.

Policymakers must first recognize the wide-ranging value of BGI for urban resilience and facilitate citizen involvement from the outset of the planning process. Long-term success depends on two crucial factors: how communities and policymakers perceive the urgency of climate-related threats and the willingness of society to invest time, resources and collaborative effort in planning, implementing and sustaining BGI (Lamond & Everett, 2019; O'Donnell et al., 2021; Puppim de Oliveira et al., 2022).

Recent evidence also suggests that participatory planning should not be seen as a "consultative formality", but rather as a structural part of implementation. Successful BGI programs are characterized by anticipatory engagement, in which communities are involved in governance design and ongoing maintenance (Puppim de Oliveira et al.,2022). Deely et al. (2020) further identify lack of community empowerment and trust as obstacles that limit the acceptability and long-term sustainability of nature-based solutions.

3.2.4 Environmental, Land Use and Technical Challenges

Beyond institutional and financial constraints, environmental, spatial and technical factors may significantly hinder the implementation of BGI. A major challenge is land use competition, especially in dense urban areas where space is scarce and development pressures are intense (Gunawardena et al., 2017; Chaurasia et al., 2024). Integrating BGI into existing systems – such as water infrastructure, roads and heritage zones – requires navigating complex spatial configurations, often with limited retrofit potential.

Multi-functionality adds further complexity. BGI elements typically serve ecological, recreational, educational and aesthetic functions. While this versatility is desirable, it can lead to competing demands and management tensions, especially in highly programmed spaces like urban parks (Deely et al., 2020; Koohestani et al., 2025). Cultural institutions, events and everyday recreation coexist in these parks, creating pressure to balance public expectations with ecological integrity (Wassenhoven, 2018).

Technical barriers are equally important. The design, operation and long-term maintenance of BGI demand interdisciplinary expertise and robust knowledge of hydrology, plant performance and soil conditions (Thorne et al., 2015; Drosou et al., 2019). Yet many planning authorities lack the technical capacity and standardized guidelines to ensure performance and resilience. Uncertainties over BGI effectiveness, particularly under future climate scenarios, further inhibit confident investment (O'Donnell et al., 2017; Koohestani et al., 2025).

Moreover, the multi-functionality of BGI can generate a need for functional trade-offs that complicate both design and governance. For instance, a wetland designed for habitat creation may also be promoted as a public park space. However, heavy foot traffic or recreational use (e.g., dog walking, sports) can degrade sensitive vegetation zones or interfere with water quality functions. Without integrated frameworks to reconcile such overlapping or conflicting demands, projects remain vulnerable to fragmentation and inefficiencies (Lamond & Everett, 2019; O'Donnell et al., 2021).

In sum, spatial constraints, multifunctional pressures and technical uncertainty converge to shape a demanding operational environment for BGI. Meeting these challenges requires systemic thinking, capacity building and long-term coordination across sectors and disciplines.

4. RESULTS

4.1 Euro-Mediterranean Experience in Implementing Blue-Green Infrastructure

The implementation of BGI across the Euro-Mediterranean region has become increasingly significant in response to growing urban and climatic pressures (MedECC, 2020; European Environment Agency, 2022; IPCC, 2022). Cities in this region, such as Athens, Madrid and Rome, share common environmental challenges – including extreme summer temperatures (Vassiliades et al., 2023), the urban heat island effect (Vardopoulos et al., 2024), biodiversity loss (D'Agata, et al., 2024), drought (Doukas et al., 2023), urban pollution (Alaimo et al., 2022) and high levels of evapotranspiration (Maialetti et al., 2024). These shared pressures call for integrated solutions that combine ecological restoration with sustainable water management, making large-scale BGI projects a critical tool for enhancing urban resilience and quality of life.

To gain an insight into effective BGI planning and execution in Mediterranean cities, two emblematic case studies have been selected: Madrid Río Park in Madrid, Spain and Appia Antica Regional Park in Rome, Italy. These two examples of already completed and functional parks are set against the ATMP in Athens, Greece – an evolving BGI initiative currently in the process of implementation. The selected case studies meet specific criteria, e.g. about contextual conditions complicating implementation, thus allowing for meaningful comparison with the ATMPt.

First, all three parks are located within major urban agglomerations that host activities of national significance in areas such as administration, business and transportation – activities that often contribute significantly to pollution and greenhouse gas emissions. According to Statista (2025), Madrid, Rome and Athens rank among the largest metropolitan areas in Europe by population, reinforcing their comparability in scale and function.

Second, these cities face similar climatic and environmental conditions typical of the Mediterranean region, which, in particular, is warming faster than the global average. In this context, the development of BGI elements integrating vegetated areas with water features emerges as an essential strategy for sustainable urban adaptation.

An additional consideration in selecting these cases was the reported dissatisfaction among urban residents regarding local climate action. According to a Eurobarometer survey (European Commission, 2016), the majority of respondents in all three cities express concerns over insufficient municipal efforts to address CC. Notably, 63% of Madrid residents, 60% of Athenians and 58% of Romans believe that their cities are not doing enough to combat the crisis. This shared perception reflects not only a sense of urgency, but also a critical opportunity for local governments to invest in nature-based solutions like BGI as a means of restoring public trust, enhancing urban environments and mitigating climate risks.

Given the above, the three cases offer a relevant and instructive comparison. They serve to highlight the diverse (or common?) ways that cities follow to meet comparable challenges during the processes of design, implementation and institutional integration of BGI elements.

4.2 Río in Madrid, Spain

Madrid Río Park (Figure 2) is a good example of urban regeneration, centered on the Manzanares River. By integrating green and blue infrastructure, the project has significantly improved urban climate conditions, enhanced biodiversity and provided accessible public space for residents (Madrid City Council, 2025). Spanning nearly seven kilometers, the park reconnected neighborhoods previously divided by the M-30 motorway, while forming an ecological corridor that links existing green areas (Hernández-Lamas et al., 2016; Santo-Tomás Muro & Romero, 2020).

Figure 2. Madrid Río Park's Location

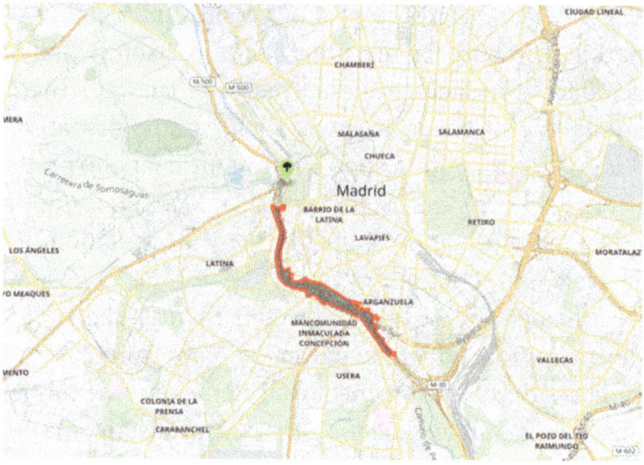

(Map data © OpenStreetMap contributors, available under the Open Database License – www.openstreetmap .org/copyright)

The project that was completed in two phases between 2005 and 2011 rerouted the M-30 underground via the Calle 30 Project and introduced new electrostatic ventilation systems to mitigate air pollution in the tunnels. In the second phase the reclaimed land was transformed into a linear park with 54 km of pedestrian paths, 30 km of cycling routes and 35 new bridges, including emblematic structures like the Arganzuela Bridge (Figure 3); additionally, some historical crossings were restored, such as the Puente de Segovia and Puente de Toledo (Urban Sustainability Exchange, 2025).

Figure 3. Arganzuela Bridge, Madrid

(Photo by Tamorlan, licensed under CC BY 3.0 – https://creativecommons.org/licenses/by/3.0/, via Wikimedia Commons)

Madrid Río employed an integrated master plan (Figure 4), combining landscape design, ecological restoration, infrastructure improvements and public amenities over an area of 2,961 hectares. The project also prioritized water quality by introducing new stormwater infrastructure and removing concrete riverbeds to support rewilding and riparian habitats (Perini, 2017).

Governance of Madrid Río fell under the City of Madrid and was supported by national funding, including allocations from Spain's Plan E stimulus program. Financial resources totaled approximately €4.4 billion, highlighting the scale of commitment and the political consensus behind the project (Madrid City Council, 2025). Ongoing operation and maintenance of the park demand sustained investment from municipal budgets and interdepartmental coordination to support landscape upkeep, water management systems and public infrastructure.

According to WEST8 (2025), one of the architectural firms involved, Madrid Río exemplifies a "mat-building" approach – an urban design strategy where green space, infrastructure and public amenities are integrated into a continuous, multi-functional layer above buried infrastructure. Despite its achievements, Madrid Río faced – and in some areas continues to face – multiple challenges.

Figure 4. Map of Madrid Río Park, Madrid, Spain

(Source: Official body – Madrid City Council / Ayuntamiento de Madrid, publicly available data via madrid.es)

Technically, tunneling beneath the Manzanares demanded complex engineering to support the park above while maintaining citywide traffic flows. The scale of the intervention caused prolonged disruption and controversy among local residents and business owners (Perini, 2017). Ecologically, while the river was partially restored, environmental groups questioned the sufficiency of habitat rehabilitation in such a constrained urban context.

Socially, early phases lacked inclusive participation, with community engagement mechanisms improving only after 2009. Criticism centered on limited transparency and insufficient involvement of residents in shaping the project's initial direction. Although later initiatives included competitions and public forums, trust had to be gradually rebuilt (Perini, 2017).

At the spatial level, local disconnections persist. Certain park zones remain poorly integrated with surrounding neighborhoods, while height differentials resulting from the buried motorway have resulted in steep slopes and access barriers. In some segments, surface-level infrastructure such as parking lots continues to fragment the landscape and limit pedestrian continuity (Bambó Naya et al., 2023).

Despite these challenges, the project was completed in 2011 and has been internationally recognized for its contribution to ecological and urban regeneration (Giannelli et al., 2018). Madrid Río represents a landmark achievement in reconfiguring urban form through nature-based solutions. Its success lies in long-term institutional coordination, technical innovation and substantial financial backing. It offers a replicable framework for integrated metropolitan infrastructure, with lessons for cities like Athens, where the ATMP remains a partially realized vision.

4.3 Appia Antica in Rome, Italy

The Appia Antica Regional Park in Rome (Figure 5) is a 4,580-hectare area integrating archaeological heritage, natural landscapes and agricultural land – making it one of Europe's largest cultural-nature reserves (Cerema, 2011).

Figure 5. Map of the Appia Antica Regional Park

(Source: Official body – Parco Regionale dell'Appia Antica, publicly available data via parcoappiaantica.it)

Centered on the ancient Appia Way, it includes sub-zones such as the Caffarella Valley, Aqueducts Archaeological Area, Tormarancia Estate and the Via Latina ruins, all connected through a 16 km perimetric network (Parco Regionale dell' Appia Antica, 2025). Since its foundation in 1988, the Park expanded twice in 2002 and 2018 (Figure 6).

Figure 6. Map of Appia Antica Regional Park's Territorial Expansions: 1988, 2002, 2018

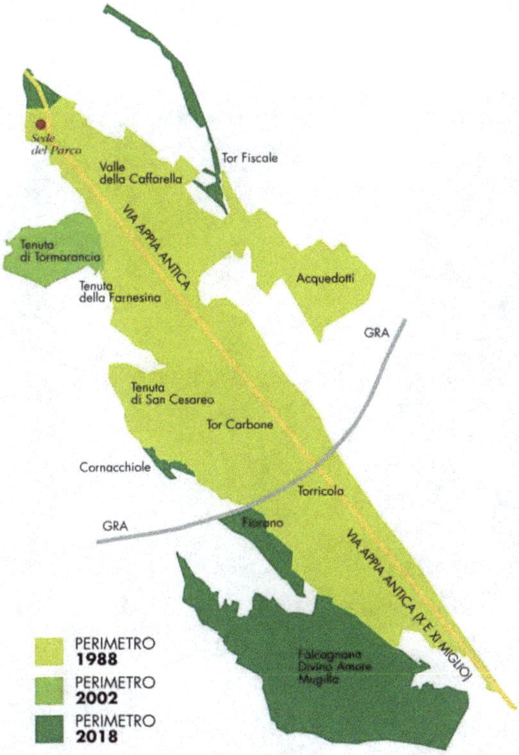

(Source: Official body – Parco Regionale dell'Appia Antica, publicly available data via parcoappiaantica.it)

Established under Regional Law No. 66 following grassroots mobilization, the Park has been managed by the Regional Authority of Lazio through an autonomous governance body – the Park Community (Comunità del Parco). This includes the President of the Metropolitan City, the Mayors of Rome, Ciampino and Marino (or their appointed delegates) and four representatives from professional associations and environmental organizations. As of today, the role of President of the Park Community is held by the Mayor of Ciampino (Parco Regionale dell' Appia Antica, 2025). This structure ensures territorial representation while enabling local government input in decision-making processes. Operational funding derives primarily from regional and national sources, complemented by EU programs and heritage grants.

The Park's terrain extends from the urban fringe near Piazza Numa Pompilio to the rural southern reaches of Frattocchie. Figure 7 shows a preserved segment of

Via Appia near Quarto Miglio, offering a visual reference to the Park's historical infrastructure. The Almone River (Figure 8), a historic watercourse with ecological and cultural value, flows through the Park, reinforcing its blue infrastructure features (Capelli, 2015).

Figure 7. Remains of the Via Appia near Quarto Miglio, Rome

(Photo by Kleuske, licensed under CC BY 3.0 – https://creativecommons.org/licenses/by/3.0/, via Wikimedia Commons)

The Park is governed by a tripartite mission: (a) conservation of archaeological and landscape assets, (b) ecological restoration of hydrogeological and biodiversity systems and (c) sustainable recreation and public education. It shelters rare plant and animal species – 43% of Rome's flora and 20% of the region's – demonstrating its role as a biodiversity hotspot (Iamonico, 2022).

Figure 8. Almone River, Rome

(Photo by Indeciso42, licensed under CC BY-SA 4.0 – https://creativecommons.org/licenses/by-sa/4.0/, via Wikimedia Commons)

Despite its achievements, the Park faces serious challenges that threaten its long-term sustainability. Environmental degradation remains a key concern, as the Almone River suffers from pollution due to urban runoff and industrial activity, diminishing its hydrological role and affecting local biodiversity (Tallini et al., 2013). Additionally, the original Roman drainage infrastructure along the Appian Way has significantly deteriorated, leading to flooding and erosion in several areas and demanding urgent restoration (Orsi, 2013).

The Park also contends with growing tourism pressures: while increased visitation has generated public interest and economic benefits, it has simultaneously placed strain on fragile ecosystems and heritage structures, particularly in sensitive archaeological zones (Brumana et al., 2023). These issues are further exacerbated by outdated management frameworks, including aging infrastructure, insufficient coordination across visitor services and limited adaptive tools to address new environmental and social demands. This confluence of environmental, infrastructural and managerial challenges underscores the need for renewed, integrated planning and long-term strategic investment.

To address these, the Park has launched digitization projects for 3D mapping of the Appian Way, aimed at enhancing conservation planning, archaeological research and public engagement (Brumana et al., 2023).

In essence, Appia Antica stands as a model of integrated BGI, balancing natural, cultural and recreational functions under a robust legal and institutional framework. Its long-term viability is tied to continued public investment, participatory governance and adaptive management – critical elements for any city seeking to replicate its success, including Athens and the ongoing development of the ATMP.

4.4 Antonis Tritsis Metropolitan Park in western Athens, Greece: A BGI element in the making

4.4.1 Western Athens: A High Social Vulnerability Urban Sub-region suffering from UHI

Western Athens is a geographical unit covering an area of 66.8 sq. km. and bordered by the National Highway Athens-Thessaloniki and the mountains Aigaleo and Poikilo (Figure 9). The population of western Athens, according to the 2021 census data, amounts 475,809 inhabitants, with a population density of 7,126.08 inhabitants per sq km (Hellenic Statistical Authority, 2021).

Figure 9. Location of Western Athens within the Attica Region, Greece

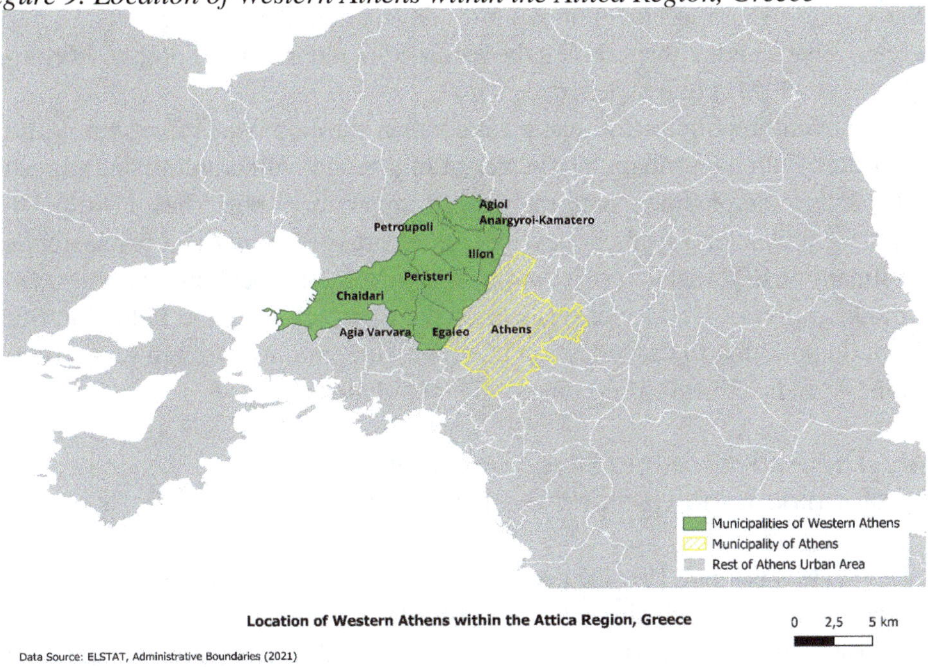

(Source: Created using QGIS – Data Source: ELSTAT – Hellenic Statistical Authority – Editing and design: Authors)

In the past, the area attracted industrial and manufacturing firms and provided cheap housing for the working class, resulting in the development of residential areas next to industrial/manufacturing complexes. The area has been stigmatized gradually by environmental degradation, spatial and social isolation and a 'cross the river' mentality (referring to flooding of Kifisos River from time-to-time). For decades, the area of western Athens remained on the margins, first serving as a place for unwelcome but necessary land uses for the urban core and later being ignored and left in a condition of spatial isolation. From the 1950s and 1960s, during the period of internal migration from rural areas to urban centers, particularly to the Greater Athens region, new waves of low-income and marginalized social groups turned to the area to find low-cost permanent residence (Giokas, 2019).

Compared to the other departments of the metropolis, western Athens presents a high composite and cumulative vulnerability index on the basis of poverty, illiteracy, population aging and population density (Delladetsimas et. al, 2015).

As a result of the predominant pattern of production of the built and urban environment since the '50's and 60's, central and western Athens face a significant problem of thermal discomfort, mainly due to the high building density, lack of greenery and other cooling areas and heavy traffic in a road network suffering from poor ventilation. The urban heat island (UHI) phenomenon is particularly intense in these areas, where summer temperatures during both day and night can be 5 to 8 degrees Celsius higher than those to be recorded in the northern suburbs (Keramitsoglou et al., 2011; Polyzos, 2016).

Consequently, policies for sustainable urban development in western Athens should start from the primary challenges of high social vulnerability and exposure to CC hazards, among them extremely high temperatures exacerbated by the UHI.

A critical element of this policy-making effort is planning and expansion/ establishment of BGI that can offer a range of benefits to local communities: from the modification of UHI, improvement of the micro-climate and reduction of atmospheric pollution to the reduction of social vulnerability through promotion of the long-term health and well-being of the population.

4.4.2 The Park's Path to Blue-Green Infrastructure: Transforming Land Use and the Landscape

The ATMP (Figure 10), located in western Athens within the administrative boundaries of the Municipality of Ilion and bordering Agioi Anargyroi-Kamatero, is the largest metropolitan park in the capital, covering 913 acres. It constitutes a rare natural enclave within the densely built urban fabric, distinguished by an extensive system of artificial lakes, vegetation and open space, which together support a wide range of flora and fauna (Antonis Tritsis Metropolitan Park, 2025).

Figure 10. Location of the Antonis Tritsis Metropolitan Park

(Image from Google Earth © Google, 2025. Used under Google Maps/Earth Additional Terms of Service – for illustrative purposes only)

The transformation of the site into a BGI component is rooted in its historical evolution. Originally a royal estate used for agricultural purposes, the area hosted orchards, gardens and facilities such as the Othonian Courtyard (Figure 11) and the Queen's Tower (Figure 12) – a Gothic-style building designed by French architect Boulanger in 1854, now part of a privately owned estate functioning as a winery (Biris, 2006; Antonis Tritsis Metropolitan Park, 2017). After a period of state ownership and partial privatization, the estate was damaged during the German occupation (1941–1944). Out of the original 2,500 acres of the Queen's Tower estate, some sections were gradually allocated to public welfare institutions and others remained under private ownership. Eventually, a portion of 913 acres was reserved and transferred to the Park Management Authority in 1993 for the specific purpose of the creation of a metropolitan park in alignment with the goals of the Athens Master Plan of 1985 (Government Gazette of Greece, 1985).

Figure 11. View of the Othonian Courtyard - Former stables

(Source: Kavouras' Personal Archive)

Figure 12. View of the Queen's Tower

(Photo by EurovisionGR, licensed under CC BY-SA 4.0 – https://creativecommons.org/licenses/by-sa/4.0/, via Wikimedia Commons)

Planning and management of the site were undertaken by the Athens Master Plan Organization with the goal of creating an environmental awareness park for the everyday needs of local residents (Figure 13). This initiative was based on a study commissioned by the Organization and developed by a consortium of private consulting firms (Athens Master Plan Organization, 1995).

Figure 13. Topographic Map of Antonis Tritsis Metropolitan Park

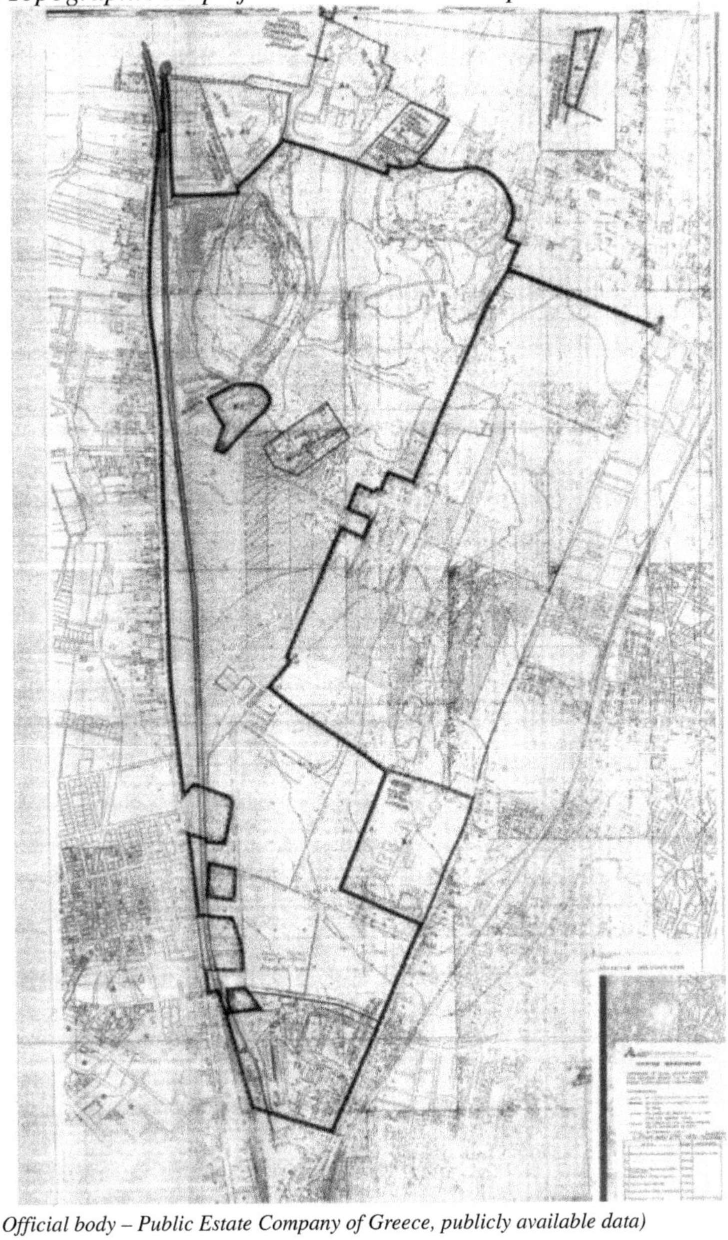

(Source: Official body – Public Estate Company of Greece, publicly available data)

Between 1992 and 2003, substantial European funding (covering about 75% of the cost) supported the development of core infrastructure and green elements (Stamoulidis & Stafylas, 2017). A key feature of this phase was a set of six artificial lakes and a connecting canal network, which replaced older seasonal waterholes and gradually evolved into a vital wetland ecosystem (Antonis Tritsis Metropolitan Park, 2017).

Unfortunately, however, the original master plan was not faithfully implemented and the Athens Organization, as a Legal Entity of Public Law, was abolished in 2012 as a result of the structural reforms that were imposed on the country after the Greek fiscal and economic crisis that started in 2009.

In a related technical report prepared by the lead partner of the consortium that originally designed the park, significant issues affecting its current operation are raised. The report was commissioned by the Hellenic Ornithological Society, which has been voluntarily active in the park and sought the firm's support in reshaping the development plan for the ATMP.

In light of the serious organizational and operational challenges facing the park, the report highlights as central issue the then institutional dysfunction and financial hardship of the Park's Administrative and Management Authority. In particular, the report emphasizes the incapacity of the managing body to perform basic maintenance duties and to implement or complete the original plans for the Park's functions and infrastructure. Instead, it had approved new land uses and functions not foreseen in the original design, many of which are incompatible with the Park's intended character or are considered entirely arbitrary (Papagiannis, T., & Associates S.A., 2008).

However, the ATMP had and still has the potential to become a major BGI element for western Athens, because of its features that are potential contributors to several aspects of urban resilience (Figure 14). Its extensive natural features enhance the city's capacity to mitigate and adapt to CC. Vegetated areas improve rainwater infiltration and contribute to flood risk reduction. The artificial lakes retain water and support biodiversity, while vegetation limits soil evaporation and erosion. Tree cover mitigates the urban heat island (UHI) effect and provides essential shading reducing ambient surface temperatures.

Figure 14. Panoramic View of the Antonis Tritsis Metropolitan Park.

Adapted from a photo by Emmanouil Koutras, licensed under CC BY-SA 4.0 – https://creativecommons.org/licenses/by-sa/4.0/, via Wikimedia Commons.

As highlighted in the literature, land use planning that includes such infrastructure is crucial for disaster preparedness and climate resilience (Sapountzaki & Dandoulaki, 2015). Indeed, the ATMP's strategic location and physical capacity make it appropriate as a hub for emergency response and civil protection, providing space for shelter, emergency vehicle access and training. The Region of Attica explicitly acknowledges the park's role in regional evacuation planning (Region of Attica, 2025).

The park hosts 389 plant species – 311 of which are native – and offers habitats ranging from pine and eucalyptus groves to agricultural stretches, including 60 acres of olive trees and 110 acres once used for peanut cultivation (Antonis Tritsis Metropolitan Park, 2017). These support diverse wildlife. The park's lakes alone sustain 198 bird species, including both resident and migratory (Latsoudis et al., 2023), as well as amphibians, reptiles, fish and mammals. Recognized by the Hellenic Botanical Society as a 'biotope island' within the urban fabric (Vasilopoulou, 2017), the ATMP serves both ecological and educational functions. Recommendations for ecological enhancement include the installation of bird shelters, creation of wetlands, management of invasive species and visitor education programs (Latsoudis et al., 2017).

The park also plays a central role in environmental education and public awareness. Its facilities host workshops on ecology, biodiversity and sustainability, offering outdoor recreational and learning opportunities to diverse social groups. Numbers of visitors (15,000–16,000 each weekend) indicate the park's relevant supra-local appeal (Antonis Tritsis Metropolitan Park, 2023). The above activities are coordinated by the Park Management Authority, which operates under the supervision of the Attica Region, in accordance with the institutional framework (Government Gazette of Greece, 2002). The Management Board is composed of members appointed by the Regional Governor, the Mayors of the municipalities involved and representatives of civil society, as specified in the framework.The park's ecosystem is inseparably linked to its water system. Of the six artificial lakes, five are located in the northern zone and one in the south. These are connected by a canal network and supported by a hydraulic installation that enables water circulation across the park (Figure 15). The lakes function as water retention basins and contribute to both vegetation support and microclimatic balance. Their aquatic vegetation – Including watercress, wattle and bitter-grass – provides habitat and food source for various faunal groups, reinforcing the role of the ATMP as a living urban ecosystem.

Figure 15. The Interconnected System of Artificial Lakes in the ATMP's Management Authority

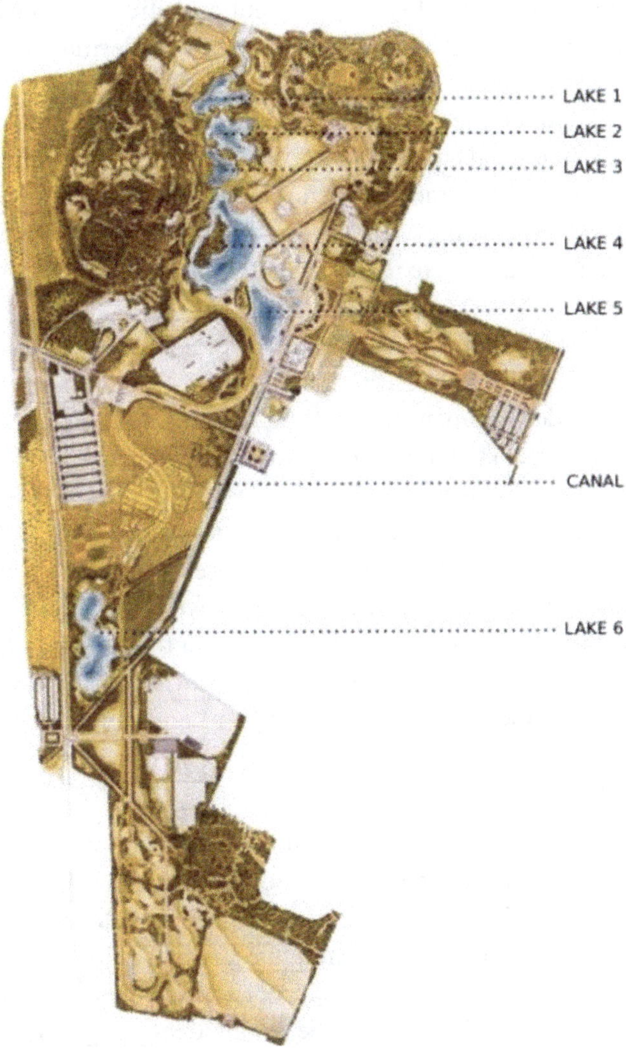

(Source: Adapted from a map of Athens Master Plan Organization, 1995, publicly available data via parkotritsis.gr)

4.4.3 Challenges for the Park Realization Process

4.4.3.a Governance and Administration Obstacles

Despite the ATMP's establishment in 2002 with a clearly defined legal mandate to promote environmental awareness and manage the park's development on sustainable principles (Government Gazette of Greece, 1985), the realization of its objectives has remained incomplete.

Persistent governance instability, inconsistent priorities and above all, failure to update and enforce the master plan have systematically undermined implementation efforts.

Fragmented responsibilities among different levels of government, particularly due to overlapping municipal and regional competencies, have resulted in disjointed decision-making and administrative inertia. This is evident in the case of ATMP, where unclear roles between the Management Authority, the Region of Attica and adjacent municipalities have led to recurring gaps in maintenance and infrastructure management.

Frequent changes in leadership and inconsistent strategic vision have delayed vital initiatives, such as habitat restoration and infrastructure modernization. The absence of a phased development framework, which would ensure continuity and define long-term goals, contributes to operational inefficiencies and a lack of clear strategic direction.

Although a Park Management Authority exists under the jurisdiction of the Region of Attica, its role remains largely executive in nature, lacking the strategic capacity to lead integrated planning or to coordinate effectively the multiple stakeholders involved. This institutional limitation did not allow for the Park to emerge as a cohesive metropolitan priority. In contrast to the cases of Madrid and Rome, the ATMP is characterized by fragmented institutional support and limited policy continuity.

4.4.3.b Financial Barriers

The Park's budget relies primarily on a monthly revenue stream drawn from the proportional allocation of the Central Autonomous Resources designated for the municipalities of Attica. This contribution functions as a form of environmental compensation, reflecting the environmental burden caused by urban activity in the region. The amount is set by joint ministerial decision and may be revised every three years, depending on the stabilization of the Park's operating costs and the degree of financial self-sufficiency achieved through its activities and the use of allocated public assets (Government Gazette of Greece, 2016). However, this framework has been proven insufficient to sustain even basic operations, let alone support long-term development. In practice, reliance on emergency national or EU funding for ad hoc purposes – such as temporary repairs, isolated infrastructure upgrades or one-off events – has led to short-term interventions with limited strategic impact, perpetuating an approach that fails to deliver coherent results.

Public funding – especially from municipal budgets – remains the Park's sole financial pillar, exposing it to broader macroeconomic and political fluctuations. The absence of a long-term financial plan or self-financing mechanisms (e.g., through services, concessions, or public-private partnerships) leaves its future viability questionable.

The comparison with Madrid Río and Appia Antica reveals the weakness of this approach. In both foreign cases, multilevel governmental support was mobilized effectively through broad coalitions of civic actors. In contrast, the ATMP has never been embraced as a collective urban vision nor did it mobilize civic movements to the extent required for continuous public pressure and political prioritization.

4.4.3.c Difficulties in Public Engagement

The ATMP has not succeeded in generating the type of public ownership or participatory engagement observed in other Euro-Mediterranean BGI examples. The absence of a strong grassroots support base has limited the park's visibility in public discourse, reducing pressure on institutions to prioritize its development. Although educational and environmental initiatives are present, they remain fragmented and lack an institutional continuity to build broader public engagement or civic stewardship.

More critically, there is no institutionalized participatory framework connected to the park's planning, monitoring, or governance. No citizen advisory council, structured consultation process, or co-management mechanism currently exists. Community participation is typically limited to sporadic cultural or environmental events, without mechanisms for sustained civic involvement or local agenda-setting.

The gap in participatory governance is not solely a matter of a top-down political-administrative system; it also reflects the wider sociopolitical dynamics of the area of western Athens, an area historically underrepresented in urban planning priorities and often lacking structured community organizations.

4.4.3.d Land Use Conflicts and Environmental Drawbacks due to Climate Change

Chronic water shortages remain one of the most pressing environmental challenges facing the ATMP. Water demand consistently exceeds available supply from internal sources – six boreholes and five wells – while the critical Eschatia well, located outside the park and owned by the Agioi Anargyroi–Kamatero Municipality, saw a dramatic drop in capacity following upstream flood control works that disrupted its feeding stream (Katsafadou, 2017). During the 2017–2018 hydrological year, transfers to the park fell by 91%, from 34 m^3/day to only 3 m^3/day (Antonis Tritsis Metropolitan Park, 2019).

In response to critical shortages, emergency measures have been employed, including water deliveries by the Greek Army (Figure 16), following evaporation losses and infrastructure failures. Meanwhile, plans for a wastewater treatment facility were abandoned due to strong local opposition (EYDAP, 2018), making water recycling impossible and forcing reliance on deep groundwater pumping – an ecologically unsustainable practice.

Figure 16. Water Carriers of the Greek Army Carry Water to the 5th Lake

(Source: Kavouras' Personal Archive)

Furthermore, parts of the park, especially the southern section, remain occupied by incompatible land uses, such as municipal depots and heavy vehicle parking facilities (Belavilas, 2017), forming a physical and functional barrier between northern and southern areas (Figure 17). These uses, although vital for the Metropolitan Region, undermine the ecological character of the park.

Connectivity with other urban green sites or corridors and peri-urban greenery also remains weak. Despite its proximity to major green and natural landmarks – e.g. the Nea Philadelphia Park and Mount Poikilo – the park is not effectively integrated into a coherent BGI system. While car access is convenient, the lack of reliable public transport connections and inadequate pedestrian and cycling infrastructure reduce the Park's attraction. The potential of the nearby railway station Pyrgos Vasilissis remains untapped due to low frequency and poor accessibility (Tsigdinos, 2017).

Figure 17. The "Barrier" of the Southern Section: Waste Trucks and Heavy Vehicle Parking Facilities Operated by the Municipalities of Ilion and Agioi Anargyroi–Kamatero

(Image taken using Google (2025), via Google Earth – Adapted by Authors)

However, the study by Polyzos et al. (2014) has envisioned the ATMP as a central node located in a wider metropolitan green corridor, potentially linking the Park with abandoned railway routes and significant urban-cultural landmarks (Figure 18). Nevertheless, realization of such visions requires a societal culture that is supportive of spatial planning, a more efficient public administration, multi-stakeholder alliances and public-private partnerships to facilitate large investments and overcome the problems of rigid land use land ownership status.

Figure 18. Network of Green and Cultural Routes in Metropolitan Athens

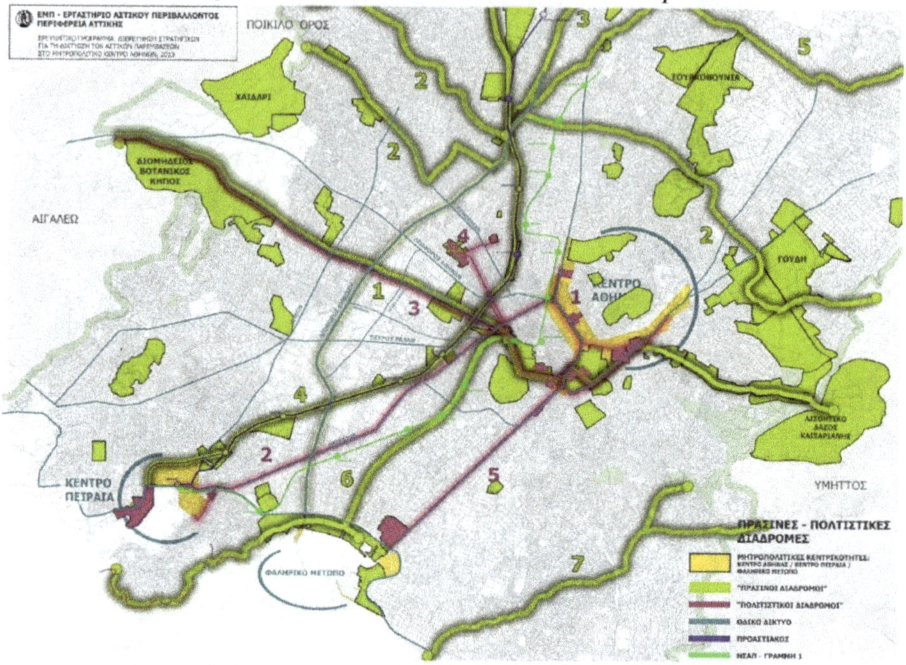

Source: Adapted from a map of Polyzos, Balabanidis & Triantis (2014), (used with permission)

4.4.4 Stakeholders' Proposals and Perspectives

A qualitative investigation was conducted through eight semi-structured interviews with representatives of the stakeholders in order to deepen understanding of the challenges and potential solutions related to the ATMP. These included two local government officials, two environmental experts, two community representatives and two urban planning professionals/academics. The interviews took place between January and March 2025, lasted between 45 and 75 minutes and followed a flexible question guide that allowed for in-depth exploration of emerging themes.

The collected data were analyzed thematically focusing on perceptions of major challenges, stakeholder roles and proposed interventions. Recurring perceptions, problem framings and solution proposals were coded and grouped according to the four challenge categories identified in section 3.2: (a) governance and multi-level coordination, (b) financial limitations and funding tools, (c) public perception, engagement and participation and (d) environmental, land use and technical challenges. At the same time, attention was paid to capturing distinct perspectives emerging from each stakeholder group.

The analysis revealed both convergence and divergence in stakeholder priorities. While all interviewees recognized the lack of a master plan and the chronic funding instability as central problems, their interpretations of causes and proposed solutions varied depending on their institutional roles and experiences. For example, environmental experts emphasized ecological degradation and biodiversity loss and the respective underlying causes, whereas municipal officials focused more on the inadequacy of operational budgets and institutional fragmentation. Community representatives voiced dissatisfaction with park neglect, safety concerns and limited access to decision-making. Academic and planning professionals stressed the lack of an overall plan, strategic continuity and the need for integrated participatory frameworks.

Table 2 below summarizes the key themes and proposals, classified by stakeholder group and challenge category:

Table 2. Summary of Stakeholder Perspectives and Proposed Interventions by Challenge Category

Challenge	Stakeholder Group	Main Concerns	Proposed Interventions
Governance and Multi-Level Coordination	Local government officials	Fragmented responsibilities; lack of a unified authority; political discontinuity	Institutional coordination; establishment of a permanent ATMP management body and a constantly updated Master and Implementation Plan
	Environmental experts	Absence of a long-term ecological strategy; weak enforcement of regulations	Establishment of a governance structure with environmental monitoring roles and ecological task forces
	Community representatives	Lack of transparency and public involvement	Participatory decision-making; citizen advisory councils
	Academics/planners	Policy fragmentation; lack of an overall plan and the ad hoc and fragmented planning interventions	Introduction of a multi-level governance framework providing for oversight mechanisms

continued on following page

Table 2. Continued

Challenge	Stakeholder Group	Main Concerns	Proposed Interventions
Financial Limitations and Funding Tools	Local government officials	Reliance on unstable municipal funding; insufficient state support	National/EU funding framework; Municipal takeover of park functions (e.g. green management)
	Environmental experts	Funding gaps in ecosystem maintenance	Link ecosystem services to funding mechanisms; pilot EU-funded projects
	Community representatives	Lack of capacity to fund basic infrastructure and safety measures	Strengthening public funding; integration of a public employment programme
	Academics/planners	Inadequate financial model; lack of long-term planning	Mixed financing model (public-private-partnerships); park self-financing roadmap
Public Perception, Engagement and Participation	Local government officials	Limited citizen participation; passive and indifferent public	Public outreach strategies; digital feedback tools
	Environmental experts	Low environmental literacy among users	Community education programs; school engagement
	Community representatives	Lack of consultation; sense of exclusion	Co-design workshops; volunteer groups
	Academics/planners	Weak culture of participatory planning	Institutionalization of participatory tools in planning processes
Environmental, Land Use and Technical Challenges	Local government officials	Incompatible uses (e.g., parking of waste trucks); water shortages	Gradual relocation of disruptive uses; development of resilient water systems
	Environmental experts	Loss of biodiversity; drought-induced stress	Reintroduction of native species; infrastructure for water reuse
	Community representatives	Pollution, dry lakes, unsafe areas	Lake system restoration; investment on infrastructure
	Academics/planners	Urban pressure on park boundaries; poor regional integration	Metropolitan green corridor development; integration into long-term spatial planning

4.5 Lessons from Madrid and Rome Transferable to Athens

The cases of Madrid Río Park and the Appia Antica Regional Park offer valuable insights into how BGI can be successfully integrated into complex urban environments. Despite their distinct contexts – one centered on riverfront regeneration, the other on landscape preservation and cultural heritage – both parks share strategic approaches that can inform the unfinished efforts to consolidate the ATMP in Athens as a fully functional BGI element. Four key lessons stand out as transferable and potentially catalytic for the Athenian context:

1. Institutional Continuity and Coordinated Governance.

Both Madrid and Rome benefited from institutional frameworks that ensured coordination among various levels of government (municipal, regional, national). In the case of Madrid, a clearly phased master plan allowed different agencies to operate under a unified vision, while Rome's Appia Antica Park emerged from a continuous grassroots pressure and was subsequently protected by regional legislation and supported by public agencies.

In contrast, the ATMP suffers from fragmented governance, a lack of continuity and the absence of a binding master plan. A transferable lesson is the creation of a stable, multi-level governance framework backed by legislative tools that ensure long-term planning, inter-agency coordination and institutional stability.

2. Strategic Use of Public Investment and Diverse Financing Mechanisms.

Madrid Río was enabled through substantial public investment – particularly through Spain's national stimulus plan – while also integrating infrastructure upgrades (e.g., burying the M-30 ring road) within a broader environmental regeneration agenda. Appia Antica leveraged EU and national conservation funds while tapping into cultural heritage grants as well.

The lesson for ATMP is dual: first, strategically aligning environmental goals with broader infrastructure or heritage projects can unlock funding synergies; second, a mixed funding model involving municipal, regional, EU and private contributions is crucial to overcoming reliance on unstable current funding status.

3. Engagement of Civic Society and Public Participation

In both Madrid and Rome cases, civil society played a central role. In Rome, community advocacy was central to the establishment of the park and continues to play a significant role in its management. In Madrid, although early planning lacked deep community involvement, subsequent engagement processes improved project acceptance and civic ownership.

In the case of ATMP, lack of community involvement and limited participatory mechanisms have proved to be obstacles to Park's development. The Park's visibility and institutional capacity could be significantly improved by the introduction of workshops for co-design, citizen advisory bodies and environmental education campaigns.

4. Integrated Design and Ecological-Cultural Connectivity

Both Madrid Río and Appia Antica exemplify the value of integrating ecological, recreational and cultural functions into a coherent spatial narrative. Madrid's river-cityscape model and Appia's merging of green infrastructure with archaeological heritage offer compelling examples of multifunctional public space with strong place identity.

In the case of Athens, the existing natural and historical features of the ATMP – such as the Royal Estate remnants and the artificial lakes – create a strong basis for establishing the Park simultaneously as an ecological and a cultural landmark. An integrated design approach that supports this dual identity, while also aiming to functionally connect the ATMP with other nearby urban and peri-urban green spaces, could significantly strengthen its role as a keystone node in a fragmented and under-connected metropolitan BGI system, especially by reinforcing linkages to Mount Poikilo, Nea Philadelphia Park and former railway corridors. This is particularly relevant for Western Athens, a densely built and environmentally vulnerable area, where the Park could act as a central element in a fragmented ecological network.

While context-specific adaptations are necessary, the key principles of institutional continuity, diversified funding, civic participation and integrated design offer a strategic blueprint for rethinking the ATMP's implementation pathway. Although foreign models cannot be directly replicated, Athens can draw inspiration from these practices to mobilize local potential, align planning efforts and ensure the long-term viability of the park as a core element of metropolitan resilience.

5. DISCUSSION

The implementation of BGI in Mediterranean cities encounters a series of complex challenges, which – as analyzed in Section 3.2 – can be grouped into four main categories: governance and multi-level coordination, financial sustainability, public engagement and environmental/spatial constraints. The three examined cases (Madrid Río, Appia Antica, ATMP) confirm the theoretical grounding of these obstacles, while also offering illustrative examples of successful or problematic management practices.

1. Governance and Multi-Level Coordination: The Madrid case highlights the importance of a cohesive and ambitious master plan, supported by clear regulatory mechanisms and intergovernmental cooperation. Rome, by contrast, illustrates the power of civic mobilization and institutional protection through regional legislation. The ATMP fails to secure such continuity, as it is marked

by administrative instability, overlapping mandates and a lack of coordination. This reflects the theoretical insights of Deely et al. (2020) and Chaurasia et al. (2024) who emphasize the role of integrated institutional frameworks and horizontal coordination in effective BGI implementation.

2. Financing and Economic Viability: Literature highlights the vulnerability of BGI initiatives to fragmented and short-term funding arrangements (Deely et al., 2020; Drosou et al., 2019; Puppim de Oliveira et al., 2022; Chaurasia et al., 2024). The substantial cost of Madrid Río (approximately €4.4 billion) was made possible through national investment (Plan E), underscoring the importance of strategically integrating infrastructure and ecological regeneration. Appia Antica leveraged EU and national cultural heritage funding. In the case of ATMP, initial development was partially supported through European funds, particularly for landscape rehabilitation and infrastructure. However, over time, the park's operation became increasingly dependent on local government funding and ad hoc state transfers, without securing a stable or self-sustaining financial model. As Yiannakou & Salata (2018) note, the absence of multi-level and innovative funding models is a structural weakness in BGI delivery.
3. Public Participation and Social Legitimacy: Public involvement is not only desirable but essential for BGI legitimacy according to literature (O'Donnell et al., 2021; Lamond & Everett, 2019). Appia Antica was born from bottom-up mobilization, where environmental NGOs and local residents resisted speculative development and framed the park as a public good. In Madrid, despite initial shortcomings, later phases included public consultations that enhanced the project's acceptance. The ATMP, by contrast, lacks structured participatory mechanisms, weakening its social legitimacy. This confirms the importance of embedding participatory approaches in planning processes, particularly in high-vulnerability areas such as western Athens (Delladetsimas et al., 2015).
4. Environmental, Spatial and Technical Challenges: All three cases face significant technical and environmental hurdles. Madrid's ecological regeneration of the Manzanares River involved not only hydrological interventions but also riparian habitat restoration, mitigating urban heat and promoting biodiversity. Appia Antica's water system remains vulnerable, yet targeted restoration efforts have been introduced. The degradation of the Almone River in Rome and the effects of climate variability on the lakes of ATMP reflect the urgent need for adaptive water management (Gunawardena et al., 2017; Ferreira et al., 2024). Conflicting land uses – such as the presence of garbage truck depots in the southern part of ATMP – highlight the difficulties of integrating BGI into densely built environments.

All in all, the most important lesson learnt from theoretical suggestions and empirical case studies is that at least in the Euro-Mediterranean region successful BGI implementation requires:

- institutional continuity and coordinated governance,
- blended and diversified funding mechanisms,
- active community engagement and
- an integrated planning vision that combines ecological, social and cultural dimensions of space.

The comparative analysis of the three cases clearly shows that large-scale BGI projects in Mediterranean urban contexts – such as those in Madrid, Rome and Athens – are shaped by a set of structural challenges that underscore the complexity of such undertakings. The ATMP as an incomplete BGI case allows for a more detailed understanding of institutional, financial, environmental, spatial and participatory barriers delaying its evolution into a mature BGI element.

Finally, the critical review of the Madrid and Rome examples provides useful potentially transferable guidance and tools for improving the feasibility and long-term sustainability of the ATMP. While this study does not advocate for mechanical replication of foreign models, it does support their critical adaptation to other contexts presenting spatial, environmental, cultural or institutional similarities. In this regard, Athens cannot simply copy external examples, but it can draw strategic lessons from them and translate those into spatial and governance practices tailored to the local context.

REFERENCES

Akbari, H., & Kolokotsa, D. (2016). Three decades of urban heat islands and mitigation technologies research. *Energy and Building, 133*, 834–842. DOI: 10.1016/j.enbuild.2016.09.067

Alaimo, L. S., Ciommi, M., Vardopoulos, I., Nosova, B., & Salvati, L. (2022). The medium-term impact of the COVID-19 pandemic on population dynamics: The case of Italy. *Sustainability (Basel), 14*(21), 13995. DOI: 10.3390/su142113995

Almaaitah, T., Appleby, M., Rosenblat, H., Drake, J., & Joksimovic, D. (2021). The potential of blue-green infrastructure as a climate change adaptation strategy: A systematic literature review. *Blue-Green Systems, 3*(2), 223–248. DOI: 10.2166/bgs.2021.016

Antonis Tritsis Metropolitan Park. (2017). *Discovering Antonis Tritsis Metropolitan Park: Education for sustainability – Educational material* [in Greek]. Antonis Tritsis Metropolitan Park.

Antonis Tritsis Metropolitan Park. (2023, March 23). Antonis Tritsis Metropolitan Park of environmental and educational activities and social economy development management authority. Verde Tec [in Greek]. https://www.verde-tec.gr/el/readText.asp?textID=806

Antonis Tritsis Metropolitan Park. (2025, March 20). Description [in Greek]. Antonis Tritsis Metropolitan Park. http://www.parkotritsis.gr/perigrafh/

Antonis Tritsis Metropolitan Park Management Authority. (2019). *Actions and perspectives* [in Greek]. Antonis Tritsis Metropolitan Park.

Athens Master Plan Organization. (1995). Environmental Awareness Park of the Queen's Tower Estate: Stage A – Phase B [Study]. Consortium of consultants: Thymios Papayannis and Associates S.A., Ioannis Dokoumetzidis, Ioannis Kolliniatis, Ioannis Alavanos / Christos Athanasopoulos G.P., O.T.M./TEPE, Technical Study Group.

Bambó Naya, R., De la Cal Nicolás, P., Medina, C., Ezquerra, I., Monclús, J., & García-Pérez, S. (2023). Quality of public space and Sustainable Development Goals: Analysis of nine urban projects in Spanish cities. *Frontiers of Architectural Research, 12*(3), 477–495. DOI: 10.1016/j.foar.2023.01.002

Belavilas, N. (2017, June 7–9). Antonis Tritsis Metropolitan Park: Assessment of the current situation – Strategy, directions, proposals [Conference presentation, in Greek]. Scientific and Environmental Three-Day Event, Ilion, Greece. Antonis Tritsis Metropolitan Park.

Biris, K. (2006). The toponyms of the city and the surroundings of Athens [in Greek]. Publications of the Archaeological Bulletin.

Brumana, R., Quilici, S., Oliva, L., Previtali, M., Gabriele, M., & Stanga, C. (2023). Multi-sensor HR mass data models toward multi-temporal-layered digital twins: Maintenance, design and XR informed tour of the multi-stratified Appian Way (PAAA). *Sensors (Basel), 23*(20), 8556. DOI: 10.3390/s23208556 PMID: 37896649

Capelli, G. (2015). La Marrana dell'acqua Mariana: Un corso d'acqua al servizio dei Papi [in Italian]. Acque Sotterranee - Italian Journal of Groundwater, 4(4), 79-82. DOI: 10.7343/as-135-15-0162

Cerema. (2011). Factsheet No. 01b: Rome, a green city – The example of Appia Antica Regional Park. République française – Ministère de l'Environnement. https://paysages-territoires-transitions.cerema.fr/IMG/pdf/blue_green_infra_rome.pdf

Chaurasia, M., Patel, K., Singh, R., Bhadouria, R., & Rao, K. S. (2024). Integrating Blue-Green as next-generation urban infrastructure in developing countries. In Joshi, P. K., Rao, K. S., Bhadouria, R., Tripathi, S., & Singh, R. (Eds.), *Blue-Green infrastructure for sustainable urban settlements*. Springer., DOI: 10.1007/978-3-031-62293-9_3

Chiesura, A. (2004). The role of urban parks for the sustainable city. *Landscape and Urban Planning, 68*(1), 129–138. DOI: 10.1016/j.landurbplan.2003.08.003

D'Agata, A., Vardopoulos, I., & Sateriano, A. (2024). Interdisciplinary insights into sustainable development and socio-environmental change in Mediterranean landscapes. In *Socio-enviromental change: the path toward sustainability* (pp. 1–18). Nova Science Publishers.

Deely, J., Hynes, S., Barquín, J., Burgess, D., Finney, G., Silió, A., Álvarez-Martínez, J. M., Bailly, D., & Ballé-Béganton, J. (2020). Barrier identification framework for the implementation of blue and green infrastructures. *Land Use Policy, 99*, 105108. DOI: 10.1016/j.landusepol.2020.105108

Delladetsimas, P. M., Sapountzaki, K., & Chalkias, C. (2015). Adaptations to the economic crisis: Reallocation of vulnerability and production of new chronic risks. In Maloutas, T., & Spyrellis, S. (Eds.), *Social atlas of Athens: Digital compendium of texts and visual material.*, DOI: 10.17902/20971.39

Doukas, Y. E., Salvati, L., & Vardopoulos, I. (2023). Unraveling the European agricultural policy sustainable development trajectory. *Land (Basel), 12*(9), 1749. DOI: 10.3390/land12091749

Douvas, L. (2007). Leisure parks: Their contribution to tourism development. Administration and Economy, 143–156.

Drosou, N., Soetanto, R., Hermawan, F., Chmutina, K., Bosher, L., & Hatmoko, J. U. D. (2019). Key factors influencing wider adoption of blue–green infrastructure in developing cities. *Water (Basel), 11*(6), 1234. DOI: 10.3390/w11061234

Escrivà Saneugenio, F., Salvati, L., D'Agata, A., & Vardopoulos, I. (2024). From traditional mosaics to urban encroachment: the intrinsic landscapes along urban-rural sustainable interfaces. In *Sprawl and the city: realizing a sustainable Mediterranean urbanization* (pp. 57–82). Nova Science Publishers.

Escrivà Saneugenio, F., Sateriano, A., Scarpitta, D., Salvia, R., Quaranta, G., D'Agata, A., Postigliola, M., & Vardopoulos, I. (2024). Sustainable land management strategies and policies amidst climate- and human-caused land degradation complexities. In *Socio-enviromental change: The path toward sustainability* (pp. 97–120). Nova Science Publishers.

European Commission. (2016). Flash Eurobarometer 419: Quality of life in European cities 2015. https://ec.europa.eu/regional_policy/en/information/publications/studies/2016/quality-of-life-in-european-cities-2015

European Commission. (2025). Green infrastructure. https://environment.ec.europa.eu/topics/nature-and-biodiversity/green-infrastructure_en

European Environment Agency. (2011). Green infrastructure and territorial cohesion: The concept of green infrastructure and its integration into policies using monitoring systems (Report No. 18/2011). https://www.eea.europa.eu/en/analysis/publications/green-infrastructure-and-territorial-cohesion

European Environment Agency. (2022). Urban adaptation in Europe: How cities and towns respond to climate change (EEA Report No. 12/2022). https://www.eea.europa.eu/publications/urban-adaptation-in-europe

European Parliament. (2020). The importance of urban and green infrastructure - European Year of Greener Cities 2022 (Texts adopted, P9_TA(2020)0241). Official Journal of the European Union. https://www.europarl.europa.eu/doceo/document/TA-9-2020-0241_EN.html

EYDAP. (2018). Technical sheet: Construction of compact MBR unit in Antonis Tritsis Park [in Greek]. https://parkotritsis.gr/wp-content/uploads/Ereynhtiko-Yliko/EreynesKaiMeletes/Gnwm.%20ektheseis/Techniko%20Deltio_Compact%20Monada%20MBR_Eydap_2018.pdf

Ferreira, J., Santos, D., & Campos, L. (2024). Blue-green infrastructure in view of Integrated Urban Water Management: A novel assessment of an effectiveness index. *Water Research, 257*, 121658. DOI: 10.1016/j.watres.2024.121658 PMID: 38696983

Giannelli, A., Giuffrida, S., & Trovato, M. (2018). Madrid Río Park: Symbolic values and contingent valuation. *Valori e Valutazioni, 21*, 75–85.

Giokas, K. (2019). The urban heat island phenomenon in western metropolitan Athens: Urban planning tools and natural methods to deal with it [Master's thesis, in Greek]. National Technical University of Athens. http://oldwww.arch.ntua.gr/sites/default/files/project/14369_/gkiokas_konstantinos.pdf

Glantzi, A., & Ramfou, B. (2012). Urban green spaces [Lecture notes, in Greek]. National Technical University of Athens. https://dspace.lib.ntua.gr/xmlui/bitstream/handle/123456789/6594/Glantzi.pdf

Government Gazette of Greece. (1985). The master plan and environmental protection programme of the wider Athens region [FEK A' 18/1985, in Greek].

Government Gazette of Greece (2002). Establishment, organization and operation of a legal entity under private law named “Organization for the Administration and Management of the Antonis Tritsis Environmental Awareness Park” at the Vassilissis Tower site [FEK A' 172/2002, in Greek].

Government Gazette of Greece (2016). New support scheme for electricity generation stations from Renewable Energy Sources and High Efficiency Cogeneration of Heat and Power – Provisions for the legal and functional unbundling of supply and distribution sectors in the natural gas market and other provisions [FEK A' 149/2016, in Greek]

Gunawardena, K. R., Wells, M. J., & Kershaw, T. (2017). Utilising green and bluespace to mitigate urban heat island intensity. *The Science of the Total Environment, 584–585*, 1040–1055. DOI: 10.1016/j.scitotenv.2017.01.158 PMID: 28161043

Hellenic Statistical Authority. (2021). 2021 Population-Housing Census. https://www.statistics.gr/en/2021-census-pop-hous

Hernández-Lamas, P., Gavilán, A., & Bernabeu-Larena, J. (2016). Parks and roads build the cities: The M-30 and Madrid-Río project, building landscape. In Territory, V. C. (Ed.), *Back to the sense of the city: International monograph book* (pp. 415–428). Centre de Política de Sòl i Valoracions. DOI: 10.5821/ctv.8121

Howard, E. (1965). Garden cities of tomorrow. MIT Press. (Original work published 1902) https://www.ipcc.ch/report/ar6/wg2/

Hubbart, J. (1994). *Reuniting man and nature: The need for natural landscapes in urban areas*. Macalester College.

Iamonico, D. (2022). Biodiversity in urban areas: The extraordinary case of Appia Antica Regional Park (Rome, Italy). *Plants*, *11*(16), 2122. DOI: 10.3390/plants11162122 PMID: 36015425

Imbrenda, V., Rontos, K., Salvati, L., Maialetti, M., Vardopoulos, I., Alhussen, A., & Ciaschini, C. (2025). Traditional landscapes in Mediterranean areas: environmental protection vs economic growth. In Ciaschini, C., & Imbrenda, V. (Eds.), *Local development and socio-environmental systems. Experiences and practices toward sustainability in Mediterranean Europe* (pp. 19–38). River Publishers. DOI: 10.1201/9788770047616-2

IPCC. (2022). *Climate change 2022: Impacts, adaptation and vulnerability. Contribution of Working Group II to the Sixth Assessment Report of the Intergovernmental Panel on Climate Change*. Cambridge University Press.

Kaplan, R. (1985). The analysis of perception via preference: A strategy for studying how the environment is experienced. *Landscape Planning*, *12*(2), 161–176. DOI: 10.1016/0304-3924(85)90058-9

Kassios, K., & Perperidou, D. (2005, March 18). The detergent contribution of urban green in the city's atmosphere [Conference presentation, in Greek]. Technical Chamber of Greece Conference “Air quality in urban areas: New data and perspectives”, Athens, Greece.

Katsafadou, K. (2017). Listing report of hydrolytic sources in Antonis Tritsis Metropolitan Park [in Greek]. Antonis Tritsis Metropolitan Park. https://parkotritsis.gr/wp-content/uploads/Ereynhtiko-Yliko/EreynesKaiMeletes/Gnwm.%20ektheseis/Katsafadou_2017.pdf

Kaur, R., & Gupta, K. (2022). Blue-green infrastructure (BGI) network in urban areas for sustainable stormwater management: A geospatial approach. *City and Environment Interactions*, *16*, 100087. DOI: 10.1016/j.cacint.2022.100087

Keramitsoglou, I., Kiranoudis, C. T., Ceriola, G., Weng, Q., & Rajasekar, U. (2011). Identification and analysis of urban surface temperature patterns in Greater Athens, Greece, using MODIS imagery. *Remote Sensing of Environment*, *115*(12), 3080–3090. DOI: 10.1016/j.rse.2011.06.014

Koohestani, S. S., Mukheibir, P., Wakefield-Rann, R., & Santamouris, M. (2025). Adopting a socio-technical perspective on the challenges and barriers in transitioning to Blue-Green Infrastructure (BGI). *Blue-Green Systems*, *7*(1), 79–94. DOI: 10.2166/bgs.2025.011

Kuo, F. E., Bacaicoa, M., & Sullivan, W. C. (1998). Transforming inner-city landscapes: Trees, sense of safety, and preference. *Environment and Behavior*, *30*(1), 28–59. DOI: 10.1177/0013916598301002

Lamond, J., & Everett, G. (2019). Sustainable blue-green infrastructure: A social practice approach to understanding community preferences and stewardship. *Landscape and Urban Planning*, *191*, 103639. DOI: 10.1016/j.landurbplan.2019.103639

Latsoudis, P., Beriatos, G., & Probonas, N. (2017). Antonis Tritsis Environmental Information and Awareness Park's avifauna: Species-habitat management proposals [in Greek]. Hellenic Ornithological Society. https://parkotritsis.gr/wp-content/uploads/Ereynhtiko-Yliko/EreynesKaiMeletes/Gnwm.%20ektheseis/Latsoudis_Ornithopanida_2017.pdf

Latsoudis, P., Beriatos, G., & Probonas, N. (2023). List of bird species in the Antonis Tritsis Park [in Greek]. Hellenic Ornithological Society. https://old.ornithologiki.gr/page_list.php?lID=5&bc=1970

Loures, L., Santos, R., & Panagopoulos, T. (2007). Urban parks and sustainable development: The case study of Portimão city [Conference paper]. Proceedings of the 3rd IASME/WSEAS International Conference on Energy, Environment, Ecosystems and Sustainable Development, Agios Nikolaos, Greece. University of Algarve.

Madrid City Council. (2025). Madrid Río. https://madrides.es/en/madrid-rio/

Madrid City Council. (2025, March 17). Parque Madrid Río. https://www.madrid.es/portales/munimadrid/es/Inicio/El-Ayuntamiento/Arganzuela/Direcciones-y-telefonos/Parque-Madrid-Rio/

Maialetti, M., Halbac-Cotoara-Zamfir, R., Vardopoulos, I., & Salvati, L. (2024). Early desertification risk in advanced economies: Summarizing past, present and future trends in Italy. *Earth (Basel, Switzerland)*, *5*(4), 690–706. DOI: 10.3390/earth5040036

McPherson, G., Simpson, J. R., Peper, P. J., Maco, S. E., Gardner, S. L., Cozad, S. K., & Xiao, Q. (2005). Municipal tree resource analysis. U.S. Department of Agriculture, Forest Service, Pacific Southwest Research Station.

Mediterranean Experts on Climate and Environmental Change (MedECC). (2020). Climate and environmental change in the Mediterranean Basin – Current situation and risks for the future (First Mediterranean Assessment Report). https://www.medecc.org/first-mediterranean-assessment-report-mar1/

Melissas, D. (2009). *Urban greenery and its management by local government* [in Greek]. Institute of Local Government.

O'Donnell, E. C., Lamond, J. E., & Thorne, C. R. (2017). Recognising barriers to implementation of Blue-Green Infrastructure: A Newcastle case study. *Urban Water Journal, 14*(9), 964–971. DOI: 10.1080/1573062X.2017.1279190

O'Donnell, E. C., Netusil, N. R., Chan, F. K. S., Dolman, N. J., & Gosling, S. N. (2021). International perceptions of urban blue-green infrastructure: A comparison across four cities. *Water (Basel), 13*(4), 544. DOI: 10.3390/w13040544

Oke, T. R. (1987). *Boundary layer climates* (2nd ed.). Routledge.

Orsi, N. (2013). The engineering behind the Via Appia. Engineering Rome. https://engineeringrome.org/the-engineering-behind-the-via-appia/

Perini, K. (2017). Madrid Río, Spain: Opportunities and policies. In K. Perini & P. Sabbion (Eds.), Urban sustainability and river restoration: Green and blue infrastructure (pp. 227–234). Wiley.

Perini, K. (2017). Madrid Río, Spain: Strategies and techniques. In K. Perini & P. Sabbion (Eds.), Urban sustainability and river restoration: Green and blue infrastructure (pp. 117–126). Wiley.

Polyzos, I. (2016, October 7). Sustainable urban development policies in metropolitan Athens: Urban avenues and metropolitan rings [Conference presentation, in Greek]. Metropolitan Interventions in West Athens Conference: Antonis Tritsis Park & Mount Egaleo-Poikilo, Peristeri, Greece.

Polyzos, I., Balabanidis, D., & Triantis, L. (2014). Sustainable urban development policies in metropolitan Athens: The strategy of integrated urban interventions - urban avenues/green routes/cultural routes [in Greek]. Region of Attica - National Technical University of Athens.

Puppim de Oliveira, J. A., Bellezoni, R. A., Shih, W., & Bayulken, B. (2022). Innovations in Urban Green and Blue Infrastructure: Tackling local and global challenges in cities. *Journal of Cleaner Production, 362*, 132355. DOI: 10.1016/j.jclepro.2022.132355

Region of Attica. (2025). Places of refuge for citizens in the event of an earthquake [in Greek]. http://www.patt.gov.gr/site/index.php?option=com_content&view=article&id=6562&Itemid=36

Regionale dell'Appia Antica, P. (2025). Parco Regionale dell'Appia Antica [in Italian]. https://www.parcoappiaantica.it

Ribeiro Hoffmann, A., Sandrin, P., & Doukas, Y. E. (Eds.). (2024). *Climate change in regional perspective. European Union and Latin American initiatives, challenges, and solutions*. Cham, Switzerland: United Nations University Series on Regionalism, Vol. 27, Springer Nature. DOI: 10.1007/978-3-031-49329-4

Santo-Tomás Muro, R., & Rodríguez Romero, E. J. (2020). City thresholds: The role of urban green infrastructures in Madrid. *ZARCH*, *14*(14), 176–187. DOI: 10.26754/ojs_zarch/zarch.2020144289

Sapountzaki, K., & Dandoulaki, M. (2015). *Hazards and disasters: Concepts and tools for evaluation, protection, and management* [in Greek]. Association of Greek Academic Libraries.

Simoniello, T., Lanfredi, M., Coluzzi, R., D'Emilio, M., Salvati, L., Vardopoulos, I., Imbrenda, V., & Samela, C. (2025). Climate-driven dynamics in the Mediterranean agroecosystems: a pre-2007 economic crisis satellite-based sustainability analysis. In *Environmental Sustainability and Global Change* (pp. 57–68). Elsevier., DOI: 10.1016/B978-0-443-31596-1.00005-2

Spronken-Smith, R. A., Oke, T. R., & Lowry, W. P. (2000). Advection and the surface energy balance across an irrigated urban park. *International Journal of Climatology*, *20*(9), 1033–1047. DOI: 10.1002/1097-0088(200007)20:9<1033::AID-JOC508>3.0.CO;2-U

Stamoulidis, K., & Stafylas, P. (2017). The area of the Queen's Tower–Tritsis Park in Athens: Strategic management for the formation of a recognizable identity with metropolitan–supra-local reach [Conference presentation]. 2nd Panhellenic Marketing Conference & Site Branding, University of Thessaly, Larissa, Greece.

Statista. (2025). Largest urban agglomerations in Europe in 2023. https://www.statista.com/statistics/1101883/largest-european-cities/

Surrey County Council. (2023). Green and blue infrastructure: Best practice and case studies [Policy report]. https://www.surreycc.gov.uk/community/climate-change/what-are-we-doing/green-and-blue-infrastructure

Tallini, M., Di Leo, A., Rossetti, C., & Berardi, F. (2013). The Sacred Almone River of the Appian Way Regional Park in Rome: Ancient myths, a ritual link between Rome and Asia Minor and water uses in the modern age. *Water Science and Technology: Water Supply, 13*(3), 727–734. DOI: 10.2166/ws.2013.098

Thorne, C. R., Lawson, E. C., Ozawa, C., Hamlin, S. L., & Smith, L. A. (2015). Overcoming uncertainty and barriers to adoption of Blue-Green Infrastructure for urban flood risk management. *Journal of Flood Risk Management, 11*(S2), S960–S972. DOI: 10.1111/jfr3.12218

Tsigdinos, S. (2017). Aspects of the urban landscape and public space: City, nature and new technologies - Tracing accessibility in the Antonis Tritsis Metropolitan Park [Master's thesis, in Greek] National Technical University of Athens. http://oldwww.arch.ntua.gr/sites/default/files/project/11940_/opseis_tsigdinos_2017.pdf

Ulrich, R. S. (1981). Natural versus urban scenes: Some psychophysiological effects. *Environment and Behavior, 13*(5), 523–556. DOI: 10.1177/0013916581135001

Urban Sustainability Exchange. (2025). The Madrid Río project. https://use.metropolis.org/case-studies/the-madrid-rio-project

Vardopoulos, I., D'Agata, A., Escrivà Saneugenio, F., & Salvati, L. (2024). *Sprawl and the city: realizing a sustainable Mediterranean urbanization.* Nova Science Publishers., DOI: 10.52305/GTVV1244

Vardopoulos, I., Escrivà Saneugenio, F., Sateriano, A., & Salvati, L. (2024). *Homage (and criticism) to the Mediterranean city. Regional sustainability and economic resilience.* River Publishers., DOI: 10.1201/9781003517306

Vardopoulos, I., Georgiou, C., Voukkali, I., Vassiliades, C., & Salvati, L. (2024). Case-based exploration of the nexus between transformative urban megadevelopments and sustainability. *Theoretical and Empirical Researches in Urban Management, 19*(3), 5–30. https://www.jstor.org/stable/27321372

Vardopoulos, I., Santamouris, M., Zorpas, A. A., Barone, G., Italos, C., & Vassiliades, C. (2024). A comparative study on discrepancies in residential building energy performance certification in a Mediterranean context. *Buildings (Basel, Switzerland), 14*(4), 1009. DOI: 10.3390/buildings14041009

Vasilopoulou, C. (2017). Water-sensitive urban design and natural water infrastructure: The case of Antonis Tritsis Park [Master's thesis, in Greek]. National Technical University of Athens. https://dspace.lib.ntua.gr/xmlui/handle/123456789/46433

Vassiliades, C., Minterides, C., Astara, O. E., Barone, G., & Vardopoulos, I. (2023). Socio-economic barriers to adopting energy-saving bioclimatic strategies in a Mediterranean sustainable real estate setting: A quantitative analysis of resident perspectives. *Energies, 16*(24), 7952. DOI: 10.3390/en16247952

Vinci, S., Vardopoulos, I., & Salvati, L. (2023). A tale of a shrinking City? Exploring the complex interplay of socio-demographic dynamics in the recent development of Attica, Greece. *Cities (London, England), 132*, 104089. DOI: 10.1016/j.cities.2022.104089

Wang, J., Liu, J., Wang, H., & Mei, C. (2020). Approaches to multi-objective optimization and assessment of green infrastructure and their multi-functional effectiveness: A review. *Water (Basel), 12*(10), 2714. DOI: 10.3390/w12102714

Wassenhoven, L. (2018). *Hellinikon passion: The development of the former Hellinikon Airport* [in Greek]. Sakkoulas.

WEST8. (2025, March 18). Vision for the riverfront. https://www.west8.com/projects/madrid-rio/

World Health Organization. (2016). Urban green spaces and health: A review of evidence. WHO Regional Office for Europe. https://iris.who.int/handle/10665/345751

Yiannakou, A., & Salata, K. (2018). Green infrastructure: An overview of its contribution to urban areas and an investigation of the provisions in the Greek planning system [in Greek]. Aichoros: Texts on Spatial Planning, Urban Planning and Development, 1(1), 43–75. DOI: 10.26253/heal.uth.ojs.aei.2018.435

KEY TERMS AND DEFINITIONS

Biodiversity: The variety of living organisms, including plants, animals and microorganisms, their genetic characteristics and the ecological communities they form.

Blue-Green Infrastructure (BGI): A strategically planned network of water (blue) and vegetation (green) elements in urban areas designed to deliver multiple ecosystem services while enhancing climate resilience.

Climate Change Hazards (CC Hazards): The adverse physical events or conditions intensified or triggered by climate change, including extreme heat, drought, heavy rainfall, flooding, sea-level rise and biodiversity loss, which pose risks to human health, infrastructure, ecosystems and urban systems

Ecosystem Services: The benefits that people derive – directly and indirectly – from natural and semi-natural systems, including air and water purification, climate regulation, flood mitigation and opportunities for recreation and mental well-being.

Hydrological Restoration: The process of re-establishing natural water flows and retention capacities – through measures like river re-wilding, wetland creation or stormwater infrastructure – to improve water quality, reduce flood risk and support aquatic habitats.

Metropolitan Park: A large-scale urban open space that integrates recreational, ecological and cultural functions to serve an entire metropolitan region rather than a single neighborhood.

Participatory Planning: An inclusive approach to urban design and policy development in which local communities, NGOs, experts and government bodies co-create visions, identify priorities and share stewardship responsibilities throughout a project.

Resilience: The capacity of urban systems – and the communities that inhabit them – to anticipate, absorb, recover from and adapt to climate-related shocks and stresses, while maintaining essential functions and structures.

Urban Heat Island (UHI): The phenomenon by which built-up areas experience higher temperatures than surrounding rural zones, driven by factors such as reduced vegetation cover, heat-absorbing materials and anthropogenic heat emissions.

Chapter 7

From Trash to Treasure: Innovations in Waste Management Towards Sustainable Energy Solutions

Ana Ramos
https://orcid.org/0000-0003-2185-1341
INEGI, Portugal

ABSTRACT

Urban waste management is a critical global issue, driven by rapid urbanization, population growth, and increasing waste generation. The United Nations projects that, by 2050, nearly 70% of the world's population will reside in cities, making effective waste strategies vital for sustainability, public health, and economic resilience. This chapter, "From Trash to Treasure: Innovations in Waste Management towards Sustainable Energy Solutions," explores waste-to-energy (WtE) technologies and other innovative approaches to address urban waste challenges. It provides an overview of cutting-edge solutions that transform waste into resources while minimizing environmental impacts. The chapter examines how WtE, aligned with circular economy principles, helps to form the backbone of sustainable waste management. Addressing technical, environmental, and socioeconomic aspects, it offers actionable insights for implementation and highlights the role of innovation in advancing urban sustainability and the global transition to a low-carbon future.

DOI: 10.4018/979-8-3373-2327-5.ch007

1. INTRODUCTION

The Global Waste Challenge: Trends, Drivers, and Impacts

The global waste challenge is a defining issue of the 21st century, exacerbated by rapid urbanization (Vardopoulos et al., 2024), population growth (Vinci et al., 2023), and unsustainable consumption patterns (Vardopoulos et al., 2025). It represents one of the most significant threats to sustainability, with implications for public health, environmental protection, and resource efficiency. According to the World Bank, municipal solid waste (MSW) generation is projected to increase from 2.01 billion tons per year in 2016 to 3.40 billion tons by 2050, with low- and middle-income countries expected to experience the most significant increases (Kaza, Yao et al. 2018). The environmental impacts of inadequate waste management are numerous and severe (Vardopoulos et al., 2021), namely regarding consequences for both public health and the global environment, with devastating effects on marine life, biodiversity and ecosystems in general (Ozbay, Jones et al. 2021). Inadequate disposal can cause pollution of rivers, oceans and soil, exposure to hazardous materials and the release of toxic gases, which are harmful to human health (Ojuri, Ayodele et al. 2018). Improper disposal sites with standing water, such as landfills, can create conditions for the spread of malaria, waste handlers being in the forefront of the concern (Egid 2024), while sanitary (controlled) landfills are designed with engineering to minimize environmental impacts, such as contamination of soil and water resources, unlike dumps (Arjwech, Somchat et al. 2020, Egid 2024).

Implementing good waste management practices can mitigate these risks, and promote healthier societies (Loizia et al., 2021). These practices include proper waste separation, daily soiling to reduce odor and vector attraction, monitoring and control of leachate, water treatment and gas capture/flaring technologies also reducing health risks. Technologies such as mechanical, biochemical (biodigesters, composting) and thermal (incineration, energy recovery) treatments are used to reduce the negative impact of waste on the environment and human health. Solid waste management is an important part of the global sustainable development agenda, contributing to the achievement of health and well-being goals (United Nations 2016). Suitable management contributes to the reduction of greenhouse gas emissions, while propelling the conservation of natural resources, reducing the need for extraction and production of new materials. Many countries have strict regulations for waste management, and international agreements emphasize the importance of such management in mitigating climate change. Proper solid waste management is a key component of the United Nations Sustainable Development Goals (SDGs) related to sustainable cities, clean water, life on land and climate action (Ram and Bracci 2024).

Landfills are a major source of methane emissions, a potent greenhouse gas contributing to climate change. Recent assessments by the United Nations indicate that waste-related emissions have escalated, particularly in rapidly urbanizing regions, underscoring an urgent need for strategic interventions (Küfeoğlu 2024). The environmental, economic, and social impacts of this surge in waste are profound, spanning the contamination of land and waterways, heightened public health risks, and the intensification of greenhouse gas emissions, placing immense pressure on existing waste management systems and has far-reaching consequences for air, water, and soil quality. Key drivers contributing to the waste crisis include population growth, rising consumption patterns, and technological advancements that promote planned obsolescence in consumer goods, as well as mounting living standards, inadequate waste management infrastructure, and the proliferation of non-biodegradable materials such as plastics, among other entities (Hossain, Law et al. 2022). Plastic waste pollution poses a threat to marine biodiversity, disrupts ecosystems, and imposes economic costs on coastal communities, with an estimated 8 million metric tons entering oceans each year (Jambeck, Geyer et al. 2015). Recent findings from Ocean Conservancy suggest these figures may be conservative, as unregulated dumping continues to rise globally. Moreover, the adoption of single-use plastics, inefficient resource recovery systems, and poorly enforced regulations exacerbate the issue (Baechler, De Frond et al. 2024). Despite heightened awareness and initiatives to promote recycling, global plastic waste production continues to rise, with only 9% of plastic waste recycled as of 2015 (Geyer, Jambeck et al. 2017). The linear "take-make-dispose" economic model has proven unsustainable and, in response, there is growing momentum toward adopting circular economy principles, where resources are reused, repurposed, and recycled to minimize waste (Potting, Hekkert et al. 2017, Zhang, Niu et al. 2022). Recycling and reuse contribute to this, keeping materials in use for as long as possible, reducing environmental impact and minimizing waste production. The economic impacts are also significant, as poor waste management contributes to higher municipal expenses, loss of valuable materials, and diminished tourism revenue. Projections suggest that a circular economy model can generate significant global net gains (Schulze 2016).

Addressing the global waste challenge is crucial for safeguarding the environment and building resilient societies. However, these efforts alone may not suffice to address the sheer magnitude of the global waste challenge. Furthermore, the impacts of poor waste management disproportionately affect vulnerable communities, contributing to social inequities. The consequences of inadequate waste management are multifaceted. Without urgent intervention, the waste crisis will continue to undermine global efforts to achieve environmental sustainability and social well-being (Ramos and Rouboa 2020).

The Urgency of Innovative Solutions to Tackle Waste Management Crises

Waste management worldwide faces huge challenges, namely being historically, an inconsistent and inefficient process, which tends to stagnate with conventional collective disposal models, with many countries failing to meet the targets set in previous plans. Globally, a significant portion of waste is not collected or managed in controlled facilities, resulting in a considerable annual net cost, while promoting conscious consumption practices, designing more sustainable products (ecodesign) and combating waste production are compulsory (Wilson 2007, Hoornweg and Bhada-Tata 2012, Zorpas 2020). The challenge becomes more complex with the obligation to introduce new collection streams, such as food waste (Vardopoulos et al., 2025), textiles (Papamichael et al., 2023) and small household hazardous waste (Vardopoulos et al., 2023). The lack of adequate infrastructure and resources for adequate collection, transportation and treatment of solid waste leads to the accumulation of garbage in public spaces and homes (Kyramargiou and Vardopoulos, 2019), increasing the risk of disease and impacting the environment (Wilson, Rodic et al. 2015, Winkler, Jackson et al. 2017). Bio-wastes, which represents a significant part of urban waste, are also challenging due to the characteristics of food waste (moist and odors attract vectors, especially in hot climates) (Vardopoulos et al., 2025). Electronic waste is also a tricky stream in some cases, as well as hazardous waste, that requires specialized treatment. The lack of dedicated collection schemes leads to landfill clogs and contamination (Kaza, Yao et al. 2018). Relative to processes, technological dependence especially in energy recovery processes, and obstacles to financing imported equipment are among the main issues encountered, backed by the logistical inefficacy (Ramos, Berzosa et al. 2020a, Ramos and Rouboa 2022). Reverse logistics needs to be implemented more effectively, with insufficient information and awareness among the population and problems with inspection and public management.

Transformative waste management solutions are more urgent than ever. Traditional landfill-based disposal methods are no longer viable due to diminishing land availability and their adverse environmental impacts. The climate crisis amplifies this urgency (Ribeiro Hoffmann et al., 2024), as landfills emit significant quantities of methane—a potent greenhouse gas (Outapa and Roi-et 2018). Waste policies have stopped prioritizing the construction of landfills, partly due to the lack of EU funding. The economic and financing aspects, including finding a balance between waste tariff revenues and operating costs to ensure economic sustainability is also an issue (Ramos, Berzosa et al. 2020a). The ambitious target of reducing the amount of waste deposited in landfills to a maximum of 10% by 2030 in the European Union puts additional pressure on the management of the remaining fraction.

Despite the challenges, there are numerous opportunities in waste management, many of which are driven by the promotion of sustainable practices, such as rethinking systems to make them closer and more convenient, encouraging conscious consumption, ecodesign and preventing food waste; encouraging home and community composting, as well as repairing and exchange programs. Citizen involvement is key (Moustairas et al., 2022), as communication and awareness campaigns are not always reflected in the amounts of recyclables collected (Eliades et al., 2022). Lack of awareness among the population can hinder the management process and the implementation of effective public policies (Etim 2024, Kumari, Ansari et al. 2025). Communication channels must be clear and accessible.

Innovation in waste management must encompass technological, policy, and financial dimensions (Pavlides et al., 2023). Addressing the global waste crisis requires a paradigm shift from the traditional "take-make-dispose" linear economy to a circular economy, where resources are reused, repurposed, and recycled. However, recycling and waste minimization alone are insufficient to manage the sheer volume of waste generated globally. Furthermore, the ongoing climate crisis demands solutions that can mitigate greenhouse gas emissions while providing additional benefits, such as renewable energy production (Ramos and Rouboa 2020, Ramos 2025). This underscores the need for innovative solutions, such as Waste-to-Energy (WtE) technologies, which have emerged as innovative to address these challenges, as they can convert waste into usable energy while reducing the volume of residual waste (Guberman 2021, Dastjerdi, Strezov et al. 2022). Figure 1 shows the myriad of possibilities from the application of WtE.

Figure 1. Waste-to-Energy possibilities.

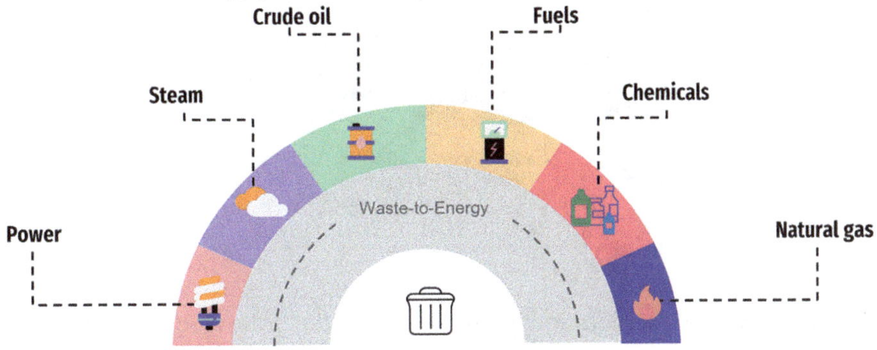

These solutions have the potential to complement recycling and composting efforts, offering a pathway toward more sustainable urban environments. Moreover, advancements in WtE technologies have addressed many historical concerns related to emissions and environmental impacts, making them a viable component of sustainable waste management strategies. In this context, it becomes imperative to explore the technological advancements, case studies, and policy frameworks that can drive the adoption of WtE solutions on a global scale.

2. OVERVIEW OF WASTE-TO-ENERGY TECHNOLOGIES

Incineration: Efficiency Improvements and Emission Control

Incineration is one of the most known thermal treatment technologies, playing a crucial role in solid waste management, especially for MSW, as it drastically reduces waste volumes. Incineration is one of the most established WtE technologies, as the combustion of waste organic and inorganic materials generates heat, which can be used for electricity production or district heating, energy-based commodities required worldwide (Neuwahl, Cusano et al. 2019). Figure 2 shows a graphical scheme of the technique. It is a physicochemical process that transforms matter by increasing its temperature, through the combustion of waste in an oxygen-rich environment. This produces steam, carbon dioxide and heat, generating hot gases and bottom ash.

Figure 2. Incineration process.

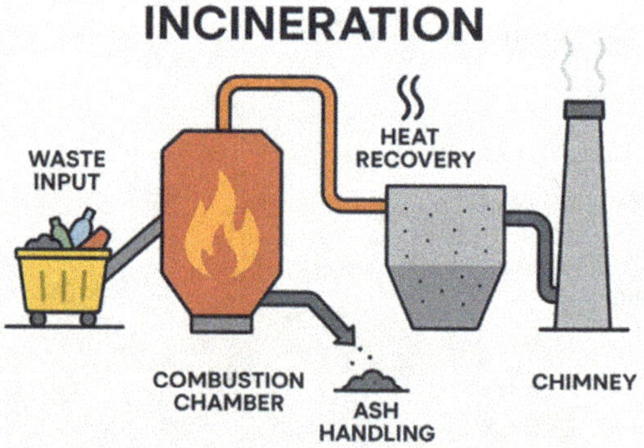

A typical waste incineration plant includes the following operations: waste storage and preparation; waste combustion in a furnace, due to rapid exothermic reactions between the waste and oxygen in the air, acting as oxidant, producing hot gases and bottom ash for disposal or revalorization; temperature decrease for the gases, often with heat recovery through steam generation (Ramos, Afonso Teixeira et al. 2018). MSW incineration plants can generate electricity by converting the heat of combustion into steam to drive a turbine. Flue gases are cooled to remove atmospheric pollutants and disposal of the waste from this treatment. Lastly, the treated gasses are dispersed into the atmosphere through a chimney.

Incineration is seen as an efficient way to reduce the volume of waste and the need for landfill space, turning what is currently a problem into a solution, since controlled combustion can generate energy (Münster and Meibom 2011, Bosmans, Vanderreydt et al. 2013). Developed countries have demonstrated the viability of incineration for waste recovery, including energy generation.

Historically, incineration plants from the 1980s and 1990s were significant sources of pollution. Nevertheless, modern incineration facilities have made significant strides in improving energy recovery efficiency and reducing emissions through advanced technologies (Tang, Dong et al. 2020). Key advancements include flue gas cleaning systems, which remove harmful pollutants from combustion gases, and selective catalytic reduction (SCR) systems, which reduce nitrogen oxide emissions. Continuous emissions monitoring systems ensure compliance with stringent environmental regulations. High-efficiency incineration plants can achieve net electrical efficiencies of up to 30%, and play a pivotal role in managing residual waste that cannot be economically recycled (Ozturk and Dincer 2020). Despite these advancements, incineration faces criticism due to concerns about air pollution and ash disposal. Bottom ash, a byproduct of incineration, can be processed and used as construction material, while fly ash requires careful handling due to its hazardous nature (Vassilev, Baxter et al. 2013).

Although the current stage of advancements has led to remarkable progress, operating waste incineration plants comes with its own set of challenges (Neehaul, Jeetah et al. 2020). These include efficiency considerations, particularly combustion efficiency and energy recovery, which can be improved through several approaches. Optimizing operational control helps to maintain stable operating conditions without major fluctuations in waste feed rate, combustion air flows, or other conditions (Ramos, Monteiro et al. 2018). Temperature, gas mixture turbulence, and gas residence time at the incineration temperature are critical to efficient combustion. Maintaining optimum temperature for instance with auxiliary burners, aids to keep the minimum required furnace temperature, especially with the heterogeneity of MSW, burning fossil fuels during start-up, shutdown, and upsets (Ramos, Monteiro et al. 2018). Controlling the amount of material fed and varying excess air helps maintain temperature (Sun, Meng et al. 2023). Excess air control is important for good burning, whereas advanced technological design can also leverage the use of improved furnace designs such as new grate systems to agitate the waste, exposing a larger surface area to air and high temperatures to aid in complete combustion (Mullinger and Jenkins 2022). Furnace design with arches and bull noses helps to extend the residence time of gases in the high temperature zone and promotes mixing with air. Proper sizing of the furnace for the amount of waste is also very important, supporting process efficiency and sufficiency. Waste prevention and preparation through screening incoming waste to the facility aids in reducing the incineration of non-combustible or pollutant-producing materials (such as batteries or other electronic debris) (Ibanez, Andres et al. 2000). While source selection to reduce ash content is impractical for most, pretreatment practices such as shredding and blending of waste can be beneficial. Regarding emissions, incineration of solid waste produces a variety of air pollutants, such as sulfur dioxide, nitrogen oxides (NOx), particulate matter, dioxins and furans. Emission control is crucial to protecting the environment and public health (National Research Council, 2000). Emissions arise from the composition of the residue, incomplete combustion and reforming reactions during gas cooling or in air pollution control equipment. Monitoring and controlling the process with computerized systems to collect and record process data, automatically control operating conditions (such as combustion temperature) by varying fuel feed and air flows (Tavares, Zsigraiová et al. 2011, Ramos, Monteiro et al. 2018). Continuous monitoring of emissions and process characteristics provides real-time feedback and facilitates the maintenance of optimal operating conditions. By implementing these and other solutions, operators can improve incineration efficiency, including energy efficiency, reducing the need for external energy sources and lowering operating costs.

There are several strategies and technologies to control emissions, such as good combustion practices, rapid gas cooling, air pollution control equipment (APCs), fugitive emission control, continuous emissions monitoring, and operator training and certification. Optimizing furnace operations improves combustion efficiency by reducing the amount of soot and the formation of incomplete combustion products such as dioxins and furans (Mullinger and Jenkins 2022). Controlling oxygen and temperature within ideal ranges, with turbulent conditions as well as promoting flue gas recirculation, minimizes the formation of NOx. Rapidly cooling flue gases to lower temperatures (typically 285°F to 300°F at the inlet) may reduce secondary formation of dioxins and furans and improve removal of mercury and acid gases. Heat recovery boilers or water quenchers are used to this outcome. Regarding APCs for particulate matter, baghouses, wet and dry electrostatic precipitators (ESPs), and inertial impact wet scrubbers (such as venturi scrubbers) are effective, while bag filters and ESPs are touted as effective for fine particles; acid gases (HCl, SO_2) require scrubbers, spray dryer absorbers and dry sorbent injection with downstream bag filters (Wang, Weller et al. 2008, Ruiz, Juarez et al. 2013). Volatile Organic Compounds (VOCs) may be controlled by thermal or catalytic oxidation and adsorption by activated carbon, being also reduced by good combustion practices. In addition to combustion modifications Nitrogen Oxides (NOx) may benefit from add-on technologies including Selective Catalytic Reduction (SCR) and Selective

Non-Catalytic Reduction (SNCR) (Mahmoudi, Baeyens et al. 2010). SCR involves injecting ammonia in the presence of a catalyst at about 600°F, whereas SNCR injects ammonia or urea into a specific temperature zone without a catalyst. Low NOx and ultra-low NOx burners are also used, while for dioxins/furans and mercury, the injection of powdered activated carbon into the gas stream prior to APC for particulate matter is widely used and effective. Activated carbon is also applied to adsorb pollutants, which are then captured in the APC. Granular activated carbon filters can also be used. The effectiveness of activated carbon injection depends on the type of carbon, dosage, temperature and gas-solid contact (Wee, Bae et al. 2022). Wet scrubbers operating at low temperatures can also be effective. Concerning fugitive emission control, in modern facilities, these vapors or particles from waste and ash handling are mitigated by designing buildings under negative pressure, drawing air from the handling areas into the combustion chamber. Enclosed ash handling areas, with air drawn through APCs, minimize worker exposure and dust leakage. External fugitive dust (e.g. from biomass piles) can be controlled by humidification, chemical stabilization or wind barriers. Continuous emissions monitoring also allows maintaining optimal operating conditions for emissions control, while operator training and certifications are essential to ensuring that furnaces and APCs operate optimally and efficiently.

Anaerobic Digestion for Biogas Production

Anaerobic digestion (AD) is another example of waste-to-energy techniques that contributes to sustainable energy goals, while supporting circular bioeconomy. It is based on the conversion of waste into renewable biogas, therefore enabling several sectors to benefit from this transformation. The global biogas production has been increasing, with a remarkable growth of ~90% between 2010 and 2019 (Piechota, Generowicz et al. 2024). Europe has been a major contributor to global production, countries such as Germany, the United Kingdom, France, Denmark and the Netherlands playing a significant role in the production and use (Gustafsson and Anderberg 2022). Over the past 40 years, facilities in the European Union and North America have developed more than in other continents, standing out for their industrial scale, energy efficiency and complexity. More recently, China and India have shown the highest consumption rate, driven by updated policies and procedures (Abanades, Abbaspour et al. 2022).

AD contributes to these biogas market needs by recovering the energy contained in biomass into a useful fuel that can be stored and transported. Biomethane production from waste recovery has increased significantly. It is estimated that the current potential for biomethane production is over 700 Mtoe worldwide (International Energy Agency 2020, International Energy Agency 2022). The applications and contribution of biogas to the national energy sector vary between countries. Global policies and programs have promoted the continuous production of biogas.

Anaerobic digestion is a biological process that decomposes organic waste in the absence of oxygen, producing biogas (primarily methane and carbon dioxide) and digestate (a nutrient-rich byproduct). Biogas can be used as a renewable energy source for electricity, heating, or as vehicle fuel after upgrading to biomethane (Pullen 2015). Figure 3 represents the general schematics for the technique.

Figure 3. Anaerobic digestion scheme.

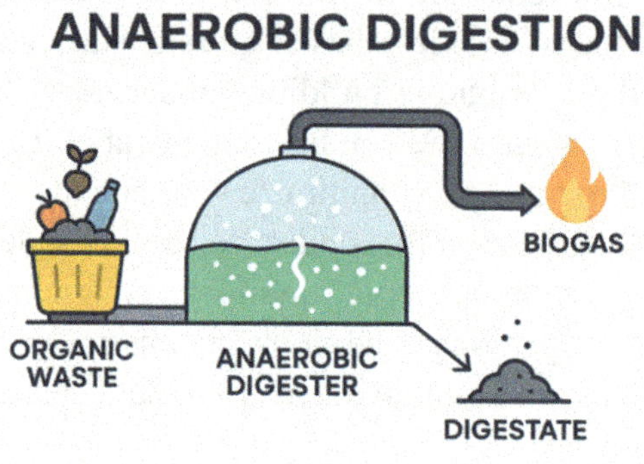

AD is an essential biological process, catalyzed by the action of anaerobic microorganisms that live in environments without molecular oxygen. It is a thermal (or, more precisely, biological treatment with energy output) technology, which enables the conversion of biodegradable organic compounds, such as MSW (specifically the organic fraction), sludge from wastewater treatment plants (WWTP), agricultural and crop residues, food industry residues, excrement and animal manure, among other materials, into biogas (Náthia-Neves, Berni et al. 2018). This complex process involves an interaction of different microorganisms, decomposing the organic matter until its final form, which consists mainly of methane (CH_4) and carbon dioxide (CO_2). It is considered a promising technology in solid waste management and effluent treatment (De Bere 2000).

The anaerobic digestion process occurs in several important stages, including hydrolysis, acidogenesis, acetogenesis and methanogenesis. Methanogenesis is the stage in which methane is produced. It is considered the rate-limiting stage in AD. Methanogenesis is carried out by methanogenic groups belonging to the domain *Archaea*, which are particularly sensitive to the process conditions and the substrate used (Wellinger, Murphy et al. 2013). There are two main pathways for methanogenesis: the direct cleavage of acetate into CH_4 and CO_2 (aceticlastic methanogenesis) and the reduction of CO_2 with hydrogen gas (hydrogenotrophic methanogenesis). The proportion of each pathway depends on the inoculum, the organic substrate and the process conditions. For the success of AD, the stable development of the methanogenic microbial community is crucial. This consortium of anaerobic microorganisms includes fermentative, syntrophic acetogenic and methanogenic bacteria. The association of different bacteria (aceticlastic and hydrogenotrophic methanogens, syntrophic acetogens, fermentative bacteria and homoacetogens) is necessary to develop methanogenic fermentation (Braga, Motteran et al. 2018). The balance between these groups increases the efficiency of biogas production and the AD process. The gaseous product of microbial activity is biogas, which composition varies depending on the type of feedstock used and the method of conducting the process. Generally, biogas consists mainly of methane (CH_4), representing 50-75%, and carbon dioxide (CO_2), representing 25-45%. Other components present in small amounts (0-2%) may include hydrogen sulphide (H_2S), hydrogen (H_2), carbon monoxide (CO), nitrogen (N_2), ammonia (NH_3) and water vapour (H_2O) (Wellinger, Murphy et al. 2013).

Unlike fossil fuels, biogas is inherently renewable as it is generated from biomass, which is a reserve of solar energy captured by photosynthesis. AD not only improves the state of a country's energy panorama but also endorses significantly the conservation of natural resources and the protection of the environment. Biogas production through AD is considered a promising renewable energy source, contributing to the current energy landscape in several ways, offering a multipurpose solution to generate the energy needed for the industrial or social sectors (Zabed, Akter et al. 2020). Applications include heat and electricity generation, cogeneration in combined heat and electricity plants (CHP), hydrogen production, use in advanced energy systems such as fuel cells, transportation fuel, usually in the form of biomethane, after treatment/purification of biogas. Biofuel production is considered a key source for achieving renewable energy contribution targets in the transport sector. AD is particularly suitable for processing food waste, agricultural residues, and other biodegradable materials (Ramos, Monteiro et al. 2022). Its integration into waste management systems can reduce methane emissions from landfills while providing a renewable energy source. In addition to biogas production, AD produces digestate, a material that can be used as a fertilizer and soil conditioner for agricultural purposes (Kassem, Hockey et al. 2020). AD also contributes to the reduction of greenhouse gas (GHG) emissions and pollution associated with waste disposal. The process allows the recycling of biomass, including nutrients, contributing to a sustainable circular economy.

The efficiency of the AD process and the resulting biogas production are influenced by several factors, including the environmental and operational conditions of the system, the nature of the substrate and the characteristics of the biodigester (Wellinger,

Murphy et al. 2013). Sudden changes in these factors can cause imbalances in the process and negatively affect performance, and may even lead to its interruption. To optimize the process, it is important to maintain stable operating conditions; select and manage the inoculum correctly, especially in the start-up phase, to avoid acidification; consider substrate pretreatments to improve biodegradability and increase biogas yield, particularly for lignocellulosic materials; optimize the design parameters of biodigesters; maintain adequate control of microbial communities; use additives, if necessary, to provide optimal conditions for bacteria; improve process monitoring and control, including management of microbial resources, to optimize productivity and reduce methane losses; consider the use of advanced technologies, such as membrane bioreactors (AnMBRs), which can increase the production quantity due to membrane specifications and high cell concentration; integrate systems (such as solar drying, heat recovery) to manage by-products and maintain optimal temperatures (Wellinger, Murphy et al. 2013). Recent advancements in membrane technologies have improved biogas purification processes, enhancing the economic feasibility of AD systems (Gkotsis, Kougias et al. 2023). Adopting AD technology can significantly reduce the environmental footprint of waste management systems while providing economic and energy benefits.

Gasification and Pyrolysis: Emerging Thermal Technologies

Gasification and pyrolysis are also techniques that enable the recovery of energy-rich products from diverse waste streams, reducing reliance on fossil fuels, facilitating the production of bio-based fuels and carbon-rich materials, also promoting carbon neutrality and resource efficiency. Gasification and pyrolysis are advanced thermal technologies that convert carbonaceous waste into synthetic gas (syngas) and bio-oil, respectively, to produce energy and other products through high-temperature processes in low-oxygen environments (Basu 2010). These technologies offer greater flexibility in feedstock handling and the potential for cleaner energy production compared to traditional incineration (Ramos, Monteiro et al. 2018). Gasification produces a syngas that can be used for electricity generation, chemical production, or as a precursor for liquid fuels. Pyrolysis yields bio-oil and biochar, which have applications in energy production and soil enhancement. Both technologies are considered cleaner than incineration due to their lower emissions and higher energy efficiency potential, and present advantages such as greater flexibility in feedstock handling and the potential for higher energy efficiencies. Figure 4 shows pyrolysis representation.

Figure 4. Pyrolysis graphical scheme.

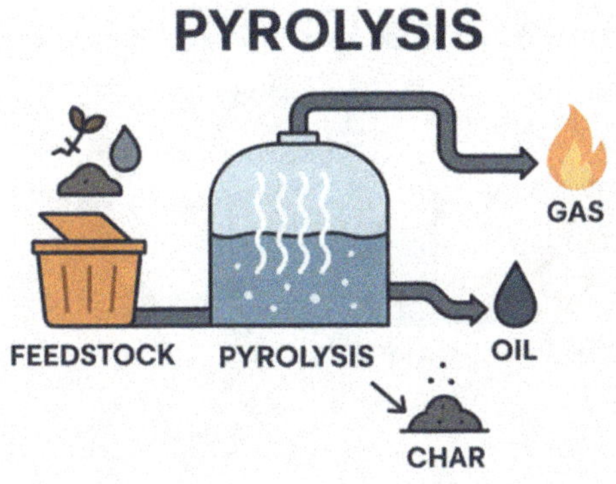

Pyrolysis is a thermal degradation (devolatization) that occurs in the absence of oxygen, generally between 150 and 700°C, releasing volatile elements and leaving a solid residue consisting of ash and char. The volatiles produced in pyrolysis are a mixture of hydrogen, CO_2, tar, CO, water vapor and light hydrocarbons. As far as products and applications are concerned, pyrolysis produces char, tar and gas, and has the potential to convert much of the energy in plastic waste into useful coal, gas and liquid oil. The gaseous products of pyrolysis (pyrolysis gas) can be a source of carbon for the production of biogas through anaerobic digestion, which requires a methanation process (Kassem, Hockey et al. 2020).

Gasification, on the other hand, is the chemical and thermal conversion of carbon-based materials into a primarily gaseous product, through partial oxidation with a gasifying agent (such as air, oxygen or steam) (Ramos, Monteiro et al. 2019). Gasification often includes pyrolysis as an initial step (Ramos, Monteiro et al. 2018). The gasification process generally involves several phases: drying (20-100°C), pyrolysis (150-700°C), oxidation (exothermic reactions that generate heat required for other reactions, producing CO_2, water and CO) and reduction (endothermic reactions that occur in the absence of O_2, producing CH_4, CO and H_2) (Basu 2010). Gasification operates at higher temperatures than pyrolysis, ranging from 800 to 1,100°C when using air as the oxidant, and up to 1,500°C with oxygen. The gasification process is generally endothermic overall, requiring heat. Figure 5 depicts a gasification general procedure.

Figure 5. Gasification process.

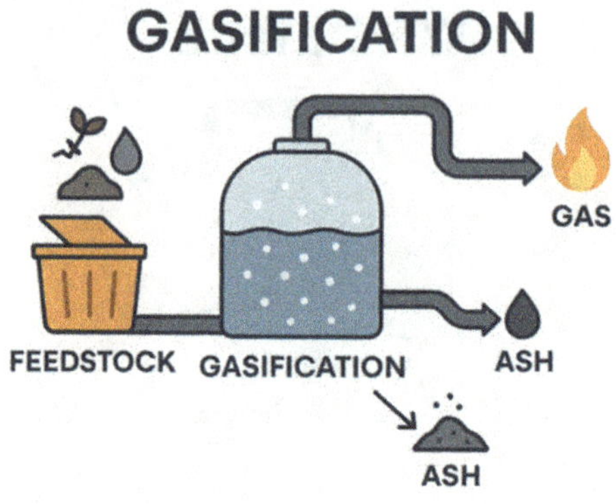

Concerning feedstocks, a wide range of waste materials can be used, including plastic waste, biomass, MSW and lignocellulosic materials. Co-gasification of biomass and plastics has also been investigated, synergy among feedstocks enhancing the outcomes of the technique (Kern, Pfeifer et al. 2013, Yang, Hu et al. 2015, Ramos, Monteiro et al. 2018, Ramos, Tavares et al. 2018, Tavares, Ramos et al. 2018). The main gaseous product of gasification is synthesis gas (syngas), which is a mixture of CO and H_2 (amidst short-chain hydrocarbons and some impurities). Syngas can be used as a direct alternative fuel or as a feedstock for the production of other chemicals, such as ammonia and methanol (Ramos and Rouboa 2020). As for other sub-products, gasification produces fly ash (including char or unreacted fuel), bottom ash, and slag.

Gasification and pyrolysis contribute significantly to the current energy landscape as part of the Waste-to-Energy strategy, which has multiple benefits such as being renewable and sustainable sources and alternatives to traditional forms of energy production, reducing dependence on petroleum derivatives. The application of solid waste gasification is expected to be effectively employed in both waste treatment and energy generation in the coming years. They allow the energy recovery of various types of waste that may not be suitable for anaerobic digestion (such as dry and less biodegradable materials or plastics). They contribute to more systematic waste management, reducing the need for landfills and pollution. These techniques also offer reduction of environmental impacts, in contrast to conventional combustion (incineration), improving the ecological performance and allowing better control of emissions, conserving the chemical energy contained in the waste (Ramos and Rouboa 2019, Ramos and Rouboa 2020). The production of syngas from waste gasification emerges as an alternative to diversify the global energy matrix, towards a low-carbon economy (Ramos, Berzosa et al. 2020b, Ramos 2025).

Despite their potential, these technologies face challenges. For gasification, tar formation is a significant hurdle, leading to operational problems and reduced gas yield (Cheng, Wu et al. 2020, De Oliveira, De Rezende et al. 2024). Pyrolysis, on the other hand, can result in a more complex product stream. For both processes, the variable composition and quality of the feedstock poses a significant challenge.

Comparative Assessment of WtE Technologies Based on Energy Efficiency, Environmental Impact, and Scalability

The efficiency of converting waste into useful energy varies between technologies. In terms of energy efficiency, the conversion of wastes through gasification is reported to be higher than for incineration, pyrolysis and biochemical technologies such as anaerobic digestion (Rafey, Prabhat et al. 2020). Factors such as waste composition and plant design affect energy efficiency. Optimizing the combustion process, recovering excess heat, and using steam for other purposes (heating/ cooling) can improve incineration efficiency. Singapore presents a case where both incineration and gasification schemes may profit from reducing the construction and operation costs, while increasing the electricity efficiency for gasification-based options would be effective to improve the economics of the project (You, Wang et al. 2016). The complexity of the technology is also related to efficiency. Pyrolysis and gasification are considered the most complex technologies, while biochemical technologies and incineration are less complex (Vukovic and Makogon 2022). In terms of scalability, complexity and economic indicators are key aspects to take into consideration (World Bank Group 2017). Gasification and pyrolysis face challenges related to their technological complexity and economic viability, especially in smaller urban contexts (Bridgwater 2003, Vreugdenhil, Zwart et al. 2009, Ahmed and Gupta 2010, Shah, Amin et al. 2023).

This complexity implies the need for intermediate-skilled labor, while incineration and anaerobic digestion require low-skilled labor. Typical plant capacities vary according to plant technicalities and local needs. Incineration may accommodate 500-2000 tons/day, gasification around 250-1000 tons/day, while pyrolysis aerobic digestion deal with circa 100-500 tons/day (Satiada and Calderon 2021). This suggests that incineration may be more scalable for very large volumes of waste, while pyrolysis and anaerobic digestion typically operate at smaller/medium scales, and gasification in the intermediate range. The applicability of technologies also affects their "scalability" in a given context. Pyrolysis and gasification are more suitable for more industrially developed urban areas, while incineration is preferable for normal urban sites and biochemical technologies for rural areas (Singh, Wadhwa et al. 2025). The variable composition of waste, including high moisture content for thermal processes and the need for extensive pretreatment for pyrolysis and gasification, pose operational challenges that can impact the scalability and economic viability of these technologies.

Regarding capital and operating costs, biochemical technologies (such as AD) are the cheapest options. Incineration is considered cheaper than other thermochemical technologies (gasification, pyrolysis) (Ramos 2025). Incineration is a commercially mature technology, while gasification and pyrolysis are not as commercially developed, although they are at various stages of technological development. There are gasification and pyrolysis projects, but incineration has been the most widely implemented.

Assessing the environmental performance of WtE technologies is crucial, often performed through Life Cycle Analysis (LCA). LCA results can vary significantly depending on accounting rules (e.g. by-product treatment and avoided landfill benefits). WtE technologies can reduce the environmental impacts of inadequate waste management practices and fossil fuel-based power generation. However, WtE technologies can also pose environmental risks if poorly implemented. The role of water-energy nexus in power plant cooling is also an issue, some plants using air as the cooling medium however, with lower levels of effectiveness (Khawaja, Alkayyali et al. 2024). Some LCA studies have indicated that anaerobic digestion and/or gasification were the best options in terms of environmental performance among the technologies compared (Hospido, Moreira et al. 2005). Incineration has shown the highest global warming potential (GWP) results among other the technologies. Gasification and pyrolysis are seen as having potential advantages, such as lower emissions, compared to conventional incineration (Ramos and Rouboa 2019). In general, WtE technologies contribute to reducing the use of landfills, mitigating associated environmental problems such as leachate and methane emissions.

In brief, while incineration effectively reduces waste volumes and generates energy, it can contribute to air pollution and greenhouse gas emissions if not properly managed, requiring advanced emission control technologies to mitigate environmental impacts. Anaerobic digestion is a promising avenue for producing renewable energy from organic waste, but its efficiency depends on feedstock quality, and managing nutrient-rich digestate presents challenges. Gasification offers high energy recovery potential and can utilize diverse waste streams, yet it faces concerns over the generation of tar and other contaminants, as well as the need for robust gas-cleaning systems. Pyrolysis presents a pathway to biofuels and carbon sequestration, but process optimization is required to ensure consistent product quality and minimize the release of volatile organic compounds. Together, these technologies represent innovative solutions to waste management and energy generation, yet they also highlight the importance of integrated approaches, regulatory frameworks, and ongoing research to maximize benefits while minimizing environmental trade-offs.

3. CASE STUDIES OF SUCCESSFUL IMPLEMENTATIONS

Cities with WtE and Other Innovative Solutions

Case studies play a crucial role in the successful implementation of WtE and other innovative waste management solutions. These serve as roadmaps, guiding decision-makers through complex planning, technical, and social challenges involved in WtE and innovative waste solutions. They provide practical, real-world insights that go beyond theory, helping policymakers, engineers, and urban planners make informed decisions. Table 1 highlights WtE key projects for the relevance of their innovations. It shows diverse initiatives illustrating the global commitment to sustainable waste management and energy production. Each city's approach reflects its unique environmental goals, technological capabilities, and community engagement strategies.

Johannesburg has implemented a landfill gas-to-energy project that captures methane emissions from five landfill sites to generate approximately 19 MW of electricity. This initiative not only reduces greenhouse gas emissions but also supplies electricity to households, contributing to the city's energy needs.

Table 1. Cities with WtE and other innovative solutions.

City	Approach	Features	Innovation	Intervention*	Project
Copenhagen, Denmark	Incineration with energy recovery	- Converts 440,000 tons of waste annually to electricity and district heating - Doubles as a public space with a ski slope and climbing wall on its roof	Integration of infrastructure and recreation Very low emissions due to advanced flue gas cleaning	2017	CopenHill
Vienna, Austria	Incineration	- Provides district heating to over 60,000 households - Designed by artist Friedensreich Hundertwasser to combine functionality and aesthetics	Community acceptance through design and transparency	2015	Spittelau plant
Tokyo, Japan	Multiple incineration plants with energy recovery	- Operates over 20 WtE plants - Integrates strict waste separation at source	High public compliance and efficient logistics	2014	Clean Authority of Tokyo

continued on following page

Table 1. Continued

City	Approach	Features	Innovation	Intervention*	Project
Pune, India	Decentralized Anaerobic digestion	- Converts organic market waste to biogas for cooking and lighting - Implemented with NGOs and municipal corporation	Community-owned and low-tech solutions for neighborhoods	2018	Maharashtra
Linköping, Sweden	Anaerobic digestion of food waste and sewage sludge	- Produces biogas to fuel city's public buses and garbage trucks	Closed-loop system linking waste and public transport	2023	Tekniska Verken
San Francisco, USA	Zero waste strategy	- Mandatory composting and recycling - Partnerships with waste service providers like Recology	Policy-driven approach with financial incentives	2022	Zero Waste I San Francisco
Kamikatsu, Japan	Zero waste initiative	- Residents separate waste into over 45 categories	Zero waste since the early 2000's; Extreme source separation with community participation	2017	Zero Waste Town
Amsterdam, Netherlands	Circular economy initiatives	- ~1/3 of MSW is incinerated to convert energy and heat - recycling a considerable amount of material and reducing carbon dioxide emissions	Combines different waste streams, including sewage treatment and incineration, achieving high efficiency (up to 95%) in conversion using CHP	2017	AEB Amsterdam
Bogotá, Colombia	Landfill Gas to Energy	- captures biogas from landfill and transforms it into electricity; one of the largest biogas systems in Colombia - capacity of 30 MW and operates 24 hours a day	1st city in Colombia to convert waste into electricity delivered to the national grid; significant reduction of CO_2 emissions; part of the revenues is credited to social investments in the surrounding communities	2014	Doña Juana Bogota
Johannesburg, South Africa	Landfill Gas to Energy	- using wastewater and methane emissions from landfills	Low-cost biogas-to-energy projects	2016	Johannesburg's Landfill Gas Project

continued on following page

Table 1. Continued

City	Approach	Features	Innovation	Intervention*	Project
Seoul, South Korea	Multiple WtE	- comprehensive waste management system - reduce landfill dependence and generate energy	Model for waste reduction and resource recovery	2019	Seoul Solution
Nairobi, Kenya	Anaerobic digestion	- community-based waste management initiatives - scalable solutions in resource-constrained settings	Local engagement and decentralized solutions	2019	Kenya Biogas Program

* Considered as the intervention that added the newest or the most innovative features

Besides the mentioned AD facilities, Nairobi is developing the Dandora Waste-to-Energy Power Station[1], a planned 45 MW facility aimed at addressing the city's waste management challenges. The project seeks to transform waste into a resource, tackling issues like waste overflow and energy deficits. Tokyo's Nerima Waste-to-Energy Plant[2] is an 18.7 MW facility that utilizes stoker incinerators to process municipal waste. This plant contributes to the city's waste management and energy production efforts. On its turn, the town of Kamikatsu has adopted a zero-waste policy, achieving an 80% recycling rate by sorting waste into 45 categories. The Kamikatsu Zero Waste Center[3] serves as a hub for education, research, and community engagement in sustainable waste practices.

Bogotá has initiated a project to transform waste into energy, aiming to process approximately 2,700 tons of waste per day by 2027. The city is also investing a portion of the profits from this initiative into social projects, demonstrating a commitment to sustainable development. San Francisco is upgrading its largest wastewater facility with Veolia's MemGas™ membrane technology[4] to convert biogas into renewable natural gas. At full capacity, the project is expected to produce enough renewable energy to offset the natural gas needs of approximately 3,800 homes annually.

Vienna's Spittelau Waste Incineration Plant processes around 270,000 tonnes of household waste annually, producing green heating and electricity. The plant is also recognized for its architectural design, integrating functionality with aesthetics. Linköping operates a waste-to-energy plant that processes municipal waste and biomass to generate electricity and heat. The facility is notable for its high efficiency and also serves as an educational site, attracting thousands of visitors annually. Similar to this, Copenhagen's Amager Bakke, also known as CopenHill, is a waste-to-energy plant that doubles as a recreational facility with a ski slope on its roof. The plant processes non-recyclable waste to produce electricity and district heating for the city. Amsterdam's AEB waste-to-energy facility processes residual waste to generate approximately 628 kWh of electricity and 208 kWh of heat per tonne of waste. The energy produced supplies the electricity grid and provides heating for about 30,000 households in the city.

It is interesting to notice that depending on the context and local needs, different aspects may be considered innovative for the particular case considered. For instance, while in Colombia the first project to convert waste into energy and delivering it into the national grid was a technological game changer, in Austria designer facilities for improved community acceptance constitute the innovation, in a more social/societal perspective. The type of feedstock also conditions the developed projects, therefore anaerobic digestion facilities are more prone to be implemented in cities with a high portion of food waste or organic feedstocks, whereas incineration is more applied for "synthetic" materials (Náthia-Neves, Berni et al. 2018, Neuwahl, Cusano et al. 2019, Kassem, Hockey et al. 2020, Ozturk and Dincer 2020).

Lessons Learned from Developing Economies

The implementation of WtE and other innovative waste management solutions in developing economies presents challenges but also shows potential. Despite thermal and biological technologies have been operating for decades in other countries and are economically viable, non-polluting and low-intermittency thermal plants, also helping to solve the problem of municipal solid waste, countries such as Brazil lags behind in this type of investment, with most of the waste illegally sent to landfills (Galvão, Alves et al. 2023). Currently, the use of WtE in Brazil is limited to the production of biogas from landfills, with few research initiatives in incineration and gasification. Incineration technology is mostly imported, as there is no local producer, being considered a risky choice that can be economically inefficient, complex and harmful to the environment and informal workers (United Nations Habitat 2010). Gasification technology in Brazil has not yet reached its full development. Despite past initiatives, new policies are needed to boost the viability of waste treatment technologies, given the lack of structured development of the circular economy in Brazil. There is significant potential for energy recovery from the approximately 230,000 tons of waste generated daily, most of which is disposed of in landfills and dumps. The case of Bogotá shows that it is possible to successfully implement WtE from landfills in a developing economy (Solano Meza, Orjuela Yepes et al. 2023). Converting biogas into electricity not only reduces emissions, but also generates revenue that can be invested in social projects, addressing the waste problem in an integrated manner (environmental, economic and social). This model can serve as an example for other cities in similar contexts.

In general, within the context of developing countries, WtE is seen as a promising green energy carrier for clean development. Strategies for implementing the WtE supply chain for circular economy systems in developing countries are being reviewed (Pan, Du et al. 2015). The assessment of WtE generation technologies in developing economies is carried out through multi-criteria sustainability analyses, simulations for energy recovery from municipal waste, and guidelines for decision-makers (Kabir and Kabir 2021). Latin America, in particular, has been the focus of studies on improving waste management strategies and the potential of biorefineries to enable circular economies (United Nations Habitat 2010, Gobierno de Colombia 2020, Ulloa-Murillo, Villegas et al. 2022). Implementing WtE and other innovative solutions presents specific challenges in developing economies, such as inefficient systems and reliance on landfills. There are technological and infrastructural limitations, financial constraints, behavioral and cultural barriers, and gaps in policy and governance.

In summary, WtE and other waste valorization technologies are essential for the transition to the circular economy and circular cities. While more developed cities such as Singapore and Amsterdam demonstrate the established and efficient use of these technologies, the case of Bogotá offers a valuable lesson for developing economies, showing that implementation is feasible and can generate environmental, economic and social benefits. Sometimes, despite the potential, countries face challenges in developing the WtE market, the suitability of technologies such as incineration and gasification, depending on the type of waste and scale. Overcoming investment obstacles and adopting policies that encourage the circular economy are crucial for the advancement of WtE in developing countries.

4. CHALLENGES AND OPPORTUNITIES

Addressing Environmental Trade-Offs: Emissions, Ash Disposal, and Community Concerns

Emerging applications such as hydrogen production from biogas reforming or through waste gasification/pyrolysis depict great potential, namely providing a clean fuel for transportation (Al-Qadri, Ahmed et al. 2024). Advanced energy systems, such as fuel cells, can be powered by biogas or syngas, offering low emissions and high efficiency for power generation and transportation (Saadabadi, Thattai et al. 2019). Another innovation is the production of lubricants from the pyrolysis and gasification of plastic waste (Sikdar, Siddaiah et al. 2020). But despite these technological advancements, WtE technologies, particularly incineration, often face public opposition due to concerns about air pollution (dioxins, heavy metals) and proximity to residential areas, and the management of toxic ash. This opposition can make it difficult to locate and license projects. Lack of public awareness about waste segregation and the circular economy can result in low participation in recycling programs and inappropriate disposal. Transparent communication and stakeholder engagement are critical to gaining public support (Di Fiore, Specht et al. 2022, Caferra, D'Adamo et al. 2023). Educational campaigns, community outreach, and involving local communities in decision-making can increase understanding, participation, and ownership. Decisions on WtE technologies should be made within the framework of a municipal waste management plan that considers the specific context of the city.

To address environmental trade-offs in innovative waste management, especially in the context of WtE solutions for sustainable energy, it is crucial to carefully manage impacts related to all these concerns. Trade-offs and choices balancing different criteria, such as sustainability (environmental, economic and social) and resilience must be adopted, tools such as LCA, Life Cycle Costing (LCC), Social-Lifecycle Analysis (S-LCA), and Techno-Economic Analysis (TEA) being used to investigate this compromise (Guinée 2016, Benoît Norris, Traverso et al. 2020, Hsu, Binyet et al. 2024, Ramos 2025).

Some key points on how to address emissions-related offsets, ash disposal, and community concerns, might be to tackle atmospheric emissions including particulate matter, nitrogen oxides, and sulfur dioxide, dioxins and heavy metals. These emissions can have environmental consequences if not properly controlled. Although consolidated in developed countries, incineration still raises doubts and discussions about environmental risks and threats to public health. An approach would be to mitigate these emissions employing advanced emissions control technologies, such as electrostatic precipitators and catalytic converters, which remove harmful pollutants from exhaust gases (World Bank Group 2017). Environmental monitoring of atmospheric emissions is an established practice, for example, in Europe. Ensuring continued effectiveness of these mitigation measures and maintaining compliance with air quality regulations would result in significant reductions in GHG emissions when compared to the landfill model. Regarding ashes that may contain residual toxins, their improper management can contaminate soil and water. Proper ash management practices, including containment and safe disposal methods, are essential to prevent environmental contamination, ash characterization being very important, to understand the most suitable valorization method (Basha, Aziz et al. 2020).

Pollution concerns may be heightened in developing countries where pollution is already a problem and environmental laws are weak, making it costly and difficult to meet international standards. The impact on the informal recycling sector ("waste pickers") is an important consideration, as incineration may affect their livelihoods, and the plan should consider this sector (Uddin, Gutberlet et al. 2020, Carenbauer 2021). The development of new WtE technologies advertised as "clean technologies" may influence the level of acceptance of incineration. Acceptance by citizens/authorities is a social criterion to be considered in the decision-making process.

Technologies and strategies need to be adapted to the local context, policies need to be incentivized to make technologies economically viable, and social inclusion needs to be ensured, particularly by considering the informal sector. Implementation can be complicated in countries with other priorities and where landfill rates are low or non-existent, making the WtE alternative seem too costly. The experience of Bogotá with the conversion of landfill biogas into electricity demonstrates that the implementation of WtE is feasible in developing economies and can generate multiple benefits (environmental, economic, social), serving as a potential model. To manage these trade-offs effectively, decisions on WtE options should be based on a short- and long-term analysis of their impacts, considering costs, technical feasibility (quality and quantity of waste), energy and environmental benefits (including GHG reduction), and social aspects (public acceptance and impact on informal workers). A methodological approach that integrates sustainability and resilience into energy transition planning is essential.

Economic Feasibility and Funding Mechanisms for WtE Projects

The economic viability varies significantly across urban contexts, being influenced by waste composition, technological maturity, regulations and economic considerations. Some technologies, such as gasification and pyrolysis, may face economic viability challenges. The high capital costs of WtE facilities pose a significant barrier. Public-private partnerships (PPPs), green bonds, and international climate finance mechanisms can help mobilize funding. Economic feasibility can be enhanced through revenue streams from electricity sales, tipping fees, and byproducts such as compost and biochar (Ramos, Afonso Teixeira et al. 2018, Ramos, Berzosa et al. 2020a, Ramos, Berzosa et al. 2020b, Ramos and Rouboa 2022).

To address the economic viability and financing mechanisms for WtE projects, it is essential to consider cost-benefit analyses, revenue streams and different ways to attract investment, especially in a context of transition towards the circular economy and sustainable energy. The economic viability of WtE projects is assessed using methodologies such as Cost-Benefit Analysis (CBA) and capital analysis (including net present value - NPV, and internal rate of return - IRR) (Ramos, Berzosa et al. 2020a, Ramos, Berzosa et al. 2020b). Incorporating uncertainty, risk and flexibility into the investment decision is also something that should be considered.

The main economic components considered include capital expenditure (CAPEX) and operational expenditures (OPEX). CAPEX are the initial costs to build the facility. These can be prohibitively high, especially in low-income countries. WtE incineration plants may have higher capital and maintenance costs than coal-fired plants of similar capacity. OPEX refer to the ongoing costs to operate and maintain the facility. Revenue derives mainly from the dumping fee paid by municipalities for waste management and from the sale of energy (electricity and/or heat). Revenues from energy sales can represent 20% to 30% of the project's total revenue (Ducharme 2010, Leme, Rocha et al. 2014). Additional revenues can come from material recovery (e.g. metals from bottom ash). Avoided costs such as avoided landfill disposal costs, also account with a balancing effect on the costs.

Comparative studies show different results for the economic viability of various WtE technologies. For instance, in Thailand, anaerobic digestion of MSW for energy recovery and fertilizer production had the highest NPV and benefit-cost ratio, making it the most economically viable option, while incineration integrated with anaerobic digestion was less economically favorable due to higher costs (Farooq, Haputta et al. 2024). In the Philippines, an economic analysis indicated incineration as the most profitable option among thermal treatments, followed by gasification and pyrolysis (Agaton, Guno et al. 2020). It was considered more optimal to invest immediately in incineration or gasification, while the investment in pyrolysis could be postponed depending on the price of electricity and the increase in the disposal tariff.

Overcoming financial constraints is essential for implementing circular economy strategies in MSW management, especially in developing countries with limited budgets. Several innovative financing mechanisms and government incentives have emerged to promote the adoption of WtE, such as the public-private partnerships, as viable solutions that attract additional financing and expertise from the private sector. In this model, the private partner finances the development and begins receiving payments after operation. PPP contracts for WtE typically involve the availability of waste management capacity and power sales agreements. However, PPPs involve a complex allocation of risks, including planning and licensing risk, which can be highly controversial due to environmental and health concerns (Castelblanco, Guevara et al. 2020). Green bonds are also a promising financial instrument to finance projects with a positive environmental impact, including waste management and WtE. Green bonds can increase the financial viability of WtE projects by providing access to capital that may not be available through traditional sources. Government incentives in the form of tax credits, grants, and fee-in tariffs encourage these projects in diverse ways. Tax credits offset costs and offer financial benefits, grants are funding programs that provide financial support, reducing financial barriers for WtE projects, while feed-in tariffs are policies that offer fixed payments for electricity generated by WtE facilities, encouraging adoption and diversifying the energy mix. Electricity generated by WtE facilities is sometimes eligible for green incentive schemes such as feed-in tariffs. Other financing policies include grants for construction of the facility, loans with good conditions, tax exemptions for operators, imposition of landfill fees, subsidies for energy prices and subsidies for tipping tariff prices.

Despite the great potential, the lack of financing and management at municipal level has been a barrier in some countries. To attract private investors, especially in developing economies, it is crucial to ensure legal certainty, transparency and a clear vision for sustainable waste management services. The creation of a regulatory framework is essential to establish the legal certainty necessary for commercially scalable projects.

In summary, the economic viability of WtE projects depends on a detailed analysis of costs and revenues, varying with the technology and the local context. Financing requires overcoming capital barriers and is facilitated by mechanisms such as PPPs, green bonds and a variety of government incentives, which are crucial to driving the transition to sustainable WtE solutions.

Policy Frameworks and Opportunities for Replicating Solutions

Supportive policy frameworks are essential to drive WtE adoption. Key elements include clear regulations for waste segregation and quality standards, incentives for renewable energy production, environmental performance monitoring and reporting requirements.

The transition to a circular economy in MSW management, which includes WtE technologies, is significantly influenced by policy frameworks and governance structures. Effective policies are essential to promote circular economy principles, facilitate sustainable waste management practices and ensure stakeholder engagement. Research indicates that policy and governance gaps exist, representing a significant barrier to the successful implementation of strategies in MSW management. Existing waste management policies are often fragmented, lacking coherence and integration with broader sustainability objectives. This fragmentation can lead to inconsistencies in practices and hinder circular initiatives (Mukwevho, Radzuma et al. 2024, Ren and Zuo 2024).

The absence of clear regulatory frameworks and enforcement mechanisms can result in non-compliance with regulations, undermining efforts to promote recycling and resource recovery. The development of comprehensive regulatory frameworks is emphasized as a crucial step to ensure sustainable and responsible WtE practices in the long term (Yukalang, Clarke et al. 2017). A supportive policy and regulatory framework are essential for the successful implementation of WtE technologies. This includes clear waste management hierarchies, incentives for renewable energy, and environmental standards that consider the full life-cycle impacts of waste treatment options.

Appropriate public policies can drive innovation and the choice of climate-friendly solutions. Ensuring legal certainty, transparency and a clear vision for sustainable waste management services is crucial, especially to attract private investors in developing economies (World Bank Group 2017). Policies need to keep pace with technological developments and promote a balance between resilient and sustainable performance of urban infrastructure. Creating a regulatory framework is essential to establish the legal certainty necessary for commercially scalable projects (related to the previous discussion on financing).

Replication of successful models requires knowledge sharing, capacity building, and the adaptation of solutions to local contexts (Jayakumar Menon and Palackal 2022). WtE technologies are seen as a potential solution that addresses the needs of waste management and energy production, contributing to the circular economy and sustainable urban development. There are several opportunities and factors that favor replication, namely the application of WtE as a waste treatment method that uses circular economy principles. There is great potential for the recovery of heat and matter from waste for energy generation. Integrating WtE with other renewable energy sources (solar, wind) and circular economy initiatives can increase the sustainability of the system (Kristia and Rabbi 2023, Lee, Lin et al. 2023, Enemuo and Ogunmodimu 2025). Adopting advanced waste sorting technologies can improve the efficiency of WtE processes. Countries such as Sweden and Japan have successfully implemented waste treatment methods using circular economy principles, such as WtE (Heshmati and Rashidghalam 2021, Rezania, Oryani et al. 2023). Highlights from Singapore show a pioneering example that has adopted incineration for energy generation, and its experiences are valuable for analyzing other cases (Tun, Palacky et al. 2020). Japan has been active in forming PPPs to export the WtE model and plants, along with expertise in waste collection, sorting and management, to parts of the Association of Southeast Asian Nations (ASEAN) region where waste infrastructure is insufficient, such as Indonesia, the Philippines and Vietnam (Hongo 2016, Hiratsuka-Sasaki and Kojima 2020, Kojima 2020). This demonstrates a model for replication.

However, replication faces challenges, such as the lack of investment and maturity of the WtE market in some regions, high capital costs, tipping fees can be low or non-existent in developing countries, making WtE comparatively more expensive than landfills. Lack of financing and management at the municipal level has been a barrier in some countries. It is necessary to analyze available technologies and the best cost-benefit of development, as feasibility varies with technology and local context.

5. VISION FOR THE FUTURE

The Role of Innovation in Achieving Sustainable Cities

Innovation will be a driving force in future waste management systems. Emerging technologies, data analytics, and digital platforms can optimize waste collection, sorting, and processing. Moreover, circular economy models, where waste is viewed as a resource, will gain prominence. Integration of WtE with renewable energy systems can further enhance sustainability (Lee, Lin et al. 2023, Hsu, Binyet et al. 2024). Figure 6 represents key aspects to take into account when planning sustainable cities.

Figure 6. Intertwining aspects to consider for the cities of tomorrow.

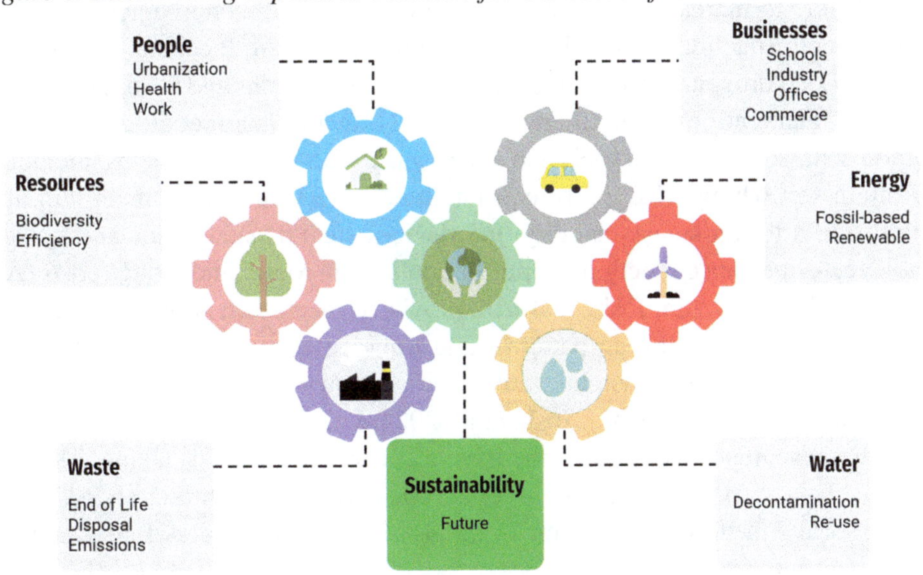

Future perspectives on waste and energy management in urban environments are intrinsically linked to the circular economy, which redefines waste as a resource. The near future anticipates a drastic increase in MSW generation worldwide, an estimated growth for global MSW generation reaching 3.4 billion tons by 2050, almost double the current figures, presenting significant environmental, social and economic challenges (Kaza, Yao et al. 2018). When combined with technological innovations, this transitioning to innovative waste management practices will drive sustainable and resilient cities.

The specific context of WtE includes the potential for increased government incentives to drive its adoption. WtE technologies will continue to play a crucial role in urban sustainability strategies, especially as cities face the dual challenges of waste management and sustainable energy production. WtE is expected to be integrated with other renewable energy sources and circular economy initiatives to increase sustainability (Lee, Lin et al. 2023, Hsu, Binyet et al. 2024). The development of comprehensive regulatory frameworks is crucial to ensure sustainable and responsible WtE practices in the long term.

The integration of technological innovations into MSW management is crucial as these technologies will not only enhance sorting and recycling processes, but can also facilitate biotechnological approaches and the implementation of digital tools that optimize waste management systems. Specific technological innovations include: emerging technologies for MSW valorization, converting waste into valuable chemicals and energy, encompassing chemical, biological, thermal, hybrid and integrated processes. For biogas, the future outlook is promising, with continued growth expected due to increasing environmental awareness and technological advances (Kougias and Angelidaki 2018, Jameel, Mustafa et al. 2024, Piechota, Generowicz et al. 2024). Hydrogen will also underscore major innovations and advances, backed by policy interventions (Boretti and Pollet 2024, Islam, Islam et al. 2024, Sadeq, Homod et al. 2024). Digital tools and smart waste management systems, including the Internet of Things (IoT), artificial intelligence (AI) and blockchain will also build bridges towards sustainability. IoT devices such as smart bins and sensors enable real-time data collection to optimize collection routes and reduce costs (Ali, Irfan et al. 2020, Kuzhin, Joshi et al. 2024). AI and machine learning can optimize collection, predict generation patterns, and improve sorting efficiency. Big data provides comprehensive insights into resource flows and system inefficiencies (Khan, Ali et al. 2024, Kuzhin, Joshi et al. 2024, Lakhouit 2025). Blockchain can increase transparency by tracking the lifecycle of waste materials (Jiang, Zhang et al. 2023, Faiz, Ninduwezuor-Ehiobu et al. 2024, Kumar, Mehra et al. 2024). AI driving sorting technologies is seen as a transformative framework for sustainable MSW management.

How Waste-to-Energy Complements Broader Climate and Sustainability Goals

WtE technologies contribute to climate goals by reducing methane emissions from landfills and providing a renewable energy source, avoiding the extraction, transportation and processing of fossil resources for that purpose. They complement other management techniques, such as recycling and composting efforts, by managing residual waste. Achieving sustainable cities requires a holistic approach that combines technological innovation, robust policy frameworks, and community engagement, and WtE will play a vital role in this transformation, offering a pathway toward resilient and sustainable urban environments (International Solid Waste Association 2024). WtE is seen as central to the development of the circular economy in cities and to the proper management of urban waste, which is vital to preventing environmental degradation. Integrating sustainability and resilience concepts into urban planning, supported by LCA methodologies, is essential (Kennedy, Pincetl et al. 2011, Pincetl, Bunje et al. 2012, Maranghi, Parisi et al. 2020). LCA helps assess the environmental performance of WtE, considering its full life cycle and the trade-offs between waste management, energy production and environmental impact, aiding in informed decision-making.

This shows how WtE complements broader climate and sustainability goals, positioning itself as a crucial solution in the transition to a circular economy and sustainable cities. Nevertheless, these techniques also contribute directly to the circular economy and sustainable waste management frameworks, namely offering a transformative approach to waste management, redefining waste as a resource and emphasizing waste minimization and resource utilization maximization, by converting non-recyclable waste into valuable resources such as energy. They are also considered a sustainable option after reduction, reuse and recycling, safely managing waste that cannot be readily recycled. On the other hand, WtE plants significantly reduce waste volumes by up to 90%, reducing the volume that would otherwise go to landfill. This reduces the need for new landfills and the associated environmental problems, such as leachate contamination (Alao, Popoola et al. 2022). In addition to energy, WtE enables the recovery of materials, such as metals, from bottom ash. The remaining ash, which represents about 10% of the initial mass of the incinerated waste, can be used as a raw material for construction, for example in the production of cement or asphalt, keeping these materials in the production cycle (Roessler, Olivera et al. 2016, Bouzar and Mamindy-Pajany 2022).

In terms of climate change mitigation, WtE contributes to the reduction of GHG emissions in several ways. By diverting waste from landfill, it avoids the emission of methane, a potent GHG produced by the decomposition of organic matter under anaerobic conditions in landfills (Chandel, Kwok et al. 2012). The energy generated by WtE can replace energy produced from fossil fuels, helping to offset the use of these and reduce the carbon footprint. Although WtE generates emissions (particulate matter, NOx, SO_2) that need to be controlled, pollutant emissions from WtE plants are considered lower than those from conventional energy conversion (O'Brien 2006).

Technological innovations are improving the efficiency of waste-to-energy conversion, such as optimizing combustion and recovering waste heat for district heating or industrial processes. Innovations such as the integration of waste heat recovery with carbon capture technologies in WtE plants can further mitigate GHG emissions and improve environmental performance, even allowing the use of captured CO_2 in applications such as horticultural greenhouses or the production of synthetic e-fuels (Dell'Aversano, Villante et al. 2024, Acampora, Grilletta et al. 2025).

6. CONCLUSIONS

WtE initiatives are recognized as drivers of technological innovation, with research efforts focused on greener and more energy-efficient solutions in response to the climate crisis. Technological innovations are improving the efficiency of biogas production and utilization processes. In addition, innovation in urban planning is needed to respond to climate change risks without exacerbating them. The development of scalable circular economy models that can be adapted to diverse urban contexts, considering socioeconomic and waste composition variations, is crucial. Integrating sustainability and resilience concepts into planning, supported by methodological life cycle assessments and consideration of trade-offs, is vital for informed decision-making and for driving climate action. WtE complements climate and sustainability objectives by redefining waste as a resource within the framework of the circular economy, mitigating climate change by reducing reliance on landfills and offsetting fossil fuel use, providing a valuable energy source and reduce environmental impacts.

In terms of environmental performance, modern incineration, with advanced technologies and good operational practices, together with efficient air pollution control systems, makes it possible to significantly reduce emissions when compared to conventional schemes. The search for efficient solutions drives technological innovation in waste treatment, aiming to reduce generation and disposal in landfills and reduce environmental risks and impacts. Anaerobic digestion is a crucial biological technology in today's energy landscape, converting various organic wastes into biogas, a versatile renewable energy source that contributes to the production of heat, electricity, and transportation fuels. Its global growth, driven by policy and technological advances, demonstrates its increasing role in sustainable waste management and the energy transition, while also offering environmental benefits and fertilizer production. Regarding more advanced WtE technologies, gasification and pyrolysis are promising thermal technologies that complement the other approaches, their development being driven by the need to manage waste sustainably and diversify energy sources, contributing to energy security, emissions reduction and a circular economy. These two techniques represent an advance in thermal waste conversion technologies, offering potential advantages in terms of lower environmental impact (compared to conventional incineration) and the production of value-added products such as synthesis gas. However, they are generally more complex and less commercially established than incineration. Its energy efficiency can be comparable or superior to incineration, depending on the configuration and the gasifying agent used. Anaerobic digestion, stands out for its low cost and good environmental performance (especially in terms of GHG), but is limited by the composition of the feedstock and typically operates at smaller scales. The selection of the most suitable WtE technology strongly depends on the specific context, including waste composition, technological capabilities and the local economic and environmental situation. Although gasification and pyrolysis are promising, they still face challenges in reaching the same maturity and scalability as incineration on a global scale.

Regarding non-technological conditions for implementation, policy frameworks are key to the success of WtE projects, which require clarity, coherence and supportive incentives. Opportunities for replication are vast, driven by the need for sustainable waste management and energy generation, and there are successful models and mechanisms such as PPPs and the use of green bonds that can facilitate scale-up, but it is crucial to overcome financial, regulatory and public acceptance barriers through a comprehensive approach.

7. FUTURE RESEARCH DIRECTIONS AND COLLABORATIVE OPPORTUNITIES

The ongoing development of advanced technologies will expand opportunities for scaling WtE solutions worldwide. Collaborative efforts among governments, research institutions, and the private sector will further enhance innovation, promote funding mechanisms, and help overcome existing barriers.

Future research directions and opportunities for collaboration in the area of innovation for waste management and sustainable energy solutions are multifaceted and crucial to achieving sustainable cities. Several areas have been identified as requiring continued research and development to drive innovation in waste management for sustainable energy production, namely:

- the development of new methods, technologies and public policies: future work needs to be directed towards a better understanding of the energy recovery of solid waste under the circular economy concept, developing new methods, new technologies and proposing new public policies.
- lifecycle thinking application: life cycle analysis (financial, environmental and social) to the energy recovery of municipal solid waste is crucial to assess the environmental performance of WtE and consider trade-offs, and its integration into planning is vital for informed decision-making. It is recommended to extend research to include a more complete assessment of social

impacts (job creation, public health effects, environmental justice considerations), as well as health risk analyses and economic impacts on income, employment, and local electricity markets.

- integrating emerging technologies into circular economy frameworks: the application of emerging technologies such as hydrothermal carbonization and plasma gasification, as well as non-thermal technologies such as fermentation and mechanical-biological treatment should be deeply explored; advanced technologies such as AI, machine learning and big data analytics offer transformative opportunities to enhance circular economy frameworks in MSW management, future research requiring a focus on developing scalable, cost-effective and easy-to-use technology solutions tailored to the needs of diverse urban contexts. AI and machine learning can optimize collection, predict generation patterns and improve sorting efficiency. Big data provides comprehensive insights into resource flows and system inefficiencies.
- integrating WtE systems with broader circular economy initiatives: exploring the potential of WtE technologies to support broader circular economy initiatives in urban environments, including material recovery (such as metals from bottom ash) and opportunities for industrial symbiosis is key. Strategies for implementing WtE supply chains for circular economy systems are another area of study underscoring this integration. It is also crucial to develop scalable circular economy models that can be adapted to diverse urban contexts, considering variations in socio-economic and waste composition.
- advancing waste-to-resource technologies: innovative waste-to-resource technologies should seek to improve the efficiency, scalability and environmental performance of these processes. Research and development should also address challenges related to feedstock variability, energy consumption and by-product management.
- dynamic modeling: future studies should incorporate dynamic modeling addressing the impact of climate change on waste generation patterns, the performance of WtE technologies in urban areas, and energy demands.
- technical and non-technical uncertainties: research should incorporate technical and non-technical uncertainties related to government policies, social acceptance, and waste management laws, so that a holistic plan is conceived towards higher degrees of sustainability in future cities.
- increasing investment in public engagement is essential to inform technology choices and promote the uptake of WtE technologies.

The transition to more sustainable waste management and energy systems requires collaboration between a variety of stakeholders. Bringing together a number of universities concerned with waste management to improve the performance of technologies, demonstrating a model for academic and research collaboration would leverage the results and findings achieved so far. Also, considering stronger collaboration between governments, the private sector and communities to widely foster the integration of circular economy principles into MSW management and to create sustainable urban systems. The private sector has an important role to play, especially given the high investment costs of some WtE technologies. Encouraging private investment requires an environment with guaranteed legal certainty, transparency and a clear vision for sustainable waste management services, with government collaboration. Developing comprehensive regulatory frameworks ensuring sustainable and responsible WtE practices in the long term requires collaboration between policymakers, industry and other stakeholders. Scientific reviews and studies can inform policymakers, practitioners and researchers, encouraging the joint exploration of innovative solutions.

In short, the future of waste and energy management in sustainable cities will be shaped by continued research into innovative technologies and effective collaboration between academia, government, the private sector and civil society. The vision for the future of waste and energy management should focus more on circular, efficient and resilient systems, driven by the need to manage the growing volume of waste and combat climate change. Innovation, in particular emerging technologies in waste treatment and systems management, is the essential driver to transform this vision into reality and build truly sustainable cities.

ACKNOWLEDGEMENT

This work is funded by national funds through the FCT – Fundação para a Ciência e a Tecnologia, I.P. and, when eligible, by COMPETE 2020 FEDER funds, under the Scientific Employment Stimulus - Individual Call (CEEC Individual) - 2021.03036. CEECIND/CP1680/CT0003; DOI - 10.54499/2021.03036.CEECIND/CP1680/CT0003(https://doi.org/10.54499/2021.03036.CEECIND/CP1680/CT0003).

REFERENCES

Abanades, S., H. Abbaspour, A. Ahmadi, B. Das, M. Ehyaei, F. Esmaeilion, M. El Haj Assad, T. Hajilounezhad, D. Jamali and A. Hmida (2022). "A critical review of biogas production and usage with legislations framework across the globe." International Journal of Environmental Science Technology in Society: 1-24.

Acampora, L., Grilletta, S., & Costa, G. J. E. (2025). The Integration of Carbon Capture, Utilization, and Storage (CCUS) in Waste-to-Energy Plants: A Review. *Energies*, *18*(8), 1883. DOI: 10.3390/en18081883

Agaton, C. B., Guno, C. S., Villanueva, R. O., & Villanueva, R. O. (2020). Economic analysis of waste-to-energy investment in the Philippines: A real options approach. *Applied Energy*, *275*, 115265. DOI: 10.1016/j.apenergy.2020.115265

Ahmed, I. I., & Gupta, A. K. (2010). Pyrolysis and gasification of food waste: Syngas characteristics and char gasification kinetics. *Applied Energy*, *87*(1), 101–108. DOI: 10.1016/j.apenergy.2009.08.032

Al-Qadri, A. A., Ahmed, U., Ahmad, N., Jameel, A. G. A., Zahid, U., & Naqvi, S. R. (2024). A review of hydrogen generation through gasification and pyrolysis of waste plastic and tires: Opportunities and challenges. *International Journal of Hydrogen Energy*, *77*, 1185–1204. DOI: 10.1016/j.ijhydene.2024.06.207

Alao, M. A., Popoola, O. M., & Ayodele, T. R. (2022). Waste-to-energy nexus: An overview of technologies and implementation for sustainable development. *Cleaner Energy Systems*, *3*, 100034. DOI: 10.1016/j.cles.2022.100034

Ali, T., Irfan, M., Alwadie, A. S., & Glowacz, A. (2020). IoT-based smart waste bin monitoring and municipal solid waste management system for smart cities. *Arabian Journal for Science and Engineering*, *45*(12), 10185–10198. DOI: 10.1007/s13369-020-04637-w

Arjwech, R., Somchat, K., Pondthai, P., Everett, M., Schulmeister, M., & Saeng-chomphu, S. (2020). Assessment of geological, hydrogeological and geotechnical characteristics of a proposed waste disposal site: A case study in Khon kaen, Thailand. *Geosciences, 10*(3), 109. DOI: 10.3390/geosciences10030109

Baechler, B. R., De Frond, H., Dropkin, L., Leonard, G. H., Proano, L., & Mallos, N. J. (2024). Public awareness and perceptions of ocean plastic pollution and support for solutions in the United States. *Frontiers in Marine Science, 10*, 1323477. DOI: 10.3389/fmars.2023.1323477

Basha, S. I., Aziz, A., Maslehuddin, M., Ahmad, S., Hakeem, A. S., & Rahman, M. M. (2020). Characterization, Processing, and Application of Heavy Fuel Oil Ash, an Industrial Waste Material – A Review. *Chemical Record (New York, N.Y.), 20*(12), 1568–1595. DOI: 10.1002/tcr.202000100 PMID: 33078557

Basu, P. (2010). *Biomass gasification and pyrolysis: practical design and theory.* Academic press.

Benoît Norris, C., Traverso, M., Neugebauer, S., Ekener, E., Schaubroeck, T., Russo Garrido, S., Berger, M., Valdivia, S., Lehmann, A., Finkbeiner, M., & Arcese, G. (2020). *Guidelines for Social Life Cycle Assessment of Products and Organizations.* Life Cycle Initiative.

Boretti, A., & Pollet, B. G. (2024). Hydrogen economy: Paving the path to a sustainable, low-carbon future. *International Journal of Hydrogen Energy, 93*, 307–319. DOI: 10.1016/j.ijhydene.2024.10.350

Bosmans, A., Vanderreydt, I., Geysen, D., & Helsen, L. (2013). The crucial role of Waste-to-Energy technologies in enhanced landfill mining: A technology review. *Journal of Cleaner Production, 55*, 10–23. DOI: 10.1016/j.jclepro.2012.05.032

Bouzar, B., & Mamindy-Pajany, Y. (2022). Manufacture and characterization of carbonated lightweight aggregates from waste paper fly ash. *Powder Technology, 406*, 117583. DOI: 10.1016/j.powtec.2022.117583

Braga, J. K., Motteran, F., Sakamoto, I. K., & Varesche, M. B. A. (2018). Bacterial and archaeal community structure involved in biofuels production using hydrothermal-and enzymatic-pretreated sugarcane bagasse for an improvement in hydrogen and methane production. *Sustainable Energy & Fuels, 2*(12), 2644–2660. DOI: 10.1039/C8SE00312B

Bridgwater, A. V. (2003). Pyrolysis and gasification of biomass and waste: proceedings of an expert meeting; Strasbourg, France, 30 September-1 October 2002, cpl Press.

Caferra, R., D'Adamo, I., & Morone, P. (2023). Wasting energy or energizing waste? The public acceptance of waste-to-energy technology. *Energy*, *263*, 10. DOI: 10.1016/j.energy.2022.126123

Carenbauer, M. G. (2021). Essential or dismissible? Exploring the challenges of waste pickers in relation to COVID-19. *Geoforum*, *120*, 79–81. DOI: 10.1016/j.geoforum.2021.01.018 PMID: 36536781

Castelblanco, G., Guevara, J., Mesa, H., & Flores, D. (2020). Risk allocation in unsolicited and solicited road public-private partnerships: Sustainability and management implications. *Sustainability (Basel)*, *12*(11), 4478. DOI: 10.3390/su12114478

Chandel, M. K., Kwok, G., Jackson, R. B., & Pratson, L. F. (2012). The potential of waste-to-energy in reducing GHG emissions. *Carbon Management*, *3*(2), 133–144. DOI: 10.4155/cmt.12.11

Cheng, L., Wu, Z., Zhang, Z., Guo, C., Ellis, N., Bi, X., Paul Watkinson, A., & Grace, J. R. (2020). Tar elimination from biomass gasification syngas with bauxite residue derived catalysts and gasification char. *Applied Energy*, *258*, 114088. DOI: 10.1016/j.apenergy.2019.114088

Dastjerdi, B., Strezov, V., Rajaeifar, M. A., Kumar, R., & Behnia, M. (2022). *Waste to Energy Technologies. Reference Module in Earth Systems and Environmental Sciences*. Elsevier.

De Bere, L. (2000). Anaerobic digestion of solid waste: State-of-the-art. *Water Science and Technology*, *41*(3), 283–290. DOI: 10.2166/wst.2000.0082 PMID: 11382003

De Oliveira, D. C., De Rezende, T. T. G., Lora, E. E. S., Venturini, O. J., & Maya, D. M. Y. (2024). Tar removal from biomass-derived syngas for hydrogen production: Oil absorption process considering brazilian sources. *International Journal of Hydrogen Energy*.

Dell'Aversano, S., Villante, C., Gallucci, K., Vanga, G., & Di Giuliano, A. (2024). E-Fuels: A Comprehensive Review of the Most Promising Technological Alternatives towards an Energy Transition. *Energies*, *17*(16), 3995. DOI: 10.3390/en17163995

Di Fiore, G., Specht, K., Rover, O. J., & Zanasi, C. (2022). Stakeholders' social acceptance of a new organic waste management policy in the city of Florianópolis (Brazil). *Journal of Cleaner Production*, *379*, 134756. DOI: 10.1016/j.jclepro.2022.134756

Ducharme, C. (2010). Technical and economic analysis of Plasma-assisted Waste-to-Energy processes M.Sc., Department of Earth and Environmental Engineering Fu Foundation of Engineering and Applied Science Columbia University September 2010 Research partially sponsored by the Earth Engineering Center, Columbia University.

Egid, B. (2024). Mosquito-borne disease, waste pickers and health equity: a co-produced study in Andhra Pradesh, India. PhD, Liverpool School of Tropical Medicine.

Eliades, F., Doula, M. K., Papamichael, I., Vardopoulos, I., Voukkali, I., & Zorpas, A. A. (2022). Carving out a niche in the sustainability confluence for environmental education centers in Cyprus and Greece. *Sustainability (Basel)*, *14*(14), 8368. DOI: 10.3390/su14148368

Enemuo, M., & Ogunmodimu, O. (2025). Transitioning the mining sector: A review of renewable energy integration and carbon footprint reduction strategies. *Applied Energy*, *384*, 125484. DOI: 10.1016/j.apenergy.2025.125484

Etim, E. (2024). Leveraging public awareness and behavioural change for entrepreneurial waste management. *Heliyon*, *10*(21), e40063. DOI: 10.1016/j.heliyon.2024.e40063 PMID: 39553649

Faiz, F., Ninduwezuor-Ehiobu, N., Adanma, U. M., & Solomon, N. O. (2024). Blockchain for sustainable waste management: Enhancing transparency and accountability in waste disposal. *Comprehensive Research and Reviews in Science and Technology*, *2*(1), 45–69. DOI: 10.57219/crrst.2024.2.1.0032

Farooq, A., Haputta, P., & Gheewala, S. H. (2024). Economic feasibility assessment of waste to energy technologies for the development of a sustainable municipal solid waste management system in Thailand. *Renewable Energy*, *233*, 121155. DOI: 10.1016/j.renene.2024.121155

Galvão, N., Alves, I. R. F. S., & Bassin, J. P. (2023). Chapter 32 - Municipal solid waste management in Brazil: overview and trade-offs between different treatment technologies. Waste Management and Resource Recycling in the Developing World. P. Singh, P. Verma, R. Singh, A. Ahamad and A. C. S. Batalhão, Elsevier: 755-772.

Geyer, R., Jambeck, J. R., & Law, K. L. (2017). Production, use, and fate of all plastics ever made. *Science Advances*, *3*(7), e1700782. DOI: 10.1126/sciadv.1700782 PMID: 28776036

Gkotsis, P., Kougias, P., Mitrakas, M., & Zouboulis, A. (2023). Biogas upgrading technologies–Recent advances in membrane-based processes. *International Journal of Hydrogen Energy*, *48*(10), 3965–3993. DOI: 10.1016/j.ijhydene.2022.10.228

Gobierno de Colombia. (2020). Bioeconomía para una Colombia Potencia viva y diversa: Hacia una sociedad impulsada por el conocimiento. Bogotá DC, Colombia.

Guberman, R. (2021). "What is Waste-to-Energy?" Retrieved 27/03/2024, 2024, from https://www.rts.com/blog/what-is-waste-to-energy/

Guinée, J. (2016). *Life Cycle Sustainability Assessment: What Is It and What Are Its Challenges? Taking Stock of Industrial Ecology. R. Clift and A. Druckman*. Springer International Publishing.

Gustafsson, M., & Anderberg, S. (2022). Biogas policies and production development in Europe: A comparative analysis of eight countries. *Biofuels, 13*(8), 931–944. DOI: 10.1080/17597269.2022.2034380

Heshmati, A., & Rashidghalam, M. (2021). Assessment of the urban circular economy in Sweden. *Journal of Cleaner Production, 310*, 127475. DOI: 10.1016/j.jclepro.2021.127475

Hiratsuka-Sasaki, A., & Kojima, M. (2020). Inter-municipal cooperation on solid waste management in Japan: Its challenges and implications for ASEAN countries. Regional Waste Management – Inter-municipal Cooperation and Public and Private Partnership. Jakarta, Indonesia. 61: 61-82.

Hongo, T. (2016). Circular economy potential and public–private partnership models in Japan. Towards a Circular Economy: Corporate

Hoornweg, D., & Bhada-Tata, P. (2012). *What a waste: a global review of solid waste management*. DOI: 10.1596/17388

Hospido, A., Moreira, T., Martín, M., Rigola, M., & Feijoo, G. (2005). Environmental evaluation of different treatment processes for sludge from urban wastewater treatments: Anaerobic digestion versus thermal processes. *The International Journal of Life Cycle Assessment, 10*(5), 336–345. DOI: 10.1065/lca2005.05.210

Hossain, S., Law, H. J., & Asfaw, A. (2022). *The waste crisis: roadmap for sustainable waste management in developing countries*. John Wiley & Sons. DOI: 10.1002/9781119811961

Hsu, H.-W., Binyet, E., Nugroho, R. A. A., Wang, W.-C., Srinophakun, P., Chein, R.-Y., Demafelis, R., Chiarasumran, N., Saputro, H., Alhikami, A. F., Sakulshah, N., & Laemthong, T. (2024). Toward sustainability of Waste-to-Energy: An overview. *Energy Conversion and Management, 321*, 119063. DOI: 10.1016/j.enconman.2024.119063

Ibanez, R., Andres, A., Viguri, J., Ortiz, I., & Irabien, J. (2000). Characterisation and management of incinerator wastes. *Journal of Hazardous Materials*, *79*(3), 215–227. DOI: 10.1016/S0304-3894(00)00268-5 PMID: 11077160

International Energy Agency. (2020). *Outlook for biogas and biomethane: Prospects for organic growth*. IEA.

International Energy Agency. (2022). *World Energy Outlook*. International Energy Agency.

International Solid Waste Association. (2024). Global waste management outlook 2024: beyond an age of waste, turning rubbish into a resource. Nairobi, Kenya: 15.

Islam, A., Islam, T., Mahmud, H., Raihan, O., Islam, M. S., Marwani, H. M., Rahman, M. M., Asiri, A. M., Hasan, M. M., Hasan, M. N., Salman, M. S., Kubra, K. T., Shenashen, M. A., Sheikh, M. C., & Awual, M. R. (2024). Accelerating the green hydrogen revolution: A comprehensive analysis of technological advancements and policy interventions. *International Journal of Hydrogen Energy*, *67*, 458–486. DOI: 10.1016/j.ijhydene.2024.04.142

Jambeck, J. R., Geyer, R., Wilcox, C., Siegler, T. R., Perryman, M., Andrady, A., Narayan, R., & Law, K. L. J. (2015). Plastic waste inputs from land into the ocean. *Science*, *347*(6223), 768–771. DOI: 10.1126/science.1260352 PMID: 25678662

Jameel, M. K., Mustafa, M. A., Ahmed, H. S., Mohammed, A., Ghazy, H., Shakir, M. N., Lawas, A. M., Mohammed, S., Idan, A. H., Mahmoud, Z. H., Sayadi, H., & Kianfar, E. (2024). Biogas: Production, properties, applications, economic and challenges: A review. *Results in Chemistry*, *7*, 101549. DOI: 10.1016/j.rechem.2024.101549

Jayakumar Menon, V., & Palackal, A. (2022). Centralized and decentralized approaches to solid waste management–a case study. *The Holistic Approach to Environment*, *12*(4), 155–164. DOI: 10.33765/thate.12.4.3

Jiang, P., Zhang, L., You, S., Van Fan, Y., Tan, R. R., Klemeš, J. J., & You, F. (2023). Blockchain technology applications in waste management: Overview, challenges and opportunities. *Journal of Cleaner Production*, *421*, 138466. DOI: 10.1016/j.jclepro.2023.138466

Kabir, Z., & Kabir, M. (2021). Solid waste management in developing countries: Towards a circular economy. Handbook of Solid Waste Management: Sustainability through Circular Economy, Springer: 1-34.

Kassem, N., Hockey, J., Lopez, C., Lardon, L., Angenent, L. T., & Tester, J. (2020). Integrating anaerobic digestion, hydrothermal liquefaction, and biomethanation within a power-to-gas framework for dairy waste management and grid decarbonization: A techno-economic assessment. *Sustainable Energy & Fuels*, *4*(9), 4644–4661. DOI: 10.1039/D0SE00608D

Kaza, S., Yao, L., Bhada-Tata, P., & Woerden, F. V. (2018). What a waste 2.0: A Global Snapshot of Solid Waste Management to 2050. Washington.

Kennedy, C., Pincetl, S., & Bunje, P. (2011). The study of urban metabolism and its applications to urban planning and design. *Environmental Pollution*, *159*(8), 1965–1973. DOI: 10.1016/j.envpol.2010.10.022 PMID: 21084139

Kern, S. J., Pfeifer, C., & Hofbauer, H. (2013). Cogasification of Polyethylene and Lignite in a Dual Fluidized Bed Gasifier. *Industrial & Engineering Chemistry Research*, *52*(11), 4360–4371. DOI: 10.1021/ie303453e

Khan, S., Ali, B., Alharbi, A. A., Alotaibi, S., & Alkhathami, M. (2024). Efficient IoT-assisted waste collection for urban smart cities. *Sensors (Basel)*, *24*(10), 3167. DOI: 10.3390/s24103167 PMID: 38794020

Khawaja, M. K., Alkayyali, K., Almanasreh, M., & Alkhalidi, A. (2024). Waste-to-energy barriers and solutions for developing countries with limited water and energy resources. *The Science of the Total Environment*, *926*, 172096. DOI: 10.1016/j.scitotenv.2024.172096 PMID: 38556009

Kojima, M. (2020). *Regional Waste Management–Inter-municipal Cooperation and Public and Private Partnership*. Economic Research Institute for ASEAN and East Asia.

Kougias, P. G., & Angelidaki, I. (2018). Biogas and its opportunities—A review. *Frontiers of Environmental Science & Engineering*, *12*(3), 1–12. DOI: 10.1007/s11783-018-1037-8

Kristia, K., & Rabbi, M. F. (2023). "Exploring the Synergy of Renewable Energy in the Circular Economy Framework: A Bibliometric Study." Sustainability (Switzerland) 15(17).

Küfeoğlu, S. (2024). Waste Management Emissions. Net Zero: Decarbonizing the Global Economies, Springer: 555-611.

Kumar, M., Mehra, R., & Saini, M. (2024). Blockchain in Waste Tracking: Enhancing Transparency and Accountability. Municipal Solid Waste Management and Recycling Technologies, IGI Global: 57-70.

Kumari, V., Ansari, A. I., & Singh, A. (2025). Empowering Urban Communities: The Role of Local Stakeholders in Solid Waste Management. Solid Waste Management for Rural Regions, IGI Global Scientific Publishing: 153-164.

Kuzhin, M. F., Joshi, A., Mittal, V., Khatkar, M., & Guven, U. (2024). *Optimizing waste management through IoT and analytics: a case study using the waste management optimization test. BIO web of conferences*. EDP Sciences.

Kyramargiou, Z., & Vardopoulos, I. (2019). Use and sustainability of the ships' waste reception facilities. Evidence from the Port of Corinth, Greece. *Journal of Management and Regional Development*, *15*(3), 93–106.

Lakhouit, A. (2025). Revolutionizing Urban Solid Waste Management with AI and IoT: A review of smart solutions for waste collection, sorting, and recycling. *Results in Engineering*, *25*, 104018. DOI: 10.1016/j.rineng.2025.104018

Lee, J., Lin, K.-Y. A., Jung, S., & Kwon, E. E. (2023). Hybrid renewable energy systems involving thermochemical conversion process for waste-to-energy strategy. *Chemical Engineering Journal*, *452*, 139218. DOI: 10.1016/j.cej.2022.139218

Leme, M. M. V., Rocha, M. H., Lora, E. E. S., Venturini, O. J., Lopes, B. M., & Ferreira, C. H. (2014). Techno-economic analysis and environmental impact assessment of energy recovery from Municipal Solid Waste (MSW) in Brazil. *Resources, Conservation and Recycling*, *87*, 8–20. DOI: 10.1016/j.resconrec.2014.03.003

Loizia, P., Voukkali, I., Zorpas, A. A., Navarro Pedreño, J., Chatziparaskeva, G., Inglezakis, V. J., Vardopoulos, I., & Doula, M. (2021). Measuring the level of environmental performance in insular areas, through key performed indicators, in the framework of waste strategy development. *The Science of the Total Environment*, *753*, 141974. DOI: 10.1016/j.scitotenv.2020.141974 PMID: 32906046

Mahmoudi, S., Baeyens, J., & Seville, J. P. K. (2010). NOx formation and selective non-catalytic reduction (SNCR) in a fluidized bed combustor of biomass. *Biomass and Bioenergy*, *34*(9), 1393–1409. DOI: 10.1016/j.biombioe.2010.04.013

Maranghi, S., Parisi, M. L., Facchini, A., Rubino, A., Kordas, O., & Basosi, R. (2020). Integrating urban metabolism and life cycle assessment to analyse urban sustainability. *Ecological Indicators*, *112*, 106074. DOI: 10.1016/j.ecolind.2020.106074

Moustairas, I., Vardopoulos, I., Kavouras, S., Salvati, L., & Zorpas, A. A. (2022). Exploring factors that affect public acceptance of establishing an urban environmental education and recycling center. *Sustainable Chemistry and Pharmacy*, *25*, 100605. DOI: 10.1016/j.scp.2022.100605

Mukwevho, P., Radzuma, P., & Roos, C. (2024). Exploring Barriers to the Effective Implementation of Integrated Waste Management Plans in Developing Economies: Lessons Learned from South African Municipalities. *Sustainability (Basel)*, *16*(22), 9643. DOI: 10.3390/su16229643

Mullinger, P., & Jenkins, B. (2022). *Industrial and process furnaces: principles, design and operation*. Butterworth-Heinemann.

Münster, M., & Meibom, P. (2011). Optimization of use of waste in the future energy system. *Energy*, *36*(3), 1612–1622. DOI: 10.1016/j.energy.2010.12.070

Náthia-Neves, G., Berni, M., Dragone, G., Mussatto, S. I., & Forster-Carneiro, T. (2018). Anaerobic digestion process: Technological aspects and recent developments. *International Journal of Environmental Science and Technology*, *15*(9), 2033–2046. DOI: 10.1007/s13762-018-1682-2

National Research Council, Commission on Life Sciences, Board on Environmental Studies, Committee on Health Effects of Waste Incineration Council (2000). Waste incineration and public health.

Neehaul, N., Jeetah, P., & Deenapanray, P. (2020). Energy recovery from municipal solid waste in Mauritius: Opportunities and challenges. *Environmental Development*, *33*, 100489. DOI: 10.1016/j.envdev.2019.100489

Neuwahl, F., Cusano, G., Gómez, B. J., Holbrook, S., & Roudier, S. (2019). *Best available techniques (BAT) reference document for waste incineration: industrial emissions directive 2010/75/EU (Integrated Pollution Prevention and Control)*. European Commission.

O'Brien, J. K. (2006). Comparison of air emissions from waste-to-energy facilities to fossil fuel power plants. *North American Waste-to-Energy Conference*. DOI: 10.1115/NAWTEC14-3187

Ojuri, O., Ayodele, F., & Oluwatuyi, O. E. (2018). Risk assessment and rehabilitation potential of a millennium city dumpsite in Sub-Saharan Africa. *Waste Management (New York, N.Y.)*, *76*, 621–628. DOI: 10.1016/j.wasman.2018.03.002 PMID: 29548830

Outapa, P., & Roi-et, V. N. (2018). Evaluation of Greenhouse Gas Emissions from Municipal Solid Waste (MSW) Management: Case Study of Lampang Municipality, Thailand. *Applied Environmental Research*, *40*(1), 46–56. DOI: 10.35762/AER.2018.40.1.5

Ozbay, G., Jones, M., Gadde, M., Isah, S., & Attarwala, T. (2021). Design and operation of effective landfills with minimal effects on the environment and human health. *Journal of Environmental and Public Health*, *2021*(1), 13. DOI: 10.1155/2021/6921607 PMID: 34531916

Ozturk, M., & Dincer, I. (2020). "An efficient waste management system with municipal solid waste incineration plant." Greenhouse Gases. *Science, Technology & Society*, *10*(4), 855–864.

Pan, S.-Y., Du, M. A., Huang, I.-T., Liu, I.-H., Chang, E., & Chiang, P.-C. (2015). Strategies on implementation of waste-to-energy (WTE) supply chain for circular economy system: A review. *Journal of Cleaner Production*, *108*, 409–421. DOI: 10.1016/j.jclepro.2015.06.124

Papamichael, I., Voukkali, I., Economou, F., Loizia, P., Demetriou, G., Esposito, M., Naddeo, V., Liscio, M. C., Sospiro, P., & Zorpas, A. A. (2023). Mobilisation of textile waste to recover high added value products and energy for the transition to circular economy. *Environmental Research*, 117716. PMID: 37995999

Pavlides, T., Vardopoulos, I., Papamichael, I., Voukkali, I., Stylianos, M., & Zorpas, A. A. (2023). Environmental sustainability assessment of excavation, construction, and demolition waste conditions and practices across Greece and Cyprus. *IOP Conference Series. Earth and Environmental Science*, *1196*(1), 012037. DOI: 10.1088/1755-1315/1196/1/012037

Piechota, G., Generowicz, N., Abd, A. A., Othman, M. R., Kowalczyk-Juśko, A., Kumar, G., & Veeremuthu, A. (2024). An introduction to biogas and biomethane. Biogas to Biomethane, Elsevier: 3-40. DOI: 10.1016/B978-0-443-18479-6.00008-9

Pincetl, S., Bunje, P., & Holmes, T. (2012). An expanded urban metabolism method: Toward a systems approach for assessing urban energy processes and causes. *Landscape and Urban Planning*, *107*(3), 193–202. DOI: 10.1016/j.landurbplan.2012.06.006

Potting, J., Hekkert, M., Worrell, E., & Hanemaaijer, A. (2017). *Circular economy: measuring innovation in the product chain*. PBL Publishers.

Pullen, T. (2015). *Anaerobic digestion-making biogas-making Energy: the earthscan expert guide*. Routledge. DOI: 10.4324/9781315770772

Rafey, A., Prabhat, K., & Samar, M. (2020). Comparison of technologies to serve waste to energy conversion. *International Journal of Waste Resources*, *10*, 372.

Ram, M., & Bracci, E. (2024). Waste Management, Waste Indicators and the Relationship with Sustainable Development Goals (SDGs): A Systematic Literature Review. *Sustainability (Basel), 16*(19), 8486. DOI: 10.3390/su16198486

Ramos, A. (2025). Integrating waste thermal conversion and lifecycle analysis for sustainable energy production: Reflecting upon environmental and economic impacts. *Sustainable Energy Technologies and Assessments, 78*, 104342. DOI: 10.1016/j.seta.2025.104342

Ramos, A., Afonso Teixeira, C., & Rouboa, A. (2018). Environmental Analysis of Waste-to-Energy—A Portuguese Case Study. *Energies, 11*(3), 548. DOI: 10.3390/en11030548

Ramos, A., Berzosa, J., Clarens, F., Marin, M., & Rouboa, A. (2020b). Environmental and socio-economic assessment of cork waste gasification: Life cycle and cost analysis. *Journal of Cleaner Production, 249*, 119316. DOI: 10.1016/j.jclepro.2019.119316

Ramos, A., Berzosa, J., Espí, J., Clarens, F., & Rouboa, A. (2020a). Life cycle costing for plasma gasification of municipal solid waste: A socio-economic approach. *Energy Conversion and Management, 209*, 112508. DOI: 10.1016/j.enconman.2020.112508

Ramos, A., Monteiro, E., & Rouboa, A. (2019). Numerical approaches and comprehensive models for gasification process: A review. *Renewable & Sustainable Energy Reviews, 110*, 188–206. DOI: 10.1016/j.rser.2019.04.048

Ramos, A., Monteiro, E., & Rouboa, A. (2022). Biomass pre-treatment techniques for the production of biofuels using thermal conversion methods – A review. *Energy Conversion and Management, 270*, 116271. DOI: 10.1016/j.enconman.2022.116271

Ramos, A., Monteiro, E., Silva, V., & Rouboa, A. (2018). Co-gasification and recent developments on waste-to-energy conversion: A review. *Renewable & Sustainable Energy Reviews, 81*, 380–398. DOI: 10.1016/j.rser.2017.07.025

Ramos, A., & Rouboa, A. (2020). Renewable energy from solid waste: Life cycle analysis and social welfare. *Environmental Impact Assessment Review, 85*, 106469. DOI: 10.1016/j.eiar.2020.106469 PMID: 32952252

Ramos, A., & Rouboa, A. (2020). Syngas production strategies from biomass gasification: Numerical studies for operational conditions and quality indexes. *Renewable Energy, 155*, 1211–1221. DOI: 10.1016/j.renene.2020.03.158

Ramos, A., & Rouboa, A. (2022). Life cycle thinking of plasma gasification as a waste-to-energy tool: Review on environmental, economic and social aspects. *Renewable & Sustainable Energy Reviews, 153*, 111762. DOI: 10.1016/j.rser.2021.111762

Ramos, A., & Rouboa, A. (2022). Life cycle thinking of plasma gasification as a waste-to-energy tool: Review on environmental, economic and social aspects. *Renewable & Sustainable Energy Reviews, 153*, 18. DOI: 10.1016/j.rser.2021.111762

Ramos, A., R. Tavares and A. Rouboa (2018). Microplastics Co-gasification with Biomass: Modelling Syngas Characteristics at Low Temperatures. Technologies and Materials for Renewable Energy, Environment and Sustainability. C. T. Salame, M. Aillerie, P. Papageorgas et al. Melville, Amer Inst Physics. 1968.

Ramos, A. T., Teixeira, C. A., & Rouboa, A. (2019). Environmental Assessment of Municipal Solid Waste by Two-Stage Plasma Gasification. *Energies, 12*(1), 137. DOI: 10.3390/en12010137

Ren, Z., & Zuo, G. (2024). Challenges of Implementing Municipal Solid Waste Separation Policy in China. *Sustainability (Basel), 16*(18), 8081. DOI: 10.3390/su16188081

Rezania, S., Oryani, B., Nasrollahi, V. R., Darajeh, N., Lotfi Ghahroud, M., & Mehranzamir, K. (2023). Review on Waste-to-Energy Approaches toward a Circular Economy in Developed and Developing Countries. *Processes (Basel, Switzerland), 11*(9), 2566. DOI: 10.3390/pr11092566

Ribeiro Hoffmann, A., Sandrin, P., & Doukas, Y. E. (Eds.). (2024). *Climate change in regional perspective. European Union and Latin American initiatives, challenges, and solutions*. Cham, Switzerland: United Nations University Series on Regionalism, Vol. 27, Springer Nature. DOI: 10.1007/978-3-031-49329-4

Roessler, J. G., Olivera, F. D., Wasman, S. J., Townsend, T. G., McVay, M. C., Ferraro, C. C., & Blaisi, N. I. (2016). Construction material properties of slag from the high temperature arc gasification of municipal solid waste. *Waste Management (New York, N.Y.), 52*, 169–179. DOI: 10.1016/j.wasman.2016.03.031 PMID: 27020344

Ruiz, J. A., Juarez, M. C., Morales, M. P., Munoz, P., & Mendivil, M. A. (2013). Biomass gasification for electricity generation: Review of current technology barriers. *Renewable & Sustainable Energy Reviews, 18*, 174–183. DOI: 10.1016/j.rser.2012.10.021

Saadabadi, S. A., Thattai, A. T., Fan, L., Lindeboom, R. E., Spanjers, H., & Aravind, P. (2019). Solid Oxide Fuel Cells fuelled with biogas: Potential and constraints. *Renewable Energy, 134*, 194–214. DOI: 10.1016/j.renene.2018.11.028

Sadeq, A. M., Homod, R. Z., Hussein, A. K., Togun, H., Mahmoodi, A., Isleem, H. F., Patil, A. R., & Moghaddam, A. H. (2024). Hydrogen energy systems: Technologies, trends, and future prospects. *The Science of the Total Environment, 939*, 173622. DOI: 10.1016/j.scitotenv.2024.173622 PMID: 38821273

Satiada, M. A., & Calderon, A. (2021). Comparative analysis of existing waste-to-energy reference plants for municipal solid waste. *Cleaner Environmental Systems, 3*, 100063. DOI: 10.1016/j.cesys.2021.100063

Schulze, G. (2016). Growth Within: A Circular Economy Vision for a Competitive Europe, Ellen MacArthur Foundation, the McKinsey Center for Business Environment: 1-22.

Shah, H. H., Amin, M., Iqbal, A., Nadeem, I., Kalin, M., Soomar, A. M., & Galal, A. M. (2023). A review on gasification and pyrolysis of waste plastics. *Frontiers in Chemistry, 10*, 38. DOI: 10.3389/fchem.2022.960894 PMID: 36819712

Sikdar, S., Siddaiah, A., & Menezes, P. L. (2020). Conversion of waste plastic to oils for tribological applications. *Lubricants (Basel, Switzerland), 8*(8), 78. DOI: 10.3390/lubricants8080078

Singh, A. K., Wadhwa, D., Tripathi, D., Pal, S., Singh, P. K., & Arora, G. (2025). Waste To Energy Technologies In Urban Setting: Comparative Analysis Of Environmental And Economic Performance. *African Journal of Biomedical Research, 28*, 1181–1189.

Solano Meza, J. K., Orjuela Yepes, D., Rodrigo-Ilarri, J., & Rodrigo-Clavero, M.-E. (2023). Comparative analysis of the implementation of support vector machines and long short-term memory artificial neural networks in municipal solid waste management models in megacities. *International Journal of Environmental Research and Public Health, 20*(5), 4256. DOI: 10.3390/ijerph20054256 PMID: 36901265

Sun, J., Meng, X., & Qiao, J. (2023). Data-driven optimal control for municipal solid waste incineration process. *IEEE Transactions on Industrial Informatics, 19*(12), 11444–11454. DOI: 10.1109/TII.2023.3246467

Tang, Y., Dong, J., Li, G., Zheng, Y., Chi, Y., Nzihou, A., Weiss-Hortala, E., & Ye, C. (2020). Environmental and exergetic life cycle assessment of incineration- and gasification-based waste to energy systems in China. *Energy, 205*, 118002. DOI: 10.1016/j.energy.2020.118002

Tavares, G., Zsigraiová, Z., Semiao, V., & da Graca Carvalho, M. (2011). Monitoring, fault detection and operation prediction of MSW incinerators using multivariate statistical methods. *Waste Management (New York, N.Y.), 31*(7), 1635–1644. DOI: 10.1016/j.wasman.2011.02.005 PMID: 21376557

Tavares, R., Ramos, A., & Rouboa, A. (2018). Microplastics thermal treatment by polyethylene terephthalate-biomass gasification. *Energy Conversion and Management, 162*, 118–131. DOI: 10.1016/j.enconman.2018.02.001

Tun, M. M., Palacky, P., Juchelkova, D., & Síťař, V. (2020). Renewable Waste-to-Energy in Southeast Asia: Status, Challenges, Opportunities, and Selection of Waste-to-Energy Technologies. *Applied Sciences (Basel, Switzerland), 10*(20), 7312. DOI: 10.3390/app10207312

Uddin, S. M. N., Gutberlet, J., Ramezani, A., & Nasiruddin, S. M. (2020). Experiencing the everyday of waste pickers: A sustainable livelihoods and health assessment in Dhaka city, Bangladesh. *Journal of International Development, 32*(6), 833–853. DOI: 10.1002/jid.3479

Ulloa-Murillo, L. M., Villegas, L. M., Rodríguez-Ortiz, A. R., Duque-Acevedo, M., & Cortés-García, F. J. (2022). Management of the Organic Fraction of Municipal Solid Waste in the Context of a Sustainable and Circular Model: Analysis of Trends in Latin America and the Caribbean. *International Journal of Environmental Research and Public Health, 19*(10), 6041. DOI: 10.3390/ijerph19106041 PMID: 35627577

United Nations (2016). Transforming our world: The 2030 agenda for sustainable development. A/RES/70/1.

United Nations Habitat. (2010). WTE industry in Latin America. Rio de Janeiro, Brazil, Habitat, UN: 73.

Vardopoulos, I., Abeliotis, K., & Lasaridi, K. (2025). A systematic informetric analysis and literature review of food waste quantification studies in the food service industry. *Sustainability (Basel), 17*(1), 103. DOI: 10.3390/su17010103

Vardopoulos, I., Chroni, C., Synani, K., Abeliotis, K., & Lasaridi, K. (2023). More than meets the eye: Uncovering the social impacts of a household hazardous waste management system in Athens. In D. F. Lekkas (Ed.), *Book of Abstracts of the 18th International Conference on Environmental Science and Technology (CEST2023), 30 August - 2 September, 2023. Athens, Greece* (p. 142). Marousi, Greece: Cosmos. ISBN: 9786188629219.

Vardopoulos, I., D'Agata, A., Escrivà Saneugenio, F., & Salvati, L. (2024). *Sprawl and the city: realizing a sustainable Mediterranean urbanization*. Nova Science Publishers.

Vardopoulos, I., Konstantopoulos, I., Zorpas, A. A., Limousi, L., Bennici, S., Inglezakis, V., & Voukkali, I. (2021). Sustainable metropolitan areas perspectives through assessment of the existing waste management strategies. *Environmental Science and Pollution Research International*, *28*(19), 24305–24320. DOI: 10.1007/s11356-020-07930-1 PMID: 32072414

Vassilev, S. V., Baxter, D., Andersen, L. K., & Vassileva, C. G. (2013). An overview of the composition and application of biomass ash. Part 2. Potential utilisation, technological and ecological advantages and challenges. *Fuel*, *105*, 19–39. DOI: 10.1016/j.fuel.2012.10.001

Vinci, S., Vardopoulos, I., & Salvati, L. (2023). A tale of a shrinking City? Exploring the complex interplay of socio-demographic dynamics in the recent development of Attica, Greece. *Cities (London, England)*, *132*, 104089. DOI: 10.1016/j.cities.2022.104089

Vreugdenhil, B., Zwart, R., & Neeft, J. P. A. (2009). *Tar formation in pyrolysis and gasification*. ECN.

Vukovic, N., & Makogon, E. (2022). Waste-to-energy generation: Complex efficiency analysis of modern technologies. *Sustainability (Basel)*, *14*(21), 13814. DOI: 10.3390/su142113814

Wang, L., Weller, C. L., Jones, D. D., & Hanna, M. A. (2008). Contemporary issues in thermal gasification of biomass and its application to electricity and fuel production. *Biomass and Bioenergy*, *32*(7), 573–581. DOI: 10.1016/j.biombioe.2007.12.007

Wee, J.-H., Bae, Y., Ahn, H., Choi, Y. O., Jeong, E., & Yeo, S. Y. (2022). Fibrous and granular activated carbon mixed media for effective gas removal as a cabin air filter. *Carbon Letters*, *32*(4), 1111–1118. DOI: 10.1007/s42823-022-00345-7

Wellinger, A., Murphy, J. P., & Baxter, D. (2013). *The biogas handbook: science, production and applications*. Elsevier.

Wilson, D. C. (2007). "Development drivers for waste management." 25(3): 198-207.

Wilson, D. C., Rodic, L., Modak, P., Soos, R., Carpintero, A., Velis, K., Iyer, M., & Simonett, O. (2015). *Global waste management outlook*. Unep.

Winkler, M. S., Jackson, D., Sutherland, D., Lim, J. M. U., Srikantaiah, V., Fuhrimann, S., & Medlicott, K. (2017). Sanitation safety planning as a tool for achieving safely managed sanitation systems and safe use of wastewater. *WHO South-East Asia Journal of Public Health, 6*(2), 34–40. DOI: 10.4103/2224-3151.213790 PMID: 28857061

World Bank Group. (2017). Converting Biomass to Energy - A Guide for Developers and Investors. Washington D.C., USA, International Finance Corporation - Energy & Water Advisory: 216.

Yang, T. H., Hu, K. M., Li, R. D., Sun, Y., & Kai, X. P. (2015). Cogasification of typical plastics and rice straw with carbon dioxide. *Environmental Progress & Sustainable Energy, 34*(3), 789–794. DOI: 10.1002/ep.12041

You, S. M., Wang, W., Dai, Y. J., Tong, Y. W., & Wang, C. H. (2016). Comparison of the co-gasification of sewage sludge and food wastes and cost-benefit analysis of gasification- and incineration-based waste treatment schemes. *Bioresource Technology, 218*, 595–605. DOI: 10.1016/j.biortech.2016.07.017 PMID: 27416510

Yukalang, N., Clarke, B., & Ross, K. (2017). Barriers to Effective Municipal Solid Waste Management in a Rapidly Urbanizing Area in Thailand. *International Journal of Environmental Research and Public Health, 14*(9), 1013. DOI: 10.3390/ijerph14091013 PMID: 28869572

Zabed, H. M., Akter, S., Yun, J. H., Zhang, G. Y., Zhang, Y. F., & Qi, X. H. (2020). Biogas from microalgae: Technologies, challenges and opportunities. *Renewable & Sustainable Energy Reviews, 117*, 21. DOI: 10.1016/j.rser.2019.109503

Zhang, Y., Niu, Y., & Zhang, T. (2022). *Introductory chapter: The overview of recent advances of sustainable waste management*. The Circular Economy-Recent Advances in Sustainable Waste Management.

Zorpas, A. A. (2020). Strategy development in the framework of waste management. *The Science of the Total Environment, 716*, 137088. DOI: 10.1016/j.scitotenv.2020.137088 PMID: 32059326

ENDNOTES

1 A Review of Nairobi's Waste-to-Energy Power Plant Initiative (2024) in Ruai – ENVACO

2 Power plant profile: Nerima Waste To Energy Plant, Japan

Chapter 8
A Structured Approach to Zero-Emission Urban Regeneration:
Integrating Planning, Innovation, and Governance for Scalable Urban Transformation

Aristotelis Vartholomaios
https://orcid.org/0000-0001-6177-1130
University of Thessaly, Greece

ABSTRACT

Cities, being at the epicenter of climate change mitigation actions, will require a deep systemic transformation to decarbonize. This cannot happen at once, but rather through a multi-tiered, deliberate, coordinated and persistent effort. This chapter presents a pyramid model as both an analytical tool and a structured approach for studying and mapping the progress across zero-emission urban regeneration projects, governance actions and policies. The model, based on socio-technical transition and multi-level governance theories, classifies these initiatives into three tiers. The first tier includes small-scale pilot projects, the second covers district-wide integrated initiatives that tackle complex cross-sectoral issues and the third involves city-wide regime transformations. The pyramid establishes a hierarchy that deals with both scaling and transformative action. It highlights the importance of innovative governance methods, effective planning and financing, and integration of social equity in zero-emission urban regenerations.

DOI: 10.4018/979-8-3373-2327-5.ch008

1. INTRODUCTION

As our planet is quickly approaching the 1.5°C global warming threshold (Tollefson, 2025), zero-emission urban regeneration stands at the threshold between ambition and necessity (Vardopoulos, 2024). Cities contribute significantly to global climate change (Ribeiro Hoffmann et al., 2024), consuming approximately two-thirds of global energy (Vardopoulos et al., 2024), and generating over 70% of CO_2 emissions (UN, 2019). While national policies and international agreements, such as the EU's "Green Deal" (EC, 2019), continue to shape overarching climate agendas (Doukas et al., 2023), the locus of meaningful implementation remains largely urban. The planning profession is thus confronted with a significant responsibility: how to retrofit, redesign, and restructure urban environments in ways that eliminate carbon emissions while enhancing their economic, social, and spatial viability. The rhetoric of urban energy transition is ubiquitous; the reality of structural transformation is far more elusive (Huovila et al., 2022).

This chapter contends that one of the primary failures of contemporary urban planning for zero carbon emissions lies in the lack of a structured approach to scaling and transformation (De Pascali & Bagaini, 2019; Droege, 2011; van den Dobbelsteen et al., 2018). The problem is not a lack of innovation (Vardopoulos et al., 2024), as cities regularly pioneer energy-efficient buildings (Vardopoulos et al., 2023), test alternative mobility solutions (Kempton et al., 2022), and experiment with decentralized energy networks (Georgiou, et al., 2025). The problem is that these often remain isolated interventions rather than cumulative, integrated processes of change (Lu et al., 2023; Wang et al., 2015). The assumption that successful pilot projects will naturally propagate themselves into wider adoption has proven misguided (Bundgaard & Borrás, 2021; Marique & Reiter, 2011; Woltering et al., 2019). Systemic transformation requires conscious organized action, as niche innovations and "small wins" tie into a overarching institutional framework of urban planning (Vardopoulos et al., 2024), design and governance (Huovila et al., 2022).

The chapter proposes a pyramid model for zero-emission urban regeneration - a conceptual framework that translates urban transition theory into a structured, policy-oriented approach. Drawing from socio-technical transition (Geels, 2019) and multi-level governance (Bache et al., 2016) theory, the model organizes urban decarbonization into three tiers: incremental innovation (Tier 1), district-level integration (Tier 2), and city-wide regime transformation (Tier 3). In doing so, it bridges a critical gap between the experimental and the systemic, guiding planners from ad-hoc interventions towards durable structural change.

The chapter proceeds as follows: First, it examines historical failures in the integration of energy and emissions in urban planning. It then presents the pyramid model, explaining in depth its three tiers: incremental projects, integrated district-level initiatives, and city-wide transformation. Then it examines case studies from experimental eco-districts to full-scale urban transitions, placing them in the proposed three-tiered hierarchy. The chapter then considers the persistent challenge of scaling and transformation. Finally, it discusses governance strategies and implementation mechanisms, including cross-sector partnerships, climate budgeting, and regulatory innovation, that can support cities in progressing through the pyramid's tiers.

2. HISTORICAL BACKGROUND

For much of the twentieth century, urban planning and energy governance developed as disjointed and largely indifferent disciplines (Jabareen, 2006). This mutual neglect eventually formed the unsustainable urban landscapes we now struggle to decarbonise. The legacy of modernist urbanism —functional zoning, car-oriented infrastructure, and sprawling suburbanisation— were conceived in a world where energy was cheap, supply seemed infinite, and emissions were of little consequence (Salvati, 2025). Cities grew on fossil fuels and became structurally dependent on them, while planning regulations, financing models, and governance structures followed note (Amado et al., 2016; Cajot et al., 2015).

It was only in the 1970s, amid the twin oil crises and early stirrings of environmental awareness, that energy became a subject of planning concern in an ad hoc and reactive manner. Fuel shortages pushed towards small-scale experimentations with passive solar design, improved insulation standards, and district heating systems. "Urban Space and Structures" by Martin and March (1972) and later "Energy, Planning and Urban Form" by Owens (1986) where two early works that turned the energy spotlight towards the urban scale, when most studies still focused on individual buildings. The planning profession, for its part, remained preoccupied with spatial order rather than energy efficiency, viewing technological interventions as the domain of engineers rather than urbanists (Amado et al., 2016).

The first serious attempts at integrating climate and energy considerations into urban planning emerge during the 1990s and 2000s. Until mid-2000s initiatives were voluntary, small-scale or situational. For example, UN's "Agenda 21" was a non-binding international agreement that encouraged municipalities to develop and share sustainability strategies. In 2007 the EU presented the "2020 Climate and Energy Package" (2009/28/EC), specifying a 20% target for reducing in Greenhouse Gas (GHG) emissions, increasing renewable energy and improving energy efficiency (Vardopoulos & Karytsas, 2019). Another cornerstone of zero-emission EU policy was the "Energy Performance of Buildings Directive" (EU/2010/31) that mandated improved building energy efficiency standards for all member states (Barone et al., 2024).

The EU has committed to full climate neutrality by 2050, adopting the European Green Deal in 2019 (Vardopoulos, 2025), and the interim "Fit for 55" legislative package (Ovaere & Proost, 2022). The EU Mission for Climate-Neutral and Smart Cities (EC, 2024) has designated 100 European cities as pilots for full decarbonisation by 2030, offering dedicated funding, regulatory support, and governance mechanisms to enable systemic transformation (Vardopoulos et al., 2023). Both within the EU and globally, similar local government cooperation networks have emerged, such as the Carbon Neutral Cities Alliance, C40 Cities and 100 Resilient Cities (Huovila et al., 2022).

Despite this progress, the structural challenges to urban decarbonisation remain formidable. An assumption during 1990s and 2000s was that flagship sustainability projects will spur a large-scale reaction from the market and that climate action will simply trickle down from policy to implementation due to market demand. Yet, despite their climate commitments, most European cities remain locked into carbon-intensive infrastructure, constrained by fiscal limitations, political frictions, bureaucratic inertia and fragmented governance (Vassiliades et al., 2023). An example is Copenhagen's 2012 pledge to become the world's first net-zero city by 2025. The ambitious aim has been pushed back indefinitely, due to underestimation of the necessity for deep systemic changes required and the capacity of existing technology (Hougaard & Christiansen, 2022).

In a similar manner, the EU's Renovation Wave Strategy (Vardopoulos & Poli, 2025), aspires to double the annual rate of energy retrofits (Schoenefeld et al., 2023), yet implementation so far has been slow. Cited reasons are the lack of adequate funding (Sonnleithner, 2021) and outdated energy calculation methodologies in national energy codes (Võsa et al., 2023). Additionally, policies and actions are often not communicated and may become targets of partisan propaganda. For example, the post-COVID-19 concept of "15-minute cities" (Voukkali et al., 2024), has become the target of far-right conspiracy theorists and climate change denialists (Marquet et al., 2025), much like Agenda-21 during the 1990s (Norton, 2014, p. 21).

Despite nearly two decades of sustainability rhetoric the underlying governance structures of most cities remain poorly equipped to advance from tactical interventions to systemic decarbonization (Sateriano et al., 2024). While decarbonization has been framed as both an urban sustainability and a resilience topic, the transition from the former to the latter is largely a question of political will and sound governance (Asprogerakas & Tasopoulou, 2024). Pilot projects, however successful, do not automatically scale; policy targets, no matter how ambitious, do not guarantee transformation (Huovila et al., 2022). Without fundamental shifts in urban governance, planning and finance, cities now face the risk of continuing to accumulate climate strategies and experimental projects, while the carbon-intensive structures of the urban economy remain largely untouched.

3. THE PYRAMID MODEL: FROM INCREMENTAL SOLUTIONS TO TRANSFORMATIVE CHANGE

The pyramid model of zero-emission urban regeneration is structured as a hierarchy of transition stages that reflect varying scales of intervention and degrees of systemic change. The pyramid structure does not indicate a necessarily linear process, but rather the stages of maturity: numerous ideas and experiments emerge at the lower levels, yet only a select few successfully progress to the highest level of integration. Each tier corresponds to a distinct focus, set of objectives, and specific challenges of implementation and governance. The model is intended to be used as an analytical lens through which local actions towards zero-emission urban regeneration can be studied and hierarchically classified. The rationale for the three-tiered pyramid is rooted in socio-technical transition theory (Geels, 2019) and multi-level governance (Bache et al., 2016).

Socio-technical transition theory offers a framework for comprehending significant transformations within societal systems, particularly those influenced by technological advancements and sustainability goals. A key component of this theory is the Multi-Level Perspective (MLP), which identifies three interconnected levels: niches, regimes, and landscapes (Geels, 2019). Niches act as protective environments for innovative technologies and practices, often insulated from mainstream influences. In contrast, regimes consist of the established configurations of institutions, infrastructures, and socio-technical practices that uphold stability and resist alterations. Finally, the landscape includes overarching structural factors such as demographic changes, geopolitical shifts, and environmental challenges, which apply pressure on both regimes and niches. Transformations occur when developments at these levels align, allowing new alternatives to confront and eventually reshape prevailing socio-technical frameworks (Geels, 2019).

The second source of inspiration, Multi-level governance (MLG), provides a framework for understanding the distribution of authority across various institutional levels (e.g. local, regional, national and international). MLG is used in scenarios where decision-making occurs across interconnected jurisdictions. This approach challenges conventional hierarchical governance models by emphasizing the role of non-state actors, policy networks, and informal coordination methods (Bache et al., 2016).

The pyramid comprises three tiers. Tier 1 represents incremental innovations, including niche applications, pilot projects, living labs, or experimental interventions (e.g. guerilla/tactical urbanism) that test new technologies or approaches. Examples include a row of solar-powered homes, a community microgrid, or an innovative street design prioritizing cyclists and pedestrians. The limited project scope is counterbalanced by the capacity to act as demonstrators of feasibility, performance, limitations, user acceptance and associated policy implications. Tier 2 integrates district initiatives at the neighborhood or district level and focuses on scaling-up and aggregating multiple projects into a more cohesive implementation vision. The focus shifts from isolated experiments to coordinated regeneration efforts (i.e. building energy retrofitting, integration of renewables and waste management at the local scale and sustainable transport solutions. Such integrated programs typically require robust partnerships between local government, the private sector and community groups.

Finally, Tier 3 represents city-wide regime transformation. This macro-scale level involves systemic reconfiguration of urban policies, infrastructures, and cultural norms to fully embed zero-emission principles. Transformative changes at this tier are characterized by strong political commitment and the introduction of new institutional frameworks, such as climate budgets or dedicated transition agencies. The ambition is not merely to expand upon the successes of lower tiers, but to fundamentally overhaul the urban system so that low-carbon solutions become the norm rather than the exception. Synthesis and cross-sectoral collaboration here are essential to overcome some of the greatest challenges regarding financing and property management, existing legislative frameworks, multi-stakeholder engagement/collaboration, social equity and practical limitations. Table 1 provides an overview of the three pyramid tiers.

Table 1. Overview of the three pyramid tiers.

Pyramid Tier	Scope & Scale	Focus and Characteristics	Illustrative Example
Tier 1: Incremental Innovations	Micro-scale projects (building or block level; single program)	Niche experiments and pilot projects introducing novel technologies or approaches. Limited in scope but high in learning value; typically driven by innovators or community initiatives.	A single zero-energy housing development or a pilot retrofit of public buildings achieving substantial energy savings.
Tier 2: Integrated District Initiatives	Meso-scale (neighborhood or district; multi-project program)	Integrated regeneration efforts combining multiple interventions (energy, transport, housing, waste, etc.) in a coordinated manner. Designed to scale up early successes and create low-carbon districts.	An eco-district redevelopment project revitalizing a grayfield/ brownfield area with zero mission buildings, renewable energy supply, sustainable transport, and circular waste systems.
Tier 3: City-wide Regime Transformation	Macro-scale (entire city or metropolitan area)	Systemic transformation embedding zero-emission objectives into city-wide policies, infrastructure, and cultural practices. Marked by robust political commitment and new institutional frameworks.	A city implementing an ambitious climate action plan that integrates all sectors (buildings, mobility, energy) and is advancing towards carbon neutrality through comprehensive measures and governance innovations.

Tier 1: Incremental Innovations

Tier 1 corresponds to the "niche" in transition theory where protected spaces allow innovative ideas to develop and mature. At Tier 1, we find the seeds of change: micro-scale interventions that often take the form of pilot projects, prototypes, or one-off experiments. Some examples of Tier 1 projects are a single building retrofitted to net-zero standards, a solar panel installation on a municipal building, an experimental street pedestrianization. They typically involve a handful of actors (sometimes just one enthusiastic department, developer, or community group) and have a limited spatiotemporal scope. The value of Tier 1 projects lies less in their immediate impact and more in the lessons they offer and the inspiration they provide. They are testbeds where new ideas can be tried relatively quickly and cheaply, revealing what works and what doesn't.

One emblematic example of a Tier 1 initiative is the Beddington Zero Energy Development (BedZED) eco-housing project in London. Completed in the early 2000s, BedZED was the UK's first large-scale zero-fossil-energy housing development. It comprises 82 homes (and some office space) designed for maximal energy efficiency and on-site renewable energy generation. Solar panels, high-insulation construction, passive solar heating, and a biofuel combined heat-and-power plant were among its features. As a self-contained project on a former brownfield site, BedZED demonstrated that near-zero-emission living was technically achievable in an urban setting. The project had strong sustainability credentials: residents enjoyed reduced energy bills and a greener environment, and the design won accolades for innovation (Chance, 2009).

However, BedZED's impact remained largely localized. It did not directly trigger a city-wide shift in London's housing construction practices, in part because it was an isolated development led by an environmentally-conscious team. In the pyramid model, BedZED can be placed in Tier 1 – an incremental innovation that provided invaluable lessons (e.g. on the importance of occupant behavior to actual energy savings, and the challenges of small-scale biomass systems)(Cherian et al., 2024). Critically, BedZED also highlighted equity questions: the development included some affordable housing units, yet the innovative nature of the project meant upfront costs were relatively high. Scaling such projects to benefit lower-income residents at a city scale remained a challenge (Chen & Pitts, 2006).

BedZED is not alone – across Europe, countless Tier 1 experiments have bloomed over recent decades. In Germany, for example, the Solar Settlement at Schlierberg, Freiburg (completed in 2006) created a 59-home housing community that produces more energy than it consumes, making it entirely emissions-free at the community level. The project could be considered ahead of its time for its net-positive energy balance. Such pilot communities, alongside projects like Copenhagen's early trials of electric buses or Barcelona's "superblock" street transformations (which started as small-area experiments), illustrate the diversity of Tier 1 approaches. Tier 1 includes both technological innovations and social experiments (e.g. community energy cooperatives). Many European cities are now home of several Tier 1 niche projects; the real question is how to connect these individual "dots" in the landscape of innovation to form a broader picture.

Tier 2: Integrated District Initiatives

Tier 2 lies on a qualitatively different scale and mode of intervention. Tier 2 initiatives take the insights and techniques proven in earlier experiments and embed them into larger, coordinated urban regeneration efforts. If Tier 1 is about experimentation and pilot projects, Tier 2 is about *scaling up*, *aggregating niches* and *broadening innovations*. Instead of a one or a few pilot buildings, we see one or more urban blocks deeply renovated as a whole; instead of one experimental mobility service, we see a city rolling out a fleet of electric buses or trams across a district; instead of an isolated PV installation we see a city-supported community energy scheme that involves hundreds of small co-owners. Planners at this stage often leverage special projects or redevelopment sites to multiply the synergistic potential of low-emission interventions.

A frequently cited example of a Tier 2 transformation is the Hammarby Sjöstad project in Stockholm. In the 1990s, Stockholm faced the task of regenerating a large brownfield waterfront area, and the city seized this opportunity to create an "eco-district" that would go beyond conventional practice. The Hammarby model integrated energy, waste, water, and transport systems for a new residential district of approximately 20,000 people. Innovations included a district heating network powered partly by renewable energy and waste heat, a vacuum waste collection system feeding an anaerobic digester to produce biogas for cooking and buses, extensive pedestrian and cycling infrastructure, and strict energy efficiency standards for buildings. The district achieved roughly 30–40% lower overall environmental impact than typical developments of its time (Pandis Iverot & Brandt, 2011).

Hammarby embodies a second-tier initiative: it scaled up from isolated green-building or green-transport practices to an entire quarter of the city using low-carbon principles. This required strong municipal leadership and multi-stakeholder collaboration – the city utility company, private developers, transport authorities, and citizen groups all played roles in execution (Pandis Iverot & Brandt, 2011). The project also enjoyed support from national policies in Sweden that encouraged renewable energy and provided funding for environmental innovation (Pandis Iveroth et al., 2013), illustrating how multi-level governance can enable Tier 2 success.

Yet Hammarby was not without its critiques. Mahzouni (2015) pointed out the "top-down" planning process that delivered in the technical aspect but neglected the socio-cultural and behavioral aspects of energy use, leading to sub-par performance over the years. The environmental success and promotion led to an increase in property values, raising concerns about the prevailing middle-class monoculture of the neighborhood (Khakee, 2007). These issues show why even at Tier 2, the social dimension must not be ignored. The benefits of sustainable districts should be made accessible to diverse urban residents, and policies (such as affordable housing quotas or community participation in planning) are needed to ensure equity.

Hammarby was not the only such endeavor in its era. Across Europe, multiple cities experimented with integrated sustainable districts around the turn of the millennium. Malmö's Western Harbor (Bo01) in Sweden, for example, was developed as a housing expo showcasing renewable energy and ecological design, and it similarly achieved substantial reductions in energy use and car dependency compared to business-as-usual (Fraker, 2013). In Germany, Freiburg's Vauban quarter – a solar-powered community built in the late 1990s on a former military base – became a notable Tier 2 success with a strong emphasis on citizen involvement and affordable green living (East, 2018; Fraker, 2013).

Such projects often emphasized on certain aspects of energy sustainability: Vauban pioneered community-led planning and car-free living; Malmö focused on architectural innovation and brownfield regeneration; Grenoble's EcoCity and Paris's Clichy-Batignolles in France integrated climate objectives with social housing and public transit improvements (Beatley, 2012; Rudden et al., 2015). Despite their differences, all these cases underscore the potential of coordinated, area-based approaches to deliver deeper carbon reductions than piecemeal efforts.

Tier 3: City-wide Regime Transformation

Advancing to Tier 3, the pyramid narrows to the pinnacle of ambition: city-wide regime transformation. At this tier, the cumulative changes from below ideally contribute to a tipping point where the city's predominant systems of energy, transport, and land-use planning undergo a fundamental reconfiguration. In theoretical terms, Tier 3 corresponds to what transitional theory would describe as a *regime shift*, where the established socio-technical regime (e.g. the fossil-fuel-based urban energy system, car-oriented transport culture, etc.) is replaced or fundamentally reoriented in favor of sustainable alternatives. This is the level of *full transitions*: the point at which low-carbon solutions cease to be marginal or experimental and instead become the new norm structuring the everyday life of the city.

From a policy and planning standpoint, Tier 3 is characterized by policy reforms, new forms of governance, large infrastructure projects and integration of climate objectives into all aspects of urban management. Some hallmarks of this stage are:

(i) the existence of financing and property management tools for deep urban retrofits,
(ii) a systematic integration of urban retrofitting in local spatial plans,
(iii) bioclimatic design codes, planning standards and possibly rating systems for urban regeneration,
(iv) the upgrade of energy grid infrastructure to handle EV charging and integration of local renewable energy,
(v) mechanisms in place to combat energy poverty, displacement of vulnerable communities and "eco-gentrification".

Achieving these requires not just technical plans but institutional transformation: cities might create new departments or agencies focused on sustainability, rewrite regulations, redirect budgets, and cultivate broad coalitions with businesses, civil society, and higher levels of government.

Oslo's zero-emission city transformation constitutes a compelling real-world case of Tier 3 actions. Over the past decade, Oslo's government has embarked on one of the most aggressive climate action strategies of any city, aiming to reduce greenhouse gas emissions by 95% by 2030 (relative to 2009 levels). Starting in 2016, Oslo integrated a carbon emissions budget into the city's financial budgeting process. The Climate Budget specifies maximum caps on allowed emissions across all municipal sectors and requires each city agency to continuously track progress at a quarterly pace (Johansen, 2022). Oslo's coordinated actions involve a city-wide, systemic approach underpinned by novel governance (the climate budget model) and multi-level coordination (e.g. leveraging Norway's renewable-heavy electricity grid and EV incentives). Oslo consciously integrates social equity in its energy transition as well. For example, revenue from road tolls is used to improve public transport while energy retrofits for low-income households are subsidized (Johansen, 2022).

The results have been tangible. Oslo's transport sector has been rapidly electrified, due to both national incentives but also municipal actions like the extensive deployment of charging stations and restrictions on fossil-fuel vehicles in the city center. Meanwhile, heating oil is being phased out in buildings, public transit and bicycle infrastructure expands and even construction sites are now mandated to use electric machinery in city-funded projects (Brebbia et al., 2017; Lind & Espegren, 2018). Oslo's Climate Agency has been established with the goal of monitoring and coordinating all these actions and engaging stakeholders. In terms of emission cuts, the city reported a significant decline of around 30% from 2009 to 2020 (Lind & Espegren, 2018).

In a similar manner, Copenhagen – often lauded for its bike culture and district heating systems – seeks to become the world's first carbon-neutral capital by 2025. The city has already cut its CO_2 emissions by roughly 75% from 2005 levels (Damsø et al., 2017; Hofstad et al., 2022). Amager Bakke, a combined heat-and-power waste-to-energy plant built in 2017 has helped massively in this transformation. What's interesting about Amager Bakke is that it's both an industrial building and a recreational facility. Its characteristic sloped roof houses a dry ski run, a hiking trail and a climbing wall, while offering a spectacular view of the city. In a sense, Amager Bakke constitutes a "counterstructure" (Gospodini, 2004); a unique techno-eco-cultural landmark that challenges preconceptions about the character of industrial sites (Bisinella et al., 2022; Nordestgaard & Arndt, 2019).

In Greece, municipalities and regions are both required to track building energy and carbon footprint and submit a comprehensive "Plan for Building Energy Efficiency" (Ministry of Environment and Energy, 2015). The plan must follow a standardized methodology for prioritizing buildings for energy renovations, crafting renovation scenarios using a cost-benefit analysis and a macro-economic analysis and delineating possible pathways for funding and completing these renovations in a reasonable time scale. The plan is required to be updated bi-annually with new real energy use data. Only local governments with an updated plan are eligible to apply for certain funding programs. The methodology is supported by tools developed in relevant INTERREG MED projects (Ministry of Environment and Energy, 2015). This is a significant departure from previous disjointed actions that were mostly constrained into situational Tier 1 projects.

In a broader sense, many European cities have now submitted or are developing a Climate City Contract (CCC). The CCC is a comprehensive local action plan tailored to each city that is both a pledge and a roadmap for reaching zero emissions by 2030 (Roversi et al., 2022). CCC's demonstrate how international, national and local funding is pooled to support actions that are prioritized according to environmental, social and economic criteria and technical limitations (Beretta & Bracchi, 2023). The document is a necessary requirement for cities participating in the "100 Climate-Neutral and Smart Cities by 2030" EU initiative and will likely form a basis for future EU funding of energy and climate actions (Nucci, 2023).

4. FROM PILOTS TO MAINSTREAM: SCALING-UP VS. TRANSFORMATION

The exploration of the three tiers of urban zero-emission regeneration in the previous section reveals that scaling is necessary but not sufficient for transformation. Scaling up implies repetition: doing more of the same action in a larger geographical area. Transformation implies a qualitative shift: a change in the very systems and cultures that underpin how the city operates. In the context of the pyramid, scaling up is about moving from Tier 1 to Tier 2 (and doing more Tier 2 projects), whereas transformation is about crossing the threshold to Tier 3 via addressing key systemic, technical, social and financial issues.

In the 2000s and 2010s, it was common to hear city leaders proudly announce the next sustainability flagship project. The implicit expectation was that if you accumulate enough of these, the city will eventually become sustainable (Bundgaard & Borrás, 2021). But we now understand that simply multiplying pilots does not automatically equate to a transformed city. Obstacles emerge in this transition, with the most common being: (i) the enormous costs of mass building energy retrofitting (Eames et al., 2014), (ii) energy grid upgrade to handle a full EV car fleet and renewables (Gill et al., 2024) and (iii) the necessary adjustments to urban form and materials to improve the urban climate and passively reduce energy demand in buildings (Vartholomaios, 2015). Furthermore, many of these actions are also resource intensive and carry a large carbon and environmental footprint on their own, something that is not often considered during action plans (Ongpeng et al., 2019). Scaling a pilot often reveals hidden interdependencies that weren't apparent at small scale.

The distinction can be illustrated with a concrete example: many European cities experimented with *low-emission zones* or *car-free days* (Tier 1 initiatives in mobility) (Nieuwenhuijsen & Khreis, 2016). Some then scaled these into permanent pedestrianized city center areas or congestion charging schemes (Tier 2-like implementations) (Soni & Soni, 2016). Sustainable Urban Mobility Plans (SUMPs) have placed such initiatives under a strategic umbrella of priorities and actions (Arsenio et al., 2016). Yet, truly transforming urban mobility towards zero-emissions requires a significant reconfiguration of the urban systems planned for maximizing car use (Tier 3) (Geels, 2019). The improvement of mass transit may not necessarily lead to a proportionate reduction in car ownership and subsequent parking demand (Huang & Chao, 2014). Similarly, EV cars will not reduce congestion but may create additional issues that must be resolved proactively. Without adequate interventions EV's may put an extraordinary burden on the public grid during peak hours (Nutkani & Khan, 2023). In high-density neighborhoods with prevailing on-street parking a jungle of charging cables hanging from apartment windows and balconies may become commonplace if left unregulated (Mlindelwa et al., 2024). Such "mundane" issues may be overlooked during Tier 1 and Tier 2 projects but become apparent only when systemic change is underway.

A common pattern has been the proliferation of pilots without mainstreaming (Bundgaard & Borrás, 2021). The observed "pilotitis" (Bouteligier, 2014) is not because pilots fail technologically, but because the leap to widespread adoption runs into non-technical barriers (lack of sustained funding, legislative barriers, institutional inertia etc.). The pyramid model explicitly prompts practitioners to ask: How will this Tier 1 project feed into Tier 2? How will this Tier 2 initiative help trigger Tier 3? What can be achieved merely via scaling and what through substantial transformative action?

Both scaling and transformative actions are needed: without some scaling, we never reach enough impact; without transformation, we cannot improve a fundamentally unsustainable system. The pyramid model's virtue is that it incorporates both: it recognizes scaling (Tier 1 to 2) as a necessary step but keeps focus on the transformative outcome (Tier 3). The challenge for practitioners and policymakers is to remain cognizant of the gap between a well-executed project and a reformed system. Bridging that gap demands conscious strategies, such as designing pilots with replication in mind, building networks of knowledge sharing between projects, timing policy changes to coincide with demonstration successes, and maintaining a long-term vision amidst short-term wins.

One city that has been attempting to bridge this gap is Freiburg in Germany. Freiburg's Vauban district was built in the 1990s as a prime example of sustainable living (Coates, 2013). In the decades since, the city has progressively integrated Vauban's principles into city-wide policies – for example, adopting stringent energy-efficiency rules for *all* new buildings and expanding renewable district heating systems – gradually moving towards a situation where what was once radical in Vauban now becomes the new standard for all Freiburg (East, 2018; Hamiduddin & Daseking, 2014). This slow but deliberate mainstreaming required continuous political commitment over decades and a civic culture supportive of green innovation. Freiburg's journey illustrates that scaling to transformation is possible but demands sustained effort, not just one-off leaps.

By contrast, consider the ambitious attempt to leap straight to Tier 3 from scratch: Masdar City in the United Arab Emirates. Launched in the late 2000s as a master-planned zero-carbon city in the desert, Masdar initially aimed to incorporate all the hallmarks of Tier 3 at once (100% renewable energy, car-free design, zero-energy buildings) (Crot, 2013). In practice, Masdar never fully achieved its original vision; over time, several of its zero-carbon goals (like banning cars entirely or powering the city exclusively with on-site renewables) were scaled back. A decade on, Masdar evolved into more of a clean-tech business hub than a full city, with much of its land still undeveloped and key sustainability features diluted (Crot, 2013; Cugurullo, 2013). The Masdar case suggests that trying to "jump" directly to Tier 3 by pumping capital into technological solutions can be problematic if the social pillar of sustainability is ignored.

Common barriers to advancing from one tier to the next can be categorized into several intertwined domains:

- **Institutional and Regulatory Barriers:** Rigid regulations at the national or local level and outdated planning and design standards may impede implementation. For example, in Greece urban layout plans determine which streets are pedestrian. Once ratified, the layout plan requires a lengthy and bureaucratic process to update it, which is the case for additional pedestrianizations. Furthermore, planning standards, until the latest update currently under review, followed a car-centric hierarchical street network definition, with little consideration for "shared streets" (i.e. streets without asphalt that allow cars to pass through, but pedestrians have priority). The proposed standards include a "calm traffic street" specifically to cover this previous regulatory gap.
- **Economic and Financial Barriers:** Many Tier 1 and 2 projects rely on grant funding, subsidies, or one-off investments that are not sustainable at mass scale. A city may afford to construct one showcase green building with external funding, but not a hundred without rethinking financing models. The benefits of zero-emission solutions (e.g. reduced pollution, climate mitigation) often accrue to society at large, while the costs may be front-loaded on particular actors (developers, utilities, buyers). This misalignment makes

scaling difficult and may give rise to "eco-gentrification" if left to market forces alone. Cities have begun to explore tools like green bonds, public-private investment partnerships, and even local carbon pricing to fund wider implementation (Zhang, 2024). Land value capture (LVC) is another often overlooked mechanism. In LVC a portion of the betterment value generated by urban regeneration is captured through taxation or development fees to finance public land acquisition, development of public infrastructure and even implement social equity policies (Canelas & Noring, 2022).

- **Technical and Infrastructure Barriers:** Some solutions cannot scale without enabling infrastructure that extends beyond the project site. Electric vehicles, as a niche, cannot become the norm without a city-wide charging network and grid upgrades; a district heating pilot cannot expand without a network of pipes and energy centers. These infrastructure investments often require coordination at higher levels of government or significant capital expenditure, making the jump to Tier 3 challenging unless there is a strategic plan and adequate funding for them.
- **Behavior, culture and property barriers:** Transformative change ultimately means people in the city live, work, and move differently. Even if the technical systems are ready, inhabitants must embrace new routines (like sorting waste for circular systems, reducing private car daily usage for shared mobility options, accepting the need for a higher degree of cooperation). Resistance, apathy, conflicting interests or simple habits can decelerate or even dismantle progress (Fu et al., 2024; Gazheli et al., 2015). Deep engagement, education, and sometimes incentives or regulations that ensure transparency in joint ventures between many small property owners are needed so that what was novel becomes the new normal (Murray & Landry, 2020). As many European cities are dense and populated by many small individual property owners, new tools that promote collective action and reach consensus will likely become essential for complex urban regeneration projects.
- **Power and Interest Barriers:** Finally, incumbent industries and interest groups that benefit from the status quo may resist change (Hess, 2014). A fossil-fuel power utility might lobby against the spread of solar neighborhood microgrids that reduce their sales; real estate developers might push back on stringent green building requirements for fear of higher costs or longer approval times (Fastenrath & Braun, 2018). Overcoming such resistance requires political will, coalition-building, and greater citizen participation in the planning and design process (de Gooyert et al., 2016; Gazheli et al., 2015).

Failed or stalled transitions often exemplify one or more of these barriers. In the next section, we examine how original approaches to governance can help surmount some of these challenges. We will see that new forms of governance –from climate budgeting to cross-sector partnerships– are emerging as key enablers for cities to climb the pyramid and achieve sustained transitions.

5. GOVERNANCE INNOVATIONS AND MULTI-LEVEL COORDINATION

Urban planning and urban regeneration are continuous, non-linear and conflict-ridden political processes (Yiannakou, 2020). If technology and design provide the tools for zero-emission urban regeneration, governance provides the rules and processes by which those tools are deployed politically. A recurrent theme in this discussion of tiers and transitions is that traditional governance arrangements often act as a brake on progress. Hierarchical bureaucracies, sectoral departmentalization, and rigid planning procedures are commonly cited barriers to urban decarbonization (Bunning, 2014; Tozer, 2018). However, around the world, we are witnessing the emergence of prototype governance models tailored to facilitate urban energy transitions. This section explores several such initiatives and analyses how they address the challenges identified earlier towards Tier 3 transformations.

One such governance innovation in recent years is Oslo's climate budget that has already been described in the previous section. The climate budget provides a quantitative accountability mechanism, forcing coordination across sectors while enhancing transparency. Another governance model gaining traction is that of integrated urban transition agencies (Nevens et al., 2013; Wamsler, 2015). These are dedicated bodies tasked with monitoring, coordinating and driving climate and energy transitions across all city departments and stakeholders. An early example was the London Climate Change Agency established in 2005 which sought to leverage public-private partnerships (PPPs) to deliver low-carbon projects across the capital (Schroeder & Bulkeley, 2009).

Another example is the Berlin Energy Agency (BEA), founded in the 1990s as a PPP. BEA was not constrained to an advisory role, but actively coordinated projects. Energy performance contracts (EPCs) were often signed between building owners and Energy Service Companies (ESCos). EPCs bind the payment terms to the actual improvement in building energy performance (Hufen & de Bruijn, 2016). BEA played a significant role in determining energy use goals for EPCs, overcoming financial barriers and upgrading over a thousand public buildings in Berlin(Shonder et al., 2010). The semi-autonomous nature of the agency allowed it to act entrepreneurially, brokering deals between the city, ESCos and banks for the interest of the municipality (Polzin et al., 2016).

The key lesson here is that an agency, task force or public entity in general with a clearly focused mission can concentrate expertise and act as a mediator and a champion of urban decarbonization. Such agencies are multidisciplinary by nature; speaking the language of engineers, urban planners, designers and development specialists, liaising between local government, industry and capital. Such body, however, requires stable support and integration with a broader (ideally nationwide) policy framework to avoid becoming a disconnected "island" of innovation.

Another trend is the use of multi-stakeholder consortia to govern and finance urban regeneration projects. Similar models have emerged where municipalities collaborate with private firms, community organizations, and sometimes regional or national agencies to pool resources and align objectives. The Bristol City Leap initiative in the UK is one recent example (Griffith & Nicholls, 2024) of a public-private joint venture set up to prioritize investments in low-carbon energy infrastructure over a decade. The city of Bristol brought together private sector partners through a competitive process. The partners will share the investment costs and technical expertise to develop renewable energy infrastructure and perform building energy retrofitting.

Another promising governance model is community driven. While "conventional" PPPs bridge the public and private sectors, local residents are often left out of the decision-making processes (Hewitt et al., 2019). One example of a community-driven model of governance is the municipal energy cooperative (Meister et al., 2020). In several European countries, local governments have encouraged or partnered with citizen cooperatives to accelerate renewables deployment and energy savings (Biresselioglu et al., 2021). For example, Germany's energy transition has been bolstered by over a thousand energy cooperatives where citizens jointly own solar farms, wind turbines, or district heating systems – often with municipalities providing seed support or favorable policies (Krug et al., 2022). In these cooperatives, residents buy shares and become co-owners of local renewable energy projects, sharing in both the decision-making and the financial returns.

Such initiatives effectively create Public-Private-People Partnerships (PPPPs), distributing urban regeneration implementation among several stakeholders. European legislation now recognizes "renewable energy communities", something that has been translated into national legislations in member states (Biresselioglu et al., 2021). Merely legislating to allow the existence of such communities is not sufficient for them to grow. The adoption of the energy cooperatives model can be accelerated by supportive local governance actions and active engagement of experts and facilitators. Where these prerequisites are met, district heating, solar energy and wind power cooperatives are becoming more commonplace, allowing individuals to participate even with little financial resources or technical expertise (Krug et al., 2022).

Multi-level governance is another critical facet of enabling Tier-3 urban transitions. City-level innovation often depends on the support – or at least the enabling environment – set by higher levels of government (regional, national or supranational). For example, in many countries, building energy codes are nationwide, which could constrain how far a city can go in mandating efficient or zero-carbon buildings (Hoppe & Lulofs, 2008). Some pioneering cities worked around this by negotiating with national authorities for pilot status or special exemptions to try stricter codes in certain districts (essentially creating Tier 2 enclaves with advanced standards). When these pilots succeeded, they provided evidence for the national government to later raise standards country-wide (thus feeding into broader transformation) (Tambach & Visscher, 2012). This was observed in the Netherlands, where experiments with natural gas-free neighborhoods in cities informed the national decision to phase out gas in all new housing (Looijen Bader & Oostra, 2022). Similarly, several German cities' experiences with energy-efficient "passive house" buildings contributed to strengthening of federal energy codes (Schepelmann, 2018).

EU's actions have had a significant role in assisting local initiatives. For example, the EU Covenant of Mayors initiative, encouraged cities to establish a concrete action plan towards climate-resilience (Haupt, 2018). INTERREG programs have provided funding for cross-border and regional projects, allowing cities to collaborate and share solutions on integrated urban development (Cosmi et al., 2017; Davidson et al., 2019). Therefore, another important aspect of zero-emission governance is advocacy and networking – cities joining forces in networks like C40, ICLEI, NetZeroCities, Rockefeller Foundation's 100 Resilient Cities and the Covenant of Mayors to collectively push for the regulatory and funding frameworks that empower local action.

6. INTEGRATING PLANNING MECHANISMS AND FINANCIAL TOOLS

Governance innovations will falter if they are not supported by appropriate planning mechanisms and financing strategies. To reach Tier 3 of the pyramid model formal planning tools –from land-use plans to mobility plans– must function as an integrated strategy rather than a set of isolated policies. In many cities, one finds a masterplan drawn up by urban planners, a separate climate action plan by environment and energy experts and a SUMP by transport engineers. Too often, these documents only loosely reference each other. The result can be contradictions on the ground as plans and strategies do not align (Crilly & Lemon, 2018).

To truly support zero-emission regeneration, planning instruments need to be reconfigured and harmonized. This means that the various mandatory tools, such as Strategic Environmental Assessments (SEAs) and SUMPs, should be aligned into a single coherent strategic process. In practice, SEAs have sometimes been treated as tick-box addendums, rather than steering documents (Campeol, 2020). A more effective approach would be to use SEA proactively: integrate climate and energy criteria into the core of plan-making. For example, when developing a new city plan, the SEA process could enforce climate and energy budgeting to guide, rather than just audit the planning process. Similarly, SUMPs, in the best cases, connect transport planning with land use planning, urban form regulation and broader environment, and social goals (Gavanas et al., 2016; Pozoukidou et al., 2017). Yet, some cities treat the SUMP as merely a transport department's manifesto that is disjointed from other plans (Hartl et al., 2024; Jordová & Brůhová-Foltýnová, 2021).

One important planning instrument that needs emphasis in zero-emission regeneration is the city's regulatory power over development control and financing. Land value capture (LVC) and other real estate tools can enable and finance urban regeneration if used wisely (Bloom, 2024; Canelas & Noring, 2022; Huston & Lahbash, 2018). When a city invests in urban regeneration, it usually increases land and property values in the vicinity. If unchecked, this can lead to gentrification and windfall gains for private landowners, while the city struggles to fund further improvements. LVC include several mechanisms -from development impact fees and betterment levies to public land leasing- to rectify this by reclaiming some of the value created by public action for public benefit (Medda, 2012). This is already commonplace in some form: for instance, many cities charge developers fees or ask for in-kind contributions (like affordable housing units or infrastructure upgrades) as conditions for granting planning permission for large projects. To support zero-emission goals, these practices can be honed and expanded. A few examples illustrate the potential:

- **Tax Increment Financing (TIF)**: A city designates a regeneration area (say a low-carbon district retrofit) and uses future increases in property tax revenues (due to rising values) to finance upfront the improvements in that area. Essentially, the project pays for itself over time via the value it generates (Squires & Lord, 2012).
- **Betterment and Infrastructure levies**: Cities like São Paulo have at times charged property owners a fee proportional to the value uplift from new infrastructure, explicitly to fund those projects (Nobre, 2023). This aligns costs and benefits and can fund expensive Tier 2/Tier 3 investments.
- **Incentives**: Another tool is to grant extra development rights (like higher building density) in exchange for significant green features or contributions (Cease et al., 2019; Greene et al., 2017; Zanon & Verones, 2013). For instance, a city could allow a developer to build a taller building if it is certified zero-carbon and if the developer funds nearby public infrastructure.

Alternatively, incentives could be used to encourage a large number of property owners to collaborate towards a complex urban project, such as the total regeneration of one or more urban blocks.

7. SOCIAL EQUITY IN ZERO-EMISSION URBAN REGENERATION

This chapter has interwoven the themes of equity and social justice with technology, policy, and design. The significance of this integration cannot be overstated: zero-emission urban regeneration actions will fail to gain the necessary momentum if they are perceived as exacerbating socioeconomic inequalities. Urban political ecology consistently reminds us that environmental interventions are almost never neutral. They redistribute resources, risks, and benefits, and thus are inherently political (Heynen, 2017). What does this mean in practical terms for the pyramid model?

Early Tier-1 experiments often set the tone for later scaling: they can either debunk the myth that sustainability is a "luxury" by thriving in low-income contexts, or they can reinforce that myth by only occurring in affluent enclaves. Prioritizing low-income neighborhoods and social housing for Tier-1 actions can help in tackling energy poverty. At Tier 2 and Tier 3, equity considerations become concrete in spatial form. The choice of where to implement an eco-district or a low-carbon zone is an equity choice. Will the city focus only on already gentrifying inner-city areas and make them even more attractive (potentially displacing existing residents)? Or will it also invest in regenerating neglected quarters, tying climate action to improvements in housing quality, public health, and local jobs in those areas?

To avoid "eco-gentrification" and the associated displacement of low-income residents, cities are experimenting with anti- safeguard policies such as community land trusts or stringent affordable housing requirements as part of green regeneration projects (Kim & Eisenlohr, 2022; Levine et al., 2022). For instance, Paris identified specific "priority neighborhoods" for building retrofits that overlap with social housing concentrations (Coscia et al., 2020). In Portland, Oregon, the "right to return" gives priority to subsidized housing to people that were displaced after urban regeneration (Mandal, 2018). The CCCs being developed by EU mission cities, for example, include stakeholder commitment sections which could be used to formalize community benefits and participation in projects (Foulds et al., 2023).

Procedural justice matters: involving a broad public in decision-making helps identify these issues early and build acceptance. The aforementioned example of citizens' climate assemblies is relevant. When a demographically diverse group of residents deliberates and produces recommendations, those tend to highlight equity concerns (e.g. an assembly might support pedestrianizing neighborhood streets only if lost car parking spaces are somehow recovered) (Ross et al., 2021). Cities like Barcelona have pioneered such assemblies, and the approach is spreading (Amorim-Maia et al., 2023). Participatory budgeting is another avenue: giving communities real power to decide a portion of the city's budget can lead to investments that reflect diverse priorities (Epting, 2020). Several zero-emission living labs have been carried out in European cities with encouraging results (Bulkeley et al., 2019; Engels et al., 2019; Giannouli et al., 2018; Voytenko et al., 2016) as situational Tier-1 projects. The timing is now appropriate to move towards a more enduring and systematic integration of such initiatives in formal urban planning procedures and governance.

Ultimately, achieving zero-emissions should not come at the cost of increased socioeconomic inequality. Effective communication and targeting of urban regeneration policies should adapt to the socio-demographics of the city (Vogiatzi et al., 2018). Instead of framing them as "we need to reduce emissions by X", a better narrative (backed by concrete action) is "we will improve your neighborhood, your housing, and your transit, and in doing so, cut emissions". When regeneration is done in partnership with communities, it can enhance social capital and trust in government, which in turn accelerates climate action in a virtuous cycle.

8. DISCUSSION

Urban planners and policymakers today face the dual challenge of scale and scope when considering urban regeneration. Zero-emission urban regeneration is not simply about city beautification or economic uplift, but also about increasing the climate resilience of cities and reducing their impact on the planet. The task requires scaling up what works while widening the scope, eventually leading to systemic change. The presented pyramid model offers a structured roadmap to navigate this challenge.

The pyramid model can be used both as an exploratory lens and a diagnostic tool. The concept of tiers can be employed by local governments and planners to assess their current portfolio of initiatives and identify gaps. For instance, a city might find it has many Tier 1 pilots but no Tier 2 programs to aggregate and build upon them – suggesting a need to create intermediating structures or scaling strategies. The model classifies plans, actions and projects aiding in the development of effective urban regeneration strategies. It also helps in discussing actions, as speaking about "pilot projects" (Tier 1), "integrated initiatives" (Tier 2) and "systemic change" (Tier 3) allows for more precise communication.

The case studies presented at each tier level have provided a real-world basis for the theoretical model. These examples also provide evidence that progression through tiers, as suggested by the model, is plausible. The chapter demonstrated how BedZED and similar projects (like the Solar Settlement in Freiburg) incubated ideas that informed larger schemes. How Hammarby, Vauban, and other eco-districts showed integration in practice and influenced city policies. Finally, how cities like Oslo and Copenhagen are beginning to turn towards transformative change through innovations in governance.

Additionally, the contrast between scaling and transformation was brought out by these examples. A key lesson is that accumulating pilot projects does not equal systemic change as more complex issues arise when considering a wider scope of implementation. Institutional, financial, technical, behavioral barriers to such change were identified and possible routes for addressing them were proposed. The tiered model opens the path for comparative analyses across different urban contexts. For urban practitioners, the cases underscore the importance of context. While the paths to zero-emission cities may differ, understanding the current maturity of actions, policies and projects along the continuum of change can be vital for decision making.

In terms of implementation, the pyramid model suggests a policy-oriented approach that is somewhat akin to a roadmap. The model could be integrated into the current climate and energy action plans to establish a clearer relation between tiers and how projects are interconnected. Such implementation would ensure that no tier is neglected: that there is always a pipeline of innovation, mechanisms for scaling, and efforts to institutionalize change. It would also explicitly allocate roles: community groups and entrepreneurs might be empowered and funded to generate Tier 1 ideas; PPPs or cooperatives could be formed for Tier 2 delivery; and municipal authorities would focus on Tier 3 frameworks and regulations. Some cities may already be implicitly doing this, yet the pyramid model renders such multi-tier thinking explicit, transparent and deliberate.

There are, of course, limitations to this framework. Reality is messy, and not every project or action will fit neatly into a tier. Furthermore, changes in government might lead to regressions and backpedaling of policies and priorities, dismantling and scaling back what has been previously built. The model simply highlights *what needs to happen* for success, and where the weak links might be. Ultimately, it is a way to simplify and clarify the complex process of urban transformation.

9. CONCLUSIONS

Moving forward, one might ask: how does this approach align with broader urban agendas, like smart cities, resilience, or post-pandemic recovery? Interestingly, it aligns well. Smart city projects often start as pilots (Tier 1) and need integration with other strategic goals. Zero-emission urban regeneration pilot projects often have a strong "smart city" component for that reason. Similarly, climate resilience requires layering social and environmental goals within an ever-changing landscape of challenges. The COVID-19 pandemic, for example, challenged our idea of daily commuting between work and home and has demonstrated a greater need for resilient public spaces, active personal mobility and well-ventilated buildings. This demand can accelerate the transition from Tier 1 and 2 to Tier 3 transformative actions.

The journey from theory to practice is often considered a challenge, and rightfully so. This chapter has attempted to make that journey a bit more navigable for urban regeneration in the zero-emission context. The chapter has attempted to delineate a path that is both visionary and pragmatic. The pyramid model encourages us to persevere methodically: nurture the seeds, cultivate the garden, and ultimately reshape the landscape. From Haussmann's Paris boulevards to the car-centric urban renewals of the 20th century, cities have demonstrated the capacity to reinvent themselves in response to new challenges or visions.

Now, in the 21st century, the climate emergency compels us to reinvent cities once more. This time cities will need to make a profound transition towards zero-emissions to ensure the survival of our civilization as global climate change progresses. Zero-emission urban regeneration is an immense undertaking, but with frameworks like the one discussed, it becomes a series of manageable and interconnected actions. As planners and citizens, our task is to connect those actions into a coherent transformation, building the pyramid one level at a time, until the future sustainable and resilient city becomes a reality.

REFERENCES

Amado, M., Poggi, F., & Amado, A. R. (2016). Energy efficient city: A model for urban planning. *Sustainable Cities and Society, 26*, 476–485. DOI: 10.1016/j.scs.2016.04.011

Amorim-Maia, A. T., Anguelovski, I., Chu, E., & Connolly, J. J. T. (2023). Governing intersectional climate justice: Tactics and lessons from Barcelona. *Environmental Policy and Governance*. Advance online publication. DOI: 10.1002/eet.2075

Arsenio, E., Martens, K., & Di Ciommo, F. (2016). Sustainable urban mobility plans: Bridging climate change and equity targets? *Research in Transportation Economics, 55*, 30–39. DOI: 10.1016/j.retrec.2016.04.008

Asprogerakas, E., & Tasopoulou, A. (2024). From sustainability to resilience: Tracing the transition path for spatial planning policy in Greece. *European Spatial Research and Policy, 31*(1), 29–52. DOI: 10.18778/1231-1952.31.1.04

Bache, I., Bartle, I., & Flinders, M. (2016). Multi-level governance. In *Handbook on Theories of Governance* (pp. 486–498). Edward Elgar Publishing., https://www.elgaronline.com/edcollchap/edcoll/9781782548492/9781782548492.00052.xml DOI: 10.4337/9781782548508.00052

Barone, G., Vardopoulos, I., Attia, S., & Vassiliades, C. (2024). Optimizing energy-efficient building renovation: Integrating double-skin façades with solar systems in the Mediterranean landscape. *Energy Reports, 12*, 2933–2945. DOI: 10.1016/j.egyr.2024.08.032

Beatley, T. (Ed.). (2012). *Green Cities of Europe*. Island Press/Center for Resource Economics., DOI: 10.5822/978-1-61091-175-7

Beretta, I., & Bracchi, C. (2023). Climate-neutral and Smart Cities: A critical review through the lens of environmental justice. *Frontiers in Sociology, 8*, 1175592. DOI: 10.3389/fsoc.2023.1175592 PMID: 37854356

Biresselioglu, M. E., Limoncuoglu, S. A., Demir, M. H., Reichl, J., Burgstaller, K., Sciullo, A., & Ferrero, E. (2021). Legal Provisions and Market Conditions for Energy Communities in Austria, Germany, Greece, Italy, Spain, and Turkey: A Comparative Assessment. *Sustainability (Basel), 13*(20), 20. Advance online publication. DOI: 10.3390/su132011212

Bisinella, V., Nedenskov, J., Riber, C., Hulgaard, T., & Christensen, T. H. (2022). Environmental assessment of amending the Amager Bakke incineration plant in Copenhagen with carbon capture and storage. *Waste Management & Research, 40*(1), 79–95. DOI: 10.1177/0734242X211048125 PMID: 34585637

Bloom, A. (2024). Public land, value capture, and the rise of speculative urban governance in post-crisis London. *Environment and Planning A. Environment & Planning A, 56*(6), 1771–1786. DOI: 10.1177/0308518X231199701

Bouteligier, S. (2014). Urban Sustainability Pilot Projects: Fit or Misfit between Challenge and Solution? In *Scale-sensitive Governance of the Environment* (pp. 188–202). John Wiley & Sons, Ltd. DOI: 10.1002/9781118567135.ch12

Brebbia, C. A., Longhurst, J., Marco, E., & Booth, C. (2017). *Sustainable Development and Planning IX*. WIT Press.

Bulkeley, H., Marvin, S., Palgan, Y. V., McCormick, K., Breitfuss-Loidl, M., Mai, L., von Wirth, T., & Frantzeskaki, N. (2019). Urban living laboratories: Conducting the experimental city? *European Urban and Regional Studies, 26*(4), 317–335. DOI: 10.1177/0969776418787222

Bundgaard, L., & Borrás, S. (2021). City-wide scale-up of smart city pilot projects: Governance conditions. *Technological Forecasting and Social Change, 172*, 121014. DOI: 10.1016/j.techfore.2021.121014

Bunning, J. (2014). Governance for regenerative and decarbonised eco-city regions. *Renewable Energy, 67*, 73–79. DOI: 10.1016/j.renene.2013.11.041

Cajot, S., Peter, M., Bahu, J.-M., Koch, A., & Maréchal, F. (2015). Energy Planning in the Urban Context: Challenges and Perspectives. *Energy Procedia, 78*, 3366–3371. DOI: 10.1016/j.egypro.2015.11.752

Campeol, G. (Ed.). (2020). *Strategic Environmental Assessment and Urban Planning: Methodological Reflections and Case Studies*. Springer International Publishing., DOI: 10.1007/978-3-030-46180-5

Canelas, P., & Noring, L. (2022). Governmentalities of land value capture in urban redevelopment. *Land Use Policy, 122*, 106396. DOI: 10.1016/j.landusepol.2022.106396

Cease, B., Kim, H., Kim, D., Ko, Y., & Cappel, C. (2019). Barriers and incentives for sustainable urban development: An analysis of the adoption of LEED-ND projects. *Journal of Environmental Management, 244*, 304–312. DOI: 10.1016/j.jenvman.2019.04.020 PMID: 31128335

Chance, T. (2009). Towards sustainable residential communities; the Beddington Zero Energy Development (BedZED) and beyond. *Environment and Urbanization, 21*(2), 527–544. DOI: 10.1177/0956247809339007

Chen, B., & Pitts, A. (2006). A Socio-Technical Perspective to Enact Zero Emissions / Energy Development in China. *Proceedings of PLEA2006 - The 23 Rd Conference on Passive and Low Energy Architecture.* PLEA2006 - The 23 rd Conference on Passive and Low Energy Architecture, Geneva. http://web5.arch.cuhk.edu.hk/server1/staff1/edward/www/plea2018/plea/2006/Vol1/PLEA2006_PAPER142.pdf

Cherian, B. A., Tokbolat, S., & Rodrigues, L. (2024). Evaluating Thermal Comfort in Zero-Energy Developments: A Case Study of BedZED. *The 1st International Conference of Net Zero Carbon Built Environment.* https://nottingham-repository.worktribe.com/output/39163737

Coates, G. J. (2013). The sustainable urban district of vauban in freiburg, germany. *International Journal of Design & Nature and Ecodynamics, 8*(4), 265–286. DOI: 10.2495/DNE-V8-N4-265-286

Coscia, C., Mukerjee, S., Palmieri, B. L., & Quintanal Rivacoba, C. (2020). Enhancing the Sustainability of Social Housing Policies through the Social Impact Approach: Innovative Perspectives form a "Paris Affordable Housing Challenge" Project in France. *Sustainability (Basel), 12*(23), 9903. Advance online publication. DOI: 10.3390/su12239903

Cosmi, C., Salvia, M., Di Leo, S., Pietrapertosa, F., & Loperte, S. (2017). Interregional Cooperation as a Key Tool for the Achievement of Strategic-Energy and Climate Targets: The Experience of the INTERREG IVC RENERGY and SEE RE-SEEties Projects. In Bisello, A., Vettorato, D., Stephens, R., & Elisei, P. (Eds.), *Smart and Sustainable Planning for Cities and Regions: Results of SSPCR 2015* (pp. 335–352). Springer International Publishing., DOI: 10.1007/978-3-319-44899-2_20

Crilly, M., & Lemon, M. (2018). The Misalignment of Policy and Practice in Sustainable Urban Design. In M. Dastbaz, W. Naudé, & J. Manoochehri (Eds.), *Smart Futures, Challenges of Urbanisation, and Social Sustainability* (pp. 69–87). Springer International Publishing. DOI: 10.1007/978-3-319-74549-7_5

Crot, L. (2013). Planning for Sustainability in Non-democratic Polities: The Case of Masdar City. *Urban Studies (Edinburgh, Scotland), 50*(13), 2809–2825. DOI: 10.1177/0042098012474697

Cugurullo, F. (2013). How to Build a Sandcastle: An Analysis of the Genesis and Development of Masdar City. *Journal of Urban Technology, 20*(1), 23–37. DOI: 10.1080/10630732.2012.735105

Damsø, T., Kjær, T., & Christensen, T. B. (2017). Implementation of local climate action plans: Copenhagen – Towards a carbon-neutral capital. *Journal of Cleaner Production, 167*, 406–415. DOI: 10.1016/j.jclepro.2017.08.156

Davidson, K., Coenen, L., & Gleeson, B. (2019). A Decade of C40: Research Insights and Agendas for City Networks. *Global Policy, 10*(4), 697–708. DOI: 10.1111/1758-5899.12740

de Gooyert, V., Rouwette, E., van Kranenburg, H., Freeman, E., & van Breen, H. (2016). Sustainability transition dynamics: Towards overcoming policy resistance. *Technological Forecasting and Social Change, 111*, 135–145. DOI: 10.1016/j.techfore.2016.06.019

De Pascali, P., & Bagaini, A. (2019). Energy Transition and Urban Planning for Local Development. A Critical Review of the Evolution of Integrated Spatial and Energy Planning. *Energies, 12*(1), 1. Advance online publication. DOI: 10.3390/en12010035

Doukas, Y. E., Salvati, L., & Vardopoulos, I. (2023). Unraveling the European agricultural policy sustainable development trajectory. *Land (Basel), 12*(9), 1749. DOI: 10.3390/land12091749

Droege, P. (2011). *Urban Energy Transition: From Fossil Fuels to Renewable Power*. Elsevier.

Eames, M., Dixon, T., Lannon, S., Hunt, M., Laurentis, C. D., Marvin, S., Hodson, M., Guthrie, P., & Georgiadou, M. (2014). *Retrofit 2050: Critical challenges for urban transitions*. Cardiff University. https://centaur.reading.ac.uk/36187/1/critical%20challenges%20briefing-March%202014.pdf

East, M. (2018). Rising Cities: Continuity, Innovation and Deliberation of Vauban District of Freiburg, Germany. *Ecocycles, 4*(2), 58–74. DOI: 10.19040/ecocycles.v4i2.106

EC. (2019). *The European Green Deal*. EU Publications.

EC. (2024, December 3). *Climate-neutral and smart cities*. https://research-and-innovation.ec.europa.eu/funding/funding-opportunities/funding-programmes-and-open-calls/horizon-europe/eu-missions-horizon-europe/climate-neutral-and-smart-cities_en

Engels, F., Wentland, A., & Pfotenhauer, S. M. (2019). Testing future societies? Developing a framework for test beds and living labs as instruments of innovation governance. *Research Policy, 48*(9), 103826. DOI: 10.1016/j.respol.2019.103826

Epting, S. (2020). Participatory Budgeting for Environmental Justice. *Ethics, Policy & Environment, 23*(1), 22–36. DOI: 10.1080/21550085.2020.1746003

Fastenrath, S., & Braun, B. (2018). Ambivalent urban sustainability transitions: Insights from Brisbane's building sector. *Journal of Cleaner Production, 176*, 581–589. DOI: 10.1016/j.jclepro.2017.12.134

Foulds, C., Valkenburg, G., Ryghaug, M., Suboticki, I., Skjølsvold, T. M., Korsnes, M., & Heidenreich, S. (2023). Implementing Mission-oriented Experiments: Recommendations on Epistemic Inclusion for City Stakeholders Working in Climate Change Initiatives. *Journal of City Climate Policy and Economy, 2*(1), 55–76. DOI: 10.3138/jccpe-2022-0014

Fraker, H. (2013). *The Hidden Potential of Sustainable Neighborhoods*. Island Press/ Center for Resource Economics., DOI: 10.5822/978-1-61091-409-3

Fu, X., Liu, G., Wu, H., Zhuang, T., Huang, R., Yuan, F., & Zhang, Y. (2024). Dissecting behavioral inertia in shaping different resident participation behaviors in neighborhood regeneration: A quantitative behavioral experiment. *Environmental Impact Assessment Review, 109*, 107632. DOI: 10.1016/j.eiar.2024.107632

Gavanas, N., Pozoukidou, G., & Verani, E. (2016). Integration of LUTI models into sustainable urban mobility plans (SUMPs). *European Journal of Environmental Sciences, 6*(1), 11–17. DOI: 10.14712/23361964.2016.3

Gazheli, A., Antal, M., & van den Bergh, J. (2015). The behavioral basis of policies fostering long-run transitions: Stakeholders, limited rationality and social context. *Futures, 69*, 14–30. DOI: 10.1016/j.futures.2015.03.008

Geels, F. W. (2019). Socio-technical transitions to sustainability: A review of criticisms and elaborations of the Multi-Level Perspective. *Current Opinion in Environmental Sustainability, 39*, 187–201. DOI: 10.1016/j.cosust.2019.06.009

Georgiou, C., Maialetti, M., Zorpas, A. A., Salvati, L., & Vardopoulos, I. (2025). Policy and planning for "zero-net energy" real estate development in rural and ex-urban districts. In *Environmental Sustainability and Global Change* (pp. 291–303). Elsevier., DOI: 10.1016/B978-0-443-31596-1.00020-9

Giannouli, I., Tourkolias, C., Zuidema, C., Tasopoulou, A., Blathra, S., Salemink, K., Gugerell, K., Georgiou, P., Chalatsis, T., Christidou, C., Bellis, V., Vasiloglou, N., & Koutsomarkos, N. (2018). A methodological approach for holistic energy planning using the living lab concept: The case of the prefecture of Karditsa. *European Journal of Environmental Sciences, 8*(1), 14–22. DOI: 10.14712/23361964.2018.3

Gill, A., T, R., Upadhyay, R., & Gautam, A. A. (2024). Strategic Grid Integration of Renewable Using Electric Vehicles for Smart Cities Stability. *E3S Web of Conferences, 540*, 10024. DOI: 10.1051/e3sconf/202454010024

Gospodini, A. (2004). Urban morphology and place identity in European cities: Built heritage and innovative design. *Journal of Urban Design*, *9*(2), 225–248. DOI: 10.1080/1357480042000227834

Greene, M., Mora, R. I., Figueroa, C., & Waintrub, N., & de D. Ortúzar, J. (. (2017). Towards a sustainable city: Applying urban renewal incentives according to the social and urban characteristics of the area. *Habitat International*, *68*, 15–23. DOI: 10.1016/j.habitatint.2017.03.004

Griffith, R., & Nicholls, M. (2024). *Financing UK place- based climate action: From Westminster to Cumberland.* Grantham Research Institute on Climate Change and the Environment, London School of Economics and Political Science.

Hamiduddin, I., & Daseking, W. (2014). Community-based planning in Freiburg, Germany: The case of Vauban. In *Community Action and Planning* (pp. 237–258). Policy Press. https://bristoluniversitypressdigital.com/edcollchap/book/9781447315193/ch013.xml

Hartl, R., Harms, P., & Egermann, M. (2024). Towards transformation-oriented planning: What can sustainable urban mobility planning (SUMP) learn from transition management (TM)? *Transport Reviews*, *44*(1), 167–190. DOI: 10.1080/01441647.2023.2239497

Haupt, W. (2018). European Municipalities Engaging in Climate Change Mitigation and Adaptation Networks: Examining the Case of the Covenant of Mayors. In Y. Yamagata & A. Sharifi (Eds.), *Resilience-Oriented Urban Planning: Theoretical and Empirical Insights* (pp. 93–110). Springer International Publishing. DOI: 10.1007/978-3-319-75798-8_5

Hess, D. J. (2014). Sustainability transitions: A political coalition perspective. *Research Policy*, *43*(2), 278–283. DOI: 10.1016/j.respol.2013.10.008

Hewitt, R. J., Bradley, N., Baggio Compagnucci, A., Barlagne, C., Ceglarz, A., Cremades, R., McKeen, M., Otto, I. M., & Slee, B. (2019). Social Innovation in Community Energy in Europe: A Review of the Evidence. *Frontiers in Energy Research*, *7*, 31. Advance online publication. DOI: 10.3389/fenrg.2019.00031

Heynen, N. (2017). Urban Political Ecology. In *International Encyclopedia of Geography* (pp. 1–9). John Wiley & Sons, Ltd., DOI: 10.1002/9781118786352.wbieg1110

Hofstad, H., Sørensen, E., Torfing, J., & Vedeld, T. (2022). Designing and leading collaborative urban climate governance: Comparative experiences of co-creation from Copenhagen and Oslo. *Environmental Policy and Governance, 32*(3), 203–216. DOI: 10.1002/eet.1984

Hoppe, T., & Lulofs, K. (2008). The Impact of Multi-Level Governance on Energy Performance in the Current Dutch Housing Stock. *Energy & Environment, 19*(6), 819–830. DOI: 10.1260/095830508785363587

Hougaard, I.-M., & Christiansen, K. L. (2022, September 13). *Net zero: Copenhagen's failure to meet its 2025 target casts doubt on other major climate plans.* The Conversation. https://theconversation.com/net-zero-copenhagens-failure-to-meet-its-2025-target-casts-doubt-on-other-major-climate-plans-190033

Huang, W.-H., & Chao, M.-C. (2014). The Impacts of the Mass Rapid Transit System on Household Car Ownership in Taipei. *[Journal of Sustainable Development of Energy, Water and Environment Systems], [2]*([2]), [191]-[207].

Hufen, H., & de Bruijn, H. (2016). Getting the incentives right. Energy performance contracts as a tool for property management by local government. *Journal of Cleaner Production, 112*, 2717–2729. DOI: 10.1016/j.jclepro.2015.10.036

Huovila, A., Siikavirta, H., Antuña Rozado, C., Rökman, J., Tuominen, P., Paiho, S., Hedman, Å., & Ylén, P. (2022). Carbon-neutral cities: Critical review of theory and practice. *Journal of Cleaner Production, 341*, 130912. DOI: 10.1016/j.jclepro.2022.130912

Huston, S., & Lahbash, E. (2018). *Land Value Capture and Tax Increment Financing: Overview and Considerations for Sustainable Urban Investment* (SSRN Scholarly Paper 3103949). Social Science Research Network. DOI: 10.2139/ssrn.3103949

Jabareen, Y. R. (2006). Sustainable Urban Forms: Their Typologies, Models, and Concepts. *Journal of Planning Education and Research, 26*(1), 38–52. DOI: 10.1177/0739456X05285119

Johansen, R. (2022). Oslo is Demonstrating Ambitious Leadership through its Climate Budget. *Journal of City Climate Policy and Economy, 1*(1), 61–64. DOI: 10.3138/jccpe-2022.1.1.0004

Jordová, R., & Brůhová-Foltýnová, H. (2021). Rise of a New Sustainable Urban Mobility Planning Paradigm in Local Governance: Does the SUMP Make a Difference? *Sustainability (Basel), 13*(11), 11. Advance online publication. DOI: 10.3390/su13115950

Kempton, Y., Salvati, L., & Vardopoulos, I. (2022). Long-term planning and development for urban and regional inclusion, safety, resilience, and sustainability. Insights from Singapore. *Region & Periphery, 14*(10), 58–78. DOI: 10.12681/rp.32607

Khakee, A. (2007). From Olympic village to middle-class waterfront housing project: Ethics in Stockholm's development planning. *Planning Practice and Research, 22*(2), 235–251. DOI: 10.1080/02697450701584402

Kim, A. M., & Eisenlohr, A. (2022). *Community Land Trusts for Sustainably Affordable Rental Housing Redevelopment: A Case Study of Rolland Curtis Gardens in Los Angeles. 24*(1). https://www.huduser.gov/portal/periodicals/cityscpe/vol23num4/community-land-trusts.html

Krug, M., Di Nucci, M. R., Caldera, M., & De Luca, E. (2022). Mainstreaming Community Energy: Is the Renewable Energy Directive a Driver for Renewable Energy Communities in Germany and Italy? *Sustainability (Basel), 14*(12), 12. Advance online publication. DOI: 10.3390/su14127181

Levine, D., Sussman, S., Ayalon, S. Y., & Aharon-Gutman, M. (2022). Rethinking Gentrification and Displacement: Modeling the Demographic Impact of Urban Regeneration. *Planning Theory & Practice, 23*(4), 578–597. DOI: 10.1080/14649357.2022.2117399

Lind, A., & Espegren, K. (2018). Mitigation of Greenhouse Gas Emissions in Urban Areas: The Case of Oslo. In Giannakidis, G., Karlsson, K., Labriet, M., & Gallachóir, B. Ó. (Eds.), *Limiting Global Warming to Well Below 2 °C: Energy System Modelling and Policy Development* (pp. 391–405). Springer International Publishing., DOI: 10.1007/978-3-319-74424-7_23

Looijen Bader, E. M., & Oostra, M. A. R. (2022). The Dutch energy transition in practice. *IOP Conference Series. Earth and Environmental Science, 1085*(1), 012019. DOI: 10.1088/1755-1315/1085/1/012019

Lu, H., Fu, Y., Xia, C., Lu, C., Wang, B., Yang, Q., & Wang, D. (2023). Low-carbon urban experiments from vision to reality: A systematic review of the literature from 2005 to 2020. *Climate Policy, 23*(8), 1058–1077. DOI: 10.1080/14693062.2023.2205371

Mahzouni, A. (2015). The 'Policy Mix' for Sustainable Urban Transition: The city district of Hammarby Sjöstad in Stockholm. *Environmental Policy and Governance, 25*(4), 288–302. DOI: 10.1002/eet.1688

Mandal, D. (2018, June 1). *How Portland's "Right to Return" is Indeed Right to Return Housing to the Underrepresented—Berkeley Political Review*. https://bpr.studentorg.berkeley.edu/2018/06/01/how-portlands-right-to-return-is-indeed-right-to-return-housing-to-the-underrepresented/

Marique, A.-F., & Reiter, S. (2011). Towards more sustainable neighbourhoods: Are good practices reproducible and extensible? A review of a few existing "sustainable neighbourhoods." *Proceedings of PLEA 2011 - 27th Conference on Passive and Low Energy Architecture*, 27–32. https://orbi.uliege.be/bitstream/2268/88755/1/PLEA2011_afm.pdf

Marquet, O., Anguelovski, I., Nello-Deakin, S., & Honey-Rosés, J. (2025). Decoding the 15-Minute City Debate: Conspiracies, Backlash, and Dissent in Planning for Proximity. *Journal of the American Planning Association*, *91*(1), 117–125. DOI: 10.1080/01944363.2024.2346596

Martin, L., & March, L. (1972). *Urban Space and Structures* (Martin, L., & March, L., Eds.). Cambridge University Press.

Medda, F. (2012). Land value capture finance for transport accessibility: A review. *Journal of Transport Geography*, *25*, 154–161. DOI: 10.1016/j.jtrangeo.2012.07.013

Meister, T., Schmid, B., Seidl, I., & Klagge, B. (2020). How municipalities support energy cooperatives: Survey results from Germany and Switzerland. *Energy, Sustainability and Society*, *10*(1), 18. DOI: 10.1186/s13705-020-00248-3 PMID: 32215236

Ministry of Environment and Energy. (2015). Energy Efficiency Plan for Buildings of Regions and Municipalities. *Υπουργείο Περιβάλλοντος και Ενέργειας*. https://ypen.gov.gr/energeia/energeiaki-exoikonomisi/ktiria/schedio-energeiakis-apodosis-ktirion-perifereion-kai-dimon/

Mlindelwa, S., Chowdhury, S. D., & Lencwe, M. J. (2024). Overview of the challenges, developments, and solutions for electric vehicle charging infrastructure. *2024 32nd Southern African Universities Power Engineering Conference (SAUPEC)*, 1–6. DOI: 10.1109/SAUPEC60914.2024.10445031

Murray, C., & Landry, C. (2020). Urban Regeneration 3.0: Realising the potential of an urban psychology. *Journal of Urban Regeneration and Renewal*, *13*(3), 231–240. DOI: 10.69554/FYIO9002

Nevens, F., Frantzeskaki, N., Gorissen, L., & Loorbach, D. (2013). Urban Transition Labs: Co-creating transformative action for sustainable cities. *Journal of Cleaner Production*, *50*, 111–122. DOI: 10.1016/j.jclepro.2012.12.001

Nieuwenhuijsen, M. J., & Khreis, H. (2016). Car free cities: Pathway to healthy urban living. *Environment International, 94*, 251–262. DOI: 10.1016/j.envint.2016.05.032 PMID: 27276440

Nobre, E. A. C. (2023). Implementing land value capture in a Global South city: Evaluation of the experience in the City of São Paulo, Brazil. *Revista Brasileira de Estudos Urbanos e Regionais, 25*(1). Advance online publication. DOI: 10.22296/2317-1529.rbeur.202327en

Nordestgaard, P. M., & Arndt, C. H. (2019). AMAGER BAKKE: A steel building with the design challenge of creating a world famous recreational roof. *ce/papers, 3*(3–4), 151–156. DOI: 10.1002/cepa.1156

Norton, R. K. (2014). Agenda 21 and Its Discontents: Is Sustainable Development a Global Imperative or Globalizing Conspiracy. *The Urban Lawyer, 46*(2), 325–360.

Nucci, C. (2023). *Città intelligenti e neutralità climatica: Verso la transizione energetica*. Istituto Lombardo - Accademia Di Scienze e Lettere - Rendiconti Di Scienze., DOI: 10.4081/scienze.2022.825

Nutkani, I., & Khan, W. (2023). Electric Vehicles Impact on Distribution Network Demand and Power Quality. *2023 IEEE International Conference on Energy Technologies for Future Grids (ETFG)*, 1–6. DOI: 10.1109/ETFG55873.2023.10407442

Ongpeng, J. M. C., Dungca, J. R., Aviso, K. B., & Tan, R. R. (2019). Minimizing the carbon footprint of urban reconstruction projects. *Journal of Cleaner Production, 240*, 118222. DOI: 10.1016/j.jclepro.2019.118222

Ovaere, M., & Proost, S. (2022). Cost-effective reduction of fossil energy use in the European transport sector: An assessment of the Fit for 55 Package. *Energy Policy, 168*, 113085. DOI: 10.1016/j.enpol.2022.113085

Owens, S. (1986). *Energy, planning, and urban form*. Pion.

Pandis Iverot, S., & Brandt, N. (2011). The development of a sustainable urban district in Hammarby Sjöstad, Stockholm, Sweden? *Environment, Development and Sustainability, 13*(6), 1043–1064. DOI: 10.1007/s10668-011-9304-x

Pandis Iveroth, S., Vernay, A.-L., Mulder, K. F., & Brandt, N. (2013). Implications of systems integration at the urban level: The case of Hammarby Sjöstad, Stockholm. *Journal of Cleaner Production, 48*, 220–231. DOI: 10.1016/j.jclepro.2012.09.012

Polzin, F., von Flotow, P., & Nolden, C. (2016). What encourages local authorities to engage with energy performance contracting for retrofitting? Evidence from German municipalities. *Energy Policy, 94*, 317–330. DOI: 10.1016/j.enpol.2016.03.049

Pozoukidou, G., Gavanas, N., & Verani, E. (2017). Land Use Models And Sustainable Urban Mobility Plans: An Integrative Approach For Strategic Planning. *International Journal of Sustainable Development and Planning*, *12*(5), 867–882. DOI: 10.2495/SDP-V12-N5-867-882

Ribeiro Hoffmann, A., Sandrin, P., & Doukas, Y. E. (Eds.). (2024). *Climate change in regional perspective. European Union and Latin American initiatives, challenges, and solutions*. Cham, Switzerland: United Nations University Series on Regionalism, Vol. 27, Springer Nature. DOI: 10.1007/978-3-031-49329-4

Ross, A., Alstine, J. M. V., Cotton, M., & Middlemiss, L. (2021). Deliberative democracy and environmental justice: Evaluating the role of citizens' juries in urban climate governance. *Local Environment*, *26*(12), 1512–1531. DOI: 10.1080/13549839.2021.1990235

Roversi, R., Boeri, A., Pagliula, S., & Turci, G. (2022). Energy Community in Action—Energy Citizenship Contract as Tool for Climate Neutrality. *Smart Cities*, *5*(1), 1. Advance online publication. DOI: 10.3390/smartcities5010018

Rudden, P. J., O'Neill, K., McEvoy, B., & Treanor, A. (2015). Environmental sustainability of European cities. *Proceedings of the Institution of Civil Engineers. Civil Engineering*, *168*(2), 75–80. DOI: 10.1680/cien.14.00037

Salvati, L. (Ed.). (2025). *Environmental sustainability and global change*. Elsevier., DOI: 10.1016/C2023-0-52570-4

Sateriano, A., D'Agata, A., & Vardopoulos, I. (Eds.). (2024). *Socio-environmental change: the path toward sustainability*. Nova Science Publishers., DOI: 10.52305/NZKT5653

Schepelmann, P. (2018). *Governance of Low-Carbon Energy System Transitions: A Case Study from North-Rhine Westphalia, Germany. 32*. https://www.adb.org/publications/low-carbon-energy-system-transitions

Schoenefeld, J. J., Stein, B., & Renz, I. (2023). Renovating Europe: How to start and steer a wave? In *Making the European Green Deal Work*. Routledge. DOI: 10.4324/9781003246985-7

Schroeder, H., & Bulkeley, H. (2009). Global Cities and the Governance of Climate Change: What is the Role of Law in Cities. *The Fordham Urban Law Journal*, *36*(2), 313–360.

Shonder, J., Morofsky, E., Schmidt, F., Morck, O., & Himanen, M. (2010). *Best Practice Guidelines for Using Energy Performance Contracts To Improve Government Buildings* (IEA ECBCS Annex 46).

Soni, N., & Soni, N. (2016). Benefits of pedestrianization and warrants to pedestrianize an area. *Land Use Policy*, *57*, 139–150. DOI: 10.1016/j.landusepol.2016.05.009

Sonnleithner, M. (2021). New Opportunities for Increasing the Renovation Rate of Buildings. *Architecture Papers of the Faculty of Architecture and Design STU*, *26*(1), 2–9. DOI: 10.2478/alfa-2021-0002

Squires, G., & Lord, A. D. (2012). The transfer of Tax Increment Financing (TIF) as an urban policy for spatially targeted economic development. *Land Use Policy*, *29*(4), 817–826. DOI: 10.1016/j.landusepol.2011.12.007

Tambach, M., & Visscher, H. (2012). Towards Energy-neutral New Housing Developments. Municipal Climate Governance in The Netherlands. *European Planning Studies*, *20*(1), 111–130. DOI: 10.1080/09654313.2011.638492

Tollefson, J. (2025). Earth breaches 1.5 °C climate limit for the first time: What does it mean? *Nature*, *637*(8047), 769–770. DOI: 10.1038/d41586-025-00010-9 PMID: 39794429

Tozer, L. (2018). *Urban Decarbonization: Politics and Practices of Carbon Neutrality*. https://utoronto.scholaris.ca/server/api/core/bitstreams/a94786de-7403-469a-b2d0-c9c2313c81dd/content

UN. (2019, September 18). *Cities: A "cause of and solution to" climate change | UN News*. https://news.un.org/en/story/2019/09/1046662

van den Dobbelsteen, A., Roggema, R., Tillie, N., Broersma, S., Fremouw, M., & Martin, C. L. (2018). 4.6—Urban Energy Masterplanning—Approaches, Strategies, and Methods for the Energy Transition in Cities. In Droege, P. (Ed.), *Urban Energy Transition* (2nd ed., pp. 635–660). Elsevier., DOI: 10.1016/B978-0-08-102074-6.00045-0

Vardopoulos, I. (2024). Urban planning policies and tools for sustainable development: insights into adaptive reuse, smart cities, and urban metabolism. In *Socio-enviromental change: The path toward sustainability* (pp. 121–140). Nova Science Publishers.

Vardopoulos, I. (2025). Cities in flux: Waste, the shadow of circularity, and the (elusive) promise of sustainability. *Sustainability (Basel)*, *17*(2), 705. DOI: 10.3390/su17020705

Vardopoulos, I., Escrivà Saneugenio, F., Sateriano, A., & Salvati, L. (2024). *Homage (and criticism) to the Mediterranean city. Regional sustainability and economic resilience*. River Publishers., DOI: 10.1201/9781003517306

Vardopoulos, I., Giannopoulos, K., Kalli, S., & Salvati, L. (2024). From tech-centric to human-centric smart tourist destinations: innovations in real estate, urban planning, and sustainability. In *Socio-enviromental change: the path toward sustainability* (pp. 65–96). Nova Science Publishers.

Vardopoulos, I., & Karytsas, S. (2019). An exploratory path analysis of climate change effects on tourism. *Sustainable Development, Culture, Traditions Journal, Special Volume in Honor of Professor George I. Theodoropoulos,* 132–152. DOI: 10.26341/issn.2241-4002-2019-sv-13

Vardopoulos, I., Papoui-Evangelou, M., Nosovo, B., & Salvati, L. (2023). Smart 'tourist cities' revisited: Culture-led urban sustainability and the global real estate market. *Sustainability (Basel), 15*(5), 4313. DOI: 10.3390/su15054313

Vardopoulos, I., & Poli, M. (2025). Decoding Europe's "Renovation Wave" energy strategy: real estate shifts for sustainable cities. In Mitoula, R. (Ed.), *2nd Open-Air Cities International Conference: Local and Regional Sustainable Development and Urban Reconstruction, 14-16 February, 2025. Book of Abstracts* (p. 173). Athens, Greece: Open Air Cities Institute. ISBN: 9786188707030.

Vardopoulos, I., Santamouris, M., Zorpas, A. A., Barone, G., Italos, C., & Vassiliades, C. (2024). A comparative study on discrepancies in residential building energy performance certification in a Mediterranean context. *Buildings (Basel, Switzerland), 14*(4), 1009. DOI: 10.3390/buildings14041009

Vardopoulos, I., Vannas, I., Xydis, G., & Vassiliades, C. (2023). Homeowners' perceptions of renewable energy and market value of sustainable buildings. *Energies, 16*(10), 4178. DOI: 10.3390/en16104178

Vartholomaios, A. (2015). The Residential Solar Block envelope: A method for enabling the development of compact urban blocks with high passive solar potential. *Energy and Building, 99*, 303–312. DOI: 10.1016/j.enbuild.2015.04.046

Vassiliades, C., Minterides, C., Astara, O. E., Barone, G., & Vardopoulos, I. (2023). Socio-economic barriers to adopting energy-saving bioclimatic strategies in a Mediterranean sustainable real estate setting: A quantitative analysis of resident perspectives. *Energies, 16*(24), 7952. DOI: 10.3390/en16247952

Vogiatzi, C., Gemenetzi, G., Massou, L., Poulopoulos, S., Papaefthimiou, S., & Zervas, E. (2018). Energy use and saving in residential sector and occupant behavior: A case study in Athens. *Energy and Building, 181*, 1–9. DOI: 10.1016/j.enbuild.2018.09.039

Võsa, K.-V., Kurnitski, J., Felsmann, C., Meinsenbach, A., Carli, M. D., Tonon, M., & Iivonen, M. (2023). National methods fail to calculate standardized deep renovation concepts for dwellings: Benchmarking in three EU climates. *E3S Web of Conferences, 396*, 04014. DOI: 10.1051/e3sconf/202339604014

Voukkali, I., Vardopoulos, I., Papamichael, I., & Zorpas, A. A. (2024). Urban metabolism considering the "15-minute-city" concept. In *1st Open-Air Cities International Conference: Local and Regional Sustainable Development and Urban Reconstruction, 16-18 February, 2024, Kallithea, Greece. Book of Abstracts* (p. 99). Athens, Greece: Open-Air Cities Institute. ISBN: 9786188707009.

Voytenko, Y., McCormick, K., Evans, J., & Schliwa, G. (2016). Urban living labs for sustainability and low carbon cities in Europe: Towards a research agenda. *Journal of Cleaner Production, 123*, 45–54. DOI: 10.1016/j.jclepro.2015.08.053

Wamsler, C. (2015). Mainstreaming ecosystem-based adaptation: Transformation toward sustainability in urban governance and planning. *Ecology and Society, 20*(2), art30. https://www.jstor.org/stable/26270196. DOI: 10.5751/ES-07489-200230

Wang, Y., Song, Q., He, J., & Qi, Y. (2015). Developing low-carbon cities through pilots. *Climate Policy, 15*(sup1), S81–S103. DOI: 10.1080/14693062.2015.1050347

Woltering, L., Fehlenberg, K., Gerard, B., Ubels, J., & Cooley, L. (2019). Scaling – from "reaching many" to sustainable systems change at scale: A critical shift in mindset. *Agricultural Systems, 176*, 102652. DOI: 10.1016/j.agsy.2019.102652

Yiannakou, A. (2020). Urban regeneration as a perpetual planning process: Understanding the role of stakeholders in property-led regeneration projects in Greek cities. *Local Economy, 35*(2), 83–104. DOI: 10.1177/0269094220914470

Zanon, B., & Verones, S. (2013). Climate change, urban energy and planning practices: Italian experiences of innovation in land management tools. *Land Use Policy, 32*, 343–355. DOI: 10.1016/j.landusepol.2012.11.009

Zhang, Y. (2024). Role of green finance, green bonds, public private partnership, and technology innovation in carbon neutrality and sustainable development. *Heliyon, 10*(18), e37189. Advance online publication. DOI: 10.1016/j.heliyon.2024.e37189 PMID: 39328522

Chapter 9
Climate Crisis and the Tourism Industry:
The Roles of Coastal and Shipping Quality Service

Grigorios L. Kyriakopoulos
https://orcid.org/0000-0003-4875-8534
National Technical University of Athens, Greece

Justas Streimikis
Lithuanian Centre for Social Sciences, Vilnius, Lithuania

Ioannis Sebos
https://orcid.org/0000-0003-4800-6893
National Technical University of Athens, Greece

ABSTRACT

Nowadays, the multifaceted interrelations among climate change, outdoor recreation activities and tourism industry have shown an ongoing research interest. Forecasting climate changes on human health and comfort, possible shifts in preferred destinations and potential alterations in current patterns of seasonality, all offer various avenues of tourism industry. Subsequently, it is essential to capture both incremental and systemic, or even radical, types of eco-innovation. These insights have substantial policy implications at developing strategies that encourage climate change adaptation and green innovation. As a result, policymakers should prioritize the integration of green technology and climate change adaptation measures in their sustainable development agendas to foster more resilient economic growth.

DOI: 10.4018/979-8-3373-2327-5.ch009

This chapter explores the services quality in both coastal context and in shipping. Four clusters have been identified, while the emerging key concepts, the prevalent methodologies and the priority actions per field and cluster have been demonstrated.

1. INTRODUCTION

Tourism sector contributes to global GDP and employment to a level of around 10%, while 7.3 and 20.6 million jobs are related to tourism (Ribeiro Hoffmann et al., 2024) The number of tourists travelling internationally in 2008 reached 922 million showing an ongoing trend based on UNWTO forecasting (Ribeiro Hoffmann et al., 2024). This situation and characteristics of global tourism determine the pivotal role of climate as an important determinant for many types of tourism industry. In the relevant literature several metrics have been developed for the suitability of climate, as that of Mieczkowski's "Tourism Climatic Index" (TCI) that summarizes a combination of seven climate variables. Based on TCI in EU, climate resources for tourism can be analyzed while concerning about future climate changes. In such a way local-derived data from regional climate models could be utilized and compared with data of period 1961-1990 with relevant long-term scenarios forecasting the period 2071-2100 (Perch-Nielsen et al., 2010). In literature noticeable methodological tools associate climate data with tourism applications among developed economics (such as in the case study of Luxembourg) disclosing information of the various aspects of climate in tourism. Such dataset refers to the thermal comfort or stressful conditions which are further related to aesthetical and physical parameters of identifying extreme events of heat waves. Other typical variables of physiological interest are that of temperature and precipitation patterns to forecast expected climatic conditions for the future through projections among fitting regional climate models (Matzarakis et al., 2013).

About half of the total number of international tourists in Europe corresponds to arrivals of about 10% of the total continental positions in tourism industry (Vardopoulos & Karytsas, 2019). In this respect tourism is conceptually and operationally linked with climate change in a variety of ways. Subsequently to this significant position of EU tourism in the EU is the fact that EU tourism is a significant contributor of greenhouse gas (GHGs) emissions. Therefore, EU-based literature studies are concentrated on climate and carbon risks for tourism and industry (Vardopoulos & Karytsas, 2019). Climate risk has been followed by carbon risks, imposed by the dominant parameters of characterizing climate and carbon risk of EU tourism (Vardopoulos & Karytsas, 2019). Critical issues involved are the following: geographical concerns especially among countries in the former Soviet Union and former Yugoslavia; sparsely conducted research on emissions and climate-based indicators that could integrate carbon emissions, climate risks, institutional policies to be adapted and implemented at destinations in the EU-territory (Vardopoulos & Karytsas, 2019).

From a long-term perspective of tourism destinations research can be driven to impacts of rising sea levels and coastal erosion (Vardopoulos & Karytsas, 2019). Based on this theoretical overview it is anticipated to occur vast climate changes leading to accruing shifts of tourist flows, thus determining outdoor international tourism expenditure and contributing to economic implications due to future climate change by 2080s within Europe (Amelung & Moreno, 2012). Such tourist behaviour is based on the statistics between bed nights and a climate-related index of human comfort involving income and prices of travellers. Climate change can also affect the decisions of regional travelling and the management of the natural resources of local tourism interest in Europe (Amelung & Moreno, 2012) (Kyriakopoulos & Sebos, 2023).

The steps of implementing this study are the following: The literature overview of Introduction section has been followed by the theoretical background of Literature Review. In this section two fields of investigation have been identified. The one field explores the services quality in a coastal context and the other field of this theoretical coverage has focused on services quality in shipping. These fields of theory are covered in the section of Literature Review. Then, the Methods and Analysis have been presented in section 3, followed by the Results and Discussion section. The results of this study have been organized in the following six themes:

- Consumer behaviour & Green innovations
- Climate change & Eco-innovations
- Coastal sustainability shipping
- Climate shipping tourism
- Green technologies & maritime industry
- Climate change & regulations

Subsequently, the "challenges and priority actions in climate change and tourism industry" and the "challenges and priority actions of quality in coastal and shipping services" have been discussed. The study ends up with the Conclusions section. In Figure 1 there is a graphical representation depicting the logic of procedures followed in this study.

Figure 1. Study Framework.

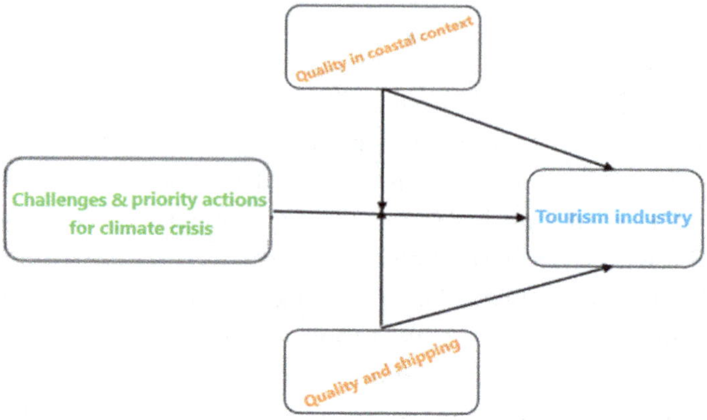

Source: Authors' own study.

Based on the study framework in Figure 1 the research objective is the investigation of those parameters that support the sustainable development of tourism industry considering the environmental dimension of it. In this context the ongoing and accelerating climate crisis is deteriorating, among others, the flows at tourism destinations, while affecting the behavioural perceptions to decide among the safest available traveling choices in national, continental, or international networks that have been developed in tourism sector. Subsequently, among the challenges and the priority actions of confronting climate crisis and regaining the trustworthiness of tourism operations, the quality services in the coastal context and in shipping are anticipated to predominately contribute in shifting the globally-threatening climate crisis into national-oriented opportunities of attracting more visitors, based on emerging standards of quality wellness and well-being. Figure 1 also graphically discloses the direct interactions between the two orange-coloured quality contexts. Both these two quality-termed contexts are the key-determinants to specify how the climate crisis (that is the start position of the graph) has been affected by the tourism industry activities and procedures (that is the end position of the graph). In an attempt to further demonstrate the crucial effect of these two quality-termed contexts, a group of six themes has been emerged in the relevant literature and are demonstrated in the results of this study.

2. LITERATURE REVIEW

2.1 Service Quality in a Coastal Context

From a societal context of analysis it was introduced an evaluation index system that contains 35 indexes regarding public service quality. Data in the period 2010-2017 among 12 coastal provinces and cities of China showed that the public service quality could be assessed by the method of entropy weight order preference similarity to the ideal solution (TOPSIS) while the variables of the method included the fiscal transparency on public service quality. It was noteworthy that the public service quality among the examined coastal provinces and cities of China was relatively high, and fiscal transparency positively affected on public service quality. Fiscal transparency increase lead to quality of public services increase (Yang, 2021). Subsequently, fiscal transparency contributed to the quality of public services while improving the scale of investment and the efficiency of public services expenditure. This behaviour was attributed to the fact that fiscal transparency can increase the expenditure on public welfare services and curb official corruption. Improvement of public service quality can guide specific motivations for other countries in order to improve public service quality by increasing fiscal transparency (Yang, 2021).

Among others, service quality is a significant determinant of recreational fisheries development. In the relevant literature 16 attributes of five dimensions on the SERVQUAL (Service Quality) method have been introduced to measure the service quality of recreational fisheries (Jiang et al., 2023). Regarding the role of tourists, authors developed an evaluation system of the recreational fishery service quality of Shengshan Island and Lianjiang County in China, based on the fuzzy Kano model and importance-performance analysis. The similarities and the significant differences in the quality of recreational fishery services within the two regions have been also denoted (Jiang et al., 2023). Subsequently, the sustainable development of recreational fisheries in both regions should strive to maintain their strengths and to improve their weaknesses. The results of this analysis results can serve as a reference for similar studies on recreational fisheries issues (Jiang et al., 2023).

In the relevant literature a qualitative research utilized data collection, observation, documentation and interviews in order to identify the development of innovation strategies and the competitiveness of eco-based coastal tourism destinations that can be managed by service quality-based organizations (Udiyana et al., 2018). Data analysis techniques included the Internal Factor Analysis Strategic (IFAS), the External Factor Analysis Strategic (EFAS) and the SWOT matrix model. It was demonstrated that the internal factors concerns, where the strengths relied on natural conditions white sandy beaches, calm blue water and strong community support. Besides, main weakness were that of low entrepreneurial spirit and the quality of local human resources, as well as the preservation of the environment (Udiyana et al., 2018). External factors indicated economic-based opportunities to improve entrepreneurship and to empower local entrepreneurs to own and manage businesses, and strong synergy among stakeholders. Threats to coastal tourism destinations were the land tenure by investors, regional carrying capacity, exploitation and environmental pollution. Future-designed strategies aim to support strategies on innovation and infrastructure, being accompanied by organizational-management and service quality innovations (Udiyana et al., 2018).

From an international perspective it can be denoted the study of Thilakshan et al. (Thilakshan et al., 2023) who assessed passenger perceptions of the service quality of the Sri Lankan Railways (the Coastal Railway Line). A questionnaire survey and interviews assessed passenger satisfaction levels in terms of service quality including air conditioning, seating comfort, availability and usability of doors and windows, restroom amenities, and informational offerings. The passenger satisfaction was positively related to high levels of punctuality, cleanliness, safety, comfort, and customer service (Thilakshan et al., 2023). it is noteworthy that necessary improvements and actions can be undertaken towards customer expectation, thus, better support a viable and passenger-focused national rail network for increased welcome travel experience to all rail customers (Thilakshan et al., 2023).

Another international-driven service quality case was focused on a coastal context (Lombardini et al., 2022). These authors investigated the overall ecosystem services' provision, assuming it as a comprehensive measure of territorial environmental performances, and urbanization processes along the Italian coastal municipalities. In particular, the Multiple Ecosystem Services Landscape Index (MESLI) combined a selected set of ecosystem services into a representative synthetic index to overall environmental values: geographic, morphological, socio-economic and territorial of urbanization processes. It was proven that more intensively urbanized areas were generally characterized by lower environmental performances. On the other hand, there was reported a variability of correlation ratios suggesting that the settlement quality and morphology played a protagonist role to deplete or to protect the overall ecosystem services provision (Lombardini et al., 2022).

Another crucial human-societal issue of broad interest and applicability is that of health organizations and times -such as that of first wave of the COVID-19 pandemic at Vancouver Coastal Health, BC, - that played a crucial role during pandemic preparedness, response and recovery from mental health and substance use services (Devane et al., 2021). A quality improvement initiative was designed and showed that healthcare providers were required to be flexible while continuing to provide patient care especially among mental health and substance use field experienced the COVID-19 response at their workplace (Devane et al., 2021). The study evaluated the examined variables were that of perceptions related to communication, redeployment, safety and well-being, affirming that the affiliated-to-survey model of ADKAR (awareness, desire, knowledge, ability and reinforcement) can support change management, which informed and recommended health-care leaders to support ongoing pandemic response procedures (Devane et al., 2021).

Besides to human-societal dimension, it is also interesting to review the forecasting and anticipated effects of service quality to the coastal context via the human interventions in nature. For this, human well-being contributors in coastal regions are that of coastal recreation and water quality. These contributors can create opportunities for ecosystem management, ecological restoration, and water quality improvements to positively affect people and the environment. Yet the effect of environmental quality on human behavior is sparsely examined among coastal ecosystem service studies (Kreitler et al., 2013). The research target is to link a quantified measure of recreational visitation within ecosystem service assessments. Subsequently, a) conceptualizing environmental condition towards the variation in recreational visitation. b) prioritizing water quality improvements, are both associated with new recreational infrastructure as parks or the modification of existing ones. Thereafter, expected benefits from potential changes in recreational visitation and water quality improvements can be demonstrated (Kreitler et al., 2013). Similar affection between economic value of natural mangrove ecosystems with habitat quality of the relevant ecosystem services was reported in a case study at the Tangerang Coastal Area of Banten Province, Indonesia (Marlianingrum et al., 2021).

Water quality degradation is a chronic problem that weakens the resilience of a social-ecological system differently than acute disturbances, such as disease or storms. In this context in a similar study researchers developed a tailored resilience framework that applied ecosystem service concepts to coastal social-ecological systems affected by degraded water quality showning the interconnection developed between the ecological system and the social system via ecological production functions and ecosystem services and, in a broader sense, the interdisciplinary coordination of research planning with inform coastal management (Merrill et al., 2018). The social system can be transferable to other geographical settings and it feeds back to the ecological system via policies and interventions to address declining water quality (Merrill et al., 2018).

Another functionality of local aqueous habitats with water quality and broader ecosystem services is the biofiltration and biodeposition capacity of freshwater bivalves C. australis that can influence water quality by reducing phytoplankton levels through filter-feeding and altering nutrient levels through excretion and biodeposition (Buelow & Waltham, 2020). In northeast Australia the native freshwater bivalves (Corbicula australis) may help restore the water quality of coastal wetlands that receive high nutrient loads in runoff from nearby agricultural activity (Buelow & Waltham, 2020). In particular larger individuals and denser populations mean that filtration and biodeposition rates are commonly to be occurred in artificial wetlands and may lead to greater processing of nutrient rich water (Buelow & Waltham, 2020). Similarly, the population density of M. leucophaeata forms dense clusters in newly invaded areas which lead to impacts on local organisms and environment. Some of the impacts of non-native species on newly colonized areas may be positive, since M. leucophaeata invasion can promote an improvement in lagoon water quality, especially within already invaded systems where high sewage-enriched effluents are released (Neves et al., 2020).

Closing this overview of the importance to: a) maintain coastal ecology, b) resist natural disasters, and, c) improve the living environment of coastal cities, it is noteworthy mentioning those studies that have focused on coastal protection forests, so as to better understand their vital role to further serve in the endogenous functions of coastal protection forest ecosystems and the necessity to develop methods of valuing quantification methods and of identifying measures to improve their resilience as ecological indicators of sustainable management on coastal protection forest ecosystems in the future (Zhu et al., 2024).

2.2 Service Quality and Shipping

Organizational differences between consumer and business-to-business services can, subsequently, lead to varied perceptions of their service quality. However, it is a fact that container shipping services have grown tremendously with the increase in globalization, especially in the burgeoning Asia-Pacific market (Lobo, 2010), thus, shipping services gain enormous importance since they comprise of different components or sub-services; some of them are the following: sales and marketing, cargo booking, documentation, operations and claims handling. It is interesting to note that from data collected among shippers in the predominant global container-port of Singapore, personal visits and claims handling were rated the lowest (Lobo, 2010). Moreover, different characteristics have been attributed among organizational buyers whom overall satisfaction scores are differentiated from that of foreign companies that gained higher rating than local and joint venture companies (Lobo, 2010).

By examining two service quality (SQ) gaps: that of the SQ gap between types of business customers and that of the SQ gap among employee statuses of business customers, it was testified a five-factor SERVQUAL measure as the initial hypothesized model, showing that the applicability of SERVQUAL to measure the perceived SQ of customers in the shipping industry (in the case of Taiwan) was empirically rejected (Chen et al., 2009). The existence of the two hypothesized gaps was verified by the method of MANOVA (Chen et al., 2009), but since then (end of 2000s') other models, or real-world case studies, or solving-oriented approaches of local impact have been gradually developed and applied in the relevant scientific area, either in Asia (Yuen & Thai, 2015) (Pham & Yeo, 2019), or abroad (Stopka et al., 2015) (Han et al., 2018), depicting that earlier developed models are vulnerable to analytical gaps, which have been emerged from the service provider to the business customer side.

Among the most prevalent issues of service quality valuation in shipping is the exploration of the role of logistics service quality in generating shipper loyalty, considering relationship quality in the unique context of container shipping. For this reason, ocean carrier-shipper relationships can be perceived in alignment with the investigation of shippers' future intentions, making necessary the use of the same carrier. Carrier selection criteria are traditionally relied on shippers' satisfaction with the service attributes, including soft concepts such as customer loyalty and logistics service quality. In the context of maritime transport, it has been testified an array of theoretical models through on data collection by a postal questionnaire survey of 227 freight forwarders in South Korea, showing the necessity of developing a high level of logistics service quality of container shipping lines jointly with high quality of services in order to attain higher than expected levels of shippers' loyalty and satisfaction (Jiang et al., 2023).

Through a literature review, interviews and surveys Thai et al. (Thai et al., 2014) developed, refined and verified a proposed service quality (SQ) model. This methodological approach was able to understand what construe the service quality, development and validation since customers are actively seeking value from service providers for low price. In recognizing the gap of the existing maritime literature 6 SQ dimensions, which are that of Corporate Image, Customer Focus, Management, Outcomes, Personnel and Technical, and their 18 associated attributes were proven reliable to measure service quality in tramp shipping. The usefulness of customers' evaluation of their service quality as well as a benchmarking tool for continuous improvement have been limited to unavoidably small-sized data collected and to the bulk commodity context. Future research should involve larger sample sizes and a broader geographical coverage of regions for general and specialized cargoes in order to enhance its reliability (Thai et al., 2014).

Another critical issue when approaching the service quality principles at shipping is that of the e-commerce platforms provided by major global liner shipping companies (LSCs). Of particular interest is how e-service quality (e-SQ) influences e-loyalty (e-LY) and establish benchmarks for individual customers (Chao et al., 2024). In such a study authors developed a structural equation modeling framework and, then, they transformed it into a network data envelopment analysis model. Authors concluded that the e-SQ of LSCs gained a significant positive impact on the e-satisfaction (e-SA) and e-LY of their customers. Additionally, trust played a crucial role to significantly moderate the interrelation between e-SA and e-LY (Chao et al., 2024)., implying that LSCs have provided an e-SQ that exceeds the expectations of their large-scale customers (Chao et al., 2024). Through benchmark analysis it was found that for small-scale shippers, virtually no slack exists between their benchmark and actual perception. However, the slack increased as the scale of the shipper increased (Chao et al., 2024). Specific aspects of improvement are related to the architecture of e-commerce platforms, the privacy protection and the size-scale of shippers (Chao et al., 2024).

3. METHODS AND ANALYSIS

In this section a methods' overview on tourism industry, coupling with climate priorities and sustainable environmental and economic growth perspectives, revealed that the key-concepts and priority actions can be collectively presented in the form of the following four clusters: Climate Green sector, social planning, sustainable environment, as shown in Table 1.

Table 1. Key-concepts and priority actions of tourism industry, coupled with climate and sustainability targets.

Clusters	Key Concepts and Priority Actions	Reference #
Climate	Seasonal behaviour of climate changes and tourism development in Europe	(Tranos & Davoudi, 2014)
Climate	Two types of indexing: Tourism Climatic Index (TCI), Holiday Climate Index (HCI)	(Nicholls & Amelung, 2008) (Scott et al., 2016)
Climate	Interrelationship between climate change and green environment, jointly with economic growth	(Apostu et al., 2023) (Kyriakopoulos, 2021) (Streimikiene et al., 2020)
Climate	Climatological parameters that promote a user-friendly analysis	(Bonsinetto & Falco, 2013)
Green sector	Green growth strategy in Europe	(Bonsinetto & Falco, 2013)
Green sector	Future growth in tourism industry suggest an ongoing and increasing number of international tourism arrivals, making an imperative need of offering high quality tourism services to all tourist destinations	(Ribeiro Hoffmann et al., 2024)
Social planning	Including the planning, the location, and the development of new accommodations, coupling with more transportation routes and fascinating attractions	(Nicholls & Amelung, 2008)

continued on following page

Table 1. Continued

Clusters	Key Concepts and Priority Actions	Reference #
Social planning	The today climate changes severity and the limited access to nature of resources are paradigms of adopting cost-effective planning that integrates economic, social and environmental policies and strategies	(Apostu et al., 2023)
Sustainable environment	Environmental challenges: including climate change, thus, research to be oriented to innovation as a way to develop and to realize sustainable solutions. Innovation challenges: including the adoption and the understanding of the "eco-innovation" concept, coupling with the encouragement of climate change adaptation and green innovation	(Machiba, 2010) (Agan & Balcilar, 2023)

Source: Synthesis of studies was made by authors' own work.

Based on Table 1 it is noteworthy that regarding the green growth strategy among European cities, these regions have to cope with challenges of greening their economy and to nurture more sustainable patterns (Streimikiene & Kyriakopoulos, 2024). Politicians and policy-makers should envisage sustainable growth including renewables, GHGs emissions, energy efficiency and biodiversity. All of these aspects can be can support both local and regional types of economy (Bonsinetto & Falco, 2013). In this regard EU member states can benefit from the Europe 2020 Strategy. This European strategy (EU2020S) can guide a roadmap of smart, sustainable and inclusive future in a timeplan of ten years for growth and jobs. The EU2020S was designed as a European exit strategy from the global economic and financial crisis in view of new European economic governance towards achieving the goals of the green growth strategy: climate change, green economy and energy (Bonsinetto & Falco, 2013).

Another critical aspect of investigation is the seasonal behaviour of climate changes and tourism development in Europe. In such a study the impact of climate change was explored regarding the winter sport tourism in European regions (Tranos & Davoudi, 2014). In this winter season and tourism sector it was stressed out the importance of the combined effect of exposure to climate change stimuli and economic sensitivity due to the fact that seasonal tourism-stated countries are better described by the fact of Northern Europeans, having historically escaped the unfavorable weather conditions in their countries in search of the traditional "sun, sand, and sea" vacation in the Mediterranean region (Nicholls & Amelung, 2008). By employing scenarios-based sessions, peak season conditions throughout northwestern Europe can be projected to improve over the course of the coming century. Besides, the length of the peak season was also projected to increase in many regions, being extended from summer to cover also spring and autumn. It can be underscored that climate change could considerably enhance Northwestern Europe's competitive advantage relative to the Mediterranean in the summer season. Such shifts in climatic suitability could be substantially impacting on the existing operation of the European tourism industry: volumes/populations of travellers' served, spatial distributions, timing of travel within and to Europe planning of locations and transportation routes, development of new accommodations and attractions (Nicholls & Amelung, 2008). Such operational transformations can act as proactive rather than as reactive stances to achieve organizational flexibility and to (re)mobile capital investments (Nicholls & Amelung, 2008).

Climate changes and the finite regeneration of nature of resources are motivators to provoke shifts among a variety of economic, social, and environmental policies and strategies. The adoption of sustainable development principles and the advancements in circular economy are globally-challenging priority worldwide. Thus, researches should be motivated to explore the European countries' path toward sustainable development by analyzing the relationship between green environment, economic growth, and circular economy issues (Apostu et al., 2023). Specific variables of investigation that shape the interrelationship between green environment, economic growth and circular economy are that of: final energy consumption, GDP, capital gross fixed capital formation, greenhouse gas (GHGs) emissions, SOx emissions, NOx emissions, as well as generation of municipal waste per capita, being also credited by the Eurostat and the World Bank databases. In relevant studies it was reported that gross fixed capital formation and total GHGs emissions can lead to decreasing effluents of municipal waste. Contrarily, enablers of municipal waste generation were that of final energy consumption, GDP, SOx and NOx emissions (Apostu et al., 2023) (Tsimnadis & Kyriakopoulos, 2024).

Assuming the facts that a) international tourism started only in the 1950s, and, b) the 800 million arrivals benchmark was reached in 2005, then, projections of future growth in the international tourism sector suggest that arrivals will experience the same growth in 15 years as it had in the last 55 years (Ribeiro Hoffmann et al., 2024). Nevertheless, in this forecasting trend there should be an ongoing background information required about climate assessment among tourism destinations, enabling travelers to assess their climate potential/footprint. In a more detailed perspective, each one destination could be performed in terms of mean conditions of relevant climatology parameters, while these climate change conditions should be linked to varied levels of heat stress and sultriness as well as decreasing cold stress. This information forms a plausible basis for an adequate assessment, aiming to provide relevant information for different environmental planning issues especially among the growing tourism sector of EU member states (Matzarakis et al., 2013).

Future researches are prospected to manage and to plan climate change adaptation and mitigation policies towards green environment in the light of economic growth, circular economy and energy poverty alleviation, especially in the case of European countries. Therefore, existing scientific results should be more or less, directly or indirectly positioned to favor processes of green transition considering the potential of jointly circular economy initiatives and economic policy measures (Apostu et al., 2023) (Kyriakopoulos, 2021) (Streimikiene et al., 2020).

It is noteworthy that in order to confront contemporary environmental challenges including climate change, research focus should be placed on innovation as a way to materialize environmental sustainability into social justice and economic prosperous societies. Therefore, understanding the "eco-innovation" term (also mentioned as "green innovation") prerequisites its definition and operational aspects, including: target, mechanism and impact (Machiba, 2010), not undermining the fact that eco-innovation activities are very diverse covering different levels and scales. In future research practices are concentrated on technological advances, where advanced industry players can adopt complementary organizational or institutional changes such as new business models and alternative modes. For this reason, it is also important future researches to monitor the evolution of corporate activities for sustainable production and analyze good practices of broader scalability, industrial-manufacturing, and applicability (Machiba, 2010) (Agan & Balcilar, 2023).

4. RESULTS AND DISCUSSION

4.1 Results: Emerging Themes of the Quality Context

The results of the study have been organized in the form of six themes that have been literature-emerged regarding the interrelated developed among the (jointly approached) phenomena of climate change and tourism industry. These themes are presented as follows.

4.1.1 Consumer Behaviour and Green Innovations

Among the most popular determinants of exploring climate change and the contribution of tourism industry are that of "green innovation" and "green products". In the relevant literature it is noted that habitually consumers resist innovative appliances that differ from the conventional ones. In this context researches have focused on green innovation practices and on the ways under which consumers resist to green innovation products, thus, moderating their influence of environmental awareness and display pro-environmental behaviors in alignment with the diffusion of innovation and expectation theories (Khan et al., 2022). The main characteristic of such an analysis is that when differentiating the cross-sectional surveys on different pools of participants and sectors there are revealed noticeable causal associations between green innovation practices and consumer resistance to green product innovation, lack of knowledge for green product manufacturers besides to varied environmental awareness and pro-environmental behavior (Khan et al., 2022).

From an macro-economic perspective economy grows vigorously, while the environment is severely damaged from day to day, advocating the green industrial revolution and the green marketing become crucial operational policies in the area of scientific and technological industry in the 21st century (Chen et al., 2008). In this context a literature-reported case study has focused on Taiwan that is a country of the highest leading global position of TFT-LCD industry density around the world, even exceed Japan and Korea (Chen et al., 2008). This study showed that the developed research model will be proven ineffective to promote green marketing as an influential factor of LCD sales (Chen et al., 2008).

The studies in the relevant literature revealed that it is vital for firms to achieve their sustainable development goals while developing and sustaining competitive advantages in green innovation. The relevant scientific research has been concentrated on the ways of firms can induce green innovation behavior (GIB), examining external factors, whereas current literature has sparsely investigated unabsorbed slack as an internal antecedent of GIB. The critical point here is researches to provide managerial implications for firms to launch green innovation activities. In such a study the affection of capability reconfiguration: a) capability evolution and b) capability substitution have been examined in the relationship between unabsorbed slack and GIB among Chinese technologically-intensive manufacturing firms (Yang et al., 2025). The results revealed that the mediating effect of capability substitution is stronger than that of capability evolution. Besides, consumer green pressure is positively associated with the relationship between unabsorbed slack and capability evolution/substitution (Yang et al., 2025).

Another theoretical tool that can be employed to investigate the effects of consumer innovation process on green consumption behaviour among targeted consumer groups is that of planned behaviour theory. In such a study among young consumers in Iran, students in universities of Ardabil province in Iran were selected by convenience sampling method (Nouri & Emkani, 2020). It was demonstrated that consumers' innovation was positively affecting on attitude, observed behaviour control and mental norms of the consumers compared to the intention for green consumption (Nouri & Emkani, 2020). Furthermore, while the Intention for green consumption was a credible indicator of green consumption behaviour, the effect of attitude and mental norms on intention for green consumption was not confirmed (Nouri & Emkani, 2020).

In the relevant literature there is shown an emerging interest in investigating the green innovation with a broad spectrum of sectors, like that of: health sector (Sarfraz et al., 2022), ecological behavior (Waheed et al., 2020), consumer perception in green tourism through circular economy (Chen & Liu, 2025), consumer behavior and green B2B marketing (Esangbedo et al., 2024). As a conclusive statement that this theoretical coverage discloses is that knowledge of green consumer behavior plays a decisive role for conscious environmental and business decisions. The relevant studies demonstrate the determinants of green curtailment behaviors and consumer adoption of innovations that are labeled as green (eco-innovations) and then, an explanation of how these two types of green behaviors are linked to each other is provided (Jansson et al., 2010). For this reason, the survey is proven the most successful and well-known analytical tool, being capable to disclose willingness to curtail energy- and environmentally-burdened technologies and willingness to adopt a so-called environmentally friendly alternative tools or products or services that are used to identify significant determinants across the behavioral categories (Jansson et al., 2010). This analysis demonstrated those values, beliefs, norms, and habit strength in order to further determine willingness to curtail and willingness for eco-innovation adoption. Previous eco-innovation adoption is found to be a strong determinant of future willingness to adopt. In future studies a two-sided approach is realistic to be followed: Deepening into green consumer behavior and demonstrating that values, beliefs and norms not only predict low involvement post-purchase behaviors but also adoption of high involvement eco-innovations (Jansson et al., 2010).

4.1.2 Climate Change and Eco-Innovations

In this emerging theme, climate change and eco-innovation can be approached as an environmental issue jointly with a technological one. In the last two decades of research eco-innovation is placed as special emphasis topic of environmental management, encouraging companies to introduce a prevention philosophy rather than correction measures of environmental damage (Durán-Romero & Urraca-Ruiz, 2015). A relevant study was focused on the field of environmentally sound technologies due to their importance for mitigating climate change, having analyzed compiled patents of the European Patent Office (EPO) in the period 1978 - 2010. In particular, the microeconomic analysis involved the technological trajectory followed by these technologies, such as the promotion of renewable technologies to abatement policies, showed that technologies that are led by developed countries can be mainly concentrated on corrective technologies rather than preventive ones. Moreover, promising technologies could be able to at least stabilize emissions making necessary to push public policies encouraging the collaboration between basic research and industry in order climate change and priority actions to be addressed (Durán-Romero & Urraca-Ruiz, 2015).

Among the more attractive methods of investigating eco-innovation for climate change mitigation is the development of models of time series data in selected indicators. Such type of data can be obtained from the European Innovation Scoreboard

2020 (Jesic et al., 2021). in which data for the EU member countries and selected third countries were collected for the analysis, considering that the Recovery Plan for Europe envisages investing 30% of the huge budget in climate change, with the goal of zero gas emission by 2050. This ambitious plan entails investments in research and eco-innovation, making necessary their credibility towards effective long-term investment policy, climate change impact, and mitigation of consequences (Jesic et al., 2021). It is also noteworthy that financing Research and Development (R&D) is low among developing countries, but in developed EU countries, there are more investments in R&D, especially in the business sector for the period 2012–2019. Therefore, eco-innovation and self-sustainable R&D financing for climate preserving products can keep a larger share among developing vs. developed countries (Jesic et al., 2021).

In the relevant literature it was reported that climate change and eco-innovations is a theme of high regional interest. In this context it is noteworthy to mention three studies that are referring to: African countries (Ben Amara & Qiao, 2023), Indonesia energy sector (Ismail et al., 2021), and a sample of companies listed on the London Stock Exchange under the precondition whether such a company supported the Sustainable Development Goal 13 on climate action (Albitar et al., 2023). A common characteristic of these case studies is the timescale of analysis extended from early 2010s' up to the early of 2020s', covering a decade of analysis each one. Among them, a study was concentrated on the impact of carbon emissions, eco-innovation adoption, and international collaboration on economic growth and the influence of economic growth on green growth, while a generalized method of moments (GMM) for 54 African countries at the period 2010 - 2019, being allied with an imbalanced panel of 540 observations (Ben Amara & Qiao, 2023). It was reported a positive affection of carbon emissions, eco-innovation international co-operation, and economic growth. Besides, economic growth and green growth are directly analogous and powerfully affected to each other. Besides, policymakers can prioritize to the growing effect of eco-innovation to attain sustainable and greener growth and climate crisis mitigation (Ben Amara & Qiao, 2023).

In another study the energy sector is characterized as emerging at low-income countries which are responsible to reduce greenhouse gases (GHGs) under the Paris agreement, implying the need to balance their development mandate and climate change policies. In this context technological innovation is considered to be one of the key strategies to maintain the global temperature below 2 °C (Ismail et al., 2021). In Indonesia the characteristics of frugal and eco-innovation were explored to identify the appropriateness of technologies within the context of emerging and low-income regions. Economic and social benefits of low-carbon technologies are prevailing at the local level, as that of energy sector in rural Indonesia, where eco-innovation's characteristics should cascade positive impacts at a larger scale gaining socioeconomic benefits for lower-income communities while contributing to GHG emission reduction. (Ismail et al., 2021).

In the case study of companies listed on the London Stock Exchange for the period of 2014–2020 researchers grounded their study on climate governance, focusing on three proxies, namely the existence of an environmental committee, climate incentives and the existence of sustainability reports (Albitar et al., 2023). It was concluded that corporate eco-innovation can be positively associated with climate change commitment. Subsequently, firms that adopt innovative approaches are more committed to climate change, thus, it can efficiently control pollution and resource used, being able to reduce their environmental impact (Albitar et al., 2023). Firms that integrate climate change issues in governance can manage climate change risks and opportunities, especially at adopting eco-innovative technologies and integrating corporate climate change issues in governance (Albitar et al., 2023).

Since climate change represents an increasing threat to society and demands collaborative actions for changing technologies and production methods, a newly research area that attempts to bridge the gap between eco-innovation with climate change mitigation refers to policies on the basis of circular economy (CE) (Durán-Romero et al., 2020). CE targets to increase the resource use efficiency and to minimize both resource inputs and generation of waste. However, up to now we are in the early developmental stage of CE growth, thus, there is incomplete knowledge of the contribution of CE eco-innovation and, subsequently, to climate change mitigation goals,(Durán-Romero et al., 2020). This research gap necessitates a better understanding on the interrelations developed among technological, environmental, policymaking and socio-economic dimensions, making a comprehensive and integrated framework in which eco-innovation can be identified and operated (Durán-Romero et al., 2020).

4.1.3 Coastal Sustainability Shipping

In the theme of coastal sustainability shipping, it is noteworthy to mention the study of (Scarpa et al., 2019) in which the impact of ship traffic in the vicinity of navigation channels in a wide shallow body of water was investigated. The study included the navigation channels in the Venice Lagoon using a new large dataset of approximately 600 measured wake events associated to specific ships. The modern industrial port and the opening of the Malamocco–Marghera channel in the late 1960s, exerted pressure on the lagoon caused by ship traffic, raising concerns about its physical integrity and habitat survival (Scarpa et al., 2019). Large vessels passing through revealed strong impact on the shallow water areas adjacent to waterways. Besides, depression wakes from such vessels can spark water level dropdown of up to 2.45 m at the channel margin, causing unusually large distortions and degrading of the lagoon morphology, making necessary the prediction of wake amplitude on the basis of ship characteristics and motion (Scarpa et al., 2019).

Similarly, to the aforementioned case study of Venice Lagoon, it has been reported another case study in the field of coastal zone of Bangladesh. The coastal zone of Bangladesh covers 47,211 square kilometers confronting the Bay of Bengal or have close contact with the Bay. Besides, 35 million individuals who represent 29% of the populace live in the waterfront zone (Haque et al., 2021). The housing difficulties of coastal areas and the potentialities of shipping-container-housing have been investigated as a realistic solution to the construction. The research findings can support architects and other scientists to enhance the engagement of shipping containers in the coastal areas as a sustainable building material meeting further applicability at similar contexts globally (Haque et al., 2021).

In another regional-specific case study it was investigated the impact of legal regulations on China's marine economic activities, analyzing the external variables related marine industries in 25 provinces of China Marine Statistics Yearbook at the period 2010 - 2017 (Yang et al., 2020). The critical variables were found that of population size and per capita output value. These variables have positively impacted on the gross tonnage of ships and the number of berths at ports. Legal regulations can negatively impact on cost production and, as result, mediocre the volumes of port sea transport and freight (Yang et al., 2020).

Another crucial issue is that of maritime transport that has received little attention in sustainability transitions research. While the sustainability transitions are mature and heterogeneous, there is need of a more nuanced perspective on socio-technical regimes to understand how the varied conditions of novel technologies adoption can support sustainability transitions (Bergek et al., 2021). In this context regime segmentation plays an influential role on key transition conditions: a) technology maturity and fitness, b) system integration and infrastructure, and c) acceptability and legitimacy (Bergek et al., 2021). Battery-electric energy storage solutions within three segments -including coastal ferry, coastal fishing, and offshore supply of Norwegian coastal shipping- suggested that the transition process can be gradually developed (Bergek et al., 2021).

Emerging issues of literature interest that can be mentioned here are the following:

- Upscaling implementation of biofuels in the shipping sector by establishing a market with subsidies for biofuels and feed-in targets (Bach et al., 2021).
- Biodiesel and liquefied biogas (LBG) and Technological Innovation Systems (TISs) can nurture advantageous market conditions for maritime biofuel having the drawback that such a type of biofuel is expensive (Bach et al., 2021).
- Comparing to battery-electric and hydrogen solutions it is reported a positive externality of the interchangeability between fossil and biofuels. This com-

parison makes biodiesel and liquefied biogas (LBG) competitive contenders for coastal shipping (Bach et al., 2021).

- Coastal shipping is a feasible and realistic alternative to the environmental unfriendly land-based transport operations, Coastal shipping can prove to become an effective transport mode that is operating as viable alternative of other modes of transportation (Everett & Kittel, 2010).

4.1.4 Climate Shipping Tourism

In the relevant literature of climate studies through shipping tourism researchers were focused on the fact that blue economy, as an emerging concept in all over the world, is subjected to multiple interlinked activities. In such a study in the case of Bangladesh among the major activities, food production, shipping and trade, tourism, ecosystem-biodiversity, new technologies and climate change challenges are promising domains that have to be holistically considered in the context of the blue economy of Bangladesh (Rouf et al., 2019). Bangladesh was proven to maintain enormous resources of high potential to uplift the existing economy, to improve livelihoods and to substantially reduce environmental risks and ecological scarcities. Nevertheless, constrains are the lack of policy, the weak institutional or organizational structure of data or information, lack of knowledge in innovating and diversifying marine products and services. Besides, marketing strategies can be followed by continuous research, educated manpower who are committed in their skillful jobs, motivate of public awareness regarding the value of maritime security in marine and coastal environment (Rouf et al., 2019). Long- and short-term strategic plans can make decisions to transform the national productivity higher making the country a model case study of successful blue economy accomplishment (Rouf et al., 2019).

4.1.5 Green Technologies and Maritime Industry

When exploring the green technologies that are operative in maritime industry it can be signified that traditional green emissions reduction in the shipbuilding industry typically focuses on fuel types and waste disposal. However, there are studies targeted to green emissions reduction in the shipbuilding industry beyond the need of specific fuels consideration, but taking a fundamentally new approach of cutting-edge technologies to control ship weight (Zeng et al., 2023). In this context, a new generation of ship full life cycle weight control technology is that of Design Index and Ship Energy Efficiency Management Regulations that has been structured on multi-table fusion in order to achieve energy-saving and emissions reduction. This innovative weight control technology is suitable for small polar cruise ships, FPSOs, and other vessels, and has been verified to reduce fuel consumption by around 25% by controlling ship weight and optimizing design (Zeng et al., 2023).

Therefore, further investigation should be directed to weight control optimization that increases the overall reliability of ship structures, power systems, stability, navigation safety, and comfort. Besides, a substantial reduction of maintenance costs, the extended life expectancy of ships service life, the concurring reduction of accident rates, all can greatly improve the reliability and carbon reduction effect of ships inspiring a green and sustainable development of the shipbuilding industry (Zeng et al., 2023). The crucial roles of new technologies and energy-fuel savings in jointly reducing carbon emissions, pollution, and environmental impacts and promoting a sustainable "greener" economy were the research objectives and findings of other similar studies in the shipping industry: bionics and biomimetic technologies (Busch et al., 2019), state-of-the-art technologies that are available in green maritime transport, including alternative fuels (than that of heavy fuel oil and marine gas oil); hybrid propulsion systems and hydrogen technologies; digitalization of maritime sector to increase vessel efficiency; hull drag reduction technologies and carbon capture technologies (Vidović et al., 2023), intelligent shipping technology, involving data security and multi-source data such as a trajectory prediction for different climates and different ship types advancing our current knowledge on the current landscape and progression trajectory of intelligent shipping technologies (Xiao et al., 2024), integration of planning problems in green maritime transportation, as green technologies involve different planning approaches to address technological and price uncertainties. Making necessary the development of new mathematical models and optimization strategies to capture emerging and planning complexities in a realistic basis (Ben Ahmed et al., 2023). All of the aforementioned domains will support current technologies' readiness and will shape their expected development over the coming years, as key areas of future research.

Jointly with the application of newly available technologies in the green maritime sector, a broad array of issues that have been raised as conflicting requirements of different ecological, technological and economical parameters that act synergistically, include the following emerging fields of investigation:

- Governmental policies, since national governments can incentivize radical reductions in harmful emissions as priority actions and motivations against increases to economic costs of maritime transport without jeopardizing the greatest benefit: that of transport of massive quantities of freight at the lowest induced cost (Vidović et al., 2023).
- Public health and hygiene measures act as precautionary and preventive measures during pandemic, as that of COVID-19, but they are also vulnerable to

increasing maritime transport costs that could disrupt the global economy, accordingly (Vidović et al., 2023).

- Industry, this is a sector that has shifted towards a gradual decrease in emissions through the implementation of "better" transitional solutions until alternative fuels eventually become low-cost fuels (Vidović et al., 2023).
- Green ports, in this context cold Ironing (CI) is one of the key points of a green port strategy to build efficiently and economically sustainable green ports in the future (Xiao et al., 2024).

4.1.6 Climate Change and Regulations

The problem setting of climate change-regulations is that secrecy jurisdictions pave the way of the legal framework perpetuating climate change. In this context it is important to understand the dynamics of climate change by enabling capital accumulation rooted in environmental degradation, making feasible and plausible to determine how these jurisdictions are issued and effected. Climate change is obliged to address the nature and the characteristics of the legislative norms that upholds exploitative and ecocide social relationships.

In the relevant literature the secrecy jurisdictions, though operationally significant, they are under-analyzed elements of the global legislative framework that facilitates climate change (Atiles & Whyte, 2025). In the case of Caribbean, which is a geographical region that operates as the epicenter for externalizing legal liabilities and extra-legal activities, climate change is disproportionately suffering its impacts (Atiles & Whyte, 2025). It was shown that corporate organizational structure enables fossil fuel companies to shield themselves from liability, enabling them to detoxify fossil fuel assets (Atiles & Whyte, 2025). Subsequently, a supplementary typology of "mechanisms of avoidance" enabled fossil fuel companies to secure key commercial advantages and operate under the radar of regulatory constraints, making an imperative need to dismantle regulatory havens in order to build up a sustainable economy (Atiles & Whyte, 2025).

Marine pollution is termed as "pollution of the marine environment", which is the collective result of different types of emerging activities, including land-based activities, vessel-related activities, dumping at sea, atmospheric and offshore hydrocarbon exploration, seabed mining (Shi & Zhang, 2017). These types of activities are factors of marine pollution that should be transboundary in nature, harmful to human health and marine ecosystem, as well as a substrate-factor of induced climate change that affects all countries globally (Shi & Zhang, 2017). The Assessment Reports of the Intergovernmental Panel on Climate Change (IPCCs) are recently and affirmatively confirming the existence of global warming, being a certain indicator of future concern and priority actions to be undertaken against climate change of detrimental consequences to natural and human systems, all continents and across oceans. In turn, substantial and realistic reduction of GHGs emissions should be viewed as international issue that needs international responses. In this context, marine pollution and climate change is a problematic issue of international impact that necessitates global regulation by the international community (Shi & Zhang, 2017).

4.2 Discussing the Challenges and Priority Actions in Climate Change and Tourism Industry

Based on the aforementioned theoretical coverage, analysis and results, it can be argued that in recent years climate change is firmly recognized by the majority of the world's scientists and governments as an issue of extreme concern for the population nationally (Kyramargiou & Vardopoulos, 2019) and globally (Vardopoulos et al., 2024). Tourism activity is itself a contributor to the problem of climate change due to its dependency on fossil fuel consumption and, in this context, outdoor recreation and tourism are likely to be directly and indirectly impacted by such a climate change and, broadly, altering the health conditions and the carrying capacity of the surrounding natural environment (Nicholls, 2006).

The tourism industry in EU is bounded on the right to leisure. The right to tourism was proclaimed universal in the Global Code of Ethics for Tourism of the United Nations World Tourism Organization (UNWTO) that has been recognized by the Universal Declaration of Human Rights, UN 1948: Article 24. Since then, although disparities have been globally emerged and reported regarding the access to leisure environment and tourism activities, in practice these rights are unquestionably key phenomena of contemporary and industrialized societies that have global effects on international recreation and tourism (Ribeiro Hoffmann et al., 2024) Among the reported disparities are that of time-seasonal and spatial-centered perspectives. Indeed, summer Southern Europe could experience climate conditions that are less favourable to tourism than the current climate, while countries in the North could enjoy better conditions. The concurring economic effects of these changes are also sizeable, albeit difficult to assess. Most important is the tourists' temporal flexibility with respect to holiday planning since, as more as intensiveness occur at the prominence of institutional rigidities -such as school holidays- then, the wider the differences between winning and losing regions in terms of economic impact can be shown (Amelung & Moreno, 2012). Climate change on tourism demand among EU member states was implemented by conducting a long-term (2100) projection accounting for climate adaptation in terms of holiday duration and frequency (Barrios & Ibañez, 2015). It was shown that climatic change was likely to affect the relative attractiveness of EU regions for tourism activities. Moreover, mainly in the Southern EU Mediterranean regions climate condition in 2100 could under current economic conditions, lower tourism revenues for up to −0.45% of GDP per year (Barrios & Ibañez, 2015). However, in the majority of Northern European regions would gain, though modest, from altered climatic conditions, reaching up to 0.32% of GDP on an annual basis. Subsequently, changes in holiday duration can be more beneficial than the change in holiday frequency in mitigating the cost of climate change. Nevertheless, it is challenging whether the EU-based findings are likely to be conditioned by broader societal and institutional factors (Barrios & Ibañez, 2015).

From a methodological prospect novel methodology can be developed to include a spatial dimension to existing each time operational-national framework, thus, identifying those regions in which forms of seasonal-based types of sport, such as winter sport tourism, can be likely to be significantly affected by, or better adjusted to, climate change (Tranos & Davoudi, 2014). Indeed, among the mostly appreciated quantification methods to investigate the optimal or unacceptable climate conditions both generally and for specific tourism segments or activities over the last 10 years are the incorporation of the Tourism Climate Index (TCI) (being subject to substantial critique) with accurate assessment of the climatic suitability of destinations for leisure tourism, which is introduced Holiday Climate Index (HCI) (Scott et al., 2016). HCI is proven advantageous comparing to TCI, since HCI is more consistent with observed visitation patterns at urban contexts where HCI at many cities is higher than that of TCI, particularly in shoulder seasons and the winter months. The varied rating scales of HCI and the component weighting system have been based on literature-stated tourists' climatic preferences, thus, empirically demonstrating that the use of the TCI should be discontinued in favour of HCI (Scott et al., 2016).

Of utmost importance issue is the pursuing of green growth which has emerged as a strategy to address environmental concerns and to simultaneously promote economic development (Gorji & Martek, 2023). Renewable energy technologies and policies are differentiated on green growth in the built environment. At investigating 20 developed and 20 developing countries in the period 2010-2021 the inclusion of panel data estimators contained the employment of generalized least squares and generalized method, revealing that the contribution of renewable energy sectors to green growth varies between developed and developing countries. In developed countries green growth is achieved through solar, wind, and biomass sources, whereas hydroelectric capacities did not. In developing countries, the wind capacity has Stimuli of investment and required technological innovation for a sustainable built environment are also spatial- and time- sensitive (Gorji & Martek, 2023).

4.3 Discussing the Challenges and Priority Actions of Quality in Coastal and Shipping Services

The idea of attributing a service quality identity to coastal contexts is coming back from 1990s' when early studies were devoted to nonpoint sources as external threats to coastal water quality (Burroughs, 1993). In this study a program design for nonpoint source control was considered in investigating those external threats to national parks. Through an analogous problem. In this context recent federal legislation seeks to limit nonpoint sources in coastal areas, being denominated as diffuse land activities that degrade water quality. Besides, external threats are occurring outside a park boundary but also affecting the resources or purposes within the park for many decades, which are subject to federal management. In this context nonpoint sources are categorized as external threat that have to be limited and to protect parks through programs and park techniques. Park techniques rely on power, usually through legal devices, but also on "soft approaches" that utilize shared values and objectives. Whenever a soft approach fails, the manager of the protected unit is empowered to undertake an adopt a hard action in order to optimally protect the resource (Burroughs, 1993).

At this point it can be mentioned that over the course of its urban history, there are urban contexts of increasing populations along coastal areas, which support economic activities and urbanization processes, as that of the Italian peninsula in which many of Italy's large cities and metropolitan areas are located along the coast side and are adhered to strong and direct connection to each other. In recent decades there is an intense tourist activity that exerts a growing anthropocentric pressure (Lombardini et al., 2022). In Figure 2 a synthesis of all stakeholders whom responsibilities are both to provide professional tourism services in alignment with quality preservation and environmental protection, are all synergistic, not antagonistic, to each other.

Figure 2. Integrated synthesis framework on the shipping-quality-climate change-ecoinnovations nexus.

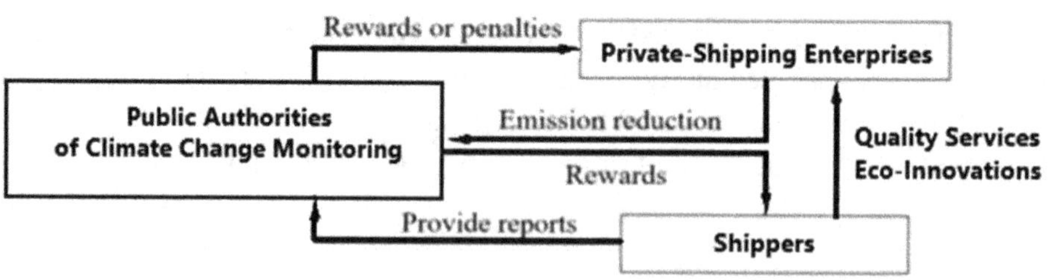

Source: Authors own study.

Based on Figure 2 it can be denoted that when local governments take an active role in regulation, they consider the associated costs and may either reward or penalize shipping companies and cargo owners based on their compliance. Governments might also receive societal accolades for their proactive stance. In cases of passive regulation, then, local governments enforce rules by incentivizing cargo owners to report any emission reduction practices by shipping companies. Shipping companies when proactively reduce emissions, then, they face higher costs but are also benefited from government incentives and additional gains from their reduction efforts. Conversely, those enterprises that are passive in their emission reduction efforts risk government fines and potential reports from cargo owners. Cargo owners who engage in supervision incur costs and additional fees but may receive rewards from actively regulating governments or by reporting non-compliant shipping companies to passively regulating governments. If cargo owners choose not a supervisor role, they only face the additional fees without any potential rewards. The relationship between all core and peripheral stakeholders is shown in Figure 2.

In such densely populated areas, it is an imperative need to assess the perception of the inhabitants on the provision of basic sanitation services and to identify demands and priorities for improving these services. Such a case is that of the Santa Rosa neighborhood (Guarujá-SP) where researches attempted to identify the demands when applying structured questionnaires in groups of blocks (Vardopoulos et al., 2021). These groups of blocks were defined being based on the social development index, in order to consider the socio-environmental complexity of the neighborhood. Perception differences are reported at the different neighborhoods showing that basic sanitation is an essential human right, and the universalization of such a type of services is challenging due to its complexity. Despite the universalization and the quality of services adherence with various social actors, the adopted policies are not always reflecting societal needs (Vardopoulos et al., 2021). Contrarily, high heterogeneity is anticipated in perceiving service provision and quality in a neighborhood level. Focusing on aqueous and coastal contexts it can be denoted that while water supply is the only service that can be assessed and valued as adequate; drainage and sanitation are among those mostly requiring improvements, while waste management has a very different perception compared to the available data (Vardopoulos et al., 2021). Especially considering the coastal context it can be argued that quality of service is primary determined by the importance of social participation in all stages of shipping and basic sanitation management. For them, it cannot be undermined the real demands of the population, making essential the designing of policies that incorporate participation results and reflect them to real world conditions (Vardopoulos et al., 2021).

5. CONCLUSIONS

From a managerial perspective this study underscores those priority areas for investments and improvements that have to be identified in order to improve services and the quality of life of the local populations. Besides, it has been stressed out that results of similar future researches can realistically contribute to the updating of municipal plans and/or regulations of local, national, or EU range of actions and regulations (Vardopoulos et al., 2021).

From a technological perspective corrosivity of the atmosphere in coastal regions is considered as a very high constraints of their tourism standards of visiting and local economy development (Castañeda-Valdés et al., 2023). Subsequently, reinforced structures are frequently showing short service life if adequate measures are not applied. Therefore, critical parameters of research interest are that of determination of time-to-corrosion-initiation (ti) and time to corrosion with induced cracking (tcc), both of which can credibly calculate the service life in buildings located in coastal cities (Castañeda-Valdés et al., 2023). Indeed, changes between the 2 aforementioned time-termed values: that of time-to-corrosion-initiation (ti) and time to corrosion with induced cracking (tcc), have been associated with concrete covering thickness, being particularly important to extremely aggressive coastal outdoor exposure sites that are located within 10 m of the shoreline. The field application in areas of such characteristics, as Havana is one of the most aggressive coastal sites in the world, include the determination of electrochemical corrosion rate in the reinforced concrete specimens. Besides to chemical, also the environmental parameters have to be determined at the sites, enabling the quantification of lowest permissible threshold levels in order the relevant coastal infrastructure to determine the concrete quality assessment before building reinforced structures exposed to extreme corrosivity and to predict a service life in very aggressive coastal sites (Castañeda-Valdés et al., 2023).

From the shipping industry perspective, it is crucial to understand that under rapidly changing environments of the international economy the shipping industry is adversely affected by survival competition. In this context, future researches should attempt to identify and to apply those shipping service quality and measures by expanding the research models of the logistics service quality into the shipping industry. In such a way shipping service can be conceptualized and developed at a scale to assess the shipping service quality. Although it is utopia to develop a perfect model, there are proposed models as useful tools of measuring the shipping service quality in coastal areas, counting and valuing the specific infrastructure and geospatial characteristics (Gi-du Kang & Kim, 2009).

Conclusively, in better identifying the key drivers of customer satisfaction in liner shipping sector future researches, a parsimonious and updated set of variables can be constructed and utilized (Yuen & Thai, 2015) (Yuen & Thai, 2017).

REFERENCES

Agan, B., & Balcilar, M. (2023). Unraveling the Green Growth Matrix: Exploring the Impact of Green Technology, Climate Change Adaptation, and Macroeconomic Factors on Sustainable Development. *Sustainability (Basel)*, *15*(11), 8530. DOI: 10.3390/su15118530

Albitar, K., Al-Shaer, H., & Liu, Y. S. (2023). Corporate commitment to climate change: The effect of eco-innovation and climate governance. *Research Policy*, *52*(2), 104697. DOI: 10.1016/j.respol.2022.104697

Amelung, B., & Moreno, A. (2012). Costing the impact of climate change on tourism in Europe: Results of the PESETA project. *Climatic Change*, *112*(1), 83–100. DOI: 10.1007/s10584-011-0341-0

Apostu, S. A., Gigauri, I., Panait, M., & Martín-Cervantes, P. A. (2023). Is Europe on the Way to Sustainable Development? Compatibility of Green Environment, Economic Growth, and Circular Economy Issues. *International Journal of Environmental Research and Public Health*, *20*(2), 1078. DOI: 10.3390/ijerph20021078 PMID: 36673838

Atiles, J., & Whyte, D. (2025). Fossil Capital in the Caribbean: The Toxic Role of "Regulatory Havens" in Climate Change. *Regulation & Governance*, *19*(2), 469–481. Advance online publication. DOI: 10.1111/rego.70001

Bach, H., Mäkitie, T., Hansen, T., & Steen, M. (2021). Blending new and old in sustainability transitions: Technological alignment between fossil fuels and biofuels in Norwegian coastal shipping. *Energy Research & Social Science*, *74*, 101957. DOI: 10.1016/j.erss.2021.101957

Barrios, S., & Ibañez, J. N. (2015). Time is of the essence: Adaptation of tourism demand to climate change in Europe. *Climatic Change*, *132*(4), 645–660. DOI: 10.1007/s10584-015-1431-1

Ben Ahmed M., Molland E., Tomasgard T. (2023). Challenges and Opportunities for Adopting Green Technologies in Maritime Transportation Planning. IFIP Advances in Information and Communication Technology, 692 AICT, 620 - 633. DOI: DOI: 10.1007/978-3-031-43688-8_43

Ben Amara, D., & Qiao, J. (2023). From economic growth to inclusive green growth: How do carbon emissions, eco-innovation and international collaboration develop economic growth and tackle climate change? *Journal of Cleaner Production*, *425*, 138986. DOI: 10.1016/j.jclepro.2023.138986

Bergek, A., Bjorgum, O., Hansen, T., Hanson, J., & Steen, M. (2021). Sustainability transitions in coastal shipping: The role of regime segmentation. *Transportation Research Interdisciplinary Perspectives, 12*, 100497. DOI: 10.1016/j.trip.2021.100497

Bonsinetto, F., & Falco, E. (2013). Analysing Italian regional patterns in green economy and climate change. Can Italy leverage on Europe 2020 strategy to face sustainable growth challenges? *Journal of Urban and Regional Analysis, 5*(2), 123–142.

Buelow, C. A., & Waltham, N. J. (2020). Restoring tropical coastal wetland water quality: Ecosystem service provisioning by a native freshwater bivalve. *Aquatic Sciences, 82*(4), 77. DOI: 10.1007/s00027-020-00747-7

Burroughs, R. H. (1993). Nonpoint sources as external threats to coastal water quality: Lessons from park service experience. *Coastal Management, 21*(2), 131–142. DOI: 10.1080/08920759309362197

Busch, J., Barthlott, W., Brede, M., Terlau, W., & Mail, M. (2019). Bionics and green technology in maritime shipping: An assessment of the effect of Salvinia air-layer hull coatings for drag and fuel reduction. *Philosophical Transactions. Series A, Mathematical, Physical, and Engineering Sciences, 377*(2138), 20180263. DOI: 10.1098/rsta.2018.0263 PMID: 30967069

Castañeda-Valdés, A., Corvo, F., Marrero-Águila, R., Fernández-Domínguez, A., & Del Angel-Meraz, E. (2023). The service life of reinforced concrete structures in an extremely aggressive coastal city. Influence of concrete quality. *Materials and Structures, 56*(1), 12. DOI: 10.1617/s11527-023-02100-4

Chao, S.-L., Yu, M.-M., & Wei, S.-Y. (2024). Ascertaining the impact of e-service quality on e-loyalty for the e-commerce platform of liner shipping companies. *Transportation Research Part E, Logistics and Transportation Review, 184*, 103491. DOI: 10.1016/j.tre.2024.103491

Chen, J. K. C., Wang, J. C. F., Tsai, W.-H., & Lin, T. S. Y. (2008). Exploring innovation model combine with the green marketing to consumer's behavior influence - A case study of a TFT-LCD company. 2008 IEEE International Conference on Industrial Engineering and Engineering Management, IEEM 2008, art. no. 4737943, 619 - 623. DOI: DOI: 10.1109/IEEM.2008.4737943

Chen, K.-K., Chang, C.-T., & Lai, C.-S. (2009). Service quality gaps of business customers in the shipping industry. *Transportation Research Part E, Logistics and Transportation Review, 45*(1), 222–237. DOI: 10.1016/j.tre.2008.02.005

Chen, T., & Liu, G. (2025). Global perspectives on environmental policy innovations: Driving green tourism and consumer behavior in circular economy. *Journal of Environmental Management*, *374*, 124138. Advance online publication. DOI: 10.1016/j.jenvman.2025.124138 PMID: 39827609

Devane, C., Saulnier, L., & Latham, L. (2021). Coastal Mental Health and Substance Use Services' COVID-19 Response: A Quality Improvement Initiative. *Nursing Leadership*, *34*(3), 40–50. DOI: 10.12927/cjnl.2021.26594 PMID: 34698013

Durán-Romero, G., López, A. M., Beliaeva, T., Ferasso, M., Garonne, C., & Jones, P. (2020). Bridging the gap between circular economy and climate change mitigation policies through eco-innovations and Quintuple Helix Model. *Technological Forecasting and Social Change*, *160*, 120246. DOI: 10.1016/j.techfore.2020.120246

Durán-Romero, G., & Urraca-Ruiz, A. (2015). Climate change and eco-innovation. A patent data assessment of environmentally sound technologies. *Innovation (North Sydney, N.S.W.)*, *17*(1), 115–138. DOI: 10.1080/14479338.2015.1011062

Esangbedo, C. O., Chang, Y., Ushie, A. M., & Hasan, M. M. (2024). Eco-Innovation and Firm Sustainability in Emerging Economies: The Role of Consumer Behavior and Green B2B Marketing. *Journal of Business-To-Business Marketing*, *31*(4), 405–420. DOI: 10.1080/1051712X.2024.2369764

Everett, S., & Kittel, C. (2010). Sustainability and Australian Coastal Shipping: Some Issues. *Australian Journal of Maritime and Ocean Affairs*, *2*(3), 82–89. DOI: 10.1080/18366503.2010.10815661

Gi-du Kang, G.-D., & Kim, Y.-D. (2009). An Analysis of the Measurement of the Shipping Service Quality. *The Asian Journal of Shipping and Logistics*, *25*(1), 41–55. DOI: 10.1016/S2092-5212(09)80011-4

Gorji, A. A., & Martek, I. (2023). The Role of Renewable Energy as a 'Green Growth' Strategy for the Built Environment. *Buildings (Basel, Switzerland)*, *13*(5), 1356. DOI: 10.3390/buildings13051356

Han, S., Cao, B., Fu, Y., & Luo, Z. (2018). A liner shipping competitive model with consideration of service quality management. *Annals of Operations Research*, *270*(1-2), 155–177. DOI: 10.1007/s10479-016-2386-y

Haque, M. O., Mohammad, F., & Rahman, R. (2021). Shipping Container Housing for Architectural Resilience in Coastal Regions: *Addressing Construction and Material Sustainability. ZEMCH International Conference*, 163 - 177.

Ismail, C., Wiropranoto, F., Takama, T., Lieu, J., & Virla, L. D. (2021). Frugal Eco-innovation for Addressing Climate Change in Emerging Countries: Case of Biogas Digester in Indonesia. Handbook of Climate Change Management: Research, Leadership. *Transformation (Durban)*, *1*, 693–719. DOI: 10.1007/978-3-030-57281-5_177

Jansson, J., Marell, A., & Nordlund, A. (2010). Green consumer behavior: Determinants of curtailment and eco-innovation adoption. *Journal of Consumer Marketing*, *27*(4), 358–370. DOI: 10.1108/07363761011052396

Jesic, J., Okanovic, A., & Panic, A. A. (2021). Net zero 2050 as an EU priroty: Modeling a system for efficient investments in eco innovation for climate change mitigation. *Energy, Sustainability and Society*, *11*(1), 50. DOI: 10.1186/s13705-021-00326-0

Jiang, X., Zhang, J., Yang, C., & Wan, R. (2023). Evaluating the service quality of insular and coastal recreational fisheries by integration of the SERVQUAL-fuzzy Kano model and importance-performance analysis. *Ocean and Coastal Management*, *243*, 106753. DOI: 10.1016/j.ocecoaman.2023.106753

Khan, M. A. S., Du, J., Malik, H. A., Anuar, M. M., Pradana, M., & Yaacob, M. R. B. (2022). Green innovation practices and consumer resistance to green innovation products: Moderating role of environmental knowledge and pro-environmental behavior. *Journal of Innovation and Knowledge*, *7*(4), 100280. Advance online publication. DOI: 10.1016/j.jik.2022.100280

Kreitler, J., Papenfus, M., Byrd, K., & Labiosa, W. (2013). Interacting Coastal Based Ecosystem Services: Recreation and Water Quality in Puget Sound, WA. *PLoS One*, *8*(2), 56670. DOI: 10.1371/journal.pone.0056670 PMID: 23451067

Kyramargiou, Z., & Vardopoulos, I. (2019). Use and sustainability of the ships' waste reception facilities. Evidence from the Port of Corinth, Greece. *Journal of Management and Regional Development*, *15*(3), 93–106. DOI: 10.54399/rbgdr.v15i3.4670

Kyriakopoulos G.L. (2021). Environmental Legislation in European and International Contexts: Legal Practices and Social Planning toward the Circular Economy. Laws 10(1), 3, 15 pp. .DOI: 10.3390/laws10010003

Kyriakopoulos G.L., Sebos I. (2023). Enhancing Climate Neutrality and Resilience through Coordinated Climate Action: Review of the Synergies between Mitigation and Adaptation Actions. Climate 11(5), 105, 27 pp. DOI: DOI: 10.3390/cli11050105

Lobo, A. (2010). Assessing the service quality of container shipping lines in the international supply chain network - shippers' perspective. *International Journal of Value Chain Management*, *4*(3), 256–266. DOI: 10.1504/IJVCM.2010.033615

Lombardini, G., Pilogallo, A., & Tucci, G. The Provision of Ecosystem Services Along the Italian Coastal Areas: A Correlation Analysis Between Environmental Quality and Urbanization (2022). Lecture Notes in Computer Science 2022, 13380 LNCS, 298 - 314. DOI: DOI: 10.1007/978-3-031-10542-5_21

Machiba, T. (2010). Eco-innovation for enabling resource efficiency and green growth: Development of an analytical framework and preliminary analysis of industry and policy practices. *International Economics and Economic Policy*, *7*(2), 357–370. DOI: 10.1007/s10368-010-0171-y

Marlianingrum, P. R., Kusumastanto, T., Adrianto, L., & Fahrudin, A. (2021). Valuing habitat quality for managing mangrove ecosystem services in coastal Tangerang District, Indonesia. *Marine Policy*, *133*, 104747. DOI: 10.1016/j.marpol.2021.104747

Matzarakis, A., Rammelberg, J., & Junk, J. (2013). Assessment of thermal bioclimate and tourism climate potential for central Europe-the example of Luxembourg. *Theoretical and Applied Climatology*, *114*(1-2), 193–202. DOI: 10.1007/s00704-013-0835-y

Merrill, N. H., Mulvaney, K. K., Martin, D. M., Chintala, M. M., Berry, W., Gleason, T. R., Balogh, S., & Humphries, A. T. (2018). A Resilience Framework for Chronic Exposures: Water Quality and Ecosystem Services in Coastal Social-Ecological Systems. *Coastal Management*, *46*(4), 242–258. DOI: 10.1080/08920753.2018.1474066 PMID: 31178625

Neves, R. A. F., Naveira, C., Miyahira, I. C., Portugal, S. G. M., Krepsky, N., & Santos, L. N. (2020). Are invasive species always negative to aquatic ecosystem services? The role of dark false mussel for water quality improvement in a multi-impacted urban coastal lagoon. *Water Research*, *184*, 116108. DOI: 10.1016/j.watres.2020.116108 PMID: 32688149

Nicholls, S. (2006). Climate change, tourism and outdoor recreation in Europe. *Managing Leisure*, *11*(3), 151–163. DOI: 10.1080/13606710600715226

Nicholls, S., & Amelung, B. (2008). Climate change and tourism in northwestern Europe: Impacts and adaptation. *Tourism Analysis*, *13*(1), 21–31. DOI: 10.3727/108354208784548724

Nouri, B. A., & Emkani, P. (2020). Effect of consumer innovation process on green consumption behaviour: An application of planned behaviour theory. *International Journal of Business Environment*, *11*(4), 347–368. DOI: 10.1504/IJBE.2020.111387

Perch-Nielsen, S. L., Amelung, B., & Knutti, R. (2010). Future climate resources for tourism in Europe based on the daily Tourism Climatic Index. *Climatic Change*, *103*(3), 363–381. DOI: 10.1007/s10584-009-9772-2

Pham, T. Y., & Yeo, G.-T. (2019). Evaluation of Transshipment container terminals' service quality in Vietnam: From the shipping companies' perspective. *Sustainability (Basel)*, *11*(5), 1503. DOI: 10.3390/su11051503

Ribeiro Hoffmann, A., Sandrin, P., & Doukas, Y. E. (Eds.). (2024). *Climate change in regional perspective. European Union and Latin American initiatives, challenges, and solutions*. Cham, Switzerland: United Nations University Series on Regionalism, Vol. 27, Springer Nature. DOI: 10.1007/978-3-031-49329-4

Rouf, M. A., Rahman, M. M., Rahman, S. M., & Ahsan, M. N. (2019). Coming stakes in the ocean: Food Production, Shipping and Trade, Tourism, Ecosystem-biodiversity, New Technologies and Climate Change Challenges in Bangladesh. *Journal of Ocean and Coastal Economics*, *6*(2), 5. DOI: 10.15351/2373-8456.1102

Sarfraz, M., Raza, M., Khalid, R., Liu, T., Li, Z., & Niyomdecha, L. (2022). Consumer Purchasing Behavior Toward Green Environment in the Healthcare Industry: Mediating Role of Entrepreneurial Innovation and Moderating Effect of Absorptive Capacity. *Frontiers in Public Health*, *9*, 823307. Advance online publication. DOI: 10.3389/fpubh.2021.823307 PMID: 35186878

Scarpa, G. M., Zaggia, L., Manfè, G., Lorenzetti, G., Parnell, K., Soomere, T., Rapaglia, J., & Molinaroli, E. (2019). The effects of ship wakes in the Venice Lagoon and implications for the sustainability of shipping in coastal waters. *Scientific Reports*, *9*(1), 19014. DOI: 10.1038/s41598-019-55238-z PMID: 31831769

Scott, D., Rutty, M., Amelung, B., & Tang, M. (2016). An inter-comparison of the Holiday Climate Index (HCI) and the Tourism Climate Index (TCI) in Europe. *Atmosphere (Basel)*, *7*(6), 80. DOI: 10.3390/atmos7060080

Shi Y., Zhang D. (2017). International regulatory responses to global challenges in marine pollution and climate change. Marine Pollution and Climate Change, 279 - 322. DOI: DOI: 10.1201/9781315119243

Stopka, O., Šimková, I., & Konečný, V. (2015). The quality of service in the public transport and shipping industry. *Nase More (Dubrovnik)*, *62*(3), 126–130. DOI: 10.17818/NM/2015/SI7

Streimikiene, D., Kyriakopoulos G.L. (2024). Climate Change Mitigation Performance in the EU Tourism Destination Sector. Journal of Tourism and Services 15(28), 268–284. . https://doi.org/DOI: 10.29036/jots.v15i28.758

Streimikiene D., Lekavičius V., Baležentis T., Kyriakopoulos G.L., Abrhám J. (2020). Climate Change Mitigation Policies Targeting Households and Addressing Energy Poverty in European Union. Energies 13(13), 3389; 24 pp. .DOI: 10.3390/en13133389

Thai, V. V., Tay, W. J., Tan, R., & Lai, A. (2014). Defining service quality in tramp shipping: Conceptual model and empirical Evidence. *Asian Journal of Shipping and Logistics*, *30*(1), 1–29. DOI: 10.1016/j.ajsl.2014.04.001

Thilakshan, T., Saranjan, P., Fernando, G. V. K., & Bandara, J. M. S. J. Analysis of Passenger Perception on Quality of Service of Railway Transportation: Coastal Railway Line, *Sri Lanka. Moratuwa Engineering Research Conference* 2023, MERCon, 420 - 425. DOI: DOI: 10.1109/MERCon60487.2023.10355413

Tranos, E., & Davoudi, S. (2014). The Regional Impact of Climate Change on Winter Tourism in Europe. *Tourism Planning & Development*, *11*(2), 163–178. DOI: 10.1080/21568316.2013.864992

Tsimnadis K., Kyriakopoulos G.L. (2024). Investigating the Role of Municipal Waste Treatment within the European Union through a Novel Created Common Sustainability Point System. Recycling 9(3), 42, 25 pp. DOI: .DOI: 10.3390/recycling9030042

Udiyana, I. B. G., Suastama, I. B. R., Astini, N. N. S., Mahanavami, G. A., Karwini, N. K., & Maretta, Y. A. (2018). Innovation strategy the development of competitifeness of eco-based coastal tourism destination, management organization and quality of services. *Journal of Environmental Management and Tourism*, *9*(4), 851–860. DOI: 10.14505//jemt.9.4(28).19

Vardopoulos, I., Giannopoulos, K., Kalli, S., & Salvati, L. (2024). From tech-centric to human-centric smart tourist destinations: innovations in real estate, urban planning, and sustainability. In *Socio-enviromental change: the path toward sustainability* (pp. 65–96). Nova Science Publishers.

Vardopoulos, I., & Karytsas, S. (2019). An exploratory path analysis of climate change effects on tourism. *Sustainable Development, Culture, Traditions Journal, Special Volume in Honor of Professor George I. Theodoropoulos,* 132–152. DOI: 10.26341/issn.2241-4002-2019-sv-13

Vardopoulos, I., Konstantopoulos, I., Zorpas, A. A., Limousi, L., Bennici, S., Inglezakis, V., & Voukkali, I. (2021). Sustainable metropolitan areas perspectives through assessment of the existing waste management strategies. *Environmental Science and Pollution Research International*, *28*(19), 24305–24320. DOI: 10.1007/s11356-020-07930-1 PMID: 32072414

Vidović, T., Šimunović, J., Radica, G., & Penga, Ž. (2023). Systematic Overview of Newly Available Technologies in the Green Maritime Sector. *Energies, 16*(2), 641. DOI: 10.3390/en16020641

Waheed, A., Zhang, Q., Rashid, Y., Tahir, M. S., & Zafar, M. W. (2020). Impact of green manufacturing on consumer ecological behavior: Stakeholder engagement through green production and innovation. *Sustainable Development (Bradford), 28*(5), 1395–1403. DOI: 10.1002/sd.2093

Xiao, G., Wang, Y., Wu, R., Li, J., & Cai, Z. (2024). Sustainable Maritime Transport: A Review of Intelligent Shipping Technology and Green Port Construction Applications. *Journal of Marine Science and Engineering, 12*(10), 1728. DOI: 10.3390/jmse12101728

Yang, D., Geng, H., & Wang, F. (2020). Marine regulations' impact on environmental sustainability in maritime shipping and coastal economic activities. *Journal of Coastal Research, 109*(sp1), 79–84. DOI: 10.2112/JCR-SI109-013.1

Yang, J., Gu, L., Liu, K., & Deng, C. (2025). Unabsorbed slack, capability reconfiguration and green innovation behavior: The moderating role of consumer green pressure. *Journal of Business and Industrial Marketing, 40*(2), 391–414. DOI: 10.1108/JBIM-02-2024-0068

Yang, Q. (2021). Fiscal Transparency and Public Service Quality Association: Evidence from 12 Coastal Provinces and Cities of China. *Journal of Risk and Financial Management, 14*(1), 13. DOI: 10.3390/jrfm14010013

Yuen, K. F., & Thai, V. V. (2015). Service quality and customer satisfaction in liner shipping. *International Journal of Quality and Service Sciences, 7*(2-3), 170–183. DOI: 10.1108/IJQSS-02-2015-0024

Yuen, K. F., & Thai, V. V. (2017). Corporate social responsibility and service quality provision in shipping firms: Financial synergies or trade-offs? *Maritime Policy & Management, 44*(1), 131–146. DOI: 10.1080/03088839.2016.1237782

Zeng, J., Yu, S., & Zhan, Y. (2023). A Novel Approach to Green Emissions Reduction and Reliability Improvement in the Maritime Industry: Multi-Table Fusion-Based Full-Life Weight Control Technology. *7th IEEE International Conference on Transportation Information and Safety, ICTIS 2023*, 1655 - 1661. DOI: DOI: 10.1109/ICTIS60134.2023.10243738

Zhu, E., Gao, H., Chen, L., Yao, J., Liu, T., & Sha, M. (2024). Interactions between coastal protection forest ecosystems and human activities: Quality, service and resilience. *Ocean and Coastal Management*, *254*, 107190. DOI: 10.1016/j.ocecoaman.2024.107190

Chapter 10
Sustainable Urban Development in Northeast India:
Integrating Traditional Ecological Knowledge With Modern Urban Resilience Frameworks Amidst Climate

Vishal Jain
https://orcid.org/0000-0003-1126-7424
School of Engineering & Technology, Vivekananda Institute of Professional Studies - Technical Campus, New Delhi, India

Archan Mitra
https://orcid.org/0000-0002-1419-3558
NITTE Institute of Communication, India

Sanchita Paul
https://orcid.org/0009-0006-6437-3991
Independent Researcher, India

ABSTRACT

Northeast India, a region rich in biodiversity and cultural heritage, is experiencing increasing challenges due to rapid urbanization, climate change, and resource scarcity. This study examines the integration of Traditional Ecological Knowledge (TEK) with modern urban resilience frameworks to promote sustainable development. Focusing on urban centers like Guwahati, Shillong, and Imphal, the research highlights the

DOI: 10.4018/979-8-3373-2327-5.ch010

role of community-driven resource management and vernacular architecture in addressing sustainability challenges. It also explores green finance mechanisms in funding urban sustainability initiatives. Additionally, the study assesses the application of GIS and AI-driven climate risk assessments to enhance urban planning. Major barriers are analyzed with policy recommendations for participatory governance. This research provides a multidimensional framework that harmonizes global ESG standards with local socio-environmental contexts, offering actionable strategies for policymakers, urban planners, and financial stakeholders.

1. INTRODUCTION

1.1 Background and Context

Northeast India, comprising the eight states of Arunachal Pradesh, Assam, Manipur, Meghalaya, Mizoram, Nagaland, Tripura, and Sikkim, is a region of immense ecological and cultural significance. This region is home to rich biodiversity, including several endemic flora and fauna species, and supports a diverse range of indigenous communities with unique traditions of environmental stewardship. However, rapid urbanization (Vardopoulos et al., 2024), unplanned infrastructure development (Salvati & Vardopoulos, 2025), and climate change (Ribeiro Hoffmann et al., 2024), pose significant threats to the sustainability of these urban centers (Singh & Singha, 2020).

The urbanization trend in Northeast India has accelerated in recent decades due to factors such as rural-to-urban migration (Escrivà Saneugenio et al., 2024), economic development initiatives (Salvati et al., 2025), and infrastructural expansion (Vardopoulos et al., 2024). According to the Census of India, the urban population of Northeast India increased from 168,000 in 1901 to 8,403,000 in 2011, and it is expected to rise to 13,454,000 by 2031, particularly in cities like Guwahati, Shillong, and Imphal, leading to increased demand for housing, transportation, and public services (Meitei & Das, 2024). However, much of this urban expansion has occurred without adequate consideration of environmental sustainability (Salvati, 2025), resulting in deforestation (Escrivà Saneugenio et al., 2024), wetland encroachment (Escrivà Saneugenio et al., 2024), and rising levels of air and water pollution (Sateriano et al., 2024).

In addition to urbanization pressures, the region faces unique climate-related challenges (Simoniello et al., 2025). Northeast India is prone to intense monsoon rains, frequent flooding, and landslides, which exacerbate vulnerabilities in both rural and urban settings (Singh & Singha, 2020). The Brahmaputra and Barak river basins experience recurring floods, causing large-scale displacement and economic losses. Rising temperatures and changing precipitation patterns further strain water resources (Vardopoulos et al., 2018), agriculture (Doukas et al., 2023), and urban infrastructure (Vardopoulos et al., 2023).

The lack of resilient urban planning frameworks exacerbates these vulnerabilities (Vardopoulos et al., 2024). Modern urbanization models in Northeast India have largely followed conventional approaches that emphasize large-scale infrastructure development without incorporating Traditional Ecological Knowledge (TEK). Indigenous communities in the region have historically practiced sustainable land-use strategies, including community-based forestry, bamboo-based housing, and agroforestry systems that enhance ecological resilience (Tiwari & Sunny, 2024). Integrating TEK with contemporary urban sustainability frameworks can offer a holistic approach to mitigating environmental risks while preserving cultural heritage.

The role of green finance mechanisms is increasingly being recognized as a crucial element in achieving sustainable urban development. Financial instruments such as green bonds, ESG-aligned investments, and climate resilience funds offer viable options for supporting sustainable infrastructure and nature-based solutions (Abhilash et al., 2023). Additionally, emerging technologies such as Geographic Information Systems (GIS), AI-driven climate risk modeling, and data-driven urban planning tools can provide critical insights for enhancing urban resilience in Northeast India (Song et al., 2023).

Given these challenges and opportunities, this research aims to explore an integrated approach that combines TEK, green finance, and modern climate resilience strategies to foster sustainable urban development in Northeast India.

1.2 Problem Statement

Despite its ecological importance, Northeast India faces a significant gap in sustainable urban planning. Cities like Guwahati, Shillong, and Imphal are expanding rapidly, often without adequate planning, leading to environmental degradation and increased climate vulnerability (Xaxa, 2019). Urban migration and unregulated development contribute to deforestation, pollution, and habitat loss, while climate change intensifies natural disasters such as floods and landslides (Elahi et al., 2024).

The rising demand for essential resources like water, energy, and waste management in urban areas further stresses the region's ecological balance (Meitei & Das, 2024). Compounding this is the underutilization of Traditional Ecological Knowledge (TEK), which has long supported sustainable practices among Indigenous communities (Xaxa, 2019). However, modern planning often ignores this knowledge.

Policy fragmentation and weak governance hinder coordinated efforts toward sustainability (Wang et al., 2024). This study explores how TEK, green finance, and emerging technologies can be integrated into a sustainable urban development framework for the region.

1.3 Research Objectives and Questions

The primary goal of this research is to develop a multidimensional framework for sustainable urban development in Northeast India by integrating Traditional Ecological Knowledge (TEK), green finance mechanisms, and modern urban resilience strategies. This research aims to answer the following questions:

- How can Traditional Ecological Knowledge (TEK) contribute to urban resilience in Northeast India?
- What role do green finance mechanisms and ESG investments play in sustainable urbanization?
- How can emerging technologies (GIS, AI-driven climate risk assessments) enhance urban resilience?
- What policy and governance barriers hinder sustainable urban development in the region?
- How can a comprehensive urban sustainability framework be developed to address these challenges?

1.4 Significance of the Study

This study contributes to the field of sustainable urbanization and climate resilience by offering a framework that integrates TEK, financial sustainability, and technological advancements. The research is particularly relevant for policymakers, urban planners, and financial stakeholders aiming to address sustainability challenges in Northeast India. It provides actionable insights into how green finance and emerging technologies can support sustainable urban development in ecologically sensitive regions. Additionally, by aligning local environmental governance with global sustainability frameworks, this study offers a model that can be replicated in other regions facing similar socio-environmental challenges.

2. LITERATURE REVIEW

2.1 Theoretical Foundations: Urban Resilience and Sustainability

The concept of urban resilience is rooted in ecological and social sciences and has evolved to encompass a multidisciplinary framework that integrates engineering, governance, and economic sustainability. Resilience, as defined by Holling (1973), refers to the ability of systems to absorb disturbances while maintaining functionality and structure. In the context of urban resilience, this means the capacity of cities to withstand and recover from environmental shocks such as floods, earthquakes, and climate change-related disruptions. The resilience approach has been widely recognized in urban planning through frameworks such as the Rockefeller Foundation's 100 Resilient Cities Initiative, which highlights the need for integrated governance, adaptive infrastructure, and community-based resilience planning (McTarnaghan et al., 2022).

Parallel to resilience, sustainability theories emphasize the necessity of long-term environmental, economic, and social equilibrium. The Triple Bottom Line (TBL) framework, introduced by Elkington (1997), conceptualizes sustainability as a balance between economic development, environmental stewardship, and social well-being. This framework is particularly relevant for Northeast India, where rapid urbanization often comes at the cost of ecological degradation and cultural displacement. Sustainable urban development, therefore, requires a shift from conventional growth-centric models to approaches that integrate local ecological knowledge, decentralized governance, and participatory urban management.

A significant extension of urban resilience thinking is the Resilience Thinking Approach proposed by Walker & Salt (2006), which underscores the importance of adaptive management and system flexibility in response to changing environmental and socio-economic conditions. This approach highlights the role of feedback mechanisms in urban systems, ensuring that policy interventions align with ecological and social realities rather than merely addressing short-term crises.

2.2 Traditional Ecological Knowledge (TEK) in Urban Sustainability

Traditional Ecological Knowledge (TEK) is the cumulative wisdom, practices, and beliefs developed by indigenous communities over generations through their interaction with the natural environment. Berkes (2012) defines TEK as a dynamic system of understanding that evolves through empirical observations, adaptive strategies, and intergenerational transmission. Its relevance in urban sustainability lies in its ability to provide time-tested solutions for climate adaptation, resource management, and disaster risk reduction, especially in regions with rich indigenous heritage such as Northeast India (Mitra & Paul, 2025).

One of the most prominent applications of TEK in sustainable urban development is its role in water conservation and flood mitigation strategies. For instance, the Zabo system of Nagaland, an indigenous method of rainwater harvesting and integrated land use, has been effective in mitigating soil erosion and ensuring year-round water availability (Vohra & Franklin, 2021; Geovetenskaper, 2023). Such community-driven solutions present valuable lessons for modern urban water management, where inefficient drainage systems and unregulated construction exacerbate flood risks.

Similarly, the vernacular architecture of Assam and Arunachal Pradesh, characterized by bamboo and stilts, showcases the potential of TEK in designing climate-resilient housing. These architectural styles have evolved to withstand high rainfall, seismic activity, and humidity, making them superior alternatives to concrete-heavy urban construction that often fails under extreme weather conditions (Lotha et al., 2024). Integrating TEK-based architectural principles into urban planning can significantly enhance sustainability while preserving cultural heritage.

Another crucial aspect of TEK is its role in agroforestry and sustainable land-use practices. The shifting cultivation systems, such as Jhum practiced in Mizoram and Manipur, have long been considered ecologically disruptive. However, recent research suggests that when managed correctly, shifting cultivation contributes to soil fertility, biodiversity conservation, and carbon sequestration (Elahi et al., 2024). Hybrid models that incorporate TEK-based agroforestry with modern land-use planning can facilitate peri-urban agricultural sustainability, ensuring food security while maintaining ecological balance (Borelli et al., 2017).

The importance of TEK in urban sustainability extends beyond environmental considerations to include community governance and decentralized resource management. Indigenous communities in Northeast India traditionally manage common resources such as forests and water bodies through collective decision-making, emphasizing long-term ecological balance over short-term economic gains. Such governance models align with contemporary sustainability discourses advocating for participatory urban planning and community-led conservation initiatives (Das et al., 2023).

2.2.1 Global and National Case Studies of TEK Integration in Urban Planning

The integration of Traditional Ecological Knowledge (TEK) into urban sustainability strategies is gaining traction globally, particularly in regions where indigenous communities have long practiced nature-based solutions to resource management and climate resilience.

In the Andean highlands of Peru, Bolivia, and Ecuador, indigenous Quechua and Aymara communities have long employed the amunas system—pre-Incan water management infrastructure designed to recharge groundwater by diverting and filtering rainwater through mountainous aquifers. In recent years, the city of Lima has revived this system in collaboration with local NGOs to mitigate water scarcity exacerbated by rapid urban growth and climate variability (Hoefsloot et al., 2024). The successful integration of this ancient system into a modern urban context highlights how TEK can support sustainable water governance and urban resilience, especially in regions facing irregular precipitation and growing water demand, much like in Northeast Indian cities such as Shillong or Guwahati.

Similarly, in Mexico City, efforts to restore the ancient chinampas—floating agricultural plots built on wetlands—are underway in the peri-urban area of Xochimilco. These floating gardens, originally developed by the Aztecs, are being revitalized not only for food production but also for biodiversity conservation, wetland restoration, and community empowerment (Nickayin et al., 2023). As a nature-based agroecological model, the chinampas offer lessons in peri-urban agriculture, water management, and heritage-based development, aligning closely with the agroforestry traditions of Northeast India such as jhum cultivation when practiced sustainably.

In Africa, urban planning in parts of Kenya has successfully integrated indigenous land-use traditions, notably the sacred Kaya forests of the Mijikenda community. These forests are culturally protected and have been incorporated into Nairobi's broader green infrastructure and heritage conservation frameworks. The recognition of indigenous conservation ethics within formal urban planning underscores the value of community stewardship in maintaining ecological balance amidst urban sprawl (Wekesae et al., 2021). This approach parallels the traditional forest management systems of communities in Nagaland and Mizoram, where collective governance over natural resources remains an enduring practice.

Traditional water harvesting systems have also played a key role in sustainable urban development across India. In Rajasthan, the revival of johads (rainwater ponds) and baolis (stepwells) in urban areas such as Alwar and Jaipur has led to enhanced groundwater recharge, improved community water access, and urban flood mitigation (Kanaujia, 2020). These water structures, rooted in centuries-old knowledge systems, have been successfully blended into modern water policy frameworks, offering valuable models for Northeast India where erratic rainfall and seasonal water scarcity pose growing challenges.

Finally, coastal indigenous communities in Mumbai, such as the Kolekars and East Indian Koli fisherfolk, contribute traditional knowledge of tidal rhythms, mangrove ecosystems, and sustainable fishing to the city's urban coastal planning efforts

(Pune, 2018). Their involvement in urban resilience initiatives, including mangrove conservation and early warning systems for storm surges, exemplifies how TEK can inform adaptive strategies in ecologically sensitive urban environments.

2.2.2 Community Participation and Participatory Governance in Urban Resilience

While Traditional Ecological Knowledge (TEK) forms the foundation of sustainable resource management in Northeast India, its practical application within urban planning frameworks often remains limited by top-down governance structures (Mitra & Paul. 2025). To bridge this gap, participatory governance models, rooted in the principles of inclusivity, decentralization, and co-production, offer a compelling approach to embed community voices into the core of urban sustainability strategies. These frameworks emphasize the democratization of planning processes, enabling local stakeholders, especially indigenous and marginalized groups, to shape development trajectories in line with their socio-ecological priorities (Bussu et al., 2022).

Communities in Northeast India possess deep-rooted knowledge systems, shaped by centuries of lived interaction with their environment. For instance, community-managed forests in Nagaland and Meghalaya operate under traditional institutional structures like the Dorbar Shnong (village council), which not only regulate land use and biodiversity conservation but also resolve conflicts and guide resource-sharing practices. These governance systems exemplify localized resilience and offer ready-made platforms for participatory urban planning (Mir et al., 2021). However, formal urban planning mechanisms, often aligned with centralized development agendas rarely recognize or integrate these decentralized forms of governance, resulting in a disconnect between grassroots practices and official policy frameworks.

Theoretical models such as Arnstein's "Ladder of Citizen Participation" (1969) and Fung's "Democracy Cube" (2006) provide useful frameworks to evaluate and enhance the level of public involvement in urban governance. Arnstein's model categorizes participation from tokenism to citizen power, encouraging planners to move beyond mere consultation toward shared decision-making and delegated authority (Loonen, 2020). In Northeast Indian cities, especially those like Shillong or Kohima where community institutions remain strong, applying such models can transform urban planning into a more inclusive, place-based process.

Emerging examples of community-led initiatives in the region illustrate the transformative potential of participatory approaches. In the Khasi Hills of Meghalaya, the Mawphlang Sacred Forest has been preserved through community stewardship that predates formal conservation laws. Urban expansion now poses a threat to these ecosystems, but local councils have begun negotiating with urban authorities to incorporate conservation zones within municipal development plans (Debnath et al., 2023). Similarly, in Guwahati, citizen collectives have mobilized around the protection of Deepor Beel, advocating for its restoration as a public commons and flood buffer zone (Mahanta et al., 2020). These instances underscore the importance of recognizing local agency in shaping resilient urban landscapes.

To institutionalize such engagement, urban governance must be restructured to create formal spaces for community participation. This could include the establishment of ward-level sustainability councils, mandatory public consultations in land-use planning, and participatory environmental impact assessments that incorporate TEK alongside scientific data. Empowering community-based organizations (CBOs) and indigenous institutions to act as co-planners, rather than passive recipients of policy, is essential for achieving socially just and ecologically grounded urban resilience (Debnath et al., 2023).

2.3 Applicability of Sustainable Urban Models to Northeast India

Sustainable urban development is central to addressing the challenges posed by climate change, population growth, and resource depletion (Vardopoulos et al., 2020). Globally, various cities have successfully implemented innovative urban development strategies, integrating resilience, sustainability, and equity into their planning processes.

Curitiba, Brazil: Integrated Public Transport and Green Urbanism

Curitiba, often recognized for its innovative urban planning, has successfully integrated public transport with ecological preservation. The city's Bus Rapid Transit (BRT) system, introduced in the 1970s, remains a global benchmark for efficient, cost-effective, and environmentally sustainable transportation (Rabinovitch, 1996). Designed to function like a subway system, Curitiba's BRT operates on exclusive bus lanes with pre-paid boarding systems, reducing congestion and carbon emissions. This model has significantly improved mobility while minimizing the urban footprint, proving that sustainable public transport can be both economically viable and environmentally friendly.

In addition to its transport innovations, Curitiba has employed green urbanism to manage urban flood risks. The city transformed flood-prone areas into public parks, such as the Barigui Park, which serves both as a flood control measure and a recreational space (De Baro, 2022). This proactive approach to ecological urbanism provides a model for Northeast Indian cities like Guwahati, which experience frequent flooding. The public transport reforms made by Curitiba's BRT model can also address congestion and inefficiencies in cities like Guwahati and Shillong.

Copenhagen, Denmark: Sustainable Mobility and Carbon Neutrality

Copenhagen is globally recognized for its commitment to sustainable mobility and carbon neutrality. Cycling infrastructure plays a pivotal role in the city's transport ecosystem, with over 40% of residents using bicycles for daily commutes (Peštová, 2021). The city's extensive bike lanes, pedestrian-friendly pathways, and incentives for non-motorized transport have contributed to reduced vehicular emissions and improved public health.

Beyond mobility, Copenhagen has established ambitious carbon neutrality targets, aiming to become the world's first carbon-neutral capital by 2025. The city integrates renewable energy sources, such as wind and solar power, into its energy grid, while also utilizing district heating and energy-efficient building designs (Petrea & Ursache, 2023). Northeast India, with its vast renewable energy potential, particularly in hydropower and wind energy, can adopt Copenhagen's model by promoting decentralized renewable energy systems and improving urban transport efficiency. Their emphasis on cycling infrastructure can be adapted to urban centers with limited space, reducing reliance on motorized transport while promoting environmental sustainability.

Freiburg, Germany: Green Urban Planning and Citizen Participation

Freiburg is often cited as a model city for sustainable urban planning, owing to its commitment to low-carbon urban development and community-driven governance. The city has implemented mixed-use zoning, promoting compact neighborhoods where residential, commercial, and recreational spaces coexist, reducing dependence on automobiles (Schmid et al., 2024). This urban design prioritizes walkability and public transport, fostering a low-emission lifestyle.

One of Freiburg's most notable initiatives is the Vauban eco-district, a community that integrates solar-powered homes, green roofs, rainwater harvesting systems, and a largely car-free environment. This approach demonstrates how integrating renewable energy solutions and sustainable housing can lead to a self-sufficient urban model. Moreover, Freiburg has actively encouraged public participation in sustainability planning, reinforcing the importance of governance structures that empower local communities (Coates, 2025). Northeast India, with its rich tradition of indigenous governance, can draw lessons from Freiburg's community-centric sustainability model, ensuring that urban development aligns with local socio-environmental contexts. The eco-district concept employed by Freiburg highlights the importance of energy-efficient housing and community-driven governance, principles that align with traditional settlement patterns in Northeast India.

Singapore: Water Management and Smart Urbanism

Singapore's approach to urban sustainability emphasizes water management and smart city innovations. Faced with water scarcity, the city-state has developed the Four National Taps strategy, ensuring water security through imported water, desalination, water reclamation (NEWater), and local catchment systems (Vodák et al., 2021). Singapore's water management policies have effectively reduced dependence on external sources, ensuring long-term resilience against climate change-induced water shortages.

In addition to its water strategies, Singapore has embraced smart urbanism, leveraging data analytics, AI-driven traffic management, and energy-efficient infrastructure through the Smart Nation Program (Shafik, 2024). This comprehensive approach underscores the role of technology in optimizing urban resource management and governance. Given the infrastructural challenges in Northeast Indian cities, integrating smart urban planning and efficient water management systems can significantly improve urban resilience and service delivery. Singapore's water management innovations, specifically, can offer solutions for Northeast Indian cities that face seasonal water scarcity, encouraging investment in rainwater harvesting, decentralized wastewater treatment, and efficient distribution systems.

2.4 The Impact of Climate Change on Northeast Indian Cities

The unique geographical and climatic conditions of Northeast India make its urban centers highly susceptible to climate change. The region experiences heavy monsoonal rainfall, which, combined with rapid urbanization and deforestation, has led to an increased frequency of floods and landslides (Das et al., 2023; Lotha et al., 2024). Northeastern cities, especially Guwahati, Dhemaji district, and other Brahmaputra adjacent cities face severe waterlogging during monsoons due to poor drainage infrastructure and unregulated urban expansion (Mandal et al., 2024). These recurring flood events cause extensive damage to housing, roads, and public utilities, exacerbating socio-economic disparities and public health risks.

Temperature variability is another growing concern. Over the past few decades, the region has witnessed rising temperatures, leading to the urban heat island (UHI) effect, particularly in densely populated cities (Lotha et al., 2024). Higher temperatures contribute to increased energy demand for cooling, further straining existing power infrastructure. Additionally, unpredictable rainfall patterns have affected water availability, intensifying drought risks and straining municipal water supplies (Khanum, 2021).

Biodiversity loss is another consequence of climate change in urban areas. Cities like Guwahati, Agartala, Dimapur, Shillong, etc., which were once surrounded by dense forests and wetlands, have seen significant reductions in green cover due to infrastructure expansion and unsustainable land-use practices (Elahi et al., 2024). The conversion of wetlands such as Deepor Beel in Assam into urban settlements has disrupted migratory bird routes and led to the degradation of aquatic biodiversity (Pandit, 2016). Rising temperatures and erratic rainfall patterns further threaten these fragile ecosystems, altering species distribution, food chains, and local biodiversity hotspots. The loss of biodiversity not only diminishes natural carbon sinks but also exacerbates climate risks such as flooding and urban heat island effects.

2.4.1 Urbanization, Climate Risks, and Policy Gaps in Northeast India

Rapid urbanization combined with intensifying climate change is creating a complex web of socio-environmental vulnerabilities across Northeast Indian cities. Urban centers such as Guwahati, Shillong, and Imphal are witnessing unprecedented land-use transformations driven by rural-to-urban migration, infrastructure expansion, and shifting economic patterns. However, this accelerated growth often occurs in ecologically sensitive zones like floodplains, hill slopes, and forested wetlands exacerbating exposure to natural disasters like flash floods, landslides, and water scarcity (Singh & Singha, 2020).

Guwahati, for instance, has experienced severe annual flooding, primarily due to the encroachment of floodplains and the degradation of the Deepor Beel wetland. Unplanned expansion into drainage basins and the obstruction of natural water channels have intensified waterlogging during monsoons. Likewise, in Shillong, rapid peri-urban sprawl has increased construction on unstable slopes, raising the risk of landslides during extreme rainfall events. The lack of slope stabilization measures, combined with deforestation in the hills, has left vulnerable populations, especially in informal settlements, at heightened risk. Meanwhile, Imphal is grappling with a rising urban heat island effect and deteriorating air quality, driven by increased vehicular emissions, population density, and reduced green cover in the city's core (Mandal et al., 2022).

A spatial analysis of these cities highlights stark disparities in resilience capacity. GIS-based mapping reveals that informal settlements in low-lying or slope-adjacent areas bear the brunt of climate-induced risks. For example, in Guwahati, nearly 45% of such settlements are situated in high-risk flood zones. These communities often lack access to climate-adaptive infrastructure, such as early warning systems, proper drainage, or reinforced housing, making them disproportionately vulnerable (Sharma et al., 2022). Moreover, socio-economic stressors such as inadequate employment, limited healthcare access, and poor waste management further compound the effects of environmental hazards.

Despite these growing threats, urban resilience policies across the region remain fragmented, underfunded, and inconsistently enforced. While national programs like the Smart Cities Mission and AMRUT have reached select urban centers, their implementation often overlooks localized environmental dynamics and indigenous governance systems. Many urban planning frameworks still adopt conventional top-down approaches that prioritize large-scale infrastructure over ecological sensitivity. There is also a glaring absence of climate vulnerability assessments in municipal development plans, leaving cities ill-prepared for extreme weather events (Sakhamuri & Sanagani, 2024).

2.5 Role of Green Infrastructure in Climate Resilience

Green infrastructure has gained prominence as an effective strategy for enhancing urban resilience. Unlike traditional grey infrastructure, which primarily relies on built solutions such as concrete flood barriers and drainage systems, green infrastructure leverages natural systems to provide ecosystem services while mitigating climate risks (Dhanya et al., 2022; Krishnakumar, 2025). In Northeast India, where rapid urban expansion has led to the depletion of natural landscapes, the integration of green infrastructure can play a pivotal role in addressing climate vulnerabilities. One significant approach involves the restoration and conservation of urban wetlands. Guwahati's Deepor Beel, a Ramsar-listed wetland, historically played a crucial role in flood mitigation and biodiversity conservation. However, encroachment and pollution have severely degraded its ecological functionality (Pandit, 2016; Elahi et al., 2024). Revitalizing such wetlands through eco-sensitive urban planning can enhance their flood absorption capacity while promoting urban biodiversity.

Another key strategy is the adoption of urban forests and green corridors. Shillong, which has been experiencing declining green cover due to unregulated construction, can benefit from afforestation initiatives within city limits. Studies show that urban forests help regulate microclimates, reduce heat island effects, and improve air quality (Adesina et al., 2024). Integrating tree-planting initiatives into municipal policies can significantly strengthen urban climate adaptation efforts. Besides, green roofs and permeable pavements are additional components of green infrastructure that can aid in climate resilience. By incorporating vegetation into rooftops, cities can lower ambient temperatures and manage stormwater more effectively (Pandey et al., 2021; Dhanya et al., 2022). In Northeast Indian cities, where traditional housing often includes sloped and tin roofs, innovative adaptations such as green terraces can be explored to increase climate resilience without compromising architectural heritage.

Additionally, nature-based solutions (NBS) offer cost-effective and sustainable ways to mitigate climate risks while enhancing urban livability. Defined as interventions that harness natural processes to address environmental challenges, NBS has been successfully implemented in cities worldwide to tackle climate vulnerabilities, especially in forming flood-resilient cities (Mabrouk et al., 2023). For Northeast Indian cities, particularly those located near the Brahmaputra basin, integrating NBS into urban planning can help build resilience against extreme weather events like floods, while preserving ecological integrity.

One exemplary case study is Singapore's ABC Waters Programme (Active, Beautiful, Clean Waters), which integrates water-sensitive urban design principles to manage stormwater through green infrastructure. By creating bio-retention swales, rain gardens, and constructed wetlands, Singapore has effectively reduced flood risks while enhancing urban aesthetics (Dhanya et al., 2022). Similar approaches can be adopted in northeastern cities like Guwahati, where improper water management exacerbates flood-related damage. The city of Rotterdam in the Netherlands also offers valuable lessons through its climate-resilient urban planning initiatives. Rotterdam's water plazas function as recreational spaces during dry periods and transform into flood retention basins during heavy rainfall (Dai et al., 2017). Given the heavy monsoonal rains in Northeast India, the concept of multi-functional urban spaces could be implemented to maximize land use efficiency while mitigating climate risks.

Community-led watershed management is another crucial aspect of nature-based urban adaptation. In the hills surrounding Shillong and Kohima, indigenous communities have practiced watershed conservation for generations, ensuring sustainable water supply and ecosystem stability (Lotha et al., 2024). Scaling up these traditional practices through municipal collaboration can provide localized solutions for urban resilience while fostering community participation.

2.6 Green Finance and ESG Investments in Urban Sustainability

The growing importance of sustainable urban development has led to an increased focus on green finance and Environmental, Social, and Governance (ESG) investments as mechanisms to drive climate-resilient infrastructure and sustainable economic growth. Green finance encompasses financial instruments and investments aimed at fostering environmental sustainability, while ESG investments ensure that economic activities align with social responsibility and governance best practices. The integration of green finance into urban sustainability policies is essential for cities aiming to reduce carbon footprints while ensuring climate adaptation and mitigation. Some of the key financial instruments include green bonds, climate resilience funds, sustainability-linked loans, and impact investments (Almazrouei & Nobanee, 2021). These financial mechanisms allow both the public and private sectors to mobilize resources toward urban sustainability goals, enhancing energy efficiency, smart mobility, and eco-friendly urban planning.

Green finance also plays a crucial role in addressing urban climate risks, particularly in regions facing extreme weather events like that of Guwahati or Shillong, where extreme monsoons often affect urban livelihood. Investment in climate-resilient infrastructure, such as flood-resistant housing, smart water management, and renewable energy grids, is critical for mitigating climate-related urban vulnerabilities. Cities worldwide have demonstrated the effectiveness of green financing models in accelerating sustainable urbanization (Adesina et al., 2024). The adoption of green finance and ESG investments can bridge the sustainability gap in Northeastern cities, providing financial resources for climate adaptation and urban resilience.

2.6.1 Global Best Practices in Green Finance and ESG Investments

Several countries have successfully implemented green finance and ESG investment models, creating urban landscapes that balance economic prosperity with environmental sustainability. One notable example is the European Green Bond Standard (EGBS), which has played a pivotal role in financing sustainable urban projects across Europe. The standard ensures transparency and credibility in green bond issuance, allowing cities to secure funding for renewable energy projects, green transport infrastructure, and energy-efficient buildings. Cities like Paris and Stockholm have leveraged green bonds to transition toward carbon-neutral urban environments, demonstrating the potential of sustainable finance in mitigating climate change (Kashif & Meo, 2024). In the United States as well, cities such as New York and San Francisco have issued green municipal bonds to finance projects that promote energy efficiency, climate adaptation, and sustainable transportation (Karpf & Mandel, 2018). The New York Green Bond Program has successfully funded initiatives such as stormwater management systems, urban reforestation programs, and energy-efficient public buildings. These investments align with the broader objectives of the Biden administration's climate resilience strategy, which emphasizes financing climate-smart cities (Mitra et al., 2024).

China has also emerged as a global leader in green finance, incorporating sustainability into its urbanization strategy. The country has established green finance pilot zones, where cities such as Shenzhen and Guangzhou implement policies that promote ESG-aligned investments, low-carbon transport, and eco-friendly urban expansion (Feng et al., 2023). Through state-backed green bonds and private sector participation, China has successfully reduced urban emissions, improved air quality, and promoted green architecture in its fast-growing metropolitan regions. Another significant leader in green finance is Singapore, a nation that has introduced ESG-driven financial policies to ensure that urban development aligns with climate resilience and social responsibility. The Sustainable Finance Action Plan, developed by the Monetary Authority of Singapore (MAS), has accelerated the growth of green loans and ESG-focused asset management (Dhanya et al., 2022). Notably, Singapore's financial institutions support investments in nature-based solutions, flood mitigation projects, and net-zero buildings, demonstrating the significance of ESG finance in securing urban sustainability.

2.6.2 Green Finance Mechanisms and Policy Applications in Northeast India

The Securities and Exchange Board of India (SEBI) issued guidelines for green bonds as early as 2017, leading to the successful issuance of bonds by public and private entities for projects related to renewable energy, clean transport, and sustainable water management (Manaktala, 2020). One notable example is the Indian Renewable Energy Development Agency (IREDA), which issued green bonds worth 1,500 crore to fund solar and wind energy projects (Pal, 2023). While such initiatives have primarily focused on larger metropolitan regions, their replication in Northeast India could support decentralized solar energy systems, micro-hydropower projects, and green building retrofits in cities like Shillong, Aizawl, and Kohima.

At the sub-national level, several Indian states are experimenting with green finance models. The state of Maharashtra, for instance, launched the country's first state-level green bond in 2022 to finance climate-resilient urban infrastructure. Similarly, Gujarat has incorporated ESG principles into its investment promotion policies, incentivizing private sector participation in sustainable development (Mukherjee, 2022). Drawing from these examples, Northeast Indian states could establish region-specific green investment funds, such as a Northeast Green Infrastructure Fund (NGIF), to channel investments into urban sustainability sectors, including electric public transport, wetland restoration, and sustainable waste management.

Public-private partnerships (PPPs) also present a valuable avenue for leveraging private capital in sustainable infrastructure projects. Urban local bodies in cities such as Guwahati and Agartala could partner with private entities and social impact investors to develop electric vehicle (EV) charging infrastructure, smart waste collection systems, and decentralized wastewater treatment plants (Kumar et al., 2022). To facilitate this, state governments must create enabling regulatory environments, including transparent project pipelines, risk-sharing frameworks, and financial de-risking instruments backed by central schemes like the Viability Gap Funding (VGF) program.

India's policy landscape is gradually aligning with global green finance trends. The National Action Plan on Climate Change (NAPCC), along with state-level climate action plans (SAPCCs), emphasizes the need for financial innovation to support low-carbon urban transitions. In addition, recent initiatives under the Reserve Bank of India's (RBI) sustainable finance roadmap, and the establishment of a sovereign green bond framework in 2022, create an ecosystem conducive to green capital mobilization (Yadav et al., 2024). These policy developments can be strategically harnessed by Northeast Indian states through capacity building, regional climate finance task forces, and green finance literacy programs for municipal authorities.

2.6.3 Adapting ESG Frameworks for Urban Sustainability in Northeast India

Environmental, Social, and Governance (ESG) frameworks are increasingly seen as essential tools for aligning investment decisions with long-term sustainability goals. However, their effective implementation in complex and diverse regions like Northeast India requires a context-sensitive adaptation. While global ESG standards, such as those developed by the Global Reporting Initiative (GRI), the Sustainability Accounting Standards Board (SASB), or the United Nations Principles for Responsible Investment (UNPRI), offer structured metrics for evaluating sustainability, these frameworks must be localized to reflect the socio-political, ecological, and governance realities of Northeast Indian cities (Teixeira Dias et al., 2023).

Urban centers in the region, such as Imphal, Shillong, and Aizawl, face unique sustainability challenges including biodiversity loss, fragile ecosystems, socio-political sensitivities, and institutional fragmentation. ESG implementation in such contexts must go beyond generic compliance and focus on regionally relevant indicators. For example, under the "Environmental" pillar, metrics such as forest cover preservation, wetland conservation, and low-carbon transport adoption should be emphasized. Under the "Social" component, indicators could include community-led planning initiatives, protection of indigenous land rights, and equitable access to water and sanitation services—areas that are highly relevant given the communal governance traditions and socio-cultural diversity in the Northeast. For the "Governance" aspect, evaluating transparency in municipal budgeting, participatory decision-making, and accountability in environmental clearances could help build trust and attract ESG-conscious investors (Singhania & Saini, 2021).

To operationalize ESG in Northeast Indian urban planning, local governments can create region-specific ESG policy roadmaps that align with both national sustainability targets (such as India's Panchamrit commitments) and international benchmarks. Collaboration with the Ministry of Housing and Urban Affairs (MoHUA), the North Eastern Council, and international development partners could facilitate capacity-building and establish clear reporting protocols. One potential mechanism is the integration of ESG metrics into Smart City project evaluations and AMRUT 2.0 schemes, thereby embedding sustainability into the performance-based funding ecosystem (Mehta et al., 2023).

2.7 Technological Interventions in Ensuring Urban Sustainability

Geospatial technology, particularly GIS, has become integral to urban planning, allowing for spatial analysis and mapping of environmental risks. GIS facilitates data-driven decision-making by enabling planners to visualize climate risks such as flood-prone zones, heat islands, and deforestation trends (Pandey et al., 2021). By integrating climate data with land-use patterns, GIS enhances climate adaptation strategies in urban centers. One notable case study is New York City's Climate Risk Assessment and Adaptation Plan, which extensively uses GIS to map sea-level rise projections, flood risks, and storm surge vulnerabilities (Nguyen, 2023). Following Hurricane Sandy, NYC integrated GIS-based risk modeling into its resilience strategy, ensuring infrastructure development aligns with climate mitigation efforts. This approach provides valuable insights for flood-prone cities in Northeast India, such as Guwahati and Imphal, where GIS-based zoning can help prevent unplanned urban expansion in high-risk areas.

In addition to GIS, AI-driven climate risk assessments have revolutionized predictive analytics in urban sustainability. AI models leverage historical and real-time climate data to forecast extreme weather events, assess infrastructure vulnerabilities, and optimize resource allocation (Song et al., 2023). The City of Los Angeles, for example, employs AI-driven heatwave prediction models to implement early warning systems and urban cooling strategies (Revi et al., 2019). These predictive capabilities can be adapted in Northeast Indian cities, where monsoonal variations, flash floods, and landslides necessitate proactive risk mitigation.

Moreover, the proliferation of big data analytics and IoT has transformed how cities approach sustainability. By processing vast amounts of urban data, cities can gain actionable insights into mobility patterns, energy consumption, air quality, and waste management. Big data enhances real-time monitoring of urban environments, ensuring cities can adapt dynamically to changing conditions. One of the most prominent examples of big data utilization is found in Amsterdam's Smart City Initiative. The city employs big data to optimize public transport, reduce energy waste, and manage traffic congestion (Ejaz et al., 2025). Sensors embedded in public infrastructure collect data on air pollution, noise levels, and pedestrian movement, allowing urban authorities to make informed policy decisions. For Northeast India, where rapid urbanization and vehicular congestion pose sustainability challenges, big data-powered traffic management systems could significantly enhance mobility and reduce carbon emissions.

The Internet of Things (IoT) also plays a pivotal role in sustainable city governance. IoT-enabled smart grids, for example, allow for efficient energy distribution and consumption monitoring (Mandal et al., 2024). In Barcelona, the IoT-driven Smart Water Management System monitors water usage, detects leaks, and optimizes distribution, reducing water wastage by 25% (Hassebo & Tealab, 2023). Such technologies could be adapted for Shillong and Agartala, where water shortages and inefficient distribution networks remain pressing issues. Additionally, IoT enhances waste management through smart bin technology. In South Korea's Seoul Metropolitan Government, smart bins use IoT sensors to detect waste levels, alerting municipal agencies for timely collection (Shin et al., 2020). Implementing such systems in Northeast Indian cities can minimize landfill overflow, reduce illegal dumping, and promote circular economy practices.

2.7.1 Technological Barriers and Solutions for Smart Urban Planning in Northeast India

While technologies such as Geographic Information Systems (GIS), artificial intelligence (AI), and big data analytics hold transformative potential for urban planning, their implementation in Northeast India faces several region-specific challenges. The deployment of these tools is often hindered by inadequate digital infrastructure, limited internet penetration in peri-urban and hilly areas, and a shortage of trained personnel capable of managing and interpreting complex spatial and environmental data. For instance, many municipalities in the region still rely on outdated paper-based planning methods, with little integration of geospatial platforms for zoning, disaster mapping, or service delivery optimization (Sigh et al., 2020). Additionally, inconsistent access to high-resolution satellite data and real-time environmental monitoring systems further constrains climate risk assessments and early warning capabilities.

Technological adoption is also hampered by the digital divide and institutional fragmentation. Urban local bodies frequently lack the financial and technical capacity to invest in modern technologies, while inter-agency coordination on data sharing remains minimal. These barriers result in siloed planning processes and missed opportunities for anticipatory governance and risk-informed infrastructure development (Esfandi et al., 2024). Without targeted intervention, the region may struggle to fully leverage digital innovations for sustainability and resilience.

2.8 Gaps and Challenges in Sustainable Urban Development

The pursuit of sustainable urban development is challenged by policy fragmentation, governance inefficiencies, socio-economic disparities, and technological barriers that hinder holistic and adaptive urban planning. In many developing regions, including Northeast India, a lack of coordination among government agencies leads to disjointed policies, ineffective land-use planning, and inefficient resource allocation. This policy fragmentation results in contradictory urban development frameworks, limiting the effectiveness of climate adaptation measures and green infrastructure initiatives (Dhanya et al., 2022; Mitra et al., 2024). Furthermore, weak institutional frameworks and bureaucratic delays exacerbate governance challenges, reducing the capacity to implement long-term sustainability projects. Without integrated policies and inter-agency cooperation, cities struggle to manage rapid urbanization while addressing climate resilience and environmental conservation.

Another major challenge is the limited financial and technical capacity to support green and nature-based urban solutions. Green finance mechanisms, such as green bonds, ESG investments, and climate resilience funds, remain underutilized due to the absence of clear regulatory frameworks and inadequate awareness among municipal bodies (Almazrouei & Nobanee, 2021). This financial gap slows the adoption of renewable energy, climate-smart infrastructure, and sustainable transport systems. Similarly, the integration of modern technologies such as GIS, AI-driven risk assessments, and IoT-enabled smart infrastructure faces barriers related to high costs, digital literacy gaps, and lack of technical expertise. These challenges prevent cities from leveraging data-driven planning and smart governance tools that could enhance urban sustainability (Elahi et al., 2024).

Socio-economic inequalities further complicate sustainability efforts, as marginalized communities often lack access to basic services, resilient housing, and sustainable infrastructure. Informal settlements in urban areas frequently experience inadequate water supply, poor waste management, and higher vulnerability to climate-related disasters. The uneven distribution of sustainability benefits exacerbates environmental injustices, necessitating inclusive urban planning approaches that integrate community participation, affordable housing, and equitable access to resources (Singh & Singha, 2020). Overcoming these challenges requires stronger governance mechanisms, financial incentives for green investments, and multi-stakeholder collaboration to ensure that sustainable urban development is both inclusive and resilient.

3. RESEARCH METHODOLOGY

3.1 Research Design

This study employs a mixed-methods research design to comprehensively examine sustainable urban development in Northeast India. The combination of qualitative and quantitative methodologies ensures a well-rounded understanding of urban resilience, policy effectiveness, and financial sustainability. The research follows an exploratory and descriptive approach, allowing for both the identification of emerging trends and a detailed examination of existing urban sustainability models.

3.2 Data Collection Methods

3.2.1 Primary Data Collection

- **Surveys and Questionnaires:** Structured and semi-structured surveys are conducted with key stakeholders, including urban planners, policymakers, environmental experts, financial institutions, and local communities. The surveys assess perspectives on urban sustainability policies, TEK (Traditional Ecological Knowledge), governance challenges, and financial investments in ESG (Environmental, Social, and Governance) initiatives.
- **Field Observations and Site Analysis:** On-site visits are conducted in green infrastructure projects, informal settlements, and flood-prone zones to document on-ground sustainability challenges. GIS-based mapping is used to assess urban vulnerabilities and spatial disparities in sustainability interventions.

3.2.2 Secondary Data Collection

- **Policy Documents and Government Reports:** The study examines urban planning frameworks, climate adaptation strategies, and green finance policies at national, state, and municipal levels. Reports from the Smart Cities Mission and AMRUT program are reviewed to assess their sustainability impacts.
- **Academic Literature and Case Studies:** A systematic review of peer-reviewed journal articles, books, and conference proceedings provides theoretical insights into sustainable urban development and governance models.
- **Financial and Investment Reports:** ESG investment trends, green bond market analyses, and international climate financing mechanisms (e.g., Green Climate Fund) are examined to evaluate their relevance in financing sustainable urban projects in Northeast India.

3.3 Data Analysis Methods

A combination of quantitative and qualitative analysis techniques is employed to interpret the collected data.

- **Quantitative Analysis:** Statistical tools such as SPSS and R are used to analyze survey data, identifying patterns in urban sustainability adoption, financial investments, and public perceptions. GIS-based spatial analysis is conducted to map climate vulnerabilities and green infrastructure distribution.
- **Qualitative Analysis:** Thematic coding of survey responses using NVivo helps identify recurring themes in policy integration, TEK adoption, and governance challenges. Comparative case study analysis provides insights into successful urban resilience models and their potential applicability in Northeast Indian cities.

3.4 Limitations of the Study

One major challenge of this research is data availability. Access to updated and publicly available government reports on climate resilience and financial investments is often limited, which restricts the depth of secondary data analysis.

While the study aims to include diverse voices, logistical and administrative barriers may hinder meaningful engagement with policymakers, financial institutions, and indigenous communities. Additionally, responses may be influenced by institutional or political biases, affecting the objectivity of primary data.

Regional variability further limits the generalizability of findings. The selected case study cities may not fully represent all urban centers in Northeast India due to significant socio-political, ecological, and cultural differences.

Lastly, limited access to high-resolution satellite imagery, proprietary tools, and the absence of standardized ESG metrics and green finance reporting frameworks present challenges in conducting robust spatial and financial assessments.

4. DATA ANALYSIS

4..1 Survey-Based Assessment of Urban Sustainability Challenges

Survey data was analyzed using SPSS and R statistical tools, focusing on key sustainability indicators such as policy effectiveness, financial investment trends, TEK integration, and governance efficiency. The survey was conducted across diverse stakeholder groups, including policymakers, environmental planners, financial experts, and community representatives.

Table 1. Summary of Stakeholder Survey Responses

Indicator	Mean Score	Standard Deviation	Stakeholder Perception
Awareness of sustainability policies	3.8	0.76	Moderate-High
Integration of TEK in urban planning	2.9	0.83	Low-Moderate
Effectiveness of green finance	3.2	0.92	Moderate
Governance transparency	2.7	1.01	Low
Investment in sustainable infrastructure	3.0	0.89	Moderate
Climate risk mitigation policies	2.6	0.95	Low

Table 1 presents key sustainability indicators based on stakeholder perceptions.

4.2 GIS-Based Spatial Analysis of Climate Vulnerabilities

GIS was utilized to map climate vulnerabilities, green infrastructure distribution, and policy effectiveness in urban resilience planning. Spatial data from urban centers in Northeast India (Guwahati, Shillong, and Imphal) were analyzed to identify high-risk zones and underutilized green infrastructure.

Table 2. GIS-Based Risk Assessment Findings

Risk Factor	Guwahati	Shillong	Imphal
Flood-Prone Areas (sq. km)	48.5	30.2	25.7
Urban Heat Island Intensity (°C)	3.8	2.5	2.1
Green Space Coverage (%)	22.4	34.7	40.2
Informal Settlements at Risk (%)	45.1	32.6	29.4

Table 2 presents key climate-related risks and their severity in three case study cities.

Figure 1. GIS Map of Climate Vulnerabilities

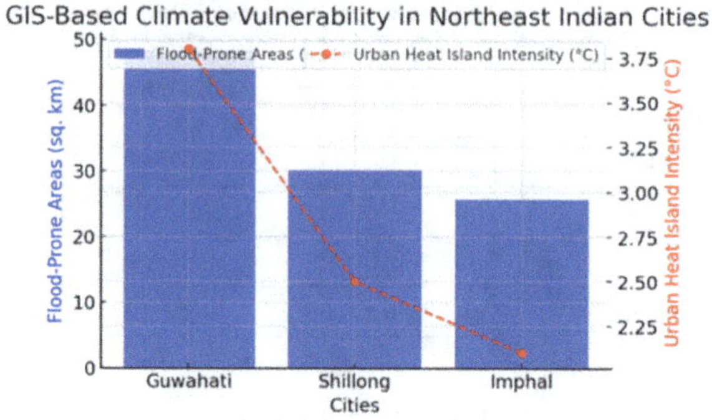

Figure 1 highlights flood-prone areas, deforestation hotspots, and infrastructure vulnerability overlays.

4.3 Thematic Analysis of Stakeholder Perspectives on Sustainability

Using NVivo software, survey responses were categorized into major themes:

- **Policy Gaps and Fragmentation:** Stakeholders identified overlapping jurisdictional issues and weak enforcement as barriers to sustainability.
- **Challenges in TEK Adoption:** Respondents noted a lack of institutional recognition of traditional ecological knowledge in urban planning policies.
- **Barriers to Green Finance:** Municipal stakeholders cited lack of awareness, regulatory gaps, and investment risks as key deterrents to ESG financing.
- **Technological Constraints:** The limited adoption of AI-driven climate risk assessments and IoT-based infrastructure management was noted as a major challenge.

4.4 Case Study Analysis of Global and Local Sustainable Urban Development Models

4.4.1 Green Finance and Climate Investments

Copenhagen's carbon neutrality goals are heavily supported by green finance instruments, including green bonds and sustainability-linked loans. While green finance mechanisms exist in the cities of Northeast India, the lack of standardized ESG frameworks and investor awareness limits large-scale adoption.

4.4.2 Sustainable Public Transport and Mobility

Curitiba's Bus Rapid Transit (BRT) system is an effective model for integrating public transport with land-use planning. In Northeast India, cities like Guwahati, Shillong, and Agartala suffer from traffic congestion and inefficient public transport infrastructure.

4.4.3 Water Management and Urban Resilience

Singapore's circular water economy utilizes rainwater harvesting, desalination, and wastewater recycling. In Northeast India, cities like Shillong and Guwahati experience water shortages due to poor conservation strategies.

4.4.4 Climate-Adaptive Infrastructure and Nature-Based Solutions

Rotterdam's flood-resilient urban planning incorporates water plazas, green rooftops, and climate-proof infrastructure. Guwahati frequently experiences flooding due to unplanned urban expansion.

Figure 2. Thematic Comparison of Global and Northeast Indian Urban Models

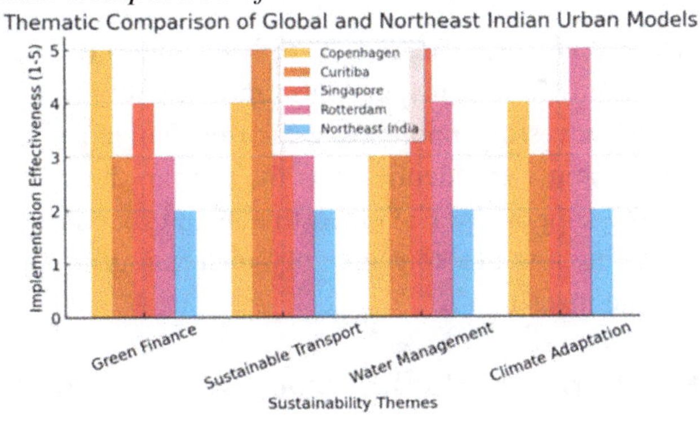

Figure 2 illustrates how different cities approach urban sustainability across key themes.

5. FINDINGS

5.1 Awareness and Perception of Urban Sustainability Policies

Survey responses indicate that awareness of urban sustainability policies is moderate to high, with a majority of respondents acknowledging the presence of such policies but questioning their effectiveness. Only 25% rated the policies as effective or highly effective, while over 40% found them moderately effective, and 35% rated them as ineffective. Lack of enforcement and inter-agency coordination were cited as the biggest policy gaps affecting urban sustainability.

5.2 Traditional Ecological Knowledge (TEK) Integration

The data reveals strong support (over 70%) for integrating TEK into modern urban planning. Respondents emphasized the importance of traditional water conservation methods, climate-resilient architecture, and indigenous land management practices. However, institutional barriers, lack of policy inclusion, and insufficient documentation were highlighted as significant challenges preventing TEK integration.

5.3 Barriers to Green Finance and ESG Investments

Findings indicate that green finance is still an underutilized mechanism in Northeast India, with only 20% of respondents familiar with ESG investments. The lack of structured ESG investment policies, weak financial incentives, and minimal private sector engagement were identified as the primary barriers. While 67% of respondents expressed support for ESG investments in urban infrastructure, actual implementation remains low due to regulatory uncertainty and limited financial literacy on sustainable investments.

5.4 Climate Resilience and Infrastructure Challenges

The survey results reflect low confidence in the resilience of urban infrastructure to climate-related disasters. Over 60% of respondents rated their city's infrastructure as weak to moderate in resilience, citing flood-prone areas, lack of climate-adaptive designs, and weak enforcement of urban resilience strategies as key concerns. Wetland restoration, urban afforestation, and nature-based drainage systems were the top solutions favored by respondents to improve climate adaptation efforts.

5.5 Public Transport Accessibility and Sustainability

Public transport accessibility remains a significant concern, with over 50% of respondents describing it as inadequate. The most demanded improvements include expanding public transit routes, introducing electric buses, and enhancing pedestrian and cycling infrastructure. Only 30% of respondents rated their public transport system as efficient, highlighting the need for policy-driven investments in sustainable urban mobility.

5.6 Governance and Stakeholder Engagement

Government transparency in urban sustainability initiatives received mixed reviews, with only 25% rating it as transparent or highly transparent, while 40% found it moderately transparent, and 35% rated it as non-transparent. Weak enforcement mechanisms, bureaucratic delays, and lack of multi-stakeholder collaboration were cited as governance challenges. Encouragingly, over 60% of respondents expressed willingness to participate in community-driven sustainability projects, highlighting a growing interest in civic engagement for sustainable development.

6. FRAMEWORK FOR SUSTAINABLE URBAN DEVELOPMENT, POLICY RECOMMENDATIONS AND STRATEGIC INTERVENTIONS

6.1 A Framework for Sustainable Urban Development

A practical framework for sustainable urbanization must bridge governance gaps, introduce enforceable policies, ensure financial feasibility, and encourage community participation. Cities in Northeast India require strategies that address immediate infrastructure gaps while preparing for long-term climate resilience. Urban planning should integrate GIS-based climate vulnerability assessments, real-time environmental monitoring, and data-driven resource management.

Key components of this framework include:

- **Enhancing Inter-Municipal Coordination:** Develop a centralized Urban Sustainability Task Force within NERUDP to ensure policy alignment across municipalities, streamline implementation, and monitor urban development progress (MoHUA, 2021).

- **Investment in Climate-Resilient Infrastructure**: Implementing mandatory environmental impact assessments (EIAs) and ensuring green-building compliance.
- **Technology-Driven Urban Solutions**: Introducing AI-powered flood warning systems, IoT-enabled waste tracking, flood monitoring, and digitalized land-use mapping (Mandal et al., 2024).
- **Financial Mechanisms for Sustainability**: Developing state-backed green bonds, tax incentives for ESG investments, and public-private partnership (PPP) models.

6.2 Policy Recommendations

Policy implementation must be systematic, time-bound, and measurable to ensure impact. A structured policy approach is categorized as follows:

Short-Term Strategies (1-3 years)

- **Enforcement of Urban Sustainability Standards**: Establish a Sustainability Code for Urban Development that mandates rainwater harvesting, energy-efficient construction, and green roofing.
- **Urban Mobility Improvement**: Introduce low-emission transport zones and mandate that 30% of public buses be converted to electric vehicles (EVs) across different cities in Northeast India, a strategy which has already been implemented in Guwahati.
- **Waste Management Pilot Programs**: Implement community-led solid waste segregation and composting projects in at least five major urban centers.

Medium-Term Strategies (3-7 years)

- **Mandatory Climate Adaptation in Infrastructure Development**: Require flood-resilient construction materials and nature-based flood control solutions, including urban wetlands and permeable roads.
- **Expansion of Green Finance Mechanisms**: Establish a Northeast Green Investment Fund (NGIF) to support solar energy, water conservation, and green public infrastructure.
- **Comprehensive Public Transport Overhaul**: Develop a Northeast Sustainable Transport Plan, including integrated smart transport hubs and expansion of metro/BRT systems.

Long-Term Strategies (7-15 years)

- **Development of Smart Climate-Resilient Cities**: Retrofit existing cities with smart energy grids, carbon-neutral buildings, and automated water recycling systems.

- **Regional Circular Economy Implementation**: Ensure that 80% of municipal waste is diverted from landfills through waste-to-energy initiatives and industrial recycling zones.
- **Sustainable Energy Transition**: 50% renewable energy penetration in urban grids through solar rooftops, micro-hydropower, and wind energy integration.

6.3 Strategic Interventions for Northeast Indian Cities

Flood and Climate Adaptation Measures: Flooding is one of the most pressing climate-related challenges in Northeast India, exacerbated by unplanned urbanization, deforestation, and encroachment on wetlands. To mitigate these risks, urban wetland protection laws should be enforced to establish buffer zones around urban rivers, preventing uncontrolled construction in flood-prone areas. Degradation of wetlands, such as Deepor Beel in Guwahati, has increased flood susceptibility. Automated stormwater drainage systems must be installed, ensuring real-time monitoring and early flood detection using geospatial AI applications (Song et al., 2023). Furthermore, integrating water retention basins and rainwater harvesting into urban planning will reduce surface runoff while replenishing groundwater sources.

Sustainable Infrastructure and Biodiversity Protection: The increasing rate of land-use change threatens urban biodiversity, necessitating stronger green infrastructure policies. One effective approach is the implementation of mandatory urban tree canopy expansion programs that require all new commercial and residential developments to allocate at least 20-30% of their land area to green spaces (Wang et al., 2024). Additionally, Ecologically Sensitive Zone (ESZ) enforcement should prohibit high-density construction in fragile ecosystems, ensuring ecological preservation. The biodiversity hotspots of Northeast India must be integrated into urban planning through adaptive management approaches (Elahi et al., 2024).

Public Transport and Mobility Transformation: To combat increasing traffic congestion and vehicular emissions, urban centers must adopt sustainable and efficient mobility solutions. The development of dedicated cycling lanes and pedestrian-friendly districts will promote non-motorized transport (NMT) while reducing pollution. Additionally, tax incentives for commuters using electric vehicles (EVs) and public transport should be introduced, targeting a 20% reduction in private vehicle dependency within five years (Singh & Singha, 2020). Furthermore, Bus Rapid Transit (BRT) corridors should be expanded to increase accessibility and reduce travel time for urban commuters.

Water Security and Sustainable Resource Management: Water scarcity is a growing concern due to erratic rainfall patterns and inefficient water management systems. Mandatory rainwater harvesting laws should require that all buildings larger than 500 square meters install rainwater collection systems, significantly improving urban water security (Vohra & Franklin, 2021). Additionally, citywide smart water metering systems should be introduced to monitor consumption and enforce tiered water pricing, reducing excessive usage and wastage. As emphasized by Lotha et al. (2024), traditional water conservation techniques from indigenous communities should be integrated into modern urban water policies.

Renewable Energy Adoption and Efficiency Enhancement: Transitioning to renewable energy is essential for sustainable urban growth. Urban authorities must mandate solar rooftop installations on at least 50% of government buildings within the next five years to reduce dependency on fossil fuels (Abhilash et al., 2023). Additionally, biogas plants should be established in all municipalities with populations exceeding 500,000, turning organic waste into clean energy and reducing landfill burdens. By leveraging smart grid technologies, cities can optimize energy distribution and improve efficiency.

Circular Economy and Sustainable Waste Management: A circular economy approach is critical for reducing urban waste generation and landfill dependency. A zero-landfill policy should be introduced in major cities, requiring segregation of plastic waste, organic waste composting, and electronic waste (e-waste) recycling (Petrea & Ursache, 2023). Furthermore, financial incentives for recycling businesses should be provided, encouraging industries to use recycled materials in construction and manufacturing. Developing waste-to-energy facilities will further contribute to urban sustainability by converting municipal solid waste into electricity (Mitra et al., 2024).

6.4 Implementation Challenges and Risk Mitigation Strategies

While the proposed interventions provide a clear roadmap for sustainable urbanization, several implementation barriers must be addressed. A proactive risk mitigation strategy will ensure the long-term success of these initiatives by tackling governance, financial, and social challenges.

Governance fragmentation remains a significant obstacle in urban sustainability efforts. Establishing a Northeast Sustainable Urbanization Task Force will enforce policy consistency across state governments, ensuring that sustainability targets are met efficiently (Meitei & Das, 2024). Furthermore, urban development monitoring dashboards should be deployed to enhance transparency and accountability, using real-time data to track the progress of urban sustainability projects (Vodák et al., 2021).

Another major limitation of sustainable urban initiatives is inadequate funding. A public-private green finance initiative should be launched to allow private investors and climate-focused financial institutions to co-finance large-scale projects

(Almazrouei & Nobanee, 2021). Additionally, performance-based incentives should be provided to municipalities meeting annual sustainability targets, rewarding cities that successfully implement green infrastructure and low-carbon transport systems (Feng et al., 2023).

In addition, public participation is crucial in shaping sustainable urban policies. Sustainability education programs should be introduced in schools, ensuring that the next generation is equipped with knowledge on climate resilience and urban sustainability (Geovetenskaper, 2023). Furthermore, community participation platforms should be established to involve residents in urban decision-making processes, ensuring that development aligns with local needs and environmental priorities (Schmid et al., 2024).

7. CONCLUSION

This research reveals the urgent need for a structured and sustainable approach to urban development in Northeast India. Rapid urbanization, climate vulnerabilities, governance inefficiencies, and financial constraints present substantial challenges to achieving urban sustainability. However, through the integration of policy-driven climate adaptation, green finance mechanisms, technology-enabled infrastructure planning, and community participation, cities in the region can transition toward resilience and sustainability. The proposed framework and strategic interventions provide a roadmap for implementing these solutions while ensuring that development aligns with ecological preservation and socio-economic inclusivity.

Looking ahead, future research and policy focus should emphasize the role of smart urban technologies in climate adaptation, including the use of AI for predictive flood management, IoT for energy-efficient city operations, and blockchain for transparent green finance tracking. Additionally, more in-depth studies on the application of Traditional Ecological Knowledge (TEK) in modern urban planning could offer insights into culturally appropriate sustainability models. Scaling up successful pilot projects in waste management, public transport electrification, and renewable energy integration will also be crucial for long-term impact. Further, fostering cross-sectoral collaboration between government bodies, private investors, and civil society organizations will strengthen the foundation for sustainable urbanization.

Ultimately, achieving sustainability in Northeast Indian cities requires a proactive, well-coordinated, and multi-stakeholder approach. By continuously refining policies and leveraging innovative financial and technological tools, urban centers in the region can emerge as models of climate-resilient and resource-efficient development.

REFERENCES

Abhilash, N., Shenoy, S. S., Shetty, D. K., Lobo, L. S., & N, S. K. (2023). Green Bond as an Innovative financial instrument in the Indian Financial Market: Insights from Systematic Literature Review Approach. *SAGE Open*, *13*(2), 21582440231178783. DOI: 10.1177/21582440231178783

Adesina, N. A., Okwandu, N. C., & Nwokediegwu, N. Z. Q. S.Abayomi Abraham AdesinaAzubuike Chukwudi OkwanduZamathula Queen Sikhakhane Nwokediegwu. (2024). Towards sustainable urban development: Conceptualizing green infrastructure and its impact on urban planning. *International Journal of Applied Research in Social Sciences*, *6*(7), 1274–1296. DOI: 10.51594/ijarss.v6i7.1266

Almazrouei, H., & Nobanee, H. (2021). *Green finance and sustainable growth.* ResearchGate.

Berkes, F. (2012). Sacred Ecology. In Routledge eBooks. DOI: 10.4324/9780203123843

Borelli, S., Conigliaro, M., Quaglia, S., & Salbitano, F. (2017). Urban and peri-urban agroforestry as multifunctional land use. In Springer eBooks (pp. 705–724). DOI: 10.1007/978-981-10-7650-3_28

Bussu, S., Bua, A., Dean, R., & Smith, G. (2022). Introduction: Embedding participatory governance. *Critical Policy Studies*, *16*(2), 133–145. DOI: 10.1080/19460171.2022.2053179

Coates, G. (2025). THE SUSTAINABLE URBAN DISTRICT OF VAUBAN IN FREIBURG, GERMANY. Ksu.

Dai, L., Wörner, R., & Van Rijswick, H. F. M. W. (2017). Rainproof cities in the Netherlands: Approaches in Dutch water governance to climate-adaptive urban planning. *International Journal of Water Resources Development*, *34*(4), 652–674. DOI: 10.1080/07900627.2017.1372273

Das, A., Gujre, N., Devi, R. J., Rangan, L., & Mitra, S. (2023). Traditional ecological knowledge towards natural resource management. In Elsevier eBooks (pp. 275–294). DOI: 10.1016/B978-0-323-90500-8.00019-1

De Baro, M. E. Z. (2022). Curitiba case study. In Cities and nature (pp. 117–162).

Debnath, J., Sahariah, D., Lahon, D., Nath, N., Chand, K., Meraj, G., Farooq, M., Kumar, P., Kanga, S., & Singh, S. K. (2023). Geospatial modeling to assess the past and future land use-land cover changes in the Brahmaputra Valley, NE India, for sustainable land resource management. *Environmental Science and Pollution Research International, 30*(49), 106997–107020. DOI: 10.1007/s11356-022-24248-2 PMID: 36418825

Dhanya, B., Ramananda, S., & Dhyani, S. (2022). Mainstreaming blue-green infrastructure in policy and planning for Urban Resilience in the Global South: Promises and Pitfalls. In *Blue-Green Infrastructure Across Asian Countries: Improving Urban Resilience and Sustainability* (pp. 499–518). Springer Singapore. DOI: 10.1007/978-981-16-7128-9_23

Doukas, Y. E., Salvati, L., & Vardopoulos, I. (2023). Unraveling the European agricultural policy sustainable development trajectory. *Land (Basel), 12*(9), 1749. DOI: 10.3390/land12091749

Ejaz, U., Ramon, W., & Olaoye, G. (2025). The Role of Big Data and AI in Smart Cities and Urban Planning.

Elahi, A., Mahanta, P., & Ahmed, N. (2024). Challenges and preservation efforts of biodiversity in Northeast India. In Journal of Emerging Technologies and Innovative Research (JETIR) [JETIR]. *Journal of Emerging Technologies and Innovative Research, 11*(4), •••.

Elkington, J. (1997). *Cannibals with Forks: The Triple Bottom Line of 21st Century Business*. Capstone.

Escrivà Saneugenio, F., D'Agata, A., Vardopoulos, I., & Salvati, L. (2024). Afforestation-deforestation interplay as a driver for urban fringe landscape change and sustainability. In *Sprawl and the city: realizing a sustainable Mediterranean urbanization* (pp. 105–128). Nova Science Publishers.

Escrivà Saneugenio, F., Salvati, L., D'Agata, A., & Vardopoulos, I. (2024). From traditional mosaics to urban encroachment: the intrinsic landscapes along urban-rural sustainable interfaces. In *Sprawl and the city: realizing a sustainable Mediterranean urbanization* (pp. 57–82). Nova Science Publishers.

Escrivà Saneugenio, F., Salvati, L., D'Agata, A., & Vardopoulos, I. (2024). From traditional mosaics to urban encroachment: the intrinsic landscapes along urban-rural sustainable interfaces. In *Sprawl and the city: realizing a sustainable Mediterranean urbanization* (pp. 57–82). Nova Science Publishers.

Esfandi, S., Tayebi, S., Byrne, J., Taminiau, J., Giyahchi, G., & Alavi, S. A. (2024). Smart cities and urban energy planning: An advanced review of promises and challenges. *Smart Cities*, *7*(1), 414–444. DOI: 10.3390/smartcities7010016

Feng, W., Bilivogui, P., Wu, J., & Mu, X. (2023). Green finance: Current status, development, and future course of actions in China. *Environmental Research Communications*, *5*(3), 035005. DOI: 10.1088/2515-7620/acc1c7

Geovetenskaper, G. U. F. (2023, June 22). Traditional ecological knowledge in agriculture for adaptation/ resilience to climate change: high mountain Asia. Greinke, L., & Mehnen, N. Urban Transformation through Public Participation. Möglichkeit und Macht Possibility and Power.

Hassebo, A., & Tealab, M. (2023). Global models of smart cities and potential IoT applications: A review. *IoT*, *4*(3), 366–411. DOI: 10.3390/iot4030017

Hoefsloot, F. I., Martínez, J., & Pfeffer, K. (2024). An emerging knowledge system for future water governance: Sowing water for Lima. *Territory, Politics, Governance*, *12*(6), 825–845. DOI: 10.1080/21622671.2021.2023365

Holling, C. S. (1973). Resilience and stability of ecological systems. *Annual Review of Ecology and Systematics*, *4*(1), 1–23. DOI: 10.1146/annurev.es.04.110173.000245

Kanaujia, K. (2020). Green Buildings: Building a Greener City, a Greener Future—An Indian Perspective. Geoecology of Landscape Dynamics, 223-241.

Kashif, U., & Meo, M. S. (2024). Case Studies in Green Bond Portfolio Management: Best Practices and Lessons Learned. In Green Bonds and Sustainable Finance (pp. 50-68). Routledge.

Khanum, F. (2021). The Northeast India in the midst of climate change.

Krishnakumar, S. (2025). Blue-Green Infrastructure for Climate Resilience: Aiding Urban Development. In Integrating Blue-Green Infrastructure into Urban Development (pp. 99-128). IGI Global Scientific Publishing.

Kumar, A., Srivastava, V., Tabash, M. I., & Chawda, D. (2022). Profitability determinants of infrastructure public private partnerships (PPPs): Empirical evidence from Indian data. *Journal of Financial Management of Property and Construction*, *27*(1), 91–111. DOI: 10.1108/JFMPC-09-2020-0062

Loonen, L. C. M. (2020). Timing the Level of Citizen Participation in Spatial Planning Processes (Master's thesis).

Lotha, T. N., Ritse, V., Nakro, V., Rudithongru, L., Hazarika, N., & Jamir, L. (2024). Climate Change Impact and Traditional Adaptation Practices in Northeast India: A Review. *Current World Environment, 19*(2), 558–575. DOI: 10.12944/CWE.19.2.5

Mabrouk, M., Han, H., Fan, C., Abdrabo, K. I., Shen, G., Saber, M., Kantoush, S. A., & Sumi, T. (2023). Assessing the effectiveness of nature-based solutions-strengthened urban planning mechanisms in forming flood-resilient cities. *Journal of Environmental Management, 344*, 118260. DOI: 10.1016/j.jenvman.2023.118260 PMID: 37354590

Mahanta, N. R., Samuel, A., & Rajput, B. (2020). Community participation and bio rights possibilities in the conservation of urban wetlands: A study of Deepor Beel (Lake), Guwahati, India. 2022 10th International Conference on Reliability, Infocom Technologies and Optimization (Trends and Future Directions) (ICRITO), 1367–1372.

Manaktala, S. (2020). Green bonds in sustainable finance: exploring the case of India. Available at *SSRN* 3644116. DOI: 10.2139/ssrn.3644116

Mandal, R., Nishant, N., Chutia, D., Aggarwal, S. P., & Sharma, B. (2024, February). LoRa Enabled IoT Sensor Framework for Monitoring Urban Flood in Guwahati City. In NIELIT's International Conference on Communication, Electronics and Digital Technologies (pp. 55-76). Singapore: Springer Nature Singapore. DOI: 10.1007/978-981-97-3604-1_5

Mandal, R. K., Alam, M. A., Prasad, R. R., & Kundra, S. (2022). Urbanization Trend In North East India: An Empirical Study. *Journal of Positive School Psychology, 6*(9), 4641–4653.

McTarnaghan, S., Morales-Burnett, J., & Marx, R. (2022). *Urban resilience: From global vision to local practice*. Urban Institute.

Mehta, M., Mehta, D., Nair, N., & Mohanty, A. (2023). *ESG assessments for cities in India*. ResearchGate.

Meitei, K. M., & Das, B. (2024). *Trends, Patterns and impacts of urbanization in North East India*. ResearchGate.

Mir, A. H., Sarma, K., & Upadhaya, K. (2021). Assessing the effectiveness of community managed forests for plant diversity conservation in Meghalaya, Northeast India. *Plant Diversity, 44*(3), 243–254. DOI: 10.1016/j.pld.2021.11.010 PMID: 35769587

Mitra, A., & Paul, S. (2025). Empowering Indigenous Communities Through Regenerative Tourism: Cultural Preservation and Socio-Economic Resilience in Northeast India. In Sousa, B. B. (Eds.), *Regenerative Tourism for Social Development* (pp. 281–310). IGI Global. DOI: 10.4018/979-8-3373-0235-5.ch012

Mitra, S., Madhuvanthi, S., & Sabumon, P. C. (2024). Integrating Green Infrastructure. Nature-Based Solutions in Achieving Sustainable Development Goals, 167.

MoHUA. (2021). *Smart Cities Mission: Government of India*. Ministry of Housing and Urban Affairs.

Mukherjee, M. (2022). *India's progress on its climate action plan–an update in early 2022*. Oxford Institute for Energy Studies.

Nguyen, K. H. (2023). Urban Adaptation and Resilience Planning for a Changing Climate: A Study of New York City (Doctoral dissertation, Rutgers The State University of New Jersey, School of Graduate Studies).

Nickayin, S. S., Jahelka, A., Ye, S., Perrone, F., & Salvati, L. (2023). Planning for just cities with nature-based solutions: Sustainability and socio-environmental inequalities in San José de Chamanga, Ecuador. *Land (Basel)*, *12*(3), 604. DOI: 10.3390/land12030604

Pal, P. (2023). Present status and future outlooks of renewable energy in India for sustainable development. A Basic Overview of Environment and Sustainable Development [Volume 2], International Academic Publishing House (IAPH), 408-433.

Pandey, R. U., Mitra, T., Wadwekar, M., Nigam, J., & Trivedi, K. (2021). Green Infrastructure as a Tool for Improving Livability of Area Based Development Projects Under Smart City Mission. Geospatial Technology and Smart Cities: ICT, Geoscience Modeling, GIS and Remote Sensing, 447-468.

Pandit, N. S.SILPIREKHA PANDIT. (2016). Transformation of urban wetlands as an effect of urban development: An analysis of Deepor Beel in Guwahati, Assam. *CREATIVE SPACE*, *4*(1), 103–114. DOI: 10.15415/cs.2016.41006

Peštová, V. (2021, November). Sustainable Urban Development: Case Study Copenhagen. In *Proceedings of the EDAMBA 2021 Conference*, Bratislava, Slovakia (Vol. 390, pp. 390-399).

Petrea, M. I., & Ursache, I. M. (2023, June). Inside the World's most sustainable smart city: lessons from Copenhagen. In Proceedings of the International Conference on "Sustainable Development of European Smart Cities"—SmartEU.

Pune, A. (2018). *Coastal Ecology and Fishing Community in Mumbai CRZ Policy*. Sustainability and Livelihoods.

Rabinovitch, J. (1996). Innovative land use and public transport policy. *Land Use Policy, 13*(1), 51–67. DOI: 10.1016/0264-8377(95)00023-2

Ribeiro Hoffmann, A., Sandrin, P., & Doukas, Y. E. (Eds.). (2024). *Climate change in regional perspective. European Union and Latin American initiatives, challenges, and solutions*. Cham, Switzerland: United Nations University Series on Regionalism, Vol. 27, Springer Nature. DOI: 10.1007/978-3-031-49329-4

Sakhamuri, H., & Sanagani, V. (2024). SDG Localization and Lessons from India: Role of NITI-Aayog. In *SDGs in the Asia and Pacific Region* (pp. 447–461). Springer International Publishing. DOI: 10.1007/978-3-031-17463-6_58

Salvati, L. (Ed.). (2025). *Environmental sustainability and global change*. Elsevier.

Salvati, L., Ciaschini, C., Postigliola, M., Ermini, B., & Vardopoulos, I. (2025). Contextual differences, contrasting trends? Sustainable development, economic performances, and divergence in land resource depletion. In *Environmental Sustainability and Global Change* (pp. 107–114). Elsevier. DOI: 10.1016/B978-0-443-31596-1.00009-X

Salvati, L., & Vardopoulos, I. (2025). Into the kaleidoscopic urbanization of Southern Europe: Modelling 'form-function' relationships through partial least squares regression. *Cities (London, England), 162*, 105933. DOI: 10.1016/j.cities.2025.105933

Sateriano, A., D'Agata, A., & Vardopoulos, I. (Eds.). (2024). *Socio-environmental change: the path toward sustainability*. Nova Science Publishers. DOI: 10.52305/NZKT5653

Schmid, B., Fricke, C., & Zengerling, C. (2024). Towards a" Freiburg Model" of Housing for the Common Good? Fostering Collaborative Housing in Urban Development. *Urban Planning, 9*, 9. DOI: 10.17645/up.8191

Shafik, W. (2024). Incorporating Artificial Intelligence for Urban and Smart Cities' Sustainability. In Maintaining a Sustainable World in the Nexus of Environmental Science and AI (pp. 23-58). IGI Global.

Sharma, S. B., Singh, A. K., & Rajawat, A. S. (2022). Flood Vulnerability and Risk Assessment with Parsimonious Hydrodynamic Modeling and GIS. In *Geospatial Modeling for Environmental Management* (pp. 3–18). CRC Press. DOI: 10.1201/9781003147107-2

Simoniello, T., Lanfredi, M., Coluzzi, R., D'Emilio, M., Salvati, L., Vardopoulos, I., Imbrenda, V., & Samela, C. (2025). Climate-driven dynamics in the Mediterranean agroecosystems: a pre-2007 economic crisis satellite-based sustainability analysis. In *Environmental Sustainability and Global Change* (pp. 57–68). Elsevier. DOI: 10.1016/B978-0-443-31596-1.00005-2

Singh, M. A., & Singha, K. (Eds.). (2020). *Understanding Urbanisation in Northeast India: issues and challenges*. Routledge. DOI: 10.4324/9781003032625

Singhania, M., & Saini, N. (2021). Systems approach to environment, social and governance (ESG): Case of Reliance industries. *Sustainable Operations and Computers*, *3*, 103–117. DOI: 10.1016/j.susoc.2021.11.003

Song, Y., Kalacska, M., Gašparović, M., Yao, J., & Najibi, N. (2023). Advances in geocomputation and geospatial artificial intelligence (GeoAI) for mapping. *International Journal of Applied Earth Observation and Geoinformation*, *120*, 103300. DOI: 10.1016/j.jag.2023.103300

Teixeira Dias, F., de Aguiar Dutra, A. R., Vieira Cubas, A. L., Ferreira Henckmaier, M. F., Courval, M., & de Andrade Guerra, J. B. S. O. (2023). Sustainable development with environmental, social and governance: Strategies for urban sustainability. *Sustainable Development (Bradford)*, *31*(1), 528–539. DOI: 10.1002/sd.2407

Tiwari, C., & Sunny, A. (2024). To the Ones Who Listen to the Mountains: Evaluating the Role of Traditional Ecological Knowledge in Conservation Efforts in the Himalayas. In *Learning 'from'and 'with'the Locals: Traditional Knowledge Systems for Environmental Sustainability in the Himalayas* (pp. 3–22). Springer Nature Switzerland. DOI: 10.1007/978-3-031-51696-2_1

Vardopoulos, I., D'Agata, A., Escrivà Saneugenio, F., & Salvati, L. (2024). *Sprawl and the city: realizing a sustainable Mediterranean urbanization*. Nova Science Publishers.

Vardopoulos, I., Escrivà Saneugenio, F., Sateriano, A., & Salvati, L. (2024). Deciphering metropolitan expansion: Settlement patterns and socio-environmental dynamics. In *Homage (and criticism) to the Mediterranean city. Regional sustainability and economic resilience* (pp. 85–104). River Publishers. DOI: 10.1201/9781003517306-4

Vardopoulos, I., Falireas, S., Konstantopoulos, I., Kaliora, E., & Theodoropoulou, E. (2018). Sustainability assessment of the agri-environmental practices in Greece. Indicators' comparative study. *International Journal of Agricultural Resources, Governance and Ecology*, *14*(4), 368–399. DOI: 10.1504/IJARGE.2018.098024

Vardopoulos, I., Ioannides, S., Georgiou, M., Voukkali, I., Salvati, L., & Doukas, Y. E. (2023). Shaping sustainable cities: A long-term GIS-emanated spatial analysis of settlement growth and planning in a coastal Mediterranean European city. *Sustainability (Basel), 15*(14), 11202. DOI: 10.3390/su151411202

Vardopoulos, I., Stamopoulos, C., Chatzithanasis, G., Michalakelis, C., Giannouli, P., & Pastrapa, E. (2020). Considering urban development paths and processes on account of adaptive reuse projects. *Buildings (Basel, Switzerland), 10*(4), 73. DOI: 10.3390/buildings10040073

Vodák, J., Šulyová, D., & Kubina, M. (2021). Advanced technologies and their use in smart city management. *Sustainability (Basel), 13*(10), 5746. DOI: 10.3390/su13105746

Vohra, K., & Franklin, M. L. (2021). Water Resources Management—An Indian Perspective. In Springer transactions in civil and environmental engineering (pp. 237–252).

Walker, B., & Salt, D. (2006). *Resilience Thinking: Sustaining Ecosystems and People in a Changing World*. Island Press.

Wang, D., Xu, P. Y., An, B. W., & Guo, Q. P. (2024). Urban green infrastructure: Bridging biodiversity conservation and sustainable urban development through adaptive management approach. *Frontiers in Ecology and Evolution, 12*, 1440477. DOI: 10.3389/fevo.2024.1440477

Wekesa, C., Ndalilo, L., & Swiderska, K. (2021). Case study for the project 'Indigenous biocultural heritage for sustainable development'.

Xaxa, A. (2019). *Tribes and urbanisation in North East India - issues and challenges*. Iitg.

Yadav, A., Anwer, N., Mahapatra, K., Shrivastava, M. K., & Khatiwada, D. (2024). Analyzing the Role of Polycentric Governance in Institutional Innovations: Insights from Urban Climate Governance in India. *Sustainability (Basel), 16*(23), 10736. DOI: 10.3390/su162310736

Chapter 11
Digital Sustainability and Corporate Digital Responsibility in ESG Adoption:
Investor Perceptions in Tunisia

Tarek Ben Noamene
https://orcid.org/0000-0001-7744-883X
Jendouba University, Tunisia

ABSTRACT

Despite the global emphasis on Environmental, Social, and Governance (ESG) principles, their adoption in Tunisia remains limited. Investor awareness is low, and corporate ESG disclosure is almost nonexistent. While regulatory measures encourage sustainable investing, ESG integration is still in its early stages. Emerging technologies like AI, blockchain, and big data offer potential solutions by enhancing transparency, sustainability assessments, and ESG reporting. However, little is known about how Tunisian investors perceive these technologies in ESG integration and investment decisions. This study examines investor perspectives within the Corporate Digital Responsibility (CDR) framework. Findings from investor interviews indicate optimism about technology's role in ESG but concerns over data accuracy, regulation, and technological expertise. Investors also stress the need for ethical technology commitments from companies.

DOI: 10.4018/979-8-3373-2327-5.ch011

1. INTRODUCTION

Environmental, Social, and Governance (ESG) factors have become fundamental considerations in investment decision-making on a global scale (Passas, 2023, 2024; Zopounidis et al., 2020). Investors have realized that firms that provide social and environmental information are not only good investments but also tend to be more financially successful in the long run (Passas, 2024; Ragazou et al., 2024). ESG factors are important determinants of investment attractiveness and business resilience, which can help companies manage their long-term risks, build their corporate reputation and align their business strategies with the expectations of society (Eccles et al., 2014). Stakeholders, including consumers, regulators and investors, are demanding more accountability and sustainability, perceiving companies that adopt ESG principles as responsible actors. ESG investing has moved from exclusionary approaches, where firms that did not meet the ESG criteria were excluded, to more strategic approaches that entail shareholder engagement and active ownership. This change has made companies to include ESG factors in their decision-making processes not only for risk avoidance but also for value creation (Park & Jang, 2021, Kagias et al., 2023; Petropoulou et al., 2024; Ragazou et al., 2022; Stavropoulou et al., 2023; Vortelinos et al., 2024; Yiannoulis et al., 2025). Today, ESG principles go beyond mere legal compliance; they serve as a means to enhance firm value while advancing broader social and environmental objectives.

Despite the growing global emphasis on ESG principles, their adoption in Tunisia remains limited. Research shows that investors in Tunisia are not very much aware of ESG frameworks (Ben Noamene, 2021; Khemiri, 2019).

Furthermore, although measures such as Law No. 2018-35, the Financial Market Council's 2021 ESG reporting guide, and the upcoming Tunindex ESG index are signs of the shift to responsible investment, ESG reporting is still in its early stages.

As of 2023, only 8 out of 75 companies listed on the Tunis Stock Exchange have ESG reports (Tunis Stock Exchange, 2024). This is mainly because traditional ESG integration strategies have not yet gained traction among companies and investors, due to limited awareness and the absence of strong regulatory incentives (Financial Market Council, 2023). Nevertheless, advanced technologies such as artificial intelligence (AI), blockchain, big data analytics, the Internet of Things (IoT) offer great opportunities to overcome the gap by enhancing ESG reporting, increasing the transparency and supporting more efficient investment decisions (Mourtzis et al., 2022).

Technology can be an important enabler of ESG by automation of sustainability evaluation, accurate data and increased investor confidence in ESG disclosures. For example, AI can carry out a real-time assessment of the environmental impact of an organization while blockchain enhances data integrity and transparency through secure ESG records (Sustainalytics, 2021; Saberi et al., 2019). These technologies not only assist organizations to meet their ESG compliance requirements but also inform investor decisions based on better and verifiable sustainability information. However, to the best of our knowledge, there is no previous work that examines how Tunisian investors perceive the role of technology in enhancing ESG disclosure. This missing knowledge raises important questions on whether investors understand the digital innovations as tools that help to enhance the corporate ESG compliance and whether they would use technology enhanced ESG information in their investment choices. Besides ESG disclosure, the ethical issues of digital transformation are growing, as companies face increasing scrutiny over the use of technology. As Tunisia is moving forward with the digitalization process, companies will be required to align their technological strategies with ethical principles; hence, the question is whether investors consider digitalization as an important corporate responsibility.

In light of these new developments, this research study aims to examine the investors' perceptions on two major elements: (1) the way they perceive the role of technology in supporting ESG integration among firms and in assisting investment choices, and (2) the way they view the responsibility of companies to use digital technologies ethically in light of the corporate digital responsibility (CDR) and digital sustainability (DS) concepts. Given Tunisia's nascent ESG ecosystem, this research provides insights into potential accelerators and barriers to adopting ESG principles.

2. LITERATURE REVIEW

2.1 The Evolving Role of ESG in Investment Strategies and Corporate Performance

ESG reporting has become an integral component of modern investment decision-making, influencing corporate transparency, risk assessment, and long-term value creation. According to research, the effectiveness of ESG disclosure depends on both regulatory mandates and governance structures alongside industry participation and investor demands (Eccles, 2014). ESG reporting affects investor confidence and corporate governance while maintaining financial system stability (Al-Hyari et al., 2023). ESG influence on investment decisions stems mainly from how regulatory frameworks establish standards for ESG disclosures. Through mandatory reporting schemes ESG information has enhanced market transparency and investor trust by delivering systematic material sustainability data.

Developed countries with standardized reporting systems show improved investor reaction when ESG integration enhances corporate accountability and aligns business strategies with long-range financial success (Christensen et al., 2021). Material ESG disclosures serve investors for evaluating corporate risks mainly in industries facing environmental and governance problems (Grewal et al., 2021).

Governance structures play a critical role in ESG disclosures because good corporate governance systems decrease investment risks while improving financial stability. Institutions that integrate ESG metrics into executive compensation plans or establish separate ESG boards attract institutional investors who see these strategies as indicators of sustainable strategic dedication (Kimbrough et al., 2024).

ESG reports that demonstrate high stability and transparency create positive stock performance conditions and reduce market volatility according to investors who see sustainability reporting efforts as recognition of their environmental initiatives (Ignatov, 2023). Even though ESG reporting follows general regulatory patterns its effects vary widely between industries because sector-specific factors determine investor expectations and business strategies. ESG initiatives serve as essential risk management tools for firms operating in environmentally sensitive industries such as energy and manufacturing to tackle legal pressures and operational interruptions while defending their reputation (Ragazou, Garefalakis, et al., 2022; Ragazou, Passas, et al., 2022; Stavropoulou et al., 2023; Vortelinos, Menegaki, et al., 2024).

According to research, businesses that proactively report environmental information achieve better financial stability and reduced capital costs supporting ideas that ESG disclosure presents companies as resilient and reliable entities (Bonetti et al., 2023). In governance-intensive industries such as finance and technology, investors focus on corporate mechanisms as main indicators of trustworthy performance. Board independence alongside anti-corruption measures and shareholder interactions serve as vital indicators that determine investor confidence (Moussa & Elmarzouky, 2024).

For instance, ESG governance structures within the banking sector have faced elevated investor focus which resulted in dedicated ESG committees becoming essential for financial risk evaluation (Rabbani et al., 2024).

The financial implications of ESG reporting remain a key topic in the literature, though findings on the financial outcomes of sustainability practices are often mixed or divergent. Governance and environmental factors tend to have positive relationships with investor confidence and long-term profitability. However, the financial impacts of social ESG initiatives vary significantly depending on the context. Buallay et al. (2022) highlight that firms spending capital on social causes see no immediate financial benefits because these initiatives demand both time and stakeholder cooperation. Brand equity along with customer loyalty and workforce productivity benefits from social ESG initiatives which ultimately drives financial sustainability. ESG role in systemic risk management has grown more significant for markets experiencing both macroeconomic uncertainty and regulatory volatility. Investors now use ESG performance as an indicator of corporate success in adapting to changing conditions since those companies with robust ESG governance structures achieve better assessments during economic turbulence (El Khoury et al., 2023). According to Abdulla and Jawad (2024) investor sentiment remains most sensitive to environmentally vulnerable region when it comes to climate-related events among climate risk indicators.

ESG Integration in Tunisia: balancing Financial Priorities and Sustainability Commitments

The evolution of ESG practices within Tunisia's investment landscape reflects a fundamental tension between financial priorities and sustainability commitments. Although ESG awareness is increasing, its incorporation into investment decisions is still influenced by local market factors, regulatory gaps, and stakeholder pressures. The existing literature shows that investor behaviour still prioritises financial performance, and that governance is the most important ESG factor that affects investment decisions (Ben Noamene, 2021). This supports the idea that ESG adoption is used as a risk management tool rather than as a long-term sustainability strategy (Khemir et al., 2019).

Financial leverage has the ability to influence the ESG disclosures of Tunisian firms, which further highlights the connection between sustainability and economic factors. The results of the study show that companies with higher leverage are more likely to have strong ESG commitments in order to meet the expectations of international investors and regulatory bodies. On the other hand, the study also shows that there is an inverse relationship between financial performance and CSR disclosure on the digital platforms, which highlights the trade-offs that Tunisian firms make between their sustainability goals and their profitability pressures (Chakroun et al., 2017). These constraints highlight the difficulties of integrating ESG practices into the Tunisian corporate culture and going beyond legal compliance and investor pressure. Governance has become the bedrock of ESG credibility in Tunisia, with good corporate governance seen as a key driver of investment. Private institutions and public institutions, as well as financial partners now consider governance mechanisms as indicators of corporate accountability, ethical leadership, and operational stability (Moussa & Elmarzouky, 2024). Those companies that have established ESG committees or clear governance structures are likely to secure the trust of investors especially in industries where governance risks are high. Nevertheless, the overall institutional environment in Tunisia, which has been influenced by socio-political changes and regulatory arbitrariness, has led to slow and uneven ESG adoption. The existing governance challenges, high youth unemployment, and social and economic inequalities remain a major drawback for building a strong ESG ecosystem in the country (Bouslah, 2015; Chakroun et al., 2017).

The evolution of ESG adoption in Tunisia reveals a complex interplay between financial imperatives, regulatory influences, and investor expectations. While there is growing awareness of ESG principles, their integration remains largely pragmatic, with governance emerging as the dominant factor in investment decision-making. This emphasis on governance aligns with risk management priorities, corporate stability, and investor confidence, highlighting the strategic role that governance mechanisms play in shaping ESG credibility. However, the limited focus on environmental and social aspects suggests that sustainability efforts are often reactive rather than proactive, driven more by external pressures than by an intrinsic commitment to long-term ESG goals.

The 2021 ESG reporting guide issued by the CMF is a good start in the right direction of promoting transparency and sustainability; however, its voluntary approach restricted its adoption. To this date, only a small number of publicly listed companies (around 10%) have started to disclose ESG-related information, which is sometimes done in a fragmented and inconsistent manner. Furthermore, cultural resistance to change, the traditional focus on short-term financial gains, and ESG literacy missing both among investors and corporate actors slow down the progress. This is also worsened by the characteristics of the Tunisian capital market which is characterized by low liquidity and investor concentration that often leads to skepticism towards ESG investments that are perceived as costly and not consistent with short-term financial results.

Moving forward, a more comprehensive ESG framework, supported by stronger institutional governance and clearer regulatory guidance, will be essential to achieving meaningful sustainability outcomes. Without deeper structural reforms, ESG adoption in Tunisia is likely to remain an instrument of risk management rather than a transformative force for sustainable development.

2.2 Managerial Theories and ESG Reporting Issues

Environmental, Social, and Governance (ESG) reporting has various managerial challenges such as, authenticity, retrospective disclosure, low customizability, limited scope, high costs, and reporting efficacy (Asif et al., 2023). The following managerial theories: Stakeholder theory (Freeman, 1984), legitimacy theory (Suchman, 1995; Archel et al., 2009), institutional theory (DiMaggio & Powell, 1983), and contingency theory (Lawrence & Lorsch, 1986) provide insights into these challenges and frameworks for tackling them. Stakeholder theory and legitimacy theory are applied to understand the stakeholder influence on ESG reporting behaviour. To achieve legitimacy, firms make disclosures that fit stakeholder demands; however, this can result in token ESG actions (Suchman, 1995; Archel et al., 2009). Many companies provide ESG information in response to the demands of investors, regulatory agencies, and customers, although they have not integrated ESG principles into their operations, thus rendering their action as symbolic compliance rather than a real initiative. This issue, known as decoupling, presents an authenticity crisis in ESG reporting where firms talk the talk but don't walk the walk. As a result, such disclosures may be viewed as performative and less reflective of actual corporate responsibility by stakeholders, leading to reputational risks and eroding investor confidence.

Institutional theory posits that firms adopt ESG reporting practices in response to normative, coercive, and mimetic pressures (DiMaggio & Powell, 1983). Standardized reporting is the result of coercive pressure, which results in ESG compliance that is driven by regulatory mandates and industry norms.

- *Coercive pressure*: Regulatory bodies force companies to adopt ESG practices to comply with regulations.
- *Normative pressure:* Professional associations and NGOs push firms to adopt sustainability initiatives.
- *Mimetic pressure:* ESG disclosure becomes a way for firms to align with their peers in the industry, rather than a means of fostering genuine sustainability efforts. This homogeneity in ESG reporting leads to increased risk of greenwashing – where firms look outwardly compliant but fail to actually implement substantive ESG initiatives, resulting in a mismatch between reported and actual sustainability performance.

Another important problem in ESG reporting is its ex-post evaluation what leads to the inability to provide accurate information about the current performance. Most ESG disclosures are based on historical information and do not always provide up to date information on the firm's sustainability performance.

According to Contingency theory, firms can broaden the scope of their disclosures by improving their information processing capacities and include ESG data from supply chain partners (upstream and downstream) (Lawrence & Lorsch, 1986). This ensures that the sustainability assessment is more holistic, since it captures not only the impact that the reporting company has on the environment and society, but also the impact that may be shifted to the company's suppliers or subcontractors. This approach is also consistent with growing stakeholder expectations for increased corporate accountability in global supply networks.

The implementation of Industry 5.0 (I5.0) technologies such as the Internet of Things (IoT), digital twins, artificial intelligence and blockchain enhances ESG reporting by allowing real time collection, analysis and verification of data. These technologies can be integrated to enable firms to move from the current retrospective ESG reporting to real time ESG disclosures that can enhance transparency and better decision making by investors. The range of ESG disclosure is usually restricted to the reporting organization and does not include significant supply chain actors. It is noteworthy that many firms practice ESG reporting of their immediate operations with no mention of the environmental and social effects happening in their value chains.

Thus, ESG reporting is influenced by various managerial challenges, such as authenticity, scope, cost, and information processing limitations. Theoretical perspectives from stakeholder theory, legitimacy theory, institutional theory, and contingency theory are useful in understanding and addressing these challenges. As the standards of transparency and disclosure are increasing with the changes in regulatory requirements and stakeholder's needs, firms are recommended to apply Industry 5.0 technologies to increase the credibility of ESG reporting, control the costs of compliance and develop the methods of sustainability performance evaluation. Further work should be done on the study of the relation between ESG reporting and new digital technologies as well as their ability to eliminate the drawbacks of the reporting process and increase corporate responsibility. Therefore, firms should shift from a conventional, rigid approach to ESG disclosure to a more flexible one based on technology to enhance stakeholder confidence, minimize greenwashing risks, and ensure that sustainability strategies are consistent with long-term value creation.

The implementation of Industry 5.0 (I5.0) technologies such as the Internet of Things (IoT), digital twins, artificial intelligence and blockchain enhances ESG reporting by allowing real time collection, analysis and verification of data. These technologies can be integrated to enable firms to move from the current retrospective ESG reporting to real time ESG disclosures that can enhance transparency and better decision making by investors. The range of ESG disclosure is usually restricted to the reporting organization and does not include significant supply chain actors. It is noteworthy that many firms practice ESG reporting of their immediate operations with no mention of the environmental and social effects happening in their value chains.

2.3 The Role of Industry 5.0 Technologies in Enhancing ESG Reporting

Environmental, Social, and Governance (ESG) reporting has gradually become a critical factor in the sustainability reporting of organizations and corporations. Nevertheless, conventional ESG reporting frameworks have some weaknesses, for instance, the problems of data accuracy, timely disclosure, flexibility of the model, cost, and effectiveness (Asif et al., 2022; Dai & Vasarhelyi, 2016). The Industry 5.0 (I5.0) technologies can be leveraged to address these challenges using automation, artificial intelligence (AI), blockchain, the Internet of Things (IoT), and digital twins to improve the quality, timeliness, and relevance of ESG reporting. The reliability of ESG reporting is mainly determined by the validity of the data collection and processing systems. Most companies use backward-looking ESG reporting which has limited capacity to capture real-time organizational performance thus missing opportunities for greater disclosure (Dai & Vasarhelyi, 2016). I5.0 technologies such as sensors and RFID help in data collection without the involvement of people, thus eliminating errors. The use of blockchain technology enhances the ESG reporting and disclosure by providing evidence of digital transactions, guaranteeing that each transaction is signed off by the network and for every corporate disclosure there is a clear trail (Coindesk, 2020). Data storage and analysis are also supported by cloud computing applications, while machine learning can detect outliers, fraud, and errors in ESG data (Asif et al., 2022). Also, drones help to lessen the reliance on people for monitoring and auditing purposes to minimize the chances of bias and errors being made. Research found that real-time and forward-looking ESG disclosure has the ability to change the way corporations disclose sustainability information. The use of digital twins provides a replica of the real-world business processes to ensure that ESG disclosures are up to date with the current performance of the organization rather than looking at the past (Saxena et al., 2022). The IoT enables the transmission of data from multiple networks in a connected network, which supports ESG monitoring and reporting to various dashboards concurrently (Yu et al., 2022). The use of clouds also improves the real time ESG disclosure by organizing large amounts of data to generate useful information and insights while the AI based models help in the ESG risk analysis and management (Verdouw et al., 2021).

Another major problem in ESG reporting is the current limited adaptability and coverage of the reports. The current ESG reports are mainly directed towards the investment community and regulators and offer the public with limited value-added information (Peloza et al., 2012). Moreover, many ESG disclosures are mainly focused on a company's direct activities and do not include its suppliers and customers in the supply chain (Hutton et al., 2001). Industry 5.0 technologies solve these problems by applying artificial intelligence, machine learning, and cloud computing to the development of ESG applications that can sort through the noise, make contextualized decisions, and present relevant ESG measures to various users (Drempetic et al., 2020). Furthermore, blockchain, IoT, and digital twins expand the ESG reporting scope by providing a full ESG impact assessment across the entire supply chain and avoiding the scenario where negative environmental or social effects are shifted to other companies (Chauhan, 2014). Thus, the increased disclosure of I5.0 technologies provides a clearer picture of a company's ESG performance and thus accountability for ESG performance.

Overall, Industry 5.0 technologies provide innovative solutions to the challenges of ESG reporting. These technologies were designed to improve authenticity, to provide real-time and prospective disclosures, to expand the scope of reporting, to reduce costs, and to enhance efficacy which in turn supports a more transparent, accurate and meaningful sustainability disclosure framework. As regulatory pressures increase and the investor expectations for ESG accountability continue to rise, the adoption of I5.0 solutions will become increasingly essential for organizations seeking to demonstrate genuine commitment to sustainable business practices.

2.4 Corporate Digital Responsibility (CDR)

Corporate Digital Responsibility CDR has emerged as a crucial framework that widens the scope of traditional Corporate Social Responsibility (CSR) to encompass responsibilities in the digital sphere. Organizations now need to manage the ethical, social, and governance challenges posed by digital transformation, as it increasingly shapes their operations and stakeholder relationships (Covucci et al., 2024).

According to Lobschat et al. (2021) CDR is a wide set of responsibilities, such as data privacy, cybersecurity, digital inclusion, and ethical AI governance. Organizations ensure the ethical handling of their digital assets and the social effects of their digital operations through CDR. The literature presents two main views on CDR. The initial focus is on the proper handling of digital technology for ethical and responsible use while meeting legal requirements and establishing transparent practices to build stakeholder trust (Joynson, 2018; Price et al., 2022). The approach follows ethical AI standards which require unbiased algorithms alongside responsible data handling and equal digital service availability. The necessity of corporate accountability emerges to safeguard digital rights while building fair digital systems (Covucci et al., 2024). The second definition sees CDR as a corporate voluntary commitment toward achieving sustainable digital transformation. Organizations should go beyond regulatory adherence by actively incorporating ethical standards into their digital strategy formation according (Herden et al., 2021; Price et al., 2022). In this line, organizations develop digital equity programs and minimize their digital footprints to ensure that technological advancements contribute positively to society. Companies that embrace CDR in this way enhance their reputations, build stakeholder trust, and reduce regulatory risks (Covucci et al., 2024; Elliott et al., 2021).

CDR remains a separate research subject from digital sustainability which produces gaps in the relationship between the two concepts. Traditional CDR frameworks concentrate on corporate governance and ethical standards, but they fail to provide substantial attention to sustainability problems which include digital pollution together with resource efficiency and digital infrastructure carbon emissions (Covucci et al., 2024; Stahl, 2024). The environmental consequences of digital transformation have led researchers to suggest that CDR should adopt sustainability principles for a more comprehensive approach (Guandalini, 2022; Covucci et al., 2024).

Corporate Digital Sustainability (CDS) framework presents a solution that combines CDR with digital sustainability through defined values, dimensions and practices (Covucci et al., 2024). The framework requires corporations to apply digital technologies responsibly while achieving sustainable development objectives. Organizations that embrace CDS can harmonize their digital governance plans with economic, social and environmental requirements to build sustainable ethical digital frameworks that last.

2.5 Digital Sustainability (DS)

Digital Sustainability (DS) is a new and developing concept that addresses how digital technologies can be used to support sustainable development in the environment, economy and society. Unlike CDR, which is mainly about the ethics and governance of corporate entities, DS explores how digital transformation can be used to achieve sustainable development in its widest sense (George et al., 2021). In various ways, scholars have defined DS for the purpose of a more detailed analysis of its effects (Covucci et al., 2024). In one interpretation, DS refers to the long-term maintenance of digital resources so that digital assets can be used in the future (Bradley, 2007; Becker et al., 2016). This interpretation is based on the digital preservation studies and focuses on the sustainability of digital infrastructures while reducing the risk of technological obsolescence. This means that digital knowledge, tools and systems will be maintained for future use and help to support the continuity of information and digital equity. From another perspective, DS is defined as the application of digital technologies to achieve sustainability goals. This view is most similar to the United Nations Sustainable Development Goals (SDGs) (Panagiotopoulos et al., 2024), especially in areas such as energy efficiency (Vardopoulos et al., 2024), circular economy (Vardopoulos et al., 2025), and carbon neutrality (George et al., 2021; Pan and Zhang, 2020). Some of the digital innovations that have been noted to be important in supporting sustainable business practices include; smart grids, AI based energy optimization and the use of blockchain for supply chain transparency. A more comprehensive perspective includes both perspectives and defines DS as a comprehensive framework that encompasses environmental, social, economic, and technological aspects (Wut et al., 2021; Abaku et al., 2023). This integrated perspective recognizes that digital transformation can be both a solution and a challenge for sustainability. On the one hand, digital tools can increase efficiency and innovation, on the other hand, they also create environmental problems such as high energy consumption (Vassiliades et al., 2024), e-waste (Vardopoulos et al., 2021), and digital inequality (Vardopoulos et al., 2023). A crucial area of focus in the DS research is the environmental effects of digital transformation. Research shows that data centers are major consumers of electricity globally and this has left many wondering about the sustainability of the digitalization process (Sharma et al., 2021). In an effort to minimize these impacts, organizations have begun to implement green IT strategies, cloud computing optimizations, and carbon offsetting strategies (Becker et al., 2018). Sustainable digital infrastructure, e-waste management and environmentally friendly software development practices have emerged as critical success factors in the pursuit of sustainable digital growth (Vardopoulos et al., 2024). Sustainability in relation to AI is also very important and can be seen in the use of AI to optimize energy use, manage resources and support digital inclusion (Elliott et al., 2021). However, issues on ethical use of AI, data protection, and equity highlight the need to establish strong regulatory measures that guarantee the correct use of AI (Lobschat et al., 2021). Experts warn that the use of AI in sustainability should not only consider the issues of efficiency but also the ethical ones, to avoid the worsening of social and economic justice (Price et al., 2022). However, the relationship between DS and CDR is still a problem, which is accompanied by the following issues: DS frameworks usually concentrate on technological and environmental components, while ethical concerns like algorithmic bias, corporate accountability, and digital governance are considered as relevant only in secondary manner. This has resulted in different strategies in both academia and corporate practice which has limited the impact of sustainable digital policies (Covucci et al., 2024). To eliminate this gap, researchers suggest that DS should be linked with ethical digital governance. The Corporate Digital Sustainability (CDS) framework is a good starting point and provides a means of incorporating sustainability into the corporate digital responsibility agenda (Covucci et al., 2024). Integrating the ethical aspects of CDR with the sustainability aspects of DS, organisations can create more comprehensive digital strategies that can support sustainability, equity and resilience. Therefore, both CDR and DS are crucial elements of the responsible use of digital technology. While CDR guarantees that corporations apply digital technologies ethically and accountably, DS guarantees that digital advancements are sustainable over time. Therefore, organizations must develop integrated strategies that combine digital ethics and sustainability in order for the advancement of technology to contribute to the overall economic, environmental, and social goals.

While ESG has been largely discussed in developed markets, research investigating the link between ESG adoption and digital technology in North African context is limited. Especially, Tunisia, a country undergoing regulatory and economic transition, has not been well covered from a scholarly perspective concerning how investors assess ESG frameworks and the enabling function of technologies like AI, blockchain, and big data. Most of the current literature is based on general sustainability practices or regulatory developments without considering investor behavior or technological integration. This chapter helps to fill this important gap by providing one of the first exploratory analyses of Tunisian investor perceptions of ESG and Corporate Digital Responsibility (CDR). It adds contextual insights to the emerging ESG discourse in under-researched regions and its findings can help to inform policy and practice in Tunisia and other emerging markets.

3. METHODOLOGY

This study uses a qualitative research approach to investigate the perception of investors towards CSR and DS in the context of ESG adoption by Tunisian companies. The study aims at investigating the expectations of investors towards the integration of the CSR and DS perspectives in the corporate ESG framework, the factors that influence their perceptions, decision-making, and trust in the ESG business models. Through the use of semi-structured interviews and thematic analysis, the study provides an in-depth understanding of investor attitudes and the main drivers of their beliefs, as well as their concerns and expectations regarding ESG adoption in Tunisia.

3.1 Study Design

The reason for choosing a qualitative approach is that it provides a detailed and meaningful understanding of the investors' perceptions of the role of CSR and DS in ESG adoption. Semi-structured interviews are used as the main data collection tool, which provides an organized framework for the study while allowing participants to develop their responses. This approach helps to ensure that the study includes both general patterns and specific experiences, thus providing a thorough understanding of how CSR and DS affect the confidence of investors in ESG adoption. A purposive sampling technique was applied to identify the participants who are involved in investment analysis and have a knowledge of ESG. The sample comprises of institutional investors, private equity managers and representatives of regulatory bodies as these stakeholders are central to the ESG investment environment in Tunisia. Their views are important in understanding how CSR and DS is perceived, the impact of the two on investment decisions and the difficulties that companies encounter in meeting the expectations of investors.

3.2 Data Collection and Procedure

To ensure that the data collected was as rich and detailed as possible, interviews were conducted at the participants' preferred locations to make the participant as comfortable as possible. Some of the participants preferred to meet the researcher face to face in their offices while the others preferred to use video conferencing to ensure flexibility. Every interview was about 70 minutes long and the questions included:

- The perceived importance of CSR and DS in ESG adoption by Tunisian companies.
- Challenges investors perceive in the integration of CSR and DS into business practices.
- The extent to which CSR and DS influence investment decisions, including sectoral variations and risk considerations.

Before each interview, a short introduction was given to explain the research objectives and to remind the participants that they were free to participate or not. The sample includes 20 managers of different investment funds in Tunisia, as well as representatives of industry associations, regulatory bodies and finance experts. This range of participants provides a thorough understanding of the ESG adoption process in the Tunisian investment environment. This composition provides a thorough analysis of ESG adoption, specifically from the perspective of investment fund managers who are key partners in defining sustainable investment strategies in Tunisia. The outcomes of this study give insights into the position of CSR and DS in the ESG adoption process in Tunisia and the convergence or divergence between investor expectations and corporate ESG strategies.

3.3 Results Analysis and Discussion

This part includes the findings of interviews with 20 investment professionals in Tunisia, based on three major areas: (a) The role of CSR and DS in ESG adoption, (b) The challenges that investors encounter in the integration of CSR and DS into business operations, and (c) The effect of CSR and DS on the investment process, with separate analysis for different sectors and risks.

3.3.1 The Perceived Importance of CSR and DS in ESG Adoption

Corporate Social Responsibility (CSR) is widely understood by investors; however, Digital Sustainability (DS) is not a much known or well-developed concept in the Tunisian investment environment. Most investors understood the significance of CSR especially in the context of governance and social responsibility but were not sure how DS relates to ESG. An investment fund manager from a private equity firm remarqued: « governance and social responsibility have always been on our agenda, but digital sustainability? This is something that we have not really examined and discussed at the strategic level yet ». This response captures a significant gap in terms of understanding and integration of DS in investment management.

Although CSR is understood as a tool for managing risks and enhancing company's image, DS is considered as a specialized topic, not a core ESG element. The reluctance to include DS into the evaluation processes indicates that investors do not yet recognize a clear financial or legal benefit of doing so. Several participants mentioned that the main advantage of CSR is that it helps to manage risks and build companies' reputation which affects investment perceptions. However, DS was mainly considered as an operational problem rather than as a strategic one. A senior portfolio manager at an institutional investment fund commented: « Governance is something that we take very seriously because it is connected with financial stability. Social factors such as the well-being of employees and diversity are becoming more important. But to be honest, I have not yet seen a situation where digital sustainability impacted an investment decision ». This statement supports the finding of Al-Hiyari et al., 2023; Ben Noamene, 2021) that in the MENA region, governance is the most important ESG factor in Tunisian investment strategies. Although social aspects are gaining importance, the absence of Digital Sustainability (DS) in investment analyses suggests that Tunisian investors primarily associate ESG with traditional financial risk measures rather than with the broader sustainability transition. This also shows the maturity difference between Tunisia and developed markets which are advancing the role of digital sustainability due to legal compliance and market demand (Covucci et al., 2024). Among the few investors familiar with the concept of Digital Sustainability (DS), discussions primarily focused on cybersecurity, data protection, and ethical AI—while overlooking the environmental impacts of digital operations. A risk analyst at an investment management firm commented: « Those companies that do not show respect to cybersecurity in the light of digital transformation are our biggest concern. In Tunisia sustainable development in the digital world—this is something we haven't really started thinking about ». This response shows some level of understanding of DS but with a greater concentration on digital risk management and not on sustainability. It appears that investors are more preoccupied with cybersecurity as a governance and operational risk but other areas such as the carbon footprint of digital structures, AI, and digital access are not well developed.

In general, the findings of the first theme indicate that while CSR—particularly governance-related aspects—is well integrated into the investment process, Digital Sustainability (DS) remains largely absent, with the focus primarily on digital safety rather than its ecological and social dimensions. This highlights the pressing need for more clearly defined ESG standards that explicitly incorporate Digital Sustainability (DS) indicators, along with guidance that helps investors understand the strategic, environmental, and social value of DS integration.

3.3.2 Challenges in the Integration of CSR and DS Into Business Practices

The most frequent issue identified was the voluntary nature of ESG reporting in Tunisia, which resulted in different disclosure practices and the absence of a universally accepted ESG framework. The quality of ESG information was also found to be diverse and not very reliable. As an ESG analyst for an asset management firm put it: « Many ESG reports in Tunisia look like advertising materials. They identify positive practices but there is almost no information on the actual results ». This is more evidence of the credibility problem with ESG reporting, which is often superficial and devoid of key performance metrics. On the other hand, in developed markets, mandatory ESG reporting standards have enhanced the reliability of ESG disclosures and, thereby, increased investor confidence (Grewal et al., 2021). This lack of definitions of ESG reporting in Tunisia means that investors have no clear understanding of the ESG risks and opportunities, which in turn leads to their inability to embed ESG into their decision-making process.

In the case of Digital Sustainability (DS), the main issue is the absence of definitions or investor expectations. Most firms are yet to go through the process of digital transformation and therefore do not have the necessary skills to develop DS strategies. An investment consultant at a venture capital firm highlighted the current limitations in DS adoption: « Companies are yet to contend with basic ESG regulations let alone think about digital sustainability, that's just not practical ». He continued, « Maybe in a few years, it will be, but for now, it's just not on their agenda ». This perspective indicates that Tunisian companies are still giving a lot of attention to the basic ESG elements, such as governance and financial disclosure, before moving on to more sophisticated sustainability areas such as DS. These factors are compounded by the low awareness of the concepts and the absence of specific guidelines from regulatory bodies, which makes DS a non-issue as far as the respondents are concerned. Presently, developed markets are addressing AI ethics, data privacy, and green IT within the ESG framework (Lobschat et al., 2021) whereas Tunisian businesses are only just starting to implement ESG, let alone digital sustainability as a critical concern.

Overall, the results of the second theme indicate that for ESG to have more internal buy-in, there is a need to provide companies with stronger financial benefits and more stringent laws and regulations.

Overall, the integration of CSR and Digital Sustainability (DS) into business practices in Tunisia remains limited and fragmented. The voluntary nature of ESG reporting, combined with inconsistent disclosures and a lack of performance-based metrics, undermines the credibility and usefulness of ESG information for investors. While governance continues to receive the most attention, the social and digital components lag significantly behind due to operational, regulatory, and awareness-related constraints. DS, in particular, suffers from a lack of definition, understanding, and institutional support, making it nearly invisible in corporate agendas. Compared to the progress seen in developed markets—where mandatory ESG standards and digital ethics are gaining traction—Tunisia is still in the early stages of embedding even foundational ESG practices. For ESG and DS to become strategic rather than symbolic, clearer regulatory frameworks, capacity-building efforts, and targeted investor education will be essential.

3.3.3 The Extent to Which CSR and DS Influence Investment Decisions

Governance is still the most important ESG factor affecting investment decisions, while CSR and DS have a more detailed and not universal impact. An Institutional Investor from the Banking Sector commented on ESG integration across sectors: « ESG, particularly governance and social impact, is a value driver for consumer goods and financial services. Industrial sectors see ESG as just the cost of doing business, not as a benefit ». This statement is consistent with the notion that ESG is considered an asset in sectors that deal with consumer goods such as banking, retail, and telecommunication. Nevertheless, in capital-intensive industries such as manufacturing and real estate, ESG is viewed as a legal enforcer rather than a strategic advisor. This gap is in line with the literature showing that ESG responses differ across sectors based on the perceived costs and benefits and the level of pressure from outside factors (Buallay, 2022). Investors also noted that ESG adoption varies significantly across sectors, with some industries showing greater initiative, while others remain reactive, engaging with ESG principles only when pushed by external forces—particularly international investors. This disparity is especially evident when it comes to the digital dimension of sustainability. As an investment manager at a regional investment fund explained: « In banking and telecom, things like governance and digital responsibility are more or less taken seriously—they kind of have to be. But in sectors like manufacturing or real estate, no one's really thinking about that unless an international investor brings it up. Otherwise, ESG just isn't on the radar. ». This finding indicates that ESG is more adopted in markets where there is high levels of regulatory pressure and consumer interaction. Organisations which do not see immediate ESG-related financial benefits or investor pressure for their sector tend to be slower in the adoption process. This is in sync with global practices where highly regulated industries like banking and technology lead the way in ESG integration, and less regulated sectors respond later (Al-Hiyari et al., 2023). Although Digital Sustainability (DS) did not affect investment decisions in any way, some investors expected that its significance will rise in the forthcoming, especially in cybersecurity and AI ethics. A senior investment analyst at a private equity firm remarqued: « For the moment, ESG is mainly about governance and social aspects. I think in a few years, digital sustainability will gain more significance, mainly for technologies ». This perspective indicates that as of now, DS is not a significant factor, but this may change as more industries become dependent on technology. As a result, AI governance, data privacy, and digital carbon footprint will likely become important ESG factors in the future. However, the reason why DS has not been included in the current mainstream investment criteria is that Tunisia is still a developing country in the ESG front yard, and especially DS is not yet part of sustainability initiatives in developed countries (Lobschat et al., 2021).

4. CONCLUSION AND FUTURE RESEARCH DIRECTIONS

The research provides important insights into how Tunisian investors perceive and incorporate ESG factors, including Corporate Social Responsibility (CSR) and Digital Sustainability (DS) into their decision-making process. The results show that while governance is the most important ESG factor, CSR is emerging as an important factor in some industries, whereas DS is not yet well explored. Investors understand the strategic importance of governance and social responsibility in consumer-oriented industries but in capital-intensive industries, ESG compliance is seen as a cost rather than an opportunity. The challenges which have been revealed – regulatory ambiguity, sectoral disparities, non-uniformity in ESG reporting, and low awareness of DS – highlight the need for a more defined ESG framework in Tunisia. Without defined disclosure guidelines and more encouragement from the regulators, companies are most likely to continue perceiving ESG as an external compliance requirement and not as a part of the company's strategy. Nevertheless, as the global ESG trends are changing, Tunisian investors may slowly start to expand the ESG focus from governance to other factors, especially due to the international market pressures and changes in the regulations.

This study has several limitations that should be acknowledged. First, the sample size was relatively small, and the respondents were 20 investment professionals, mostly from institutional funds and regulatory bodies. A larger sample that includes individual and international investors, may give a better understanding of the ESG adoption in Tunisia. Furthermore, this study focused on the Tunisian context only while some trends could be relevant to the MENA region. A comparison of the ESG adoption between Tunisia and other emerging markets can provide a better insight into the differences and similarities of the ESG trends. Another weakness is the absence of time series data because this study captures the views of investors at a particular point in time. Since ESG adoption is an on-going process that is influenced by the changes in the regulations and the global market trends, future work should try to use a longitudinal design in order to examine the changes in the ESG integration over time. Digital Sustainability is a relatively underexplored area, further work should be done on the financial and strategic implications of DS, especially in sectors where digitalization is rapidly increasing. Industry based research may provide more detailed insights into the ESG practices and the problems encountered in each industry. Finally, as Tunisia is enhancing its ESG reporting regulations, future work could assess whether the new disclosure norms help in enhancing information disclosure, investor confidence, and corporate accountability.

5. PRACTICAL IMPLICATIONS

The findings of this research are important practical implications of the changing model of ESG integration in Tunisia specifically from the perspective of investors. Since governance is the most important ESG factor in investment decisions, CSR is emerging as an important factor in certain industries, and DS is not yet fully understood, the following implications can be drawn for businesses, regulators, and policy makers.

5.1 Investor Awareness and ESG Integration

The study shows that Tunisian investors have understood that governance is the key to ESG integration, with emphasis on adherence to corporate governance, accountability and control as well as risk management. However, the perception of CSR and DS differs according to the industry sector, which shows that ESG adoption is not uniform. In consumer-related industries, social responsibility is growing as a key factor in investors' decision-making process; however, in capital-intensive industries, ESG is often considered a cost center rather than a value-creating activity. This suggests the need for specific awareness campaigns and training for investors and corporate executives to understand the long-term value of ESG compliance, beyond the compliance requirements.

5.2 Regulatory Enhancements and Policy Recommendations

This study reveals some problems including regulatory uncertainty, lack of harmonization across sectors and absence of standardized ESG disclosure. Without clear disclosure requirements and regulatory incentives, companies may continue to consider ESG as one more external compliance obligation rather than as a strategic management tool. Therefore, Tunisian regulators should:

- Issue clear and standardized ESG reporting guidelines to ensure that the reports are comparable and relevant across the various industries.
- Provide incentives such as tax incentives or easier access to funding for companies that embrace good ESG practices.
- Ensure that corporate executives and investors are well informed and trained on sustainability issues to improve their appreciation of sustainability issues.
- Promote cooperation between regulatory bodies, financial institutions and industry associations to establish the ESG integration practice.

5.3 Corporate Strategy and ESG as a Competitive Advantage

This study concludes that, within the Tunisian context, ESG adoption should be approached strategically rather than reactively. Organizations that proactively integrate governance, social responsibility, and digital sustainability into their operations are more likely to gain a competitive advantage in both local and international markets. Socially responsible practices contribute to improved brand image, customer loyalty, and product differentiation. Moreover, industrial companies should recognize the potential benefits of embedding ESG principles into their operations to enhance productivity, strengthen risk management, and support long-term financial performance.

5.4 The Role of International Market Pressures

Although current investors in Tunisia are more concerned with governance than other ESG factors, global trends and international market pressures may result in the adoption of a more comprehensive ESG approach. As investors and multinational corporations continue to incorporate sustainability into their supply chain and partnership, Tunisian companies will have to meet these standards to remain attractive to these investors. Practical measures include:

- Using internationally recognized ESG frameworks like GRI and SASB to attract foreign investors.
- Improving corporate disclosures through voluntary disclosure of ESG information even if there are no legal requirements to do so.
- Interacting with international organizations that are focused on sustainability in order to get funding, partnerships and practices.

5.5 Research and Decision Making Based on Data

This study has some limitations that suggest the need for further research to develop better ESG integration strategies in Tunisia. The limitations of the study include; the small sample size, the absence of time series data, and the use of investors' perceptions rather than real financial outcomes. Future research should: Increase the sample size to include more investors including individual investors and international investors to have a more accurate picture of the ESG adoption process. Make comparisons with other countries to see how Tunisia's ESG adoption compares with other emerging markets and identify best practices that can be adopted locally. Use longitudinal studies to determine the changes in the ESG integration over time and to capture the effects of policy changes and market trends. Examine the financial effects of ESG integration by evaluating the relationship between ESG ratings and financial performance in the Tunisian market. Explore the influence of the international investors on the ESG adoption and how their standards affect the organizational decisions.

5.6 Digital Sustainability: An Untapped Opportunity

Digital Sustainability (DS) is a currently underrepresented concept in the Tunisian context although it is gaining importance on the global business stage. As more industries become dependent on digital technologies, integrating DS into ESG reporting will be crucial to achieve long-term sustainability. To this end, there is a lack of sector-specific strategies for companies and policymakers to put into place to address this gap. These include strategies for energy-efficient digital infrastructure, strong data protection and the sustainable transformation of digital business models. Green fintech and other emerging digital solutions such as the use of blockchain for transparent ESG reporting and AI for sustainability forecasting should also be encouraged through investment. Moreover, increasing awareness on DS through the means of industry specific workshops, training and collaboration between technology firms and sustainability focused organizations is key in enabling an informed and more proactive ESG ecosystem.

5.7 Industry Based Research and ESG Strategy Development

Due to the differences in the implementation of ESG approaches across various industries, it is crucial to conduct industry-specific research and develop tailor-made ESG strategies to address sector-specific challenges and opportunities. Practical steps to advance ESG integration across sectors in Tunisia should be tailored to industry-specific needs. In the financial services industry, institutions should be encouraged to incorporate ESG factors into credit risk assessments and investment decision-making processes. Environmental sustainability can be encouraged in manufacturing and industrial organizations through energy management and sustainable supply chain management. For technology and innovation driven firms, the promotion of green IT practices and sustainable software development can help to ensure that digital transformation is sustainable in the long term. In the retail and consumer goods sector, the important steps include the strengthening of CSR initiatives, ethical sourcing and the adoption of sustainable packaging to address the current and future consumer and regulatory expectations.

This chapter offers practical implications that can help decision makers understand how investors in Tunisia perceive and incorporate ESG factors into their decisions. The results also show that there is a need for better regulatory frameworks, more education, and a proper implementation of ESG across all sectors of the economy. Although governance is currently the most important factor, CSR is also becoming relevant in consumer-oriented industries, while DS is still not fully understood. To promote the development of a more sustainable investment environment, regulators, businesses, and investors need to jointly improve ESG education, revise reporting regulations, and promote sustainability as a key to future business growth instead of treating it as a regulatory obligation.

In the light of growing international focus on ESG, Tunisia has a chance to become a pioneer in sustainable and socially responsible investment if its stakeholders actively face the identified gaps and issues. Furthermore, organizations should proactively implement ESG by incorporating it into the organization's culture rather than as an add-on. A move to sustainability innovation will lead to sustainable growth, resilience and international competitiveness. Thus, in order to meet the ESG goals, the credibility and the attractiveness of the Tunisian companies to investors and consumers will be increased. Collaboration between businesses, academia, and regulatory bodies can be fostered to share knowledge, innovate and develop a more cohesive ESG ecosystem in Tunisia.

REFERENCES

Al-Hiyari, A., Ismail, A. I., Kolsi, M. C., & Kehinde, O. H. (2023). Environmental, social and governance performance (ESG) and firm investment efficiency in emerging markets: The interaction effect of board cultural diversity. *Corporate Governance (Bradford), 23*(3), 650–673. DOI: 10.1108/CG-03-2022-0133

Asif, M., Searcy, C., & Castka, P. (2022). Exploring the role of industry 4.0 in enhancing supplier audit authenticity, efficacy, and cost effectiveness. *Journal of Cleaner Production, 331*, 129939. DOI: 10.1016/j.jclepro.2021.129939

Asif, M., Searcy, C., & Castka, P. (2023). ESG and Industry 5.0: The role of technologies in enhancing ESG disclosure. *Technological Forecasting and Social Change, 195*, 122806. DOI: 10.1016/j.techfore.2023.122806

Becker, C., Maemura, E., Moles, N., Whyte, J., & Mann, J. (2016). Assessing digital sustainability: The digital sustainability model. Benchmark DP, University of Toronto. Retrieved from https://tspace.library.utoronto.ca/handle/1807/104046

Ben Hmida Bouslah, K. (2015). Responsible investment in Tunisia. In Hebb, T., Hawley, J. P., Hoepner, A. G., Neher, A. L., & Wood, D. (Eds.), *The Routledge Handbook of Responsible Investment* (pp. 125–137). Routledge Taylor & Francis Group.

Ben Noamene, T., Halcro, K., Chaher, D., & Talib, A. (2021). Female directors and firm financial performance: The case of the FTSE 100. *Organization Development Journal, 39*(1), 59–76.

Bonetti, P., Cho, C. H., & Michelon, G. (2023). Environmental disclosure and the cost of capital: Evidence from the Fukushima nuclear disaster. *European Accounting Review, 33*(5), 1693–1721. DOI: 10.1080/09638180.2023.2203410

Bradley, K. (2007). Defining digital sustainability. *Library Trends, 56*(1), 148–163. DOI: 10.1353/lib.2007.0044

Buallay, A. (2022). Toward sustainability reporting in the MENA region: The effects on sector's performance. *Managerial Finance, 48*(8), 1137–1155. DOI: 10.1108/MF-09-2021-0422

Buallay, A., Hamdan, R., Barone, E., & Hamdan, A. (2022). Increasing female participation on boards: Effects on sustainability reporting. *International Journal of Finance & Economics, 27*(1), 111–124. DOI: 10.1002/ijfe.2141

Chakroun, R., Matoussi, H., & Mbirki, S. (2017). Determinants of CSR disclosure of Tunisian listed banks: A multi-support analysis. *Social Responsibility Journal, 13*(3), 552–584. DOI: 10.1108/SRJ-04-2016-0055

Chauhan, S., & Amit, . (2014). A relational study of firm's characteristics and CSR expenditure. *Procedia Economics and Finance, 11*, 23–32. DOI: 10.1016/S2212-5671(14)00172-5

Christensen, H. B., Hail, L., & Leuz, C. (2021). Mandatory CSR and sustainability reporting: Economic analysis and literature review. *Review of Accounting Studies, 26*(3), 1176–1248. DOI: 10.1007/s11142-021-09609-5

Coindesk, 2020. Digital voting is coming. Let's do it right [online]. Available: Coindesk. https://www.coindesk.com/markets/2020/11/10/digital-voting-is-coming -lets-do- it-right/ (Accessed May 24 2023).

Covucci, C., Confetto, M. G., Ključnikov, A., & Panait, M. (2024). Unrevealing the nexus between digital sustainability and corporate digital responsibility: A dual-track systematic literature review towards a framework of corporate digital sustainability. *Technology in Society, 79*, 79. DOI: 10.1016/j.techsoc.2024.102743

Dai, J., & Vasarhelyi, M. A. (2016). Imagineering Audit 4.0. J. Emerg. Technol. Account. 13, 1–15. Dimaggio, P.J., Powell, W.W., 1983. The iron cage revisited: Institutional isomorphism and collective rationality in organizational fields. *American Sociological Review, 48*, 147–160.

Drempetic, S., Klein, C., & Zwergel, B. (2020). The influence of firm size on the ESG score: Corporate sustainability ratings under review. *Journal of Business Ethics, 167*(2), 333–360. DOI: 10.1007/s10551-019-04164-1

Eccles, R. G., Ioannou, I., & Serafeim, G. (2014). The impact of corporate sustainability on organizational processes and performance. *Management Science, 60*(11), 2835–2857. https://www.jstor.org/stable/24550546. DOI: 10.1287/mnsc.2014.1984

El Khoury, R., Nasrallah, N., & Alareeni, B. (2023). ESG and financial performance of banks in the MENAT region: Concavity–convexity patterns. *Journal of Sustainable Finance & Investment, 13*(1), 406–430. DOI: 10.1080/20430795.2021.1929807

Elliott, K., Price, R., Shaw, P., Spiliotopoulos, T., Ng, M., Coopamootoo, K., & van Moorsel, A. (2021). Towards an equitable digital society: Artificial intelligence (AI) and corporate digital responsibility (CDR). *Society, 58*(3), 179–188. DOI: 10.1007/s12115-021-00594-8 PMID: 34149122

George, G., Merrill, R. K., & Schillebeeckx, S. J. D. (2021). Digital sustainability and entrepreneurship: How digital innovations are helping tackle climate change and sustainable development. *Entrepreneurship Theory and Practice*, *45*(5), 999–1027. DOI: 10.1177/1042258719899425

Grewal, J., Hauptmann, C., & Serafeim, G. (2021). Material sustainability information and stock price informativeness. *Journal of Business Ethics*, *171*(3), 513–544. DOI: 10.1007/s10551-020-04451-2

Guandalini, I. (2022). Sustainability through digital transformation: A systematic literature review for research guidance. *Journal of Business Research*, *148*, 456–471. DOI: 10.1016/j.jbusres.2022.05.003

Herden, C. J., Alliu, E., Cakici, A., Cormier, T., Deguelle, C., Gambhir, S., Griffiths, C., Gupta, S., Kamani, S. R., Kiratli, Y.-S., Kispataki, M., Lange, G., Moles De Matos, L., Tripero Moreno, L., Betancourt Nunez, H. A., Pilla, V., Raj, B., Roe, J., Skoda, M., … Edinger-Schons, L. M. (2021). Corporate digital responsibility: New corporate responsibilities in the digital age. Sustainability Management Forum | NachhaltigkeitsManagementForum, 29(1), 13–29.

Hutton, J. G., Goodman, M. B., Alexander, J. B., & Genest, C. M. (2001). Reputation management: The new face of corporate public relations? *Public Relations Review*, *27*(3), 247–261. DOI: 10.1016/S0363-8111(01)00085-6

Ignatov, K., & Rudolf, M. (2023). Sentimental sustainability: Does what companies say tell more than what companies do? *Financial Markets, Institutions and Instruments*, *32*(4), 221–252. DOI: 10.1111/fmii.12181

Joynson, C. (2018, August). Corporate digital responsibility: Principles to guide progress. LinkedIn. Retrieved from https://www.linkedin.com/pulse/corporate-digital-responsibility-principles-guide-progress-joynson/

Kagias, P., Sariannidis, N., Garefalakis, A., Passas, I., & Kyriakogkonas, P. (2023). Validating the Whistleblowing Maturity Model Using the Delphi Method. Administrative Sciences 2023, Vol. 13, Page 120, 13(5), 120. DOI: 10.3390/admsci13050120

Khemir, S. (2019). Perception of ESG criteria by mainstream investors: Evidence from Tunisia. *International Journal of Emerging Markets*, *14*(5), 752–768. DOI: 10.1108/IJOEM-05-2017-0172

Khemir, S., Baccouche, C., & Ayadi, S. D. (2019). The influence of ESG information on investment allocation decisions: An experimental study in an emerging country. *Journal of Applied Accounting Research*, *20*(4), 458–480. DOI: 10.1108/JAAR-12-2017-0141

Kimbrough, M. D., Wang, X., Wei, S., & Zhang, J. (2024). Does voluntary ESG reporting resolve disagreement among ESG rating agencies? *European Accounting Review*, *33*(1), 15–47. DOI: 10.1080/09638180.2022.2088588

Lobschat, L., Mueller, B., Eggers, F., Brandimarte, L., Diefenbach, S., Kroschke, M., & Wirtz, J. (2021). Corporate digital responsibility. *Journal of Business Research*, *122*, 875–888. DOI: 10.1016/j.jbusres.2019.10.006

Mourtzis, D., Angelopoulos, J., & Panopoulos, N. (2022). A literature review of the challenges and opportunities of the transition from Industry 4.0 to Society 5.0. *Energies*, *15*(17), 6276. DOI: 10.3390/en15176276

Moussa, A. S., & Elmarzouky, M. (2024). Beyond compliance: How ESG reporting influences the cost of capital in UK firms. *Journal of Risk and Financial Management*, *17*(8), 1–21. DOI: 10.3390/jrfm17080326

Pan, S. L., & Zhang, S. (2020). From fighting COVID-19 pandemic to tackling sustainable development goals: An opportunity for responsible information systems research. *International Journal of Information Management*, *55*, 102196. DOI: 10.1016/j.ijinfomgt.2020.102196 PMID: 32836647

Panagiotopoulos, P., Vardopoulos, I., Maialetti, M., Ciaschini, C., Koundouri, P., & Salvati, L. (2024). Reimagining sustainable development and economic performance indicators: A human-centric Maslow–Bossel blueprint. *Economies*, *12*(12), 338. DOI: 10.3390/economies12120338

Park, S. R., & Jang, J. Y. (2021). The impact of ESG management on investment decision: Institutional investors' perceptions of country-specific ESG criteria. *International Journal of Financial Studies*, *9*(3), 1–27. DOI: 10.3390/ijfs9030048

Passas, I. (2023). Accounting for integrity: ESG and financial disclosures: the challenge of internal fraud in management decision-making [Hellenic Mediterranean University]. DOI: 10.12681/eadd/55359

Passas, I. (2024). The Evolution of ESG: From CSR to ESG 2.0. Encyclopedia 2024, Vol. 4, Pages 1711-1720, 4(4), 1711–1720. DOI: 10.3390/encyclopedia4040112

Peloza, J., Loock, M., Cerruti, J., & Muyot, M. (2012). Sustainability: How stakeholder perceptions differ from corporate reality. *California Management Review*, *55*(1), 74–97. DOI: 10.1525/cmr.2012.55.1.74

Petropoulou, A., Angelaki, E., Rompogiannakis, I., Passas, I., Garefalakis, A., & Thanasas, G. (2024). Digital Transformation in SMEs: Pre- and Post-COVID-19 Era: A Comparative Bibliometric Analysis. Sustainability 2024, Vol. 16, Page 10536, 16(23), 10536. DOI: 10.3390/su162310536

Price, R., Dörr, S., Frick, T., Joynson, C., Merx, O., & Wade, M. (2022). The international CDR manifesto. Retrieved from https://corporatedigitalresponsibility.net/cdr-manifesto-english

Rabbani, M. R., Kiran, M., Bhuiyan, A. B., & Al-Hiyari, A. (2024). Does diversity in top management and boards affect ESG performance? Evidence from Islamic and conventional banks in the MENA region. *International Journal of Islamic and Middle Eastern Finance and Management*, *17*(6), 1244–1266. DOI: 10.1108/IMEFM-06-2024-0281

Ragazou, K., Garefalakis, A., Zafeiriou, E., & Passas, I. (2022). Agriculture 5.0: A New Strategic Management Mode for a Cut Cost and an Energy Efficient Agriculture Sector. Energies 2022, Vol. 15, Page 3113, 15(9), 3113. DOI: 10.3390/en15093113

Ragazou, K., Lemonakis, C., Passas, I., Zopounidis, C., & Garefalakis, A. (2024). ESG-driven ecopreneur selection in European financial institutions: Entropy and TOPSIS analysis. *Management Decision*, *63*(4), 1316–1345. DOI: 10.1108/MD-12-2023-2425

Ragazou, K., Passas, I., Garefalakis, A., Zafeiriou, E., & Kyriakopoulos, G. (2022). The Determinants of the Environmental Performance of EU Financial Institutions: An Empirical Study with a GLM Model. Energies 2022, Vol. 15, Page 5325, 15(15), 5325. DOI: 10.3390/en15155325

Saberi, S., Kouhizadeh, M., Sarkis, J., & Shen, L. (2019). Blockchain technology and its relationships to sustainable supply chain management. *International Journal of Production Research*, *57*(7), 2117–2135. DOI: 10.1080/00207543.2018.1533261

Saxena, A., Singh, R., Gehlot, A., Akram, S. V., Twala, B., Singh, A., Montero, E. C., & Priyadarshi, N. (2022). Technologies empowered environmental, social, and governance (ESG): An Industry 4.0 landscape. *Sustainability (Basel)*, *15*(1), 309. DOI: 10.3390/su15010309

Sharma, G. D., Reppas, D., Muschert, G., & Pereira, V. (2021). Investigating digital sustainability: A retrospective bibliometric analysis of literature leading to future research directions. *Clinical Hemorheology and Microcirculation*.

Stahl, B. C. (2024). From corporate digital responsibility to responsible digital ecosystems. *Sustainability (Basel)*, *16*(12), 4972. DOI: 10.3390/su16124972

Stavropoulou, E., Spinthiropoulos, K., Ragazou, K., Papademetriou, C., & Passas, I. (2023). Green Balanced Scorecard: A Tool of Sustainable Information Systems for an Energy Efficient Business. Energies 2023, Vol. 16, Page 6432, 16(18), 6432. DOI: 10.3390/en16186432

Sustainalytics, (2021). available at: https://www.sustainalytics.com/esg-ratings

Tunis Stock Exchange. (2024). available at: https://www.bvmt.com.tn

Vardopoulos, I., Abeliotis, K., & Lasaridi, K. (2025). A systematic informetric analysis and literature review of food waste quantification studies in the food service industry. *Sustainability (Basel), 17*(1), 103. DOI: 10.3390/su17010103

Vardopoulos, I., Giannopoulos, K., Kalli, S., & Salvati, L. (2024). From tech-centric to human-centric smart tourist destinations: innovations in real estate, urban planning, and sustainability. In *Socio-enviromental change: the path toward sustainability* (pp. 65–96). Nova Science Publishers.

Vardopoulos, I., Konstantopoulos, I., Zorpas, A. A., Limousi, L., Bennici, S., Inglezakis, V., & Voukkali, I. (2021). Sustainable metropolitan areas perspectives through assessment of the existing waste management strategies. *Environmental Science and Pollution Research International, 28*(19), 24305–24320. DOI: 10.1007/s11356-020-07930-1 PMID: 32072414

Vardopoulos, I., Papoui-Evangelou, M., Nosovo, B., & Salvati, L. (2023). Smart 'tourist cities' revisited: Culture-led urban sustainability and the global real estate market. *Sustainability (Basel), 15*(5), 4313. DOI: 10.3390/su15054313

Vardopoulos, I., Santamouris, M., Zorpas, A. A., Barone, G., Italos, C., & Vassiliades, C. (2024). A comparative study on discrepancies in residential building energy performance certification in a Mediterranean context. *Buildings (Basel, Switzerland), 14*(4), 1009. DOI: 10.3390/buildings14041009

Vassiliades, C., Kyriakides, N., Pissourios, I., & Vardopoulos, I. (2024). Sustainable architectural innovations for energy-efficient multi-storey buildings through a novel taxonomy and evaluation approach. *Strategic Planning for Energy and the Environment, 43*(4), 1195–1224. DOI: 10.13052/spee1048-5236.43415

Verdouw, C., Tekinerdogan, B., Beulens, A., & Wolfert, S. (2021). Digital twins in smart farming. *Agricultural Systems, 189*, 103046. DOI: 10.1016/j.agsy.2020.103046

Vortelinos, D., Menegaki, A., Passas, I., Garefalakis, A., & Viskadouros, G. (2024). Heterogeneous Responses of Energy and Non-Energy Assets to Crises in Commodity Markets. Energies 2024, Vol. 17, Page 5438, 17(21), 5438. DOI: 10.3390/en17215438

Vortelinos, D., Passas, I., Floros, C., & Garefalakis, A. (2024). Corporate Social Responsibility and Country Governance: An International Comparative Study Amid the COVID-19 Pandemic. International Journal of Financial Studies 2024, Vol. 12, Page 110, 12(4), 110. DOI: 10.3390/ijfs12040110

Wut, T. M., Lee, D., Ip, W. M., & Lee, S. W. (2021). Digital sustainability in the organization: Scale development and validation. *Sustainability (Basel)*, *13*(6), 3530. DOI: 10.3390/su13063530

Yiannoulis, Y., Vortelinos, D., & Passas, I. (2025). Exploring Audit Opinions: A Deep Dive into Ratios and Fraud Variables in the Athens Exchange. Accounting and Auditing 2025, Vol. 1, Page 3, 1(1), 3. DOI: 10.3390/accountaudit1010003

Yu, H., Wei, W., Li, J., & Li, Y. (2022). The impact of green digital finance on energy resources and climate change mitigation in carbon neutrality: Case of 60 economies. *Resources Policy*, *79*, 103116. DOI: 10.1016/j.resourpol.2022.103116

Zopounidis, C., Garefalakis, A., Lemonakis, C., & Passas, I. (2020). Environmental, social and corporate governance framework for corporate disclosure: A multicriteria dimension analysis approach. *Management Decision*, *58*(11), 2473–2496. DOI: 10.1108/MD-10-2019-1341

Chapter 12
The Insular Rural Landscape as a Factor of Sustainable Development

Odysseas Kopsidas
https://orcid.org/0000-0003-4900-702X
Aristotle University of Thessaloniki, Greece

ABSTRACT

The subject of this paper is the contribution of the rural landscape to the sustainable development of the island area. The research objective of this study is the protection, improvement and promotion of the rural island area with appropriate proposals to enhance environmental, economic and social resilience. The methodology chosen includes primary research of autopsy and free interviews and secondary research of literature study. International, European and national guidelines and the corresponding international and national institutional framework for landscape protection and management with a focus on rural areas are analyzed.

1. INTRODUCTION

The rural island landscape is a synthesis of the interaction of natural and man-made resources, as shaped by the long-standing occupation of the inhabitants with the productive activities of agriculture and farming (Doukas et al., 2023). The conservation and enhancement of the rural island landscape as an aesthetic resource is investigated (Pace et al., 2025). Investigate the improvement and protection of the rural landscape and the possibility of promoting it as a protected cultural resource and an element of intangible cultural heritage. Investigate whether the rural island landscape can be a resource for the development of the primary sector, in particular

DOI: 10.4018/979-8-3373-2327-5.ch012

agriculture and livestock farming, and whether the products produced can add value to the sustainable development of entrepreneurship (Vardopoulos et al., 2018). Investigate the development and promotion of the rural landscape by strengthening the agri-food sector and linking it with tourism and culture (Doukas et al., 2024).

The research questions are investigated in a specific study area, the island of Andros. The presentation of international, European and national landscape policies, their correlation with major European and national development policies as well as landscape management policy in the countryside planning, will lead to the study area focusing on the analysis of its rural landscape and ending up promoting the contribution of the rural landscape to the sustainable development of Greece.

The Mediterranean landscape of the numerous Greek islands is characterized by its diversity and unique character (Vardopoulos et al., 2024). It is the product of centuries of osmosis of elements of the special natural and cultural heritage and is considered an important characteristic for determining the island's identity (Stefani and Tsilimigas, 2023).

The management of the landscape is a structural component of sustainable development (Vardopoulos et al., 2024), and is projected as the coordinated action to ensure the preservation of the landscape, in harmony with social, economic and environmental changes (Simoniello et al., 2025), as stated in article 1 of Law 3827/25-02-2010 which ratified the European Landscape Convention (Law 3827/2010).

Rural landscapes are areas occupied by crops, pastures and natural vegetation and are the creations of rural populations, which change under the influence of the evolution of the techniques they apply, the agricultural policy, the programs of the farmers, as well as the perception they have formed the populations that live in them (Beopoulos, 2005).

The island landscape is composed of the interaction of natural resources: geomorphology, vegetation, water surfaces and anthropogenic resources, both in the material dimension of space (residential space, architecture and morphological characteristics of buildings and traditional structures (elevations of agricultural land, infrastructure such as bridges, roads, paths, etc.), as well as in the intangible (cultural heritage elements, traditions, customs, etc.) (Stefani and Tsilimigas, 2023). The inhabitants' involvement in the productive activities of agriculture and animal husbandry contributed to the formation of the island landscape, gradually building a harmonious whole between the natural and man-made environment, thus shaping the island's agricultural landscape. Its existence is a tangible proof of the potential of human dexterity and ingenuity in sculpting the rugged and rocky island terrain and providing the necessary resources for living throughout time.

2. METHODOLOGY

To prepare the paper, a primary investigation of autopsy and free interviews was conducted in the study area and a secondary investigation of a literature review. The data collection includes a study of Greek and foreign language literature, statistical data and scientific publications as well as free interviews with a representative of a public body of the local government and residents of the island. Specifically: Collection, recording and evaluation of general guidelines, international, European, national and regional texts on the rural landscape. Collection, recording and evaluation of the most specific directions, international, European, national and regional texts for the study area The description and evaluation of the existing situation in the study area and the policy for the protection and promotion of the landscape, especially the rural landscape, is done through international and European and national strategic texts, studies and policies (developmental and spatial) and a study of the relevant national and supranational legislation. The synthesis of primary and secondary information results in conclusions and propositions.

3. IMPLEMENTATION

Insular rural landscapes, characterized by their unique geographical isolation and distinct ecosystems, play a critical role in achieving sustainable development. These landscapes, found on islands around the world, are shaped by their limited spatial resources, rich biodiversity, and deeply rooted cultural traditions. Due to their geographic constraints, these areas often require innovative strategies to balance economic development, environmental conservation, and social well-being.

The sustainable development of insular rural landscapes hinges on their ability to preserve natural resources while fostering resilient communities and sustainable livelihoods. These landscapes are home to fragile ecosystems that are highly sensitive to external pressures such as climate change, overexploitation, and urbanization. At the same time, they often host indigenous knowledge systems, agricultural practices, and traditional architecture that reflect centuries of adaptation to their unique environments.

This intersection of vulnerability and cultural richness offers both challenges and opportunities. By adopting approaches such as sustainable agriculture, eco-tourism, renewable energy solutions, and integrated land-use planning, insular rural areas can become models for sustainable development. Moreover, these landscapes often serve as biodiversity hotspots and carbon sinks, contributing significantly to global environmental goals.

Understanding the dynamics of insular rural landscapes and their potential for sustainability requires a multidisciplinary approach. This involves examining the interplay between their natural, economic, and sociocultural dimensions, as well as fostering collaboration between local communities, policymakers, and researchers. As islands are increasingly on the frontlines of global environmental change, the sustainable management of their rural landscapes can provide valuable lessons for other regions striving for balance between development and conservation.

The insular rural landscape is not merely a passive setting but an active agent of sustainable development, offering innovative pathways to address the interconnected challenges of our time. The landscape is directly linked to sustainable development as the concept of landscape as well as the concept of sustainable development involves the same space-time interconnection of the past and the present in direct reflection with future perspectives and needs. The 2030 Agenda of the United Nations (UN) sets the global charter and guidelines for sustainable development with a time horizon of 2030. The 2030 Agenda promotes the integration of all three dimensions of sustainable development – social, environmental and economic – in all sectorial policies, while promoting the interconnection and coherence of SDG-related policy and legislative frameworks (General Secretariat of Legal and Parliamentary Affairs). For Greece, the SDGs provide an important opportunity for recovery and transition to a new development model, with a balance between the three pillars: economic, social and environmental (ypen.gov.gr, 2020).

Of the 17 Sustainable Development Goals defined in the Agenda, those related to the rural landscape are: Goal two (2) for zero poverty and hunger directly links the fight against poverty and hunger with the sustainability of the agricultural sector which it characterizes as a key solution for development. The implementation of resilient agricultural practices that increase productivity ensures the provision of sustainable food production systems, while contributing to the preservation of ecosystems by enhancing adaptive capacity to climate change and extreme weather events (UN Regional Information Centre). The traditional agricultural landscape, especially the island landscape, has a centuries-old tradition in the use of gentle and therefore sustainable farming practices (fallowing, crop rotation) and therefore the management goes hand in hand with the specific goal of the Agenda.

Goal fifteen (15) for life on land seeks, among other things, the sustainable use of terrestrial ecosystems, preventing the phenomenon of desertification and land degradation, while maintaining the biodiversity of ecosystems. (UN Regional Information Centre). The loss of arable land due to desertification is affecting the agricultural sector and heightening the risk of a global food crisis. Characteristically, due to drought and desertification, twelve million hectares are lost every year (twenty-three hectares every minute) in place of which twenty million tons of cereals could have been cultivated (UN Regional Information Center). The preservation of traditional rural landscape elements is an important tool in preventing the phenomenon of desertification. The terraces, as shale structures, contribute to the retention of water while at the same time helping to concentrate the nutrients of the earth. The maintenance and utilization of terraces as elements of the rural island landscape can work effectively in controlling erosion and subsequent land abandonment (Beopoulos and Papadopoulos, 2003).

Goal twelve (12) for responsible production and consumption focuses on sustainable tourism, as a factor in the development of jobs and the promotion of local culture and local products. (UN Regional Information Centre). Ecotourism is the most suitable model of tourism for the visitor's information and experiential contact with the elements of the landscape each region, because it learns to respect the past as an integral part of a sustainable future. All the above objectives converge on the primary need to inform and raise awareness of the individual as part of the social whole for the concept and manifestations of sustainable development.

The Strategic Plan of the Common Agricultural Policy (C.P.P) for the period 2023-2027 is the basic policy text for the development of the primary sector and the rural areas of the Country, for a more resilient, green and digital Agriculture, in harmony with the respective priorities of the strategies of the European Green Deal. Through his interventions, he attempts to signal the shift to a new productive model for Greek Agriculture oriented towards social, economic and environmental sustainability, with an emphasis on the age-related revitalization of rural areas and the promotion of young entrepreneurship (Strategic Plan of the Common Agricultural Policy 2023-2027).

To effectively integrate insular rural landscapes into sustainable development frameworks, targeted and inclusive strategies must be adopted. Below are key implementation measures that address environmental, economic, and social dimensions while fostering resilience and sustainability.

Environmental Conservation and Resource Management

1. Preservation of Biodiversity: Establish protected areas and community-led conservation initiatives to safeguard endemic species and fragile ecosystems. Utilize biodiversity monitoring programs to assess ecological health.
2. Sustainable Agriculture: Promote agroecological practices that enhance soil health, reduce water usage, and minimize chemical inputs. Encourage crop diversification and the cultivation of indigenous plant species adapted to local conditions.
3. Climate Adaptation Strategies: Implement reforestation, coastal restoration, and sustainable water management systems to mitigate the impacts of climate change, such as rising sea levels and extreme weather events.

Promoting Green and Circular Economies

1. Eco-Tourism Development: Design eco-tourism programs that highlight cultural and natural heritage while minimizing environmental impacts. Collaborate with local communities to create sustainable tourism experiences that generate income and encourage conservation.
2. Renewable Energy Integration: Invest in renewable energy infrastructure such as solar, wind, and micro-hydropower systems to reduce reliance on imported fossil fuels and enhance energy security.
3. Circular Economy Practices: Encourage waste reduction, recycling, and the reuse of materials to manage the limited resources typical of islands efficiently. Support local enterprises that utilize waste as a resource.

Sociocultural Engagement and Education

1. Community Involvement: Involve local populations in decision-making processes to ensure that development initiatives align with community needs and respect traditional knowledge. Empower women and marginalized groups to participate in governance and resource management.
2. Education and Capacity Building: Provide training in sustainable practices, entrepreneurship, and environmental stewardship to build local capacity for managing resources and driving development.
3. Cultural Heritage Preservation: Protect and celebrate the cultural heritage of rural communities, including traditional architecture, crafts, and agricultural techniques, as integral elements of sustainable development.

Policy and Governance Frameworks

1. Integrated Land-Use Planning: Develop land-use plans that balance economic activities, environmental conservation, and urban development. Utilize geospatial data and participatory mapping to ensure equitable resource allocation.
2. Incentives for Sustainability: Introduce policies such as tax incentives, subsidies, or grants to encourage sustainable practices in agriculture, tourism, and energy production.
3. Regional and Global Cooperation: Engage in knowledge-sharing networks with other island regions to exchange best practices, access funding opportunities, and address transboundary challenges like marine conservation and climate adaptation.

Monitoring and Evaluation

1. Establish key performance indicators (KPIs) to assess the social, economic, and environmental impacts of development programs. Conduct regular evaluations to refine strategies and ensure alignment with sustainability goals.
2. Integrate citizen science initiatives to engage local communities in data collection and foster a sense of ownership over conservation and development projects.

4. DISCUSSION

The comparative analysis of international, European, national, regional development and spatial policy texts highlight the general framework for the value of the landscape and in particular the dynamic correlation of the rural landscape with the primary sector and sustainable tourism for the study area. The protection and preservation of the rural landscape contributes to the sustainable use of ecosystems and the development of sustainable tourism, is in agreement with the Sustainable Development Goals 12 and 15 of the UN. At the same time, the restoration of the landscape and the support of green infrastructure networks are actions on which the Territorial Agenda 2030 of the EU focuses in order to achieve one of its two primary goals for a 'Green Europe'.

The dynamic function of the rural landscape, agri-food and tourism is a common trend in development policies. At the national level, in the Strategic Plan of the Common Agricultural Policy (2023-2027) the integration of the elements of the rural landscape is foreseen as a model of the CAP for granting subsidies to the producer, proposing as an intervention the continuation of cultivation on the terraces.

The agri-food sector takes the lead in the National Smart Specialization Strategy (2021-2027) as a development priority sector, while in the National Tourism Action Plan "2030" agritourism, ecotourism and culinary tourism are included among complementary tourism products that enhance the visitor's experience. At the level of spatial planning policy, in the General Framework for Spatial Planning and Sustainable Development (Government Gazette 128/03-07-2008) the restoration and promotion of cultural heritage and the landscape is set as a more specific objective, while in another part of the Government Gazette it recognizes the agricultural sector as an important "manager" of the rural landscape.

In the Regional Spatial Planning with the revision of the Regional Spatial Frameworks, general policies for the sustainable management of the landscape and specific quality objectives are integrated after the definition of landscape zones at the Regional level. The function of the three components, rural landscape, agri-food and tourism becomes more evident in the development and spatial directions given for the study area. Experiential tourism, agri-food and green activities are three priority axes of the Smart Specialization Strategy for the South Aegean Region that is clearly related to the protection and utilization of the rural island landscape. In the Strategy, the produced agri-food products of the Region stand out for: (a) the quality and (b) gentle cultivation techniques.

5. CONCLUSION

The study contributes to the redefinition of the value of the rural island landscape as an iconic cultural and aesthetic resource and at the same time as an important productive activity of the primary sector and tourism, proposing their sustainable orientation based on environmental economic and social resilience. It introduces proposals for the protection of the rural landscape in the study area and its use for the alternative management of agriculture and tourism. The suggestions that follow are based on the conclusions and the removal of threats and disadvantages and highlighting the strengths and opportunities of the SWOT analysis.

In this context, the proposals for the preservation and utilization of the rural landscape answer the research questions of this paper by documenting its value in the social, economic and environmental sustainability of the study area: (a) as an aesthetic resource, (b) as a protected cultural resource and element of intangible cultural heritage, (c) as a resource for the development of the primary sector and the sustainable development of entrepreneurship, (d) as a resource for the interconnection of the rural landscape with the agri-food sector, tourism and culture. The planning and implementation of public policy for the insular rural landscape must be approached with the aim of not only preserving it but highlighting it as a comparative development advantage.

The insular rural landscape holds immense potential as a driver of sustainable development, embodying a delicate balance between ecological integrity, cultural heritage, and economic resilience. These landscapes, with their unique geographical and ecological characteristics, serve as living laboratories for innovative practices that integrate environmental conservation with human well-being.

Preserving and utilizing the natural and cultural assets of insular rural areas requires holistic strategies that respect their limited resources and vulnerabilities. By fostering sustainable practices such as eco-tourism, renewable energy adoption, and climate-adaptive agriculture, these regions can address challenges like biodiversity loss, climate change, and economic instability while empowering local communities. Moreover, their role as biodiversity hotspots and carbon sinks underscores their global importance in combating environmental degradation.

Ultimately, the sustainable development of insular rural landscapes depends on inclusive governance, community engagement, and interdisciplinary collaboration. By valuing the traditional knowledge of local populations and integrating it with modern sustainability frameworks, these landscapes can become beacons of innovation and resilience. As the world faces interconnected environmental and social challenges, the lessons learned from these unique settings can provide invaluable guidance for achieving global sustainability goals.

In conclusion, the insular rural landscape is not only a critical factor in the pursuit of sustainability but also a testament to the power of harmonizing human activity with the natural world. By investing in their protection and sustainable development, we can safeguard these irreplaceable landscapes for future generations while contributing to a more equitable and sustainable planet.

REFERENCES

Beopoulos, N., Papadopoulos, A. (2003). "Desertification as Human Absence and Sterility of Places: An Introduction to a Controversial Phenomenon".

Beopoulos, N. (2005). Diploma thesis School of Architectural Engineering E.M.P. Athens." "Agriculture produces landscapes." Mercouri Estate.

Doukas, Y. E., Salvati, L., & Vardopoulos, I. (2023). Unraveling the European agricultural policy sustainable development trajectory. *Land (Basel)*, *12*(9), 1749. DOI: 10.3390/land12091749

Doukas, Y. E., Vardopoulos, I., & Petides, P. (2024). Challenging the status quo: a critical analysis of the common agricultural policy's shift towards sustainability. In Climate change in regional perspective. European Union and Latin American initiatives, challenges, and solutions. (pp. 115–134). Cham, Switzerland: United Nations University Series on Regionalism, Vol. 27, Springer Nature. DOI: 10.1007/978-3-031-49329-4_8

Pace, L., Galella, M., Ciaschini, C., Sateriano, A., Cudlin, P., Vardopoulos, I., Rontos, K., & Salvati, L. (2025). Environmental-economic dimensions of land-use transformations: exploring the Mediterranean rural sustainable development. In *Environmental Sustainability and Global Change* (pp. 81–93). Elsevier., DOI: 10.1016/B978-0-443-31596-1.00007-6

Simoniello, T., Lanfredi, M., Coluzzi, R., D'Emilio, M., Salvati, L., Vardopoulos, I., Imbrenda, V., & Samela, C. (2025). Climate-driven dynamics in the Mediterranean agroecosystems: a pre-2007 economic crisis satellite-based sustainability analysis. In *Environmental Sustainability and Global Change* (pp. 57–68). Elsevier., DOI: 10.1016/B978-0-443-31596-1.00005-2

Stefani, F., & Tsilimigas, G. (2023). *Landscape in spatial planning: a critical evaluation of the treatment of the degraded landscapes of the Greek island area in regional spatial planning*. Aichoros.

Vardopoulos, I., Escrivà Saneugenio, F., Sateriano, A., & Salvati, L. (2024). *Homage (and criticism) to the Mediterranean city. Regional sustainability and economic resilience*. River Publishers., DOI: 10.1201/9781003517306

Vardopoulos, I., Falireas, S., Konstantopoulos, I., Kaliora, E., & Theodoropoulou, E. (2018). Sustainability assessment of the agri-environmental practices in Greece. Indicators' comparative study. *International Journal of Agricultural Resources, Governance and Ecology*, *14*(4), 368–399. DOI: 10.1504/IJARGE.2018.098024

Vardopoulos, I., Maialetti, M., Scarpitta, D., & Salvati, L. (2024). Spatially explicit analysis of landscape structures, urban growth, and economic dynamics in Metropolitan regions. *Urban Science (Basel, Switzerland)*, *8*(4), 150. DOI: 10.3390/urbansci8040150

Chapter 13

Building Heritage Resilience in the Age of Climate Change: A Discussion on Cultural Preservation and the Future of Conservation

Tanima Ray
https://orcid.org/0009-0004-1074-3470
Birla Institute of Technology Mesra, Ranchi, India

Satyaki Sarkar
https://orcid.org/0000-0002-5161-2344
Birla Institute of Technology Mesra, Ranchi, India

Swagata Payra
https://orcid.org/0000-0003-2398-4658
Birla Institute of Technology Mesra, Ranchi, India

ABSTRACT

Heritage sites are facing growing threats every day due to the ever-increasing, abrupt changes brought on by climate change, like extreme weather conditions. It is important to take urgent measures to protect these valuable sites from the damage caused by these environmental shifts. This discussion explores how climatic solutions can be integrated into heritage conservation by different measures. Important parameters related to climate change and heritage conservation have been identified as a basis for understanding their impacts. It also gives significance to risk assessment and prioritization as brought about by multi-hazard frameworks through

DOI: 10.4018/979-8-3373-2327-5.ch013

systematic approaches to the identification and mitigation of cultural assets that are most vulnerable. Community involvement in the construction of means of adaptation is important because it suggests that conservation initiatives are sustainable and culturally relevant. These results open pathways to future research to focus on the development of adaptive strategies and further refinement of technological applications.

1. INTRODUCTION

1.1 Climate Change and Heritage Preservation

Managing cultural infrastructure in heritage towns is of more importance than ever because climate change (Ribeiro Hoffmann et al., 2024), has greater effects on our urban environments (Nicolini, 2024) (Tianchen Dai, 2022). Careful strategy is necessary for the successful implementation of these valuable places so that they may be safeguarded and their resistance to environmental stresses increases, as stated (Ababneh, 2016). The UNESCO World Heritage Convention provides a much-needed framework for this effort, promoting the identification, protection, and preservation of cultural and natural heritage worldwide.

Evolving transportation infrastructure and increased accessibility are placing strain on heritage towns, often leading to accelerated degradation of their cultural and environmental assets.

To address this issue, (Yu, Safarov, Wang, Buzrukova, & Janzakov, 2023) proposed moderate control on the development speed of transportation networks which is necessary, particularly in ecologically and culturally fragile regions. This approach ensures that infrastructure growth aligns with the preservation of heritage, balancing urbanization with environmental sustainability. Cities and towns can easily protect their cultural heritage by monitoring the rate of development while improving accessibility. It makes towns grow steadily and sustainably while at the same time retaining the distinctive cultural identity that defines them.

Archaeological heritage is preserved through controlled excavations and conservation methods that minimize site damage to portray each location's history and culture (Kurdi, 2017). Since these sites must be preserved appropriately, the communities must take part in their protection as well; this also calls for proper recording of discoveries and knowledge enhancement of the public (T K, 2024).

The approaches for cultural heritage preservation under climate change and community expectations underline the importance of integrating environmental issues with cultural conservation (Hiswara, 2023). For example, a heritage town experiencing greater flooding because of climate change would put in place adaptive methods to safeguard major monuments, assuring tourist safety and accessibility. Simultaneously, the Cultural Infrastructure Need and Assessment in Heritage Towns identifies the various types of infrastructure that are required to support cultural activities, such as museums and community centres, along with surveys and community consultations, which help to assess public interest and involvement (Getzner, 2024). For identifying the best locations for new cultural infrastructure, land suitability analysis is necessary and establish that developments meet accessibility and ecological standards while contributing to the town's character. A new cultural hub or other varied cultural infrastructures that serve locals and tourists while maintaining historic landscapes might be suggested by looking at land use patterns, for instance (Caprioli, 2021).

Newly developed logistics create a breakthrough potential for walled towns of Mediterranean port cities, which represent one of the most prestigious destinations for cultural renaissance. (Anna Maria Colavitti, 2019) have examined the opportunities and risks of regeneration strategies focused on heritage conservation, drawing from case studies of four Mediterranean port cities that have pursued UNESCO World Heritage status for their walled towns. It underlines a multi-layered approach towards the historical, architectural, and urban value of such places while requiring the strategy for the preservation of their authenticity with minimal negative impact from gentrification (Vardopoulos et al., 2024).

Government agencies, experts in conservation, local communities, and international organizations must collectively take response measures toward the increasing susceptibility of historical landmarks to environmental degradation (Vardopoulos et al., 2021).

The first requirements include defining tangible and intangible cultural resources possessed by heritage towns and identifying their population's ties to the historic past. The resilience of the built environment could be further maintained by local governments and conservation bodies, giving priority to the integration of climate-responsive designs into the structural and architectural planning of heritage assets (Tsilika & Vardopoulos, 2022).

The United Nations' Sustainable Development Goals (SDGs), particularly Goal 11.4, emphasize the protection of the world's cultural and natural heritage as part of building sustainable cities. Climate change presents significant challenges to this goal, requiring integrated responses that combine innovation, governance, and climate justice.

This chapter begins by contextualizing the challenges faced by heritage towns under climate stress. It then explores technological innovations, such as GIS, in managing cultural infrastructure. Subsequent sections examine governance, community participation, and institutional mechanisms, followed by case studies that illustrate adaptive strategies, and attempting to formulate a framework for climate-responsive heritage conservation.

2. CLIMATE CHANGE IMPACTS ON HERITAGE SITES

2.1 Vulnerabilities of Heritage Sites in the Face of Climate Change

Since global warming is at an accelerating rate, cultural heritage sites, in particular those identified in heritage towns, are also experiencing extreme perils that cause threats to their historical integrity and to identify the community around them. Unique architectural structures, as well as aged, cobbled roads and buildings, symbolize the real tapestry for cultural expression, showing stories of successive generations (Vardopoulos, 2022). This story should be protected despite climate change impacts. Anthropogenic pressures -including urban expansion and industrialization- are now affecting these primarily historically significant locations (Vardopoulos et al., 2024).

These are socio-economic challenges rather than purely environmental ones (Sateriano et al., 2024). Climate change erodes the very base of these heritage towns, bringing about the erosion of historical buildings, loss of artefacts, and, in the worst-case scenario, the erasure of cultural identities. It is the cracking of the once proud facades of old European cities and flooding the streets in the coastal towns of Southeast Asia.

Rising temperatures are pushing up storms, hurricanes, and heat waves in intensity, all of which cause damage due to their devastating effects on frail structures. For example, ancient buildings near coastal areas are susceptible to destruction by saltwater corrosion, while heat waves may cause buildings made from traditional material to crack faster and degrade quickly.

Geospatial technology has emerged as one of the valuable instruments to analyze how climate change makes heritage vulnerable. Such approaches also facilitate profound spatial analysis; the outcome provides detailed information in relation to planning, hence forming sharper, fact-driven protection methods. For example, with GIS, we can develop risk maps that predict the effects of rising sea levels, flooding, and soil erosion on historical buildings. This kind of information enables decision-makers to take proactive steps before issues escalate.

Ecological risk assessments are essential for understanding the complex interaction of natural and anthropogenic factors with such vulnerable sites. In WHSs, it allows one to integrate diversified datasets, such as meteorological and hydrological conditions, alongside very detailed historical records.

Recent research confirms the seriousness of the issue. (Giglio, Frontera, Malara, & Armocida, 2024) says that the built environment in heritage towns is increasingly threatened by climate change. According to them, the degradation of materials, including stone and wood, can be accelerated by changes in temperature, moisture, and pollution levels. These changes are both physically damaging to the buildings and culturally important structures. Adverse weather conditions drastically transform the face of historic towns by altering the aesthetic identity and depriving their ability to attract tourists—who rely so much on being authentic and remaining connected to history. This calls for specific strategies unique to the site, taking account of the environment and architectural expression at each given heritage site.

2.2 Global Efforts and Local Strategies for Heritage Preservation

With all of these rising issues in mind, global initiatives and local actions have been designed to safeguard cultural heritage sites from climatic change effects. Several international organizations, such as UNESCO and ICOMOS, are actively involved in their work to raise awareness about the importance of climatic adaptation for various heritage sites.

Governments and heritage organizations are adopting innovative methods of conservation at the local level in order to preserve historical structures (Vardopoulos, 2023). Apart from geospatial technology (Vardopoulos et al., 2023), other such methods are being adopted, like energy retrofitting (Barone et al., 2024), which helps in saving the physical and environmental qualities of heritage buildings. For example, green roofs can be added along with better insulation and energy-efficient windows that could protect buildings from extreme temperatures and reduce carbon footprints (Alessandra Gandini, 2016).

In addition, conservationists are increasingly using susceptibility analysis to evaluate the structural and material integrity of historical buildings. (Marvin Ravan, 2023) tells us how different environmental factors, such as humidity, wind velocity, and soil conditions, affect the stability of cultural heritage structures. Experts can better prioritize conservation efforts by identifying areas of vulnerability, such as buildings located near fault lines or those exposed to high levels of air pollution.

Some traditional construction techniques are now being revived to safeguard cultural heritage sites from the threats of modernity. Developed centuries ago, such techniques often are more suitable for local environmental conditions and may be more long-lasting. For example, in areas prone to earthquakes, heritage buildings could be reinforced with timber and stone materials, which are less rigid than concrete and hence would be less likely to collapse during tremors. It could be built along the coastal lines using natural construction materials and this can ensure that the effects of rising sea levels reduce waterlogging.

In addition to these pragmatic steps, education and social mobilization are essential factors in the long-term sustenance of cultural heritage preservation. Local populations must be involved in the conservation processes since they best know the historical and cultural value of such sites. (Muthukumar, 2017) calls for the education of new generations on the importance of monuments and heritage sites. Information passed down through the ages is necessary to ensure that future generations will be well-placed to protect and appreciate cultural legacies. The Clean India Mission also seeks to sensitize society towards the damage pollution causes to cultural landmarks, nudging community-driven conservation. They engage the local community and mainly people who may be considerably affected by environmental protection and heritage conservation. Thus, they ensure sustainability in preservation efforts.

Now, India has been keen on the preservation of the monument sites such as the Taj Mahal. According to Muthukumar, the roles played by ASI and UNESCO help in safeguarding these monumental buildings. For instance, through buffer zones established between the Chola Temples and the encroaching urban sprawl and industrialization by collaborating with the local stakeholders, the ASI safeguards these important monument sites. It is responsible for guarding the physical structures of the temples and ensuring that the cultural values that they hold remain intact for generations.

3. INTEGRATING CLIMATE DATA INTO CONSERVATION EFFORTS

3.1 The Role of GIS in Heritage Conservation

Climate data must be integrated into the protection of heritage. The use of a GIS-based geodatabase will combine the flood hazard and vulnerability values for cultural heritage sites. This will give an understanding of specific risks that such sites are facing which guides the approaches to conservations. For example, the flood risk analysis of the Castile and León BICs is based on three major matrices: one is the hazard itself, the other deals with vulnerability, and the third deals with estimating the overall flood risk associated with the cultural sites (Arrighi, Brugioni, Castelli, Franceschini, & Mazzanti, 2018). These matrices give a framework for flood risk analysis that encompasses both the frequency and severity of hazards. Therefore, the researchers and conservationists are not only able to analyze current risks but also to forecast future threats that flooding might present according to the projections on climate. By integrating the matrices, heritage managers are capable of identifying sites with the highest risk and creating intervention measures to be applied, such as strengthening flood defences, improving drainage systems, or changing site access in order to reduce exposure. This is also a technique for monitoring so that conservation efforts remain dynamic and responsive in changing conditions over time.

3.2 Flood Risk Assessment and Vulnerability Matrices

The integration of geospatial tools provides a more refined understanding of the wider impacts of climate-related risks. The use of GIS technology can better map areas with multiple hazards, including floods, landslides, and coastal erosion and be used to make comprehensive risk maps. The aggregation of several layers of hazard data with vulnerability assessments allows for making more detailed site-specific recommendations regarding the protection of cultural assets. Mapping, on the other hand, can also be seen as a decision-making tool for urban planners and policymakers in this respect. Furthermore, heritage professionals can make future strategies for protecting heritage much better.

3.3 Advanced Tools for Climate-Responsive Heritage Protection

Risk Mapping Tool for Cultural Heritage Protection, developed through the Interreg Central Europe Projects ProteCHt2save and STRENCH, is to help policymakers in the preservation of heritage at risk from extreme weather events (Interreg Central Europe ProteCHt2save, Accessed October 2024). Using data from 12 climate models, historical records, and Earth Observation insights, the tool is able to give clear projections of climate impacts. With 12 different climate models, the tool is thus able to provide an all-encompassing view of the potential risks facing cultural heritage sites across multiple regions, thus enhancing the predictive capacity for climate-related hazards. This tool also offers insight into past climate variability and its correlation with site vulnerability and ensures that both current and historical climate trends inform conservation efforts. It simulates various climatic scenarios in order to appraise the extent of possible implications on cultural heritage, allowing policymakers to prioritize those parts most vulnerable further and then define appropriate mitigation mechanisms.

3.4 3D Digital Technology and Photogrammetry in Heritage Conservation

More revolutionized dealings with historical artefacts are 3D digital technology and photogrammetry. In fact, by using 3D scanning and photogrammetry, high-resolution models of monuments and artifacts can be developed which capture even the smallest details in their structure and condition. In this regard, the technology has the capability of recording heritage sites in real-time and providing a very good source for monitoring and analysis in the future. Photogrammetry, for example, uses high-resolution images to produce three-dimensional models. In doing so, it gives information about the physical state of sites and the materials used in construction. Based on these digital models, one is able to ascertain damage potential from environmental stressors such as wind, rain or temperature changes. Researchers also look at how these models can be integrated with qualitative weathering models that track the development of materials over time and simulate the future behaviour of these materials during future events of weathering (Coletti, 2024). These ideas offer a more dynamic conservation approach, where data is updated continuously, and long-term heritage site health can be better monitored and managed.

3.5 Impact of Climate Change on Material Degradation

In addition, research may be focused on the analysis of the influence of climate change on certain materials, like stone, through the application of damage indexes to express the degree of degradation caused by environmental factors (Randazzo, et al., 2020). Damage indexes allow for quantifying the amount of degradation that has occurred and make it possible to compare damage rates at various locations. For incorporating these indexes, heritage managers have a clear outline of the potential effects that are brought by any specific weather phenomenon and climatic condition to work on specific conservation efforts for varying materials. Through this, even materials that seem to be on the edge may be easily isolated and conservation techniques applied in an attempt to preserve them before severe damage. These technologies, along with the GIS-based risk assessments, give a powerful framework for understanding and mitigating the impacts.

3.6 Weathering Models and Digital Simulations for Heritage Preservation

Incorporating 3D digital modelling and simulations of weathering into conservation procedures greatly improves monitoring and prediction of the impacts on cultural heritage. Because such models can easily be updated on a regular basis and shared cross-platform, this also offers accessible means for collaborative work among networks of international conservationists, researchers, and policymakers. The continued development of these technologies holds the promise of increasing the precision of climate risk assessments, supporting more informed conservation decision-making and better safeguarding cultural heritage in the face of climate change.

3.7 Collaboration and Global Networks in Climate Risk Management

From previous research and studies within the discipline, some critical parameters that would describe the particular climatic factors that affect cultural heritage sites have been established. Among them are the variation in temperature, patterns in precipitation, the strength of winds, flood danger, and extreme weather events occurrence. By analytically reviewing all these factors, researchers should build up a more holistic understanding of the dynamics of climate affecting cultural heritage sites and develop techniques to mitigate the risks linked to climate change.

3.8 Selection of Parameters for Climate Risk Assessment

Based on literature reviews and other research work concerning the effects of climate change on heritage conservation, a number of parameters that should be considered to evaluate the impact of climate on cultural heritage sites have been determined. In an effort to create universally generalizable parameters that should ideally affect every site under general climatological factors rather than its site characteristics at this moment, only specific parameters were considered. Therefore, key parameters that included temperature variation, variability in precipitation patterns and wind intensity, possibility of flooding risk, along frequency of extreme weather conditions were chosen as these greatly influence the physical and environmental conditions of cultural heritage assets. The above factors were considered because they possess the potential for material degradation or structural damage and environmental stress, which may adversely impact a heritage site.

The parameters were chosen to give a comprehensive view of the climate dynamics that affect cultural heritage preservation. In addition to environmental factors, vulnerability indicators like the risk of flooding and the potential for extreme weather events were included to provide a more complete view of the potential threats these sites may face in the future. Table 1 outlines the parameters, which form the basis of the study's analytical framework.

Table 1. List of Parameters

Sr. No.	Parameters for Thematic Areas	Sub-Parameters	Reference
	Climatological Aspects		
1.	**Rainfall**	Magnitude, Frequency, Excessive, Low, Intensity, Median raindrop diameter, wind-driven rain, Intense, Heavy, Rain amount, Rain pH, Rain drop size, runoff coefficient, and water depth.	(Giuseppe Forino, 2016), (Reuter, Hellmund, Moßgraber, & Hertweck, 2023), (Marius Mosoarca, 2017), (Bethune Carmichael, 2017), (Riccardo Cacciotti, 2021), (Coletti, 2024), (Giglio, Frontera, Malara, & Armocida, 2024), (A. Agapiou, 2015), (Shanwen Zheng, 2018)
2.	**Carbon Emissions**	Reduce, Reduced due to the use of traditional materials	(Alexandra Keller, 2016)
3.	**Greenhouse Gas Emissions**	Acts and Laws to reduce 80% by 2050, Increased	(Erica Avrami, 2021), (Erin Seekamp, 2020)
4.	**Humidity**	Humidity Cycle, Wetting and Drying Cycles, increased, relative humidity, Low, Fluctuations.	(Giuseppe Forino, 2016), (Giglio, Frontera, Malara, & Armocida, 2024)
5.	**Embodied Energy**	Evaluation	(Webb, 2017)
6.	**Topography/Urban Characteristics**	Exposure factor (α), Local topography, Features	(Marius Mosoarca, 2017), (Baldwin, 2022)
7.	**Building Age**	70 and more in years	(Erica Avrami, 2021)
8.	**Meteorological Measurements**	Air temperature, Air Speed-spot measurements, Range of Temperature, Low, High, Monthly Mean Temperature, Variation, Extreme, Atmospheric Pressure-Sensitivity error, Response time	(Sombat Prajonsant, 2018), (Alessandra Gandini, 2016), (Erin Seekamp, 2020), (Abdelrazek Elnaggar, 2024), (Mohammed ElAdl, 2022), (B. Pioppi, 2019)
9.	**Extreme Weather Events**	Probability of Occurrence, Frequency, Duration, Intensity, Storm damage, Storm Surge, Increased precipitation, Landslide, Heat waves, Saltwater Intrusion, Coastal flooding, earthquakes, tsunamis, landslides, wind-generated waves, flood, fire, drought periods, Differential Subsidence	(Giuseppe Forino, 2016), (Sandra Fatoric L. E., 2020), (C. Michael Hall, 2015), (Bethune Carmichael, 2017), (Marvin Ravan, 2023), (Giglio, Frontera, Malara, & Armocida, 2024), (E Asprogerakas A. G., 2021), (Danilo Marco Campanaro, 2015)
10.	**Air Pollution**	Concentration of Sulphur dioxide, Concentration of Nitric Acid, Carbon Dioxide, Particulate matter, Carbon fractions of particulate matter, rise in PM 2.5	(Alexandra Keller, 2016), (A. Agapiou, 2015), (Abdelrazek Elnaggar, 2024), (Baghel, 2022), (Ameneh Ahmadi, 2017)

continued on following page

Table 1. Continued

Sr. No.	Parameters for Thematic Areas	Sub-Parameters	Reference
	Climatological Aspects		
11.	**Coastal Sedimentation/ Erosion**	Erosion, Elevated deposition rates, Dune Pressure, Increased fluvial erosion	(Jen Heathcote, 2017), (Marvin Ravan, 2023), (Georgian Ionu PURCAREATA, 2015), (Ameneh Ahmadi, 2017), (Sandra Fatoric E. S., 2017)
12.	**Sea-level Rise**	High Tides, Flooding	(C. Michael Hall, 2015), (Erin Seekamp, 2020), (E Asprogerakas A. G., 2021)
13.	**Material Degradation**	Weight Loss of Material per Unit Surface	(Coletti, 2024)
14.	**Flooding**	Overtopping of Rainwater Goods, Type- Flash floods or gradual, Water depth, flow depths, flow velocities, flood sediment, flood time lag, Magnitude	(Jen Heathcote, 2017), (Sandra Fatoric E. S., 2017), (Julio Garrote, 2020), (E Asprogerakas A. G., 2021), (Abdelaziz Elfadaly, 2018)

Parameters for this study have been selected systematically to ensure that the most relevant and impactful factors are included within the research framework. This helps for a deeper understanding of potential impacts on heritage sites from climate change, with a more precise and holistic assessment of risks. The present study, using a multifaceted approach, a combination of qualitative and quantitative methods, provides us with a better understanding of how the factors affecting climatology have an impact on the preservation and management of heritage towns.

4. CHALLENGES IN MANAGING CLIMATE-RESPONSIVE HERITAGE CONSERVATION

4.1 Challenges in Assessing Climate Change Risks to Heritage Sites

Climate change is becoming a highly pressing matter for researchers and restorers in the world. It started being treated as a relatively far-off issue in terms of its time scale, but its effects are being felt now with great intensity and frequency. Heritage sites in urban settings are among the most vulnerable. (Laura Quesada-Ganuza, 2021) noted a significant gap in research on the impact of climate change on historic urban areas, which are very often filled with buildings and landmarks that are hundreds of years old and deeply embedded in local identity and history.

One of the greatest challenges in dealing with climate-responsive conservation is that the methods for assessing risks are not adequate, and current approaches often focus too much on the physical damage to heritage structures, flooding, erosion, and extreme weather events, among others, while neglecting all other impacts. For example, social vulnerabilities are rarely factored into any estimation of risk-which includes local communities displaced by rising sea levels or economic impacts on tourism-dependent regions. Normal consideration is given to local traditions, oral histories, or communal rituals relating to heritage sites to be considered the intangible aspects of cultural heritage, losing view to the fact that such intangible aspects of cultural heritage can be just as vulnerable to climate change as the physical structure with which it is associated.

The lack of any overarching framework for the assessment of climate risks stands out as the main barrier in order to develop effective preservation strategies. According to (Bertolin, 2019), we require a more sophisticated data-driven methodology that transcends the present forms of physical damage assessments. New approaches have to think in much more complex terms, reflecting the relationship between the built environment, local communities, and climatic change. The integration of several fields, including geography, sociology, economics, and engineering, holds the hope to create correct assessments of risks when it comes to cultural heritage vulnerabilities. At this pace, such integrated risk assessment will increasingly form the foundation through which our conserved activities shall be led by effective, efficient, and sustainable strategies as a result of acceleration in climate change.

To effectively address these risks, a multidisciplinary approach is essential. After all, understanding which site is more vulnerable to climate change impacts will determine the strategies to be used since the increase in temperature and other extreme weather conditions may impact different structures, landscapes, and cultural practices in different ways. While much attention remains on physical preservation, intangible cultural heritage such as local knowledge, traditions, and rituals potentially stand to lose or change because of climate pressures.

4.2 Balancing Preservation with Climate Adaptation

This is one of the most significant challenges in climate-responsive heritage conservation, as it has to balance preserving the cultural and historical integrity of a site with the implementation of necessary climate adaptation measures. There is a growing need to adapt heritage sites as climate change brings new risks to heritage sites. These changes often require altering the original structure, appearance, and form of the site. Therefore, this creates a balancing act between preserving and adapting is especially clear in the historic city areas that are home to some of the world's most cherished cultural landmarks.

Modern climate-resilient technologies, that is energy-efficient systems, flood barriers, and insulation, frequently clash with traditional construction techniques and aesthetic values that are central to cultural heritage. This is a problem for conservationists who would like to keep the heritage site 'genuine' but ensure that the heritage sites are ready to face the irreversible effects of global climatic change. For a scenario, putting up modern HVAC systems in some ancient buildings or strengthening walls by using contemporary materials may serve the purpose of sheltering against the forces of nature. However, such acts of intrusion will alter the building to an extent. This dilemma underlines the broader challenge of adapting heritage sites to modern needs without compromising their cultural and historical significance.

It is not the conservation of what already exists; it is permitting heritage sites to change while staying culturally intact. Therefore, a need for novel solutions that embrace traditional building skills but integrate these with modern technologies that support sustainability lies here. Another thing is that these solutions must survive the future alterations of climate because risks and impacts will continue evolving. The challenge, therefore, is to design strategies that respond to both current conditions and future uncertainties in climate patterns.

As (Wang & Fouseki, 2025) discuss in their exploration of time rupture and urban heritage preservation, the concept of time rupture captures the disruption of historical continuity brought about by rapid urbanization, which often leads to the erosion of local identity and a diminished sense of place. We find in the case of Dongjiadu in Shanghai the complicated interplay between modern urban development and cultural heritage, underlining the challenges and possibilities that this dynamic offers. As part of these adaptive and context-sensitive strategies, approaches go beyond traditional, static conservation approaches, pressing for the melding of modern growth with the continuation of cultural memory. Therefore, by adopting this multifaceted approach, time rupture will be effectively treated so that heritage will remain an alive, relevant element of cities.

4.3 Engaging Local Communities in Heritage Conservation

From ancestral dances in South America, cultural practices serve as a testament to resilience and history. In many European villages, storytelling traditions have been shared across generations, shaping the community's collective memory and keeping their unique identity alive. By passing stories along within oral traditions through performances in various village festivals and festivities, stories breathe new life as they cultivate an original bond of old to their audience. Just as in the interactive projection mapping in the Mastic Museum, it actually brings the uniqueness of the village of Mastic to the fore while on Chios Island to connect local people with cultural values.

One critical challenge in climate-responsive heritage conservation relates to the issue of engaging the local communities within the process. Much of heritage conservation focuses on the physical preservation of the structures, but at least equally as important is the social and cultural significance. Cultural heritage is not about buildings but about people, traditions, and practices woven around these places. In other words, no conservation strategy can be sustainable long term if the concerns and perceptions of local communities are not considered.

According to (Crowley, 2022), "heritage conservation efforts often focus on the material aspects of heritage, such as restoring architecture or preventing environmental damage, but are oblivious to the significance of intangible cultural heritage." Local traditions, knowledge, and community practices linked to heritage sites are critical aspects that define the identity of a place. Simply keeping the physical nature of conservation without focusing on its intangible counterpart will risk the loss of the same to climate change.

According to (Oladeji, 2022), the sustainability of cultural heritage is closely tied to local communities' active participation in conservation and management, as shown by a study conducted in Ondo and Ekiti States, South Western Nigeria. It highlights the essence of community participation in heritage management through volunteerism, attending cultural events, and skill acquisition.

The local communities should be involved in the conservation process to design appropriate and context-specific strategies, taking into consideration their cultural and societal contexts. However, it is quite challenging to involve the local communities. They feel that their voice is not heard in the course of conservations, particularly when decisions are taken by people from outside, not understanding the context of the area. Platforms must be developed to permit conservationists, residents, and other stakeholders to enter a dialogue with one another, thereby leading to conservation strategies that preserve not only the physical integrity of heritage sites but also cultural traditions and community values as determinants for giving meaning to such sites.

4.4 Integrating Traditional Techniques with Modern Urban Design

Integration of traditional building techniques with modern urban design is one of the most challenging aspects of climate-responsive heritage conservation. Most of the historical buildings were conceived and designed based on climate resilience. They relied on passive cooling systems, natural ventilation, and local materials that suited the environmental context. In contrast, modern urban design places other needs as a priority, including energy efficiency, building codes, and aesthetics. As (Kapoor, 2024) notes, traditional and modern methods of urban planning are sometimes at odds. A good example is how some modern buildings depend on air conditioning and insulation for energy efficiency while others may depend on natural ventilation and thick walls for the same effect. The older passive strategies have been adopted according to the local climate and conditions and can be more sustainable than some of the modern approaches.

This requires a solid understanding of traditional building techniques while also finding ways to integrate modern technologies.

4.5 Breaking Down Barriers in Cross-Disciplinary Collaboration

Effective climate-responsive heritage conservation requires collaboration with various experts. Communication and collaboration from various disciplines, however, can prove to be one of the major barriers to success in successful conservation (Curran, 2021). Researchers in heritage science, architecture, urban planning, sociology, and environmental science have different languages, models, and methodologies, which can easily create confusion and hinder progress. While the conservationist may study the restoration of buildings, the environmental scientist perhaps measures long-term sustainability in local ecosystems. Each approach is valuable, but in the absence of effective coordination and communication, it can also result in broken and disjointed efforts. Interdisciplinary collaborations must be encouraged with a culture of fruitful collaboration. This will be achieved through the designing of joint projects on research, workshops, and conservation, bringing together people from different fields.

To successfully break down barriers in cross-disciplinary collaboration, it is essential to promote a shared understanding of the value of each discipline in addressing complex heritage conservation challenges. The communication channels should be kept open and on board, and through joint workshops and collaboration platforms, the stakeholders should be involved at the decision-making level, and their expertise should be used up to the fullest limit. This develops effective and sustainable strategies that promote long-term resilience in heritage conservation efforts.

4.6 Institutional and Policy Barriers to Effective Implementation

In implementing climate adaptation strategies for heritage conservation, one major challenge will be institutional and policy barriers. Several policy frameworks about heritage conservation have not yet come of age to deal with the realities of climate change. Laws and regulations on heritage conservancy may be more focused on preserving heritage in its pristine state and do not provide the flexibility to change due to new environmental realities. Even government agencies and heritage organizations work in silos, with less coordination between departments or between local, national, and international bodies. This creates a lack of alignment that will make it tough to implement adaptation measures for climate change and also delay much-needed action.

Therefore, new policies on heritage conservation have to take precedence that is aligned with a more climate resilient view. What it calls for, therefore, is a shift in thought. What had been earlier thought of as cultural heritage as preserved, exactly how they are found now, has to shift towards embracing a dynamic aspect, where they would be adaptable in response to challenges brought on by climate factors. Institutional reform and coordination between all government levels and heritage organizations are also important to facilitate an enabling environment for climate-responsive heritage conservation.

4.7 Critical Shortcomings

(E Asprogerakas A. G., 2021) has talked about institutional barriers such as unclear policies, fragmented governance, and poor coordination which hinder effective climate adaptation for cultural heritage, especially archaeological sites, and that while some frameworks exist, gaps in implementation, crisis management, and monitoring remain critical challenges.

Other considerable factors would be:

1. **Economic Feasibility:** Limited research on the long-term economic viability of adaptive technologies for heritage sites, especially in lower-income and resource-poor regions.
2. **Institutional Inertia and Fragmented Governance:** Lack of detailed studies on overcoming bureaucratic resistance and improving coordination among different institutions and governance levels.
3. **Absence of Standardized Metrics:** No widely accepted frameworks or tools exist to measure the effectiveness of climate adaptation strategies specifically for heritage conservation.
4. **Limited Integration of Digital Technologies:** Emerging digital tools like GIS, remote sensing, and digital archives are underutilized in heritage management, particularly in resource-constrained areas.
5. **Need for Context-Specific Research:** Insufficient investigation into how adaptive strategies can be tailored to diverse cultural, environmental, and socio-economic contexts globally.

5. PROPOSED FRAMEWORK FOR CLIMATE-RESPONSIVE HERITAGE CONSERVATION

5.1 Integrating Climate Science into Heritage Conservation Planning

Heritage climatology is an emerging interdisciplinary field that studies the relationship between climate and heritage conservation. According to (Brimblecombe, 2024), it is an applied science that allows for the prediction and quantification of the effects of climate change pose on cultural heritage. This discipline integrates climate data, climate parameters, and projections to inform decisions in heritage management. It addresses both extreme events like floods or storms and the slower, cyclic climate processes, so it enables a much greater understanding of long-term threats. However, even though the significance of climate impact on heritage is well recognized, Brimblecombe says an integrated, systematic approach is in its infancy. Of course, researchers with backgrounds in heritage conservation and climate science must also work together to define the frameworks that will connect their areas of expertise. Still, without such an organized approach-the significant importance of which cannot be overstated- more challenges arise. Harmonization between both disciplines is paramount in ensuring proper management of climatic risks as well as improving the resilience of heritage conservation with regard to climatic changes like global warming.

Adding climate science to conservation planning involves more than the integration of data; it includes understanding the changeable nature of climate impacts. Heritage sites, being located mostly in vulnerable areas such as coastal zones, mountains, and urban sprawls of cities, are particularly vulnerable to climate-related risks. These accelerated temperatures shifted precipitation patterns and increased frequent extreme weather can speed up the physical degradation process- from the erosion of building materials to the flooding of archaeological sites. Ultimately, by working in tandem with climate scientists and heritage professionals, the conservation strategies would be more responsive not only to the threats existing today but also to those that will emerge tomorrow.

5.2 Risk Assessment and Prioritization

Heritage conservation in the light of growing impacts from climate change forms the foundation based on effective risk assessment. In this regard, (Sevieri, 2020) suggests the development of a multi-hazard risk prioritization framework for estimating the vulnerabilities of cultural heritage assets. The framework for using a multilevel rapid-visual survey (RVS) form for gathering data has been designed specifically to assess seismic and wind-related risks associated with heritage buildings. On the other hand, this framework exceeds mere physical vulnerability and includes the abstract elements upon which cultural value, local importance, and even the social contexts of each site can be drawn. With consideration given to both tangible and intangible elements, the framework comes closer to providing a comprehensive way of understanding the overall resilience of a site. It was effective in 25 heritage buildings tested in Iloilo City, Philippines. Such testing revealed important lessons regarding how various heritage structures should be ranked for risk management. Improvement in the framework for the future would thus entail a more refined determination of wind vulnerability as extreme events continue to grow with the change in climate. This would significantly improve risk management, especially in those coastal and windy areas where cultural heritage sites come under increasing pressure (Adetunji, 2024).

5.3 Adaptive Management and Flexibility

Heritage sites' adaptability is the key to their management. Adaptive management, according to (Della Spina, 2021), plays a crucial role in response to changing conditions over time. Combining multi-criteria analysis with discounted cash flow analysis, we understand that an innovative approach to optimal decision-making may help prioritize investment choices in recovering and reusing historical public heritage. The hybrid approach will balance the heritage sites' historical and cultural value with the monetary and practical constraints of their management in a contemporary context. Such studies also indicate that a public-private partnership is necessary for the sustainable management of heritage assets. By broadening governmental support, public-private partnerships can make heritage conservation more financially viable so that heritage sites will be revitalized and maintained in the long run. Moreover, Della Spina's research suggests that sustainability should lead to heritage management; hence, any conservation effort would be economically viable and environmentally conscious. This helps in the attainment of the more general goals of urban regeneration and adaptive reuse of historic buildings as functional structures.

As climate change continues to alter the conditions under which heritage sites operate, conservation strategies must remain flexible and adaptable. Adaptive management is key in providing a framework on which to respond to unexpected challenges rather than rigid preservation practices that fail to account for ongoing environmental changes. The time has never been more crucial to embrace flexibility in heritage conservation because of ever-quickening climate change and the creation of new approaches able to accommodate evolving risks and opportunities (Montuori, 2024) .

5.4 Sustainable Conservation Practices

It is actually in terms of heritage sustainability that resources have become scarcer and with a continuously changing climate, it threatens the viability of preservation efforts. According to (Frasca, 2024), estimating energy demand in collection facilities is one of the major necessities while proposing sustainable conservation practices where it aims to find cost-effective solutions that do not compromise the integrity of the cultural asset by assessing the energy needs of heritage buildings, particularly in collection facilities. Adjusting microclimate thresholds can serve as a low-cost, high-impact strategy for energy conservation in heritage buildings. This approach not only supports the achievement of sustainable development goals, especially SDG 11.4, which involves the protection of cultural heritage, but it also provides a way to strike a balance between conservation needs and environmental and economic goals. Focusing on energy-efficient strategies will help heritage conservation to be more in line with global sustainability efforts. Frasca's study adds to a growing body of knowledge regarding the ways in which microclimate alterations are capable of significantly saving energy consumption while it is the cheapest and environmentally friendlier choice to preserve the cultural heritage over time.

(Kayan B.A., 2017) also discusses sustainable conservation practices in his study on "Green Maintenance" methodologies, especially for heritage buildings made of laterite stone in Melaka, Malaysia. The study centres on low-carbon repair techniques, which minimize CO_2 emissions by quantifying embodied carbon expenditure. The study concludes that using repair techniques with longevity and low maintenance, such as stone replacement, is the most sustainable option for heritage conservation. This method has conveyed the long-term sustainability benefit, as structures in need of less maintenance over time reduce the carbon footprint associated with frequent repairs. While the article would point to the need for further research into the refinement of the process of laterite stone repair through its life cycle, it would then mean that it is quite possible to implement eco-friendly methods within heritage conservation methods, especially those involving resources that become taken into careful consideration.

(Santosuosso, 2022) highlights flexibility as a key aspect of risk management and sustainable conservation practices. The study suggests a sequential approach for developing a risk index (RI) to guide conservation interventions on cultural heritage assets prior to decisions on bounded rationality in preference over perfect optimization. Using a 17th-century church in Rome, which houses Bernini's *Ecstasy of Saint Teresa*, as a case study, the methodology allows managers to create an adaptable risk index by estimating its determinants under varying levels of certainty, risk, or uncertainty. This approach allows heritage managers to make informed decisions even with partial data and promotes the early identification of risks before irreversible damage occurs. However, further studies can improve its precision with better data sets and structured questionnaires.

In Venice, Italy, innovative measures such as the MoSE flood barriers and the strategic management of inlet closures have been proposed to mitigate the impact of rising water levels and ensure the continuity of maritime activities. (Cavallaro, 2017) explores a numerical approach aimed at minimizing interference between the MoSE flood barriers and maritime traffic during flooding events in Venice. By leaving one of the three inlets open during events with non-extreme water levels, vessel transit can continue while still maintaining acceptable water levels in the lagoon. The study finds that partial closure of the MoSE is viable for events with water levels under 130 cm MZPS but requires adjustments to the current closure criteria. Therefore, this method can be applied in similar coastal areas, though further field tests are needed to confirm its effectiveness.

5.5 Collaboration with Global Networks

The challenges of heritage conservation in the urban environment are complex, and international cooperation needs to be realized to overcome the problems. Public authorities and governance structures play diverse roles in their respective regions with respect to setting up conservation practices (Holma, 2021). They have researched five city case studies, both the Global North and South: London, Paris, Milan, Beirut, and Rio de Janeiro, to examine how local priorities and political dynamics shape decisions around heritage conservation. Cases are selected to reflect the tension between the intrinsic cultural value of heritage and imperatives of urban growth that usually place economic development ahead of preservation.

Spatial regulations and heritage designations are usually region-specific and may be the product of regional political climates, economic priorities, and the capacity for enforcement. In essence, successful heritage conservation can only be achieved if all competing priorities are struck and heritage sites are protected for posterity; this is possible only through international collaboration where the experiences and lessons from one region inform best practices for heritage conservation in another. Hence, Holma's work presents the urgent global cooperation involving heritage professionals with policymakers towards further sustainable, flexible, and sensitive conservation strategies within the heritage in our cities.

Global networks have played a much more extensive role than just urban conservation. According to (Heravi, 2021), "the cultural heritage sector has always been interested in developing knowledge and metadata sharing and interoperability played an extended role." The fact that they have adopted linked data technology more and more within the sector is evidence of this change. In her study, Heravi proved that universities and national institutions lead in the implementations of linked data projects, therefore providing a facility through which it will be possible for them to share resources as well as accelerate the research they have been doing. Smaller ones may struggle because of cost and technical restrictions. Although the principles of linked data itself, if adopted by the cultural heritage sector, include collaboration, discoverability, and knowledge sharing, it is very promising to enhance the efficiency of heritage conservation overall in many countries.

This is from the fact that, over recent years, this role has exceeded just heritage site preservation to address shared knowledge as well as a need for the joint problem. International forums and conferences now give room for presenting best practices in case studies or research findings, among other issues such as cross-border partnerships. Networking has further established mutual understanding creating an avenue in which cities would have opportunities facing similar challenges towards collaborating with solutions. Such partnerships can convert global wisdom into local practices and strategies considering regional cultural, social, and economic contexts with regard to their sustainable heritage preservation in the long term. Collaboration among heritage professionals and global networks has become both strategically necessary as well as the necessary mechanism in itself for conservation practices to gain maximum effectiveness everywhere around the globe.

5.6 Economic and Institutional Mechanisms

Economic mechanisms such as heritage adaptation funds, green bonds, and tax incentives for conservation investments can enhance the financial sustainability of climate-responsive strategies. Governments can collaborate with international development agencies to mobilize climate finance specifically targeted at heritage preservation. (Casey, 2019) argues that institutional barriers within the National Park Service hinder effective climate change adaptation for coastal cultural resources. (Alexandrakis, 2019) talks about cultural heritage sites that generate significant socioeconomic benefits—especially through tourism, and that protecting them from climate-induced natural hazards requires integrated strategies that inform policy and raise public awareness. Sustainable economic resilience for climate-vulnerable small island developing states can be achieved by integrating inclusive governance, urban heritage conservation, and strategic public-private partnerships to support debt repayment and urban regeneration (Allam, 2019).

Simultaneously, institutional mechanisms like the creation of integrated heritage-climate task forces, inter-ministerial coordination units, and decentralized conservation authorities can ensure that policy implementation is agile and context-sensitive. By embedding conservation within broader urban development and climate adaptation agendas, institutional reforms can bridge gaps between heritage science, governance, and financial planning. (Xiao Xiao, 2024)proposes a machine learning–based adaptation framework that helps coastal tourism destinations balance tangible and intangible cultural heritage preservation under fiscal constraints, effectively addressing funding challenges and enabling cost-effective, data-driven climate adaptation strategies. Cost-Effectiveness Analysis (CEA) can guide financially sound decision-making in cultural heritage preservation by evaluating conservation strategies without monetizing cultural value, using a climate-risk case at a Roman archaeological site has been described by (Revez, 2019).

6. CONCLUSION

Climate change is progressively changing our world, and impacts on the preservation of cultural heritage pose unprecedented challenges. In this respect, this discussion proceeds to identify some of the increasingly growing needs for climate-resilient approaches towards heritage preservation. It, thereby, emphasizes incorporating modern technology and, adaptive management practices and sustainable solutions for protecting the cultural site. If we observe, from sea level rise to extreme weather events, this threatens the roots of heritage towns and their cultural infrastructure, thus requiring effective management and long-term planning.

In any case, one can trace actual environmental changes using these technologies for the forecasting and prioritization of future risks actions to safeguard, both tangible as well as intangible elements from heritage sites. Another essential thing, aspect is an excellent risk assessment prioritization framework because it serves as an indispensable tool to classify cultural heritage into those with huge risks attached thereto. These frameworks should, therefore, consider three other types of threats: environmental, structural, and cultural risks associated with each location. Systematic consideration of these factors will enable policymakers to prioritize those heritage sites that need urgent action and provide corresponding resources. In this way, it addresses both immediate threats to heritage assets as well as its long-term sustainability, making it both efficient and targeted. Communities play a huge role in the conservation of heritage. Working with local populations brings ownership and a sense of responsibility. This ensures that strategies used for conservation are both technically sound and culturally relevant. It is through active involvement that the conservation of heritage towns is strengthened in a more participatory and sustainable manner. Sustainable forms of conservation strengthen long-term climate change problems as a backbone. Low carbon-repair techniques and energy-efficient solutions reduce environmental impacts in conserving heritage sites by keeping them lively for future generations. Another point would be integration with climate-resilient infrastructure to be taken in urban planning; the real nature and functional role of the heritage towns shall still be achieved through the infrastructures designed, which at the same time preserve and give way for community growth without their cultural heritage loss.

Another aspect will be that coordination between sectors becomes more important for the effective management of heritage. For this reason, conservationists, government organizations, local government authorities, and even international institutions need to convene and evolve coherent, climate-sensitive strategies to demonstrate shared responsibilities towards the preservation of cultural property. This collective effort calls for breaking silos between different disciplines and sectors and, through the exchange of knowledge and expertise, developing more adaptable solutions. Beyond this, digital tools and linked data are increasingly integrated into heritage management and open up new possibilities for global collaboration that will enhance resource sharing, research, and best practices. Heritage conservation must look forward to a future built on adaptability, as climate change is an ongoing, dynamic challenge, and the strategies to manage cultural heritage need to be flexible enough to adapt to the shifting environmental conditions, both anticipated and unforeseen. Commonly, it would demand continuous observation, risk evaluation, and sustainability for a longer period to enhance the resilience of heritage sites. Adaptation of strategies in conservation to assess and sustain the cultural asset would protect it without losing its authenticity or significance. The protection of cultural heritage in a climate-changed world demands a new perspective. Heritage is not a timeless, frozen thing; it is a living, adaptive piece of our total heritage. The systems which we will adopt to preserve these sites must be aware of the fact: to protect and to adapt in tandem. Innovation creatively combined with collaborative arrangements and flexibility will ultimately ensure that our cultural heritage remains pride, identity, and resilience for generations to come.

Addressing these impacts of climate change on cultural heritage takes more than saving buildings or monuments; it saves the very essence of what these places are known.

Ultimately, preserving cultural heritage in a climate-altered world is not merely about protecting buildings or monuments — it is about protecting identity, memory, and meaning. This calls for a holistic, inclusive, and future-oriented approach that considers environmental dynamics, community needs, and sustainable development. Through collective action, innovation, and adaptive governance, the world's cultural heritage can be protected while remaining vibrant and resilient for generations to come.

REFERENCES

Ababneh, A. (2016). eritage Management and Interpretation: Challenges to Heritage Site-Based Values, Reflections from the Heritage Site of Umm Qais, Jordan. *Archaeologies, 12*(1), 38–72. DOI: 10.1007/s11759-016-9290-6

Abdelaziz Elfadaly, W. A. (2018). Management of Cultural Heritage Sites Using Remote Sensing Indices and Spatial Analysis Techniques. *Surveys in Geophysics.*

Abdelrazek Elnaggar, M. S. B. (2024). Risk analysis for preventive conservation of heritage collections in Mediterranean museums: case study of the museum of fine arts in Alexandria (Egypt). Heritage Science.

Adetunji, O., & Daly, C. (2024). Climate Risk Management in Cultural Heritage for Inclusive Adaptation Actions in Nigeria. *Heritage, 7*(3), 1237–1264. DOI: 10.3390/heritage7030060

Agapiou, A., Lysandrou, V., Alexakis, D. D., Themistocleous, K., Cuca, B., Argyriou, A., Sarris, A., & Hadjimitsis, D. G. (2015). Cultural heritage management and monitoring using remote sensing data and GIS: The case study of Paphos area, Cyprus. *Computers, Environment and Urban Systems, 54*, 230–239. DOI: 10.1016/j.compenvurbsys.2015.09.003

Alessandra Gandini, L. G. (2016). Towards Sustainable Historic Cities: Adaptation to Climate Change Risks. The International Journal of Entrepreneurship and Sustainability Issues.

Alexandra Keller, N. C. (2016). *Resilience of historic cities and adaptation to climate change*. Urbanism. Architecture. Constructions.

Alexandrakis, G. M. (2019). Economic and Societal Impacts on Cultural Heritage Sites, Resulting from Natural Effects and Climate Change. Heritage, 279-305.

Allam, Z. &. (2019). Climate Change and Economic Resilience through Urban and Cultural Heritage: The Case of Emerging Small Island Developing States Economies. Economies, 62.

Ameneh Ahmadi, H. M. (2017). Land Suitability Evaluation to Determine the Appropriate Areas of Development: A Case Study of Hormuz Island. *Open Journal of Ecology, 7*(9), 518–534. DOI: 10.4236/oje.2017.79035

Anna Maria Colavitti, A. U. (2019). Applying the HUL approach to walled towns of Mediterranean seaport cities. Journal of Place Management and Development. Retrieved from DOI: 10.1108/JPMD-03-2018-0025

Arrighi, C., Brugioni, M., Castelli, F., Franceschini, S., & Mazzanti, B. (2018). *Flood risk assessment in art cities: The exemplary case of Florence*. Flood Risk Management.

Asprogerakas, E., A. G. (2021). The gap of cultural heritage protection with climate change adaptation in the context of spatial planning. The case of Greece. *International Conference on Environmental Design*. DOI: 10.1088/1755-1315/899/1/012022

Asprogerakas, E. A. G. (2021). The gap of cultural heritage protection with climate change adaptation in the context of spatial planning. The case of Greece. IOP Conf. Ser.: Earth Environ. Sci. IOP Publishing Ltd.

Baghel, A. (2022). GIS-based Industrial Land Suitability Analysis for locating Industrial Parks in Raipur and Nava Raipur. *IOP Conference Series. Earth and Environmental Science*, *1032*(1), 012024. DOI: 10.1088/1755-1315/1032/1/012024

Baldwin, M. F., Fox, A., Klondike, T., Hovis, M., Shear, T., Joca, L., Hester, M., & Cubbage, F. (2022). Geospatial Analysis and Land Suitability for "FloodWise" Practices: Nature-Based Solutions for Flood Mitigation in Eastern, Rural North Carolina. *Land (Basel)*, *11*(9), 1504. DOI: 10.3390/land11091504

Barone, G., Vardopoulos, I., Attia, S., & Vassiliades, C. (2024). Optimizing energy-efficient building renovation: Integrating double-skin façades with solar systems in the Mediterranean landscape. *Energy Reports*, *12*, 2933–2945. DOI: 10.1016/j.egyr.2024.08.032

Bertolin, C. (2019). Preservation of Cultural Heritage and Resources Threatened by Climate Change. *Geosciences*, *9*(6), 250. Advance online publication. DOI: 10.3390/geosciences9060250

Bethune Carmichael, G. W. (2017). Testing the scoping phase of a bottom-up planning guide designed to support Australian Indigenous rangers manage the impacts of climate change oncultural heritage sites. The International Journal of Justice and Sustainability.

Brimblecombe, P. R., & Richards, J. (2024). Applied climatology for heritage. *Theoretical and Applied Climatology*, *155*(8), 7325–7333. DOI: 10.1007/s00704-024-05059-6

Casey, A., & Becker, A. (2019). Institutional and Conceptual Barriers to Climate Change Adaptation for Coastal Cultural Heritage. *Coastal Management*, *47*(2), 169–188. DOI: 10.1080/08920753.2019.1564952

Cavallaro, L. I., Iuppa, C., & Foti, E. (2017). Effect of Partial Use of Venice Flood Barriers. *Journal of Marine Science and Engineering*, *5*(4), 58. DOI: 10.3390/jmse5040058

Coletti, C. (2024). *Climate Change Threats to Stone Cultural Heritage: State of the Art of Quantitative Damage Functions and New Challenges for a Sustainable Future*. Heritage., DOI: 10.3390/heritage7060154

Crowley, K. J., Jackson, R., O'Connell, S., Karunarthna, D., Anantasari, E., Retnowati, A., & Niemand, D. (2022). Cultural heritage and risk assessments: Gaps, challenges, and future research directions for the inclusion of heritage within climate change adaptation and disaster management. *Climate Resilience and Sustainability*, *1*(3), e45. Advance online publication. DOI: 10.1002/cli2.45

Curran, K., & Zimmermann, N. (2021). The Dynamics of Collaboration in Heritage Science. *Studies in Conservation*, *67*(1-2), 136–149. DOI: 10.1080/00393630.2021.1875175

Danilo Marco Campanaro, G. L.-M. (2015). 3D GIS for cultural heritage restoration: A 'white box' workflowDanilo. Journal of Cultural Heritage.

Della Spina, L. (2021). Cultural Heritage: A Hybrid Framework for Ranking Adaptive Reuse Strategies. *Buildings (Basel, Switzerland)*, *11*(3), 132. Advance online publication. DOI: 10.3390/buildings11030132

Erica Avrami, J. L. (2021). Energy and historic buildings: Toward evidence-based policy reform. *Journal of Cultural Heritage Management and Sustainable Development*.

Erin Seekamp, E. J. (2020). Resilience and transformation of heritage sites to accommodate for loss and learning in a changing climate. *Climatic Change*, *162*(1), 41–55. DOI: 10.1007/s10584-020-02812-4

Frasca, F. V., Verticchio, E., Bosco, E., Kuka, E., Lee, D. S.-H., Andersen, C. K., Bertolin, C., & Siani, A. M. (2024). Assessing microclimate thresholds for heritage preventive conservation to achieve sustainable and energy efficiency goals in a changing climate. *Scientific Reports*, *14*(1), 18707. Advance online publication. DOI: 10.1038/s41598-024-69395-3 PMID: 39134605

Georgian Ionu PURCAREATA, M. B. (2015). A model of land suitability general analysis for new infrastructure projects in the Barsa country (Romania). Forum Geographic.

Getzner, M. (2024). Cultural infrastructure as part of social infrastructure: perspectives of cultural policy and economics. In *A.-T. R. Getzner, Handbook of Social Infrastructure* (pp. 232–248). Edward Elgar Publishing. DOI: 10.4337/9781800883130.00024

Giglio, F., Frontera, P., Malara, A., & Armocida, F. (2024). Materials and Climate Change: A Set of Indices as the Benchmark for Climate Vulnerability and Risk Assessment for Tangible Cultural Heritage in Europe. *Sustainability (Basel)*, *16*(5), 2067. Advance online publication. DOI: 10.3390/su16052067

Giuseppe Forino, J. M. (2016). A proposed assessment index for climate change-related risk for cultural heritage protection in Newcastle (Australia). *International Journal of Disaster Risk Reduction*, *19*, 235–248. Advance online publication. DOI: 10.1016/j.ijdrr.2016.09.003

Heravi, E. D. (2021). Linked Data and Cultural Heritage: A Systematic Review of Participation, Collaboration. [JOCCH]. *Journal on Computing and Cultural Heritage*, *14*(2), 18.

Hiswara, A. A., Aziz, A. M., & Pujowati, Y. (2023). Cultural Preservation in a Globalized World: Strategies for Sustaining Heritage. *West Science Social and Humanities Studies*, *1*(03), 98–106. DOI: 10.58812/wsshs.v1i03.250

Jen Heathcote, H. F. (2017). Predicting and Adapting to Climate Change: Challenges for the Historic Environment. *The Historic Environment*, *8*(2), 89–100. Advance online publication. DOI: 10.1080/17567505.2017.1317071

Julio Garrote, A. D.-H. (2020). A Framework Proposal for Regional-Scale Flood-Risk Assessment of Cultural Heritage Sites and Application to the Castile and León Region (Central Spain). *Water (Basel)*, *12*(2), 329. Advance online publication. DOI: 10.3390/w12020329

Kapoor, L. E. (2024). Climate Responsive Strategies in Vernacular Architecture: A Comparative Analysis at Various Latitude. *International Research Journal of Multidisciplinary Scope*, *5*(4), 1047–1068. DOI: 10.47857/irjms.2024.v05i04.01637

Kayan, B.A., H. I. (2017). Green Maintenance for Heritage Buildings: Low Carbon Repair Appraisal Approach on Laterite Stones. *Chemical Engineering Transactions*, *56*, 337–342. DOI: 10.3303/CET1756057

Kurdi, S. M. (2017). Heritage conservation and urban development: A supporting management model for the effective incorporation of archaeological sites in the planning process. *Journal of Cultural Heritage*, *28*, 117–128. Advance online publication. DOI: 10.1016/j.culher.2017.05.007

Laura Quesada-Ganuza, L. G. (2021). Do we know how urban heritage is being endangered by climate change? A systematic and critical review. *International Journal of Disaster Risk Reduction*, *65*, 65. DOI: 10.1016/j.ijdrr.2021.102551

Marius Mosoarca, A. I. (2017). Failure analysis of historical buildings due to climate change. *Engineering Failure Analysis*, *82*, 666–680. DOI: 10.1016/j.engfailanal.2017.06.013

Marvin Ravan, M. J. (2023). A Vulnerability Assessment Framework for Cultural Heritage Sites: The Case of the Roman Ruins of Tróia. *International Journal of Disaster Risk Science*, *14*(1), 26–40. Advance online publication. DOI: 10.1007/s13753-023-00463-4

Michael, C., & Hall, T. B. (2015). Climate change and cultural heritage: Conservation and heritage tourism in the Anthropocene. *Journal of Heritage Tourism*.

Mohammed ElAdl, F. F. (2022). Managing microclimate challenges for museum buildings in Egypt. *Ain Shams Engineering Journal*.

Montuori, M. (2024). Bridging past and present: Cutting-edge technologies for predictive conservation of built cultural heritage. *The International Archives of the Photogrammetry, Remote Sensing and Spatial Information Sciences*, *XLVIII-4/W10-2024*, 139–145. DOI: 10.5194/isprs-archives-XLVIII-4-W10-2024-139-2024

Muthukumar, J. (2017). CONSERVATION OF HERITAGE MONUMENTS FROM ENVIRONMENTAL CHANGE: A CASE STUDY. Research Directions.

Nicolini, E. (2024). Climate change adaptation and mitigation and historic centers preservation. Underway and repeatable technological design solutions. *Cities (London, England)*, *152*, 152. DOI: 10.1016/j.cities.2024.105174

Oladeji, S. O., Grace, O., & Ayodeji, A. A. (2022). Community Participation in Conservation and Management of Cultural Heritage Resources in Yoruba Ethnic Group of South Western Nigeria. *SAGE Open*, *12*(4), 21582440221130987. Advance online publication. DOI: 10.1177/21582440221130987

Pioppi, B. I. P. (2019). Cultural heritage microclimate change: Human-centric approach to experimentally investigate intra-urban overheating and numerically assess foreseen future scenarios impact. *The Science of the Total Environment*. Advance online publication. DOI: 10.1016/j.scitotenv PMID: 31757533

Pougkakioti, V. M., Lazoglou, M., & Maistrou, E. (2023). Climate Change vs Cultural Heritage: An Adaptation Strategy for the Archaeological Site of Ancient Messene. *American Journal of Climate Change*, *12*(3), 456–488. Advance online publication. DOI: 10.4236/ajcc.2023.123021

Randazzo, L., Collina, M., Ricca, M., Barbieri, L., Bruno, F., Arcudi, A., & La Russa, M. (2020). *Damage Indices and Photogrammetry for Decay Assessment of Stone-Built Cultural Heritage: The Case Study of the San Domenico Church Main Entrance Portal*. Sustainability.

Reuter, J., Hellmund, T., Moßgraber, J., & Hertweck, P. (2023). The KERES Ontology: Protecting Cultural Heritage from Extreme Climate Events. Heritage., 10.3390.

Revez, M. J. (2019). Analysing the Cost-Effectiveness of Heritage Conservation Interventions: A Methodological Proposal within Project STORM. *International Journal of Architectural Heritage*, •••, 985–999.

Ribeiro Hoffmann, A., Sandrin, P., & Doukas, Y. E. (Eds.). (2024). Climate change in regional perspective. European Union and Latin American initiatives, challenges, and solutions. Cham, Switzerland: United Nations University Series on Regionalism, Vol. 27, Springer Nature. DOI: 10.1007/978-3-031-49329-4

Riccardo Cacciotti, A. K. (2021). Climate change-induced disasters and cultural heritage: Optimizing management strategies in Central Europe. *Climate Risk Management*.

Sandra Fatoric, E. S. (2017). Securing the Future of Cultural Heritage by Identifying Barriers to and Strategizing Solutions for Preservation under Changing Climate Conditions. *Sustainability (Basel)*, *9*(11), 2143. Advance online publication. DOI: 10.3390/su9112143

Sandra Fatoric, L. E. (2020). Realising the potential of cultural heritage to achieve climate change actions in the Netherlands. *Journal of Environmental Management*. PMID: 32791326

Santosuosso, P. (2022). A satisfaction-based model for risk indexing in cultural heritage conservation. *Journal of Cultural Heritage*, *57*, 173–183. DOI: 10.1016/j.culher.2022.08.007

Sateriano, A., D'Agata, A., & Vardopoulos, I. (Eds.). (2024). *Socio-environmental change: the path toward sustainability*. Nova Science Publishers., DOI: 10.52305/NZKT5653

Sevieri, G. G., Galasso, C., D'Ayala, D., De Jesus, R., Oreta, A., Grio, M. E. D. A., & Ibabao, R. (2020). A multi-hazard risk prioritisation framework for cultural heritage assets. *Natural Hazards and Earth System Sciences*, *20*(5), 1391–1414. DOI: 10.5194/nhess-20-1391-2020

Shanwen Zheng, B. H. (2018). Ecological Wisdom and Inspiration Underlying the Planning and Construction of Ancient Human Settlements: Case Study of Hongcun UNESCO World Heritage Site in China. *Sustainability*.

Sombat Prajonsant, V. F. (2018). *THE STUDY OF THERMAL COMFORT OF I-SAN VERNACULAR BUDDHIST HOLY TEMPLES (SIMS) IN BURIRAM PROVINCE. Humanities*. Arts and Social Sciences Studies.

T K, G. K.T K. (2024). Valorization of intangible cultural heritage through documentation: An Indian scenario. *Library Hi Tech News*, *41*(8), 10–14. Advance online publication. DOI: 10.1108/LHTN-02-2024-0031

Tianchen Dai, X. Z. (2022). A systematic review of studies at the intersection of urban climate and historical urban landscape. *Environmental Impact Assessment Review*, *97*, 106894. Advance online publication. DOI: 10.1016/j.eiar.2022.106894

Tsilika, E., & Vardopoulos, I. (2022). The FIX-up mix-up; undue façadism or adaptive reuse? Examining the former FIX brewery transformation into the National Museum of Contemporary Art in Athens. *Archnet-IJAR: International Journal of Architectural Research*, *16*(3), 688–709. DOI: 10.1108/ARCH-09-2021-0255

Vardopoulos, I. (2022). Industrial building adaptive reuse for museum. Factors affecting visitors' perceptions of the sustainable urban development potential. *Building and Environment*, *222*, 109391. DOI: 10.1016/j.buildenv.2022.109391

Vardopoulos, I. (2023). Adaptive reuse for sustainable development and land use: A multivariate linear regression analysis estimating key determinants of public perceptions. *Heritage*, *6*(2), 809–828. DOI: 10.3390/heritage6020045

Vardopoulos, I. (2024). Urban planning policies and tools for sustainable development: insights into adaptive reuse, smart cities, and urban metabolism. In *Socio-environmental change: The path toward sustainability* (pp. 121–140). Nova Science Publishers.

Vardopoulos, I., D'Agata, A., Escrivà Saneugenio, F., & Salvati, L. (2024). *Sprawl and the city: realizing a sustainable Mediterranean urbanization*. Nova Science Publishers., DOI: 10.52305/GTVV1244

Vardopoulos, I., Ioannides, S., Georgiou, M., Voukkali, I., Salvati, L., & Doukas, Y. E. (2023). Shaping sustainable cities: A long-term GIS-emanated spatial analysis of settlement growth and planning in a coastal Mediterranean European city. *Sustainability (Basel)*, *15*(14), 11202. DOI: 10.3390/su151411202

Vardopoulos, I., Tsilika, E., Sarantakou, E., Zorpas, A. A., Salvati, L., & Tsartas, P. (2021). An integrated SWOT-PESTLE-AHP model assessing sustainability in adaptive reuse projects. *Applied Sciences (Basel, Switzerland)*, *11*(15), 7134. DOI: 10.3390/app11157134

Wang, K., & Fouseki, K. (2025). Sustaining the Fabric of Time: Urban Heritage, Time Rupture, and Sustainable Development. *Land (Basel)*, *14*(1), 193. Advance online publication. DOI: 10.3390/land14010193

Webb, A. L. (2017). Energy retrofits in historic and traditional buildings: A review of problems and methods. *Renewable & Sustainable Energy Reviews*, *77*, 748–759. Advance online publication. DOI: 10.1016/j.rser.2017.01.145

Xiao Xiao, P. L. (2024). Sustainable Adaptation Planning for Cultural Heritage in Coastal Tourism Destinations Under Climate Change: A Mixed-Paradigm of Preservation and Conservation Optimization. *Journal of Travel Research*, *63*(1), 215–233. DOI: 10.1177/00472875221143479

Yu, J., Safarov, B., Wang, C., Buzrukova, M., & Janzakov, B. (2023). The Effect of Transportation Networks on Heritage Tourism and New Urbanization—Empirical Research Based on Rich Heritage Sites in a Chinese Province. Heritage, 7293–7315. Retrieved from DOI: 10.3390/heritage6120383

Compilation of References

"Case Study on Smart City Projects in India: An Analysis of Nagpur, Allahabad and Dehradun." https://dl.acm.org/doi/fullHtml/10.1145/3184558.3191522 (October 18, 2024).

"Digital Skills: Succeeding in a Digital World: Week 2: 2 | OpenLearn - Open University." https://www.open.edu/openlearn/mod/oucontent/view.php?id=105154&section=2 (October 18, 2024).

"Embracing the Future: The Significance of Open Standards in Technology - ETHAR™ | The XR Experience Platform." https://www.ethar.com/2024/01/01/embracing-the-future-the-significance-of-open-standards-in-technology/ (October 18, 2024).

"Insights Into Cities in The Digital Era: A People-Centered Approach to Smart Cities." https://www.cdpp.co.in/articles/insights-into-cities-in-the-digital-era-a-people-centered-approach-to-smart-cities (October 18, 2024).

"Public-Private Partnerships (PPPs): Definition, How They Work, and Examples." https://www.investopedia.com/terms/p/public-private-partnerships.asp (October 18, 2024).

"RKMA Market Research Reports." https://www.rkma.com/SFPL/cuim24/ (October 18, 2024).

"Smart Cities: Big Data, Civic Hackers, and the Quest for a New Utopia by Anthony M. Townsend | SMART CITIES: Big Data, Civic Hackers, and the Quest for a New Utopia." https://ssir.org/books/excerpts/entry/smart_cities_big_data_civic_hackers_and_the_quest_for_a_new_utopia (November 14, 2024).

"The United Indian | Government Sector | Strategies for Enhancing Digital Literacy in India." https://theunitedindian.com/news/blog?Digital-literacy-in-India&b=193&c=3 (October 18, 2024).

"Top 10: Smart Cities Driving Sustainability around the World | Sustainability Magazine." https://sustainabilitymag.com/top10/top-10-smart-cities-around-the-world (October 18, 2024).

"What Are the Cybersecurity Risks for Smart Cities? - Institute for Defense and Business." https://www.idb.org/what-are-the-cybersecurity-risks-for-smart-cities/ (October 18, 2024).

Ababneh, A. (2016). eritage Management and Interpretation: Challenges to Heritage Site-Based Values, Reflections from the Heritage Site of Umm Qais, Jordan. *Archaeologies, 12*(1), 38–72. DOI: 10.1007/s11759-016-9290-6

Abanades, S., H. Abbaspour, A. Ahmadi, B. Das, M. Ehyaei, F. Esmaeilion, M. El Haj Assad, T. Hajilounezhad, D. Jamali and A. Hmida (2022). "A critical review of biogas production and usage with legislations framework across the globe." International Journal of Environmental Science Technology in Society: 1-24.

Abdelaziz Elfadaly, W. A. (2018). Management of Cultural Heritage Sites Using Remote Sensing Indices and Spatial Analysis Techniques. *Surveys in Geophysics.*

Abdelrazek Elnaggar, M. S. B. (2024). Risk analysis for preventive conservation of heritage collections in Mediterranean museums: case study of the museum of fine arts in Alexandria (Egypt). Heritage Science.

Abhilash, N., Shenoy, S. S., Shetty, D. K., Lobo, L. S., & N, S. K. (2023). Green Bond as an Innovative financial instrument in the Indian Financial Market: Insights from Systematic Literature Review Approach. *SAGE Open, 13*(2), 21582440231178783. DOI: 10.1177/21582440231178783

Acampora, L., Grilletta, S., & Costa, G. J. E. (2025). The Integration of Carbon Capture, Utilization, and Storage (CCUS) in Waste-to-Energy Plants: A Review. *Energies, 18*(8), 1883. DOI: 10.3390/en18081883

Adamkiewicz, J., Kochańska, E., Adamkiewicz, I., & Łukasik, R. M. (2022). Greenwashing and sustainable fashion industry. *Current Opinion in Green and Sustainable Chemistry, 38*, 100710. DOI: 10.1016/j.cogsc.2022.100710

Addas, A. (2023). The Concept of Smart Cities: A Sustainability Aspect for Future Urban Development Based on Different Cities. *Frontiers in Environmental Science, 11*, 1241593. DOI: 10.3389/fenvs.2023.1241593

Addy, W. A., Ofodile, O. C., Adeoye, O. B., Oyewole, A. T., Okoye, C. C., Odeyemi, O., & Ololade, Y. J. (2023). *Data-driven sustainability: How fintech innovations are supporting green finance*. ResearchGate.

Adesina, N. A., Okwandu, N. C., & Nwokediegwu, N. Z. Q. S.Abayomi Abraham AdesinaAzubuike Chukwudi OkwanduZamathula Queen Sikhakhane Nwokediegwu. (2024). Towards sustainable urban development: Conceptualizing green infrastructure and its impact on urban planning. *International Journal of Applied Research in Social Sciences*, *6*(7), 1274–1296. DOI: 10.51594/ijarss.v6i7.1266

Adetunji, O., & Daly, C. (2024). Climate Risk Management in Cultural Heritage for Inclusive Adaptation Actions in Nigeria. *Heritage*, *7*(3), 1237–1264. DOI: 10.3390/heritage7030060

Agan, B., & Balcilar, M. (2023). Unraveling the Green Growth Matrix: Exploring the Impact of Green Technology, Climate Change Adaptation, and Macroeconomic Factors on Sustainable Development. *Sustainability (Basel)*, *15*(11), 8530. DOI: 10.3390/su15118530

Agapiou, A., Lysandrou, V., Alexakis, D. D., Themistocleous, K., Cuca, B., Argyriou, A., Sarris, A., & Hadjimitsis, D. G. (2015). Cultural heritage management and monitoring using remote sensing data and GIS: The case study of Paphos area, Cyprus. *Computers, Environment and Urban Systems*, *54*, 230–239. DOI: 10.1016/j.compenvurbsys.2015.09.003

Agaton, C. B., Guno, C. S., Villanueva, R. O., & Villanueva, R. O. (2020). Economic analysis of waste-to-energy investment in the Philippines: A real options approach. *Applied Energy*, *275*, 115265. DOI: 10.1016/j.apenergy.2020.115265

Ahmad, H., Yaqub, M., & Lee, S. H. (2024). Environmental-, social-, and governance-related factors for business investment and sustainability: A scientometric review of global trends. *Environment, Development and Sustainability*, *26*(2), 2965–2987. DOI: 10.1007/s10668-023-02921-x PMID: 36714213

Ahmed, I. I., & Gupta, A. K. (2010). Pyrolysis and gasification of food waste: Syngas characteristics and char gasification kinetics. *Applied Energy*, *87*(1), 101–108. DOI: 10.1016/j.apenergy.2009.08.032

Akbari, H., & Kolokotsa, D. (2016). Three decades of urban heat islands and mitigation technologies research. *Energy and Building*, *133*, 834–842. DOI: 10.1016/j.enbuild.2016.09.067

Alaimo, L. S., Ciommi, M., Vardopoulos, I., Nosova, B., & Salvati, L. (2022). The medium-term impact of the COVID-19 pandemic on population dynamics: The case of Italy. *Sustainability (Basel)*, *14*(21), 13995. DOI: 10.3390/su142113995

Alao, M. A., Popoola, O. M., & Ayodele, T. R. (2022). Waste-to-energy nexus: An overview of technologies and implementation for sustainable development. *Cleaner Energy Systems*, *3*, 100034. DOI: 10.1016/j.cles.2022.100034

Albitar, K., Al-Shaer, H., & Liu, Y. S. (2023). Corporate commitment to climate change: The effect of eco-innovation and climate governance. *Research Policy*, *52*(2), 104697. DOI: 10.1016/j.respol.2022.104697

Alessandra Gandini, L. G. (2016). Towards Sustainable Historic Cities: Adaptation to Climate Change Risks. The International Journal of Entrepreneurship and Sustainability Issues.

Alexandra Keller, N. C. (2016). *Resilience of historic cities and adaptation to climate change*. Urbanism. Architecture. Constructions.

Alexandrakis, G. M. (2019). Economic and Societal Impacts on Cultural Heritage Sites, Resulting from Natural Effects and Climate Change. Heritage, 279-305.

Al-Hiyari, A., Ismail, A. I., Kolsi, M. C., & Kehinde, O. H. (2023). Environmental, social and governance performance (ESG) and firm investment efficiency in emerging markets: The interaction effect of board cultural diversity. *Corporate Governance (Bradford)*, *23*(3), 650–673. DOI: 10.1108/CG-03-2022-0133

Ali, T., Irfan, M., Alwadie, A. S., & Glowacz, A. (2020). IoT-based smart waste bin monitoring and municipal solid waste management system for smart cities. *Arabian Journal for Science and Engineering*, *45*(12), 10185–10198. DOI: 10.1007/s13369-020-04637-w

Allam, Z. &. (2019). Climate Change and Economic Resilience through Urban and Cultural Heritage: The Case of Emerging Small Island Developing States Economies. Economies, 62.

Allen, J. G., MacNaughton, P., Satish, U., Santanam, S., Vallarino, J., & Spengler, J. D. (2016). Associations of cognitive function scores with carbon dioxide, ventilation, and volatile organic compound exposures in office workers: A controlled exposure study of green and conventional office environments. *Environmental Health Perspectives*, *124*(6), 805–812. DOI: 10.1289/ehp.1510037 PMID: 26502459

Almaaitah, T., Appleby, M., Rosenblat, H., Drake, J., & Joksimovic, D. (2021). The potential of blue-green infrastructure as a climate change adaptation strategy: A systematic literature review. *Blue-Green Systems*, *3*(2), 223–248. DOI: 10.2166/bgs.2021.016

Almazrouei, H., & Nobanee, H. (2021). *Green finance and sustainable growth*. ResearchGate.

Al-Qadri, A. A., Ahmed, U., Ahmad, N., Jameel, A. G. A., Zahid, U., & Naqvi, S. R. (2024). A review of hydrogen generation through gasification and pyrolysis of waste plastic and tires: Opportunities and challenges. *International Journal of Hydrogen Energy*, *77*, 1185–1204. DOI: 10.1016/j.ijhydene.2024.06.207

Al-Qawasmi, J., Asif, M., El Fattah, A. A., & Babsail, M. O. (2019). Water effciency and management in sustainable building rating systems: Examining variation in criteria usage. *Sustainability (Basel)*, *11*(8), 2416. Advance online publication. DOI: 10.3390/su11082416

Alqudah, M. Z., García, L. S., & Garcia-Benau, M. A. (2024). *Influence of ESG among Jordanian fintech businesses: The moderating role of sustainable finance technology*. Springer Link., DOI: 10.1007/s43621-024-00712-6

Amado, M., Poggi, F., & Amado, A. R. (2016). Energy efficient city: A model for urban planning. *Sustainable Cities and Society*, *26*, 476–485. DOI: 10.1016/j.scs.2016.04.011

Amelung, B., & Moreno, A. (2012). Costing the impact of climate change on tourism in Europe: Results of the PESETA project. *Climatic Change*, *112*(1), 83–100. DOI: 10.1007/s10584-011-0341-0

Ameneh Ahmadi, H. M. (2017). Land Suitability Evaluation to Determine the Appropriate Areas of Development: A Case Study of Hormuz Island. *Open Journal of Ecology*, *7*(9), 518–534. DOI: 10.4236/oje.2017.79035

Amorim-Maia, A. T., Anguelovski, I., Chu, E., & Connolly, J. J. T. (2023). Governing intersectional climate justice: Tactics and lessons from Barcelona. *Environmental Policy and Governance*. Advance online publication. DOI: 10.1002/eet.2075

Anand, A., & Singh, P. (2025). Interdisciplinary Studies: Issues, Perspectives, and Challenges From National Education Program 2020. *Integrating Personalized Learning Methods Into STEAM Education*, 233-260.

Anna Maria Colavitti, A. U. (2019). Applying the HUL approach to walled towns of Mediterranean seaport cities. Journal of Place Management and Development. Retrieved from DOI: 10.1108/JPMD-03-2018-0025

Antonios, P., Konstantinos, K., & Christos, G. (2023). A Systematic Review on Semantic Interoperability in the IoE-Enabled Smart Cities. *Internet ofThings : Engineering Cyber Physical Human Systems*, *22*, 100754. DOI: 10.1016/j.iot.2023.100754

Antonis Tritsis Metropolitan Park Management Authority. (2019). *Actions and perspectives* [in Greek]. Antonis Tritsis Metropolitan Park.

Antonis Tritsis Metropolitan Park. (2017). *Discovering Antonis Tritsis Metropolitan Park: Education for sustainability – Educational material* [in Greek]. Antonis Tritsis Metropolitan Park.

Antonis Tritsis Metropolitan Park. (2023, March 23). Antonis Tritsis Metropolitan Park of environmental and educational activities and social economy development management authority. Verde Tec [in Greek]. https://www.verde-tec.gr/el/readText.asp?textID=806

Antonis Tritsis Metropolitan Park. (2025, March 20). Description [in Greek]. Antonis Tritsis Metropolitan Park. http://www.parkotritsis.gr/perigrafh/

Apanaviciene, R., Vanagas, A., & Fokaides, P. A. (2020). Smart building integration into a smart city (SBISC): Development of a new evaluation framework. *Energies*, *13*(9), 2190. Advance online publication. DOI: 10.3390/en13092190

Apostu, S. A., Gigauri, I., Panait, M., & Martín-Cervantes, P. A. (2023). Is Europe on the Way to Sustainable Development? Compatibility of Green Environment, Economic Growth, and Circular Economy Issues. *International Journal of Environmental Research and Public Health*, *20*(2), 1078. DOI: 10.3390/ijerph20021078 PMID: 36673838

Aria, M., & Cuccurullo, C. (2017). Bibliometrix: An R-tool for comprehensive science mapping analysis. *Journal of Informetrics*, *11*(4), 959–975. DOI: 10.1016/j.joi.2017.08.007

Arjwech, R., Somchat, K., Pondthai, P., Everett, M., Schulmeister, M., & Saengchomphu, S. (2020). Assessment of geological, hydrogeological and geotechnical characteristics of a proposed waste disposal site: A case study in Khon kaen, Thailand. *Geosciences*, *10*(3), 109. DOI: 10.3390/geosciences10030109

Arrighi, C., Brugioni, M., Castelli, F., Franceschini, S., & Mazzanti, B. (2018). *Flood risk assessment in art cities: The exemplary case of Florence*. Flood Risk Management.

Arsenio, E., Martens, K., & Di Ciommo, F. (2016). Sustainable urban mobility plans: Bridging climate change and equity targets? *Research in Transportation Economics*, *55*, 30–39. DOI: 10.1016/j.retrec.2016.04.008

Asif, M., Searcy, C., & Castka, P. (2022). Exploring the role of industry 4.0 in enhancing supplier audit authenticity, efficacy, and cost effectiveness. *Journal of Cleaner Production*, *331*, 129939. DOI: 10.1016/j.jclepro.2021.129939

Asif, M., Searcy, C., & Castka, P. (2023). ESG and Industry 5.0: The role of technologies in enhancing ESG disclosure. *Technological Forecasting and Social Change, 195*, 122806. DOI: 10.1016/j.techfore.2023.122806

Asprogerakas, E. A. G. (2021). The gap of cultural heritage protection with climate change adaptation in the context of spatial planning. The case of Greece. IOP Conf. Ser.: Earth Environ. Sci. IOP Publishing Ltd.

Asprogerakas, E., A. G. (2021). The gap of cultural heritage protection with climate change adaptation in the context of spatial planning. The case of Greece. *International Conference on Environmental Design*. DOI: 10.1088/1755-1315/899/1/012022

Asprogerakas, E., & Tasopoulou, A. (2024). From sustainability to resilience: Tracing the transition path for spatial planning policy in Greece. *European Spatial Research and Policy, 31*(1), 29–52. DOI: 10.18778/1231-1952.31.1.04

Athens Master Plan Organization. (1995). Environmental Awareness Park of the Queen's Tower Estate: Stage A – Phase B [Study]. Consortium of consultants: Thymios Papayannis and Associates S.A., Ioannis Dokoumetzidis, Ioannis Kolliniatis, Ioannis Alavanos / Christos Athanasopoulos G.P., O.T.M./TEPE, Technical Study Group.

Atiles, J., & Whyte, D. (2025). Fossil Capital in the Caribbean: The Toxic Role of "Regulatory Havens" in Climate Change. *Regulation & Governance, 19*(2), 469–481. Advance online publication. DOI: 10.1111/rego.70001

Bache, I., Bartle, I., & Flinders, M. (2016). Multi-level governance. In *Handbook on Theories of Governance* (pp. 486–498). Edward Elgar Publishing., https://www.elgaronline.com/edcollchap/edcoll/9781782548492/9781782548492.00052.xml DOI: 10.4337/9781782548508.00052

Bach, H., Mäkitie, T., Hansen, T., & Steen, M. (2021). Blending new and old in sustainability transitions: Technological alignment between fossil fuels and biofuels in Norwegian coastal shipping. *Energy Research & Social Science, 74*, 101957. DOI: 10.1016/j.erss.2021.101957

Badhwar, A., Islam, S., Tan, C. S. L., Panwar, T., Wigley, S., & Nayak, R. (2024). Unraveling green marketing and greenwashing: A systematic review in the context of the fashion and textiles industry. *Sustainability (Basel), 16*(7), 2738. DOI: 10.3390/su16072738

Baechler, B. R., De Frond, H., Dropkin, L., Leonard, G. H., Proano, L., & Mallos, N. J. (2024). Public awareness and perceptions of ocean plastic pollution and support for solutions in the United States. *Frontiers in Marine Science, 10*, 1323477. DOI: 10.3389/fmars.2023.1323477

Baghel, A. (2022). GIS-based Industrial Land Suitability Analysis for locating Industrial Parks in Raipur and Nava Raipur. *IOP Conference Series. Earth and Environmental Science, 1032*(1), 012024. DOI: 10.1088/1755-1315/1032/1/012024

Baldi, F., & Pandimiglio, A. (2022). The role of ESG scoring and greenwashing risk in explaining the yields of green bonds: A conceptual framework and an econometric analysis. *Global Finance Journal, 52*, 1044028322000138. DOI: 10.1016/j.gfj.2022.100711

Baldwin, M. F., Fox, A., Klondike, T., Hovis, M., Shear, T., Joca, L., Hester, M., & Cubbage, F. (2022). Geospatial Analysis and Land Suitability for "FloodWise" Practices: Nature-Based Solutions for Flood Mitigation in Eastern, Rural North Carolina. *Land (Basel), 11*(9), 1504. DOI: 10.3390/land11091504

Bambó Naya, R., De la Cal Nicolás, P., Medina, C., Ezquerra, I., Monclús, J., & García-Pérez, S. (2023). Quality of public space and Sustainable Development Goals: Analysis of nine urban projects in Spanish cities. *Frontiers of Architectural Research, 12*(3), 477–495. DOI: 10.1016/j.foar.2023.01.002

Barone, G., Vardopoulos, I., Attia, S., & Vassiliades, C. (2024). Optimizing energy-efficient building renovation: Integrating double-skin façades with solar systems in the Mediterranean landscape. *Energy Reports, 12*, 2933–2945. DOI: 10.1016/j.egyr.2024.08.032

Barrios, S., & Ibañez, J. N. (2015). Time is of the essence: Adaptation of tourism demand to climate change in Europe. *Climatic Change, 132*(4), 645–660. DOI: 10.1007/s10584-015-1431-1

Basdekidou, V., & Papapanagos, H. (2023). Blockchain technology adoption for disrupting FinTech functionalities: A systematic literature review for corporate management, supply chain, banking industry, and stock markets. *ResearchGate*. https://www.researchgate.net/publication/383969110_Blockchain_Technology_Adoption_for_Disrupting_FinTech_Functionalities_A_Systematic_Literature_Review_for_Corporate_Management_Supply_Chain_Banking_Industry_and_Stock_Markets

Basha, S. I., Aziz, A., Maslehuddin, M., Ahmad, S., Hakeem, A. S., & Rahman, M. M. (2020). Characterization, Processing, and Application of Heavy Fuel Oil Ash, an Industrial Waste Material – A Review. *Chemical Record (New York, N.Y.), 20*(12), 1568–1595. DOI: 10.1002/tcr.202000100 PMID: 33078557

Basu, P. (2010). *Biomass gasification and pyrolysis: practical design and theory.* Academic press.

Beatley, T. (Ed.). (2012). *Green Cities of Europe*. Island Press/Center for Resource Economics., DOI: 10.5822/978-1-61091-175-7

Becker, C., Maemura, E., Moles, N., Whyte, J., & Mann, J. (2016). Assessing digital sustainability: The digital sustainability model. Benchmark DP, University of Toronto. Retrieved from https://tspace.library.utoronto.ca/handle/1807/104046

Belavilas, N. (2017, June 7–9). Antonis Tritsis Metropolitan Park: Assessment of the current situation – Strategy, directions, proposals [Conference presentation, in Greek]. Scientific and Environmental Three-Day Event, Ilion, Greece. Antonis Tritsis Metropolitan Park.

Ben Ahmed M., Molland E., Tomasgard T. (2023). Challenges and Opportunities for Adopting Green Technologies in Maritime Transportation Planning. IFIP Advances in Information and Communication Technology, 692 AICT, 620 - 633. DOI: DOI: 10.1007/978-3-031-43688-8_43

Ben Amara, D., & Qiao, J. (2023). From economic growth to inclusive green growth: How do carbon emissions, eco-innovation and international collaboration develop economic growth and tackle climate change? *Journal of Cleaner Production*, *425*, 138986. DOI: 10.1016/j.jclepro.2023.138986

Ben Hmida Bouslah, K. (2015). Responsible investment in Tunisia. In Hebb, T., Hawley, J. P., Hoepner, A. G., Neher, A. L., & Wood, D. (Eds.), *The Routledge Handbook of Responsible Investment* (pp. 125–137). Routledge Taylor & Francis Group.

Ben Noamene, T., Halcro, K., Chaher, D., & Talib, A. (2021). Female directors and firm financial performance: The case of the FTSE 100. *Organization Development Journal*, *39*(1), 59–76.

Benoît Norris, C., Traverso, M., Neugebauer, S., Ekener, E., Schaubroeck, T., Russo Garrido, S., Berger, M., Valdivia, S., Lehmann, A., Finkbeiner, M., & Arcese, G. (2020). *Guidelines for Social Life Cycle Assessment of Products and Organizations*. Life Cycle Initiative.

Beopoulos, N. (2005). Diploma thesis School of Architectural Engineering E.M.P. Athens." "Agriculture produces landscapes." Mercouri Estate.

Beopoulos, N., Papadopoulos, A. (2003). "Desertification as Human Absence and Sterility of Places: An Introduction to a Controversial Phenomenon".

Beretta, I., & Bracchi, C. (2023). Climate-neutral and Smart Cities: A critical review through the lens of environmental justice. *Frontiers in Sociology*, *8*, 1175592. DOI: 10.3389/fsoc.2023.1175592 PMID: 37854356

Bergek, A., Bjorgum, O., Hansen, T., Hanson, J., & Steen, M. (2021). Sustainability transitions in coastal shipping: The role of regime segmentation. *Transportation Research Interdisciplinary Perspectives, 12*, 100497. DOI: 10.1016/j.trip.2021.100497

Berkes, F. (2012). Sacred Ecology. In Routledge eBooks. DOI: 10.4324/9780203123843

Bertolin, C. (2019). Preservation of Cultural Heritage and Resources Threatened by Climate Change. *Geosciences, 9*(6), 250. Advance online publication. DOI: 10.3390/geosciences9060250

Bethune Carmichael, G. W. (2017). Testing the scoping phase of a bottom-up planning guide designed to support Australian Indigenous rangers manage the impacts of climate change oncultural heritage sites. The International Journal of Justice and Sustainability.

Bibri, S. E., Huang, J., Jagatheesaperumal, S. K., & Krogstie, J. (2024). The Synergistic Interplay of Artificial Intelligence and Digital Twin in Environmentally Planning Sustainable Smart Cities: A Comprehensive Systematic Review. *Environmental Science and Ecotechnology, 20*, 100433. DOI: 10.1016/j.ese.2024.100433 PMID: 38831974

Biresselioglu, M. E., Limoncuoglu, S. A., Demir, M. H., Reichl, J., Burgstaller, K., Sciullo, A., & Ferrero, E. (2021). Legal Provisions and Market Conditions for Energy Communities in Austria, Germany, Greece, Italy, Spain, and Turkey: A Comparative Assessment. *Sustainability (Basel), 13*(20), 20. Advance online publication. DOI: 10.3390/su132011212

Biris, K. (2006). The toponyms of the city and the surroundings of Athens [in Greek]. Publications of the Archaeological Bulletin.

Bisinella, V., Nedenskov, J., Riber, C., Hulgaard, T., & Christensen, T. H. (2022). Environmental assessment of amending the Amager Bakke incineration plant in Copenhagen with carbon capture and storage. *Waste Management & Research, 40*(1), 79–95. DOI: 10.1177/0734242X211048125 PMID: 34585637

Bloom, A. (2024). Public land, value capture, and the rise of speculative urban governance in post-crisis London. *Environment and Planning A. Environment & Planning A, 56*(6), 1771–1786. DOI: 10.1177/0308518X231199701

Bonetti, P., Cho, C. H., & Michelon, G. (2023). Environmental disclosure and the cost of capital: Evidence from the Fukushima nuclear disaster. *European Accounting Review, 33*(5), 1693–1721. DOI: 10.1080/09638180.2023.2203410

Bonsinetto, F., & Falco, E. (2013). Analysing Italian regional patterns in green economy and climate change. Can Italy leverage on Europe 2020 strategy to face sustainable growth challenges? *Journal of Urban and Regional Analysis*, *5*(2), 123–142.

Borelli, S., Conigliaro, M., Quaglia, S., & Salbitano, F. (2017). Urban and peri-urban agroforestry as multifunctional land use. In Springer eBooks (pp. 705–724). DOI: 10.1007/978-981-10-7650-3_28

Boretti, A., & Pollet, B. G. (2024). Hydrogen economy: Paving the path to a sustainable, low-carbon future. *International Journal of Hydrogen Energy*, *93*, 307–319. DOI: 10.1016/j.ijhydene.2024.10.350

Bosmans, A., Vanderreydt, I., Geysen, D., & Helsen, L. (2013). The crucial role of Waste-to-Energy technologies in enhanced landfill mining: A technology review. *Journal of Cleaner Production*, *55*, 10–23. DOI: 10.1016/j.jclepro.2012.05.032

Bouteligier, S. (2014). Urban Sustainability Pilot Projects: Fit or Misfit between Challenge and Solution? In *Scale-sensitive Governance of the Environment* (pp. 188–202). John Wiley & Sons, Ltd. DOI: 10.1002/9781118567135.ch12

Bouzar, B., & Mamindy-Pajany, Y. (2022). Manufacture and characterization of carbonated lightweight aggregates from waste paper fly ash. *Powder Technology*, *406*, 117583. DOI: 10.1016/j.powtec.2022.117583

Bradley, K. (2007). Defining digital sustainability. *Library Trends*, *56*(1), 148–163. DOI: 10.1353/lib.2007.0044

Bradu, P., Biswas, A., Nair, C., Sreevalsakumar, S., Patil, M., Kannampuzha, S., Mukherjee, A. G., Wanjari, U. R., Renu, K., Vellingiri, B., & Gopalakrishnan, A. V. (2022). AV. Recent advances in green technology and Industrial Revolution 4.0 for a sustainable future. *Environmental Science and Pollution Research International*, *30*(60), 1–32. DOI: 10.1007/s11356-022-20024-4 PMID: 35397034

Braga, J. K., Motteran, F., Sakamoto, I. K., & Varesche, M. B. A. (2018). Bacterial and archaeal community structure involved in biofuels production using hydrothermal-and enzymatic-pretreated sugarcane bagasse for an improvement in hydrogen and methane production. *Sustainable Energy & Fuels*, *2*(12), 2644–2660. DOI: 10.1039/C8SE00312B

Brebbia, C. A., Longhurst, J., Marco, E., & Booth, C. (2017). *Sustainable Development and Planning IX*. WIT Press.

Bridgwater, A. V. (2003). Pyrolysis and gasification of biomass and waste: proceedings of an expert meeting; Strasbourg, France, 30 September-1 October 2002, cpl Press.

Brimblecombe, P. R., & Richards, J. (2024). Applied climatology for heritage. *Theoretical and Applied Climatology, 155*(8), 7325–7333. DOI: 10.1007/s00704-024-05059-6

Brumana, R., Quilici, S., Oliva, L., Previtali, M., Gabriele, M., & Stanga, C. (2023). Multi-sensor HR mass data models toward multi-temporal-layered digital twins: Maintenance, design and XR informed tour of the multi-stratified Appian Way (PAAA). *Sensors (Basel), 23*(20), 8556. DOI: 10.3390/s23208556 PMID: 37896649

Buallay, A. (2022). Toward sustainability reporting in the MENA region: The effects on sector's performance. *Managerial Finance, 48*(8), 1137–1155. DOI: 10.1108/MF-09-2021-0422

Buallay, A., Hamdan, R., Barone, E., & Hamdan, A. (2022). Increasing female participation on boards: Effects on sustainability reporting. *International Journal of Finance & Economics, 27*(1), 111–124. DOI: 10.1002/ijfe.2141

Buelow, C. A., & Waltham, N. J. (2020). Restoring tropical coastal wetland water quality: Ecosystem service provisioning by a native freshwater bivalve. *Aquatic Sciences, 82*(4), 77. DOI: 10.1007/s00027-020-00747-7

Bulkeley, H., Marvin, S., Palgan, Y. V., McCormick, K., Breitfuss-Loidl, M., Mai, L., von Wirth, T., & Frantzeskaki, N. (2019). Urban living laboratories: Conducting the experimental city? *European Urban and Regional Studies, 26*(4), 317–335. DOI: 10.1177/0969776418787222

Bundgaard, L., & Borrás, S. (2021). City-wide scale-up of smart city pilot projects: Governance conditions. *Technological Forecasting and Social Change, 172*, 121014. DOI: 10.1016/j.techfore.2021.121014

Bunning, J. (2014). Governance for regenerative and decarbonised eco-city regions. *Renewable Energy, 67*, 73–79. DOI: 10.1016/j.renene.2013.11.041

Burroughs, R. H. (1993). Nonpoint sources as external threats to coastal water quality: Lessons from park service experience. *Coastal Management, 21*(2), 131–142. DOI: 10.1080/08920759309362197

Busch, J., Barthlott, W., Brede, M., Terlau, W., & Mail, M. (2019). Bionics and green technology in maritime shipping: An assessment of the effect of Salvinia air-layer hull coatings for drag and fuel reduction. *Philosophical Transactions. Series A, Mathematical, Physical, and Engineering Sciences, 377*(2138), 20180263. DOI: 10.1098/rsta.2018.0263 PMID: 30967069

Bussu, S., Bua, A., Dean, R., & Smith, G. (2022). Introduction: Embedding participatory governance. *Critical Policy Studies, 16*(2), 133–145. DOI: 10.1080/19460171.2022.2053179

Butt, A., Khan, S. N., Zakir, M. M., Rana, W., & Laila, A. (2021). Firms greenwashing practices and consumers' perception: Role of marketing and non-marketing external stake holders in firms greenwashing practices. *Indian Journal of Economics and Business, 20*(3), 689–706.

Caferra, R., D'Adamo, I., & Morone, P. (2023). Wasting energy or energizing waste? The public acceptance of waste-to-energy technology. *Energy, 263*, 10. DOI: 10.1016/j.energy.2022.126123

Cagáňová, D., Cehlár, M., & Horňáková, N. (Eds.). 2024. "*Smart Cities: Importance of Management and Innovations for Sustainable Development.*" DOI: 10.1007/978-3-031-56533-5

Cajot, S., Peter, M., Bahu, J.-M., Koch, A., & Maréchal, F. (2015). Energy Planning in the Urban Context: Challenges and Perspectives. *Energy Procedia, 78*, 3366–3371. DOI: 10.1016/j.egypro.2015.11.752

Campeol, G. (Ed.). (2020). *Strategic Environmental Assessment and Urban Planning: Methodological Reflections and Case Studies.* Springer International Publishing., DOI: 10.1007/978-3-030-46180-5

Canelas, P., & Noring, L. (2022). Governmentalities of land value capture in urban redevelopment. *Land Use Policy, 122*, 106396. DOI: 10.1016/j.landusepol.2022.106396

Capelli, G. (2015). La Marrana dell'acqua Mariana: Un corso d'acqua al servizio dei Papi [in Italian]. Acque Sotterranee - Italian Journal of Groundwater, 4(4), 79-82. DOI: 10.7343/as-135-15-0162

Capgemini (2022) 'Consumer perspectives on sustainability: Trends and insights', *Capgemini Research Institute*.

Carenbauer, M. G. (2021). Essential or dismissible? Exploring the challenges of waste pickers in relation to COVID-19. *Geoforum, 120*, 79–81. DOI: 10.1016/j.geoforum.2021.01.018 PMID: 36536781

Casey, A., & Becker, A. (2019). Institutional and Conceptual Barriers to Climate Change Adaptation for Coastal Cultural Heritage. *Coastal Management, 47*(2), 169–188. DOI: 10.1080/08920753.2019.1564952

Castañeda-Valdés, A., Corvo, F., Marrero-Águila, R., Fernández-Domínguez, A., & Del Angel-Meraz, E. (2023). The service life of reinforced concrete structures in an extremely aggressive coastal city. Influence of concrete quality. *Materials and Structures*, *56*(1), 12. DOI: 10.1617/s11527-023-02100-4

Castelblanco, G., Guevara, J., Mesa, H., & Flores, D. (2020). Risk allocation in unsolicited and solicited road public-private partnerships: Sustainability and management implications. *Sustainability (Basel)*, *12*(11), 4478. DOI: 10.3390/su12114478

Cavallaro, L. I., Iuppa, C., & Foti, E. (2017). Effect of Partial Use of Venice Flood Barriers. *Journal of Marine Science and Engineering*, *5*(4), 58. DOI: 10.3390/jmse5040058

Cease, B., Kim, H., Kim, D., Ko, Y., & Cappel, C. (2019). Barriers and incentives for sustainable urban development: An analysis of the adoption of LEED-ND projects. *Journal of Environmental Management*, *244*, 304–312. DOI: 10.1016/j.jenvman.2019.04.020 PMID: 31128335

Cerema. (2011). Factsheet No. 01b: Rome, a green city – The example of Appia Antica Regional Park. République française – Ministère de l'Environnement. https://paysages-territoires-transitions.cerema.fr/IMG/pdf/blue_green_infra_rome.pdf

Chakroun, R., Matoussi, H., & Mbirki, S. (2017). Determinants of CSR disclosure of Tunisian listed banks: A multi-support analysis. *Social Responsibility Journal*, *13*(3), 552–584. DOI: 10.1108/SRJ-04-2016-0055

Chance, T. (2009). Towards sustainable residential communities; the Beddington Zero Energy Development (BedZED) and beyond. *Environment and Urbanization*, *21*(2), 527–544. DOI: 10.1177/0956247809339007

Chandel, M. K., Kwok, G., Jackson, R. B., & Pratson, L. F. (2012). The potential of waste-to-energy in reducing GHG emissions. *Carbon Management*, *3*(2), 133–144. DOI: 10.4155/cmt.12.11

Chao, S.-L., Yu, M.-M., & Wei, S.-Y. (2024). Ascertaining the impact of e-service quality on e-loyalty for the e-commerce platform of liner shipping companies. *Transportation Research Part E, Logistics and Transportation Review*, *184*, 103491. DOI: 10.1016/j.tre.2024.103491

Chatziparaskeva, G., Papamichael, I., Voukkali, I., Loizia, P., Sourkouni, G., Argirusis, C., & Zorpas, A. A. (2022). End-of-life of composite materials in the framework of the circular economy. *Microplastics.*, *1*(3), 377–392. DOI: 10.3390/microplastics1030028

Chaudhari, P., Xiao, Y., Cheng, M. M. C., & Li, T. (2024). Fundamentals, Algorithms, and Technologies of Occupancy Detection for Smart Buildings Using IoT Sensors. In Sensors (Vol. 24, Issue 7). Multidisciplinary Digital Publishing Institute (MDPI). DOI: 10.3390/s24072123

Chauhan, S., & Amit, . (2014). A relational study of firm's characteristics and CSR expenditure. *Procedia Economics and Finance*, *11*, 23–32. DOI: 10.1016/S2212-5671(14)00172-5

Chaurasia, M., Patel, K., Singh, R., Bhadouria, R., & Rao, K. S. (2024). Integrating Blue-Green as next-generation urban infrastructure in developing countries. In Joshi, P. K., Rao, K. S., Bhadouria, R., Tripathi, S., & Singh, R. (Eds.), *Blue-Green infrastructure for sustainable urban settlements*. Springer., DOI: 10.1007/978-3-031-62293-9_3

Cheliatsidou, A., Sariannidis, N., Garefalakis, A., Passas, I., & Spinthiropoulos, K. (2023). Exploring Attitudes towards Whistleblowing in Relation to Sustainable Municipalities. Administrative Sciences 2023, Vol. 13, Page 199, 13(9), 199. DOI: 10.3390/admsci13090199

Chen, B., & Pitts, A. (2006). A Socio-Technical Perspective to Enact Zero Emissions / Energy Development in China. *Proceedings of PLEA2006 - The 23 Rd Conference on Passive and Low Energy Architecture*. PLEA2006 - The 23 rd Conference on Passive and Low Energy Architecture, Geneva. http://web5.arch.cuhk.edu.hk/server1/staff1/edward/www/plea2018/plea/2006/Vol1/PLEA2006_PAPER142.pdf

Chen, J. K. C., Wang, J. C. F., Tsai, W.-H., & Lin, T. S. Y. (2008). Exploring innovation model combine with the green marketing to consumer's behavior influence - A case study of a TFT-LCD company. 2008 IEEE International Conference on Industrial Engineering and Engineering Management, IEEM 2008, art. no. 4737943, 619 - 623. DOI: DOI: 10.1109/IEEM.2008.4737943

Cheng, L., Wu, Z., Zhang, Z., Guo, C., Ellis, N., Bi, X., Paul Watkinson, A., & Grace, J. R. (2020). Tar elimination from biomass gasification syngas with bauxite residue derived catalysts and gasification char. *Applied Energy*, *258*, 114088. DOI: 10.1016/j.apenergy.2019.114088

Chen, K.-K., Chang, C.-T., & Lai, C.-S. (2009). Service quality gaps of business customers in the shipping industry. *Transportation Research Part E, Logistics and Transportation Review*, *45*(1), 222–237. DOI: 10.1016/j.tre.2008.02.005

Chen, R., & Zhang, T. (2024). Artificial intelligence applications implication for ESG performance: Can digital transformation of enterprises promote sustainable development? *Chinese Management Studies, 18*(1). Advance online publication. DOI: 10.1108/CMS-11-2023-0653

Chen, S. (2024). The influence of artificial intelligence and digital technology on ESG reporting quality. *International Journal of Global Economics and Management, 3*(1), 301–310. Advance online publication. DOI: 10.62051/IJGEM.v3n1.36

Chen, T., & Liu, G. (2025). Global perspectives on environmental policy innovations: Driving green tourism and consumer behavior in circular economy. *Journal of Environmental Management, 374*, 124138. Advance online publication. DOI: 10.1016/j.jenvman.2025.124138 PMID: 39827609

Cherian, B. A., Tokbolat, S., & Rodrigues, L. (2024). Evaluating Thermal Comfort in Zero-Energy Developments: A Case Study of BedZED. *The 1st International Conference of Net Zero Carbon Built Environment*. https://nottingham-repository.worktribe.com/output/39163737

Chiesura, A. (2004). The role of urban parks for the sustainable city. *Landscape and Urban Planning, 68*(1), 129–138. DOI: 10.1016/j.landurbplan.2003.08.003

Christensen, H. B., Hail, L., & Leuz, C. (2021). Mandatory CSR and sustainability reporting: Economic analysis and literature review. *Review of Accounting Studies, 26*(3), 1176–1248. DOI: 10.1007/s11142-021-09609-5

Coates, G. (2025). THE SUSTAINABLE URBAN DISTRICT OF VAUBAN IN FREIBURG, GERMANY. Ksu.

Coates, G. J. (2013). The sustainable urban district of vauban in freiburg, germany. *International Journal of Design & Nature and Ecodynamics, 8*(4), 265–286. DOI: 10.2495/DNE-V8-N4-265-286

Coindesk, 2020. Digital voting is coming. Let's do it right [online]. Available: Coindesk. https://www.coindesk.com/markets/2020/11/10/digital-voting-is-coming-lets-do- it-right/ (Accessed May 24 2023).

Coletti, C. (2024). *Climate Change Threats to Stone Cultural Heritage: State of the Art of Quantitative Damage Functions and New Challenges for a Sustainable Future*. Heritage., DOI: 10.3390/heritage7060154

Cone Communications. (2017). *2017 Cone Communications CSR Study*. Cone Communications.

Corona, B., Shen, L., Reike, D., Carreón, J. R., & Worrell, E. (2019). Towards sustainable development through the circular economy—A review and critical assessment on current circularity metrics. *Resources, Conservation and Recycling*, *151*, 104498. DOI: 10.1016/j.resconrec.2019.104498

Cortesi, A., Vardopoulos, I., & Salvati, L. (2022). A partial least squares analysis of the perceived impact of sustainable real estate design upon wellbeing. *Urban Science (Basel, Switzerland)*, *6*(4), 69. DOI: 10.3390/urbansci6040069

Coscia, C., Mukerjee, S., Palmieri, B. L., & Quintanal Rivacoba, C. (2020). Enhancing the Sustainability of Social Housing Policies through the Social Impact Approach: Innovative Perspectives form a "Paris Affordable Housing Challenge" Project in France. *Sustainability (Basel)*, *12*(23), 9903. Advance online publication. DOI: 10.3390/su12239903

Cosmi, C., Salvia, M., Di Leo, S., Pietrapertosa, F., & Loperte, S. (2017). Interregional Cooperation as a Key Tool for the Achievement of Strategic-Energy and Climate Targets: The Experience of the INTERREG IVC RENERGY and SEE RE-SEEties Projects. In Bisello, A., Vettorato, D., Stephens, R., & Elisei, P. (Eds.), *Smart and Sustainable Planning for Cities and Regions: Results of SSPCR 2015* (pp. 335–352). Springer International Publishing., DOI: 10.1007/978-3-319-44899-2_20

Covucci, C., Confetto, M. G., Ključnikov, A., & Panait, M. (2024). Unrevealing the nexus between digital sustainability and corporate digital responsibility: A dual-track systematic literature review towards a framework of corporate digital sustainability. *Technology in Society*, *79*, 79. DOI: 10.1016/j.techsoc.2024.102743

Crawley, D., & Aho, I. (1999). Building environmental assessment methods: Applications and development trends. *Building Research and Information*, *27*(4–5), 300–308. DOI: 10.1080/096132199369417

Crilly, M., & Lemon, M. (2018). The Misalignment of Policy and Practice in Sustainable Urban Design. In M. Dastbaz, W. Naudé, & J. Manoochehri (Eds.), *Smart Futures, Challenges of Urbanisation, and Social Sustainability* (pp. 69–87). Springer International Publishing. DOI: 10.1007/978-3-319-74549-7_5

Crot, L. (2013). Planning for Sustainability in Non-democratic Polities: The Case of Masdar City. *Urban Studies (Edinburgh, Scotland)*, *50*(13), 2809–2825. DOI: 10.1177/0042098012474697

Crowley, K. J., Jackson, R., O'Connell, S., Karunarthna, D., Anantasari, E., Retnowati, A., & Niemand, D. (2022). Cultural heritage and risk assessments: Gaps, challenges, and future research directions for the inclusion of heritage within climate change adaptation and disaster management. *Climate Resilience and Sustainability, 1*(3), e45. Advance online publication. DOI: 10.1002/cli2.45

Cugurullo, F. (2013). How to Build a Sandcastle: An Analysis of the Genesis and Development of Masdar City. *Journal of Urban Technology, 20*(1), 23–37. DOI: 10.1080/10630732.2012.735105

Cui, L., Xie, G., Qu, Y., Gao, L., & Yang, Y. (2018). Security and Privacy in Smart Cities: Challenges and Opportunities. *IEEE Access : Practical Innovations, Open Solutions, 6*, 46134–46145. DOI: 10.1109/ACCESS.2018.2853985

Cui, Z., Li, Q., & Bai, M. (2023). Does fintech improve corporate ESG performance? Evidence from China. *SSRN*. https://papers.ssrn.com/sol3/papers.cfm?abstract_id=5103526

Curran, K., & Zimmermann, N. (2021). The Dynamics of Collaboration in Heritage Science. *Studies in Conservation, 67*(1-2), 136–149. DOI: 10.1080/00393630.2021.1875175

Da Silva, K. P., & Ramirez, K. N. (2021). Green buildings: a worldwide overview on leed, breeam and green star certifications / Green buildings: uma análise do panorama mundial das certificações leed, breeam e green star. *Brazilian Journal of Development, 7*(8), 82761–82778. DOI: 10.34117/bjdv7n8-471

Dai, J., & Vasarhelyi, M. A. (2016). Imagineering Audit 4.0. J. Emerg. Technol. Account. 13, 1–15. Dimaggio, P.J., Powell, W.W., 1983. The iron cage revisited: Institutional isomorphism and collective rationality in organizational fields. *American Sociological Review, 48*, 147–160.

Dai, L., Wörner, R., & Van Rijswick, H. F. M. W. (2017). Rainproof cities in the Netherlands: Approaches in Dutch water governance to climate-adaptive urban planning. *International Journal of Water Resources Development, 34*(4), 652–674. DOI: 10.1080/07900627.2017.1372273

Damsø, T., Kjær, T., & Christensen, T. B. (2017). Implementation of local climate action plans: Copenhagen – Towards a carbon-neutral capital. *Journal of Cleaner Production, 167*, 406–415. DOI: 10.1016/j.jclepro.2017.08.156

Danilo Marco Campanaro, G. L.-M. (2015). 3D GIS for cultural heritage restoration: A 'white box' workflowDanilo. Journal of Cultural Heritage.

Das, A., Gujre, N., Devi, R. J., Rangan, L., & Mitra, S. (2023). Traditional ecological knowledge towards natural resource management. In Elsevier eBooks (pp. 275–294). DOI: 10.1016/B978-0-323-90500-8.00019-1

Dastjerdi, B., Strezov, V., Rajaeifar, M. A., Kumar, R., & Behnia, M. (2022). *Waste to Energy Technologies. Reference Module in Earth Systems and Environmental Sciences*. Elsevier.

Datta, K. (2023). AI-driven public administration: Opportunities, challenges, and ethical considerations. *ResearchGate*. https://www.researchgate.net/publication/387297421_ai-driven_public_administration_opportunities_challenges_and_ethical_considerations

Davidson, K., Coenen, L., & Gleeson, B. (2019). A Decade of C40: Research Insights and Agendas for City Networks. *Global Policy*, *10*(4), 697–708. DOI: 10.1111/1758-5899.12740

De Baro, M. E. Z. (2022). Curitiba case study. In Cities and nature (pp. 117–162).

De Bere, L. (2000). Anaerobic digestion of solid waste: State-of-the-art. *Water Science and Technology*, *41*(3), 283–290. DOI: 10.2166/wst.2000.0082 PMID: 11382003

de Freitas Netto, S. V., Sobral, M. F. F., Ribeiro, A. R. B., & Soares, G. R. D. L. (2020). Concepts and forms of greenwashing: A systematic review. *Environmental Sciences Europe*, *32*(1), 1–12. DOI: 10.1186/s12302-020-0300-3

de Gooyert, V., Rouwette, E., van Kranenburg, H., Freeman, E., & van Breen, H. (2016). Sustainability transition dynamics: Towards overcoming policy resistance. *Technological Forecasting and Social Change*, *111*, 135–145. DOI: 10.1016/j.techfore.2016.06.019

De Oliveira, D. C., De Rezende, T. T. G., Lora, E. E. S., Venturini, O. J., & Maya, D. M. Y. (2024). Tar removal from biomass-derived syngas for hydrogen production: Oil absorption process considering brazilian sources. *International Journal of Hydrogen Energy*.

De Pascali, P., & Bagaini, A. (2019). Energy Transition and Urban Planning for Local Development. A Critical Review of the Evolution of Integrated Spatial and Energy Planning. *Energies*, *12*(1), 1. Advance online publication. DOI: 10.3390/en12010035

De Veirman, M., Cauberghe, V., & Hudders, L. (2017). Marketing through Instagram influencers: The impact of number of followers and product divergence on brand attitude. *International Journal of Advertising*, *36*(5), 798–828. DOI: 10.1080/02650487.2017.1348035

Debnath, J., Sahariah, D., Lahon, D., Nath, N., Chand, K., Meraj, G., Farooq, M., Kumar, P., Kanga, S., & Singh, S. K. (2023). Geospatial modeling to assess the past and future land use-land cover changes in the Brahmaputra Valley, NE India, for sustainable land resource management. *Environmental Science and Pollution Research International, 30*(49), 106997–107020. DOI: 10.1007/s11356-022-24248-2 PMID: 36418825

Deely, J., Hynes, S., Barquín, J., Burgess, D., Finney, G., Silió, A., Álvarez-Martínez, J. M., Bailly, D., & Ballé-Béganton, J. (2020). Barrier identification framework for the implementation of blue and green infrastructures. *Land Use Policy, 99*, 105108. DOI: 10.1016/j.landusepol.2020.105108

Dell'Aversano, S., Villante, C., Gallucci, K., Vanga, G., & Di Giuliano, A. (2024). E-Fuels: A Comprehensive Review of the Most Promising Technological Alternatives towards an Energy Transition. *Energies, 17*(16), 3995. DOI: 10.3390/en17163995

Della Spina, L. (2021). Cultural Heritage: A Hybrid Framework for Ranking Adaptive Reuse Strategies. *Buildings (Basel, Switzerland), 11*(3), 132. Advance online publication. DOI: 10.3390/buildings11030132

Delladetsimas, P. M., Sapountzaki, K., & Chalkias, C. (2015). Adaptations to the economic crisis: Reallocation of vulnerability and production of new chronic risks. In Maloutas, T., & Spyrellis, S. (Eds.), *Social atlas of Athens: Digital compendium of texts and visual material.*, DOI: 10.17902/20971.39

Delmas, M. A., & Burbano, V. C. (2011). The drivers of greenwashing. *California Management Review, 54*(1), 64–87.

Dempere, J., Alamash, E., & Mattos, P. (2024). Unveiling the truth: Greenwashing in sustainable finance. *Frontiers in Sustainability, 5*, 1362051. DOI: 10.3389/frsus.2024.1362051

Devane, C., Saulnier, L., & Latham, L. (2021). Coastal Mental Health and Substance Use Services' COVID-19 Response: A Quality Improvement Initiative. *Nursing Leadership, 34*(3), 40–50. DOI: 10.12927/cjnl.2021.26594 PMID: 34698013

Devey, J., & Ferraris, C. (2023). Ensuring ethical AI: Strengthening data governance and algorithmic accountability in the digital era. *ResearchGate*. https://www.researchgate.net/publication/384628034_Ensuring_Ethical_AI_Strengthening_Data_Governance_and_Algorithmic_Accountability_in_the_Digital_Era

Dhanya, B., Ramananda, S., & Dhyani, S. (2022). Mainstreaming blue-green infrastructure in policy and planning for Urban Resilience in the Global South: Promises and Pitfalls. In *Blue-Green Infrastructure Across Asian Countries: Improving Urban Resilience and Sustainability* (pp. 499–518). Springer Singapore. DOI: 10.1007/978-981-16-7128-9_23

Di Fiore, G., Specht, K., Rover, O. J., & Zanasi, C. (2022). Stakeholders' social acceptance of a new organic waste management policy in the city of Florianópolis (Brazil). *Journal of Cleaner Production*, *379*, 134756. DOI: 10.1016/j.jclepro.2022.134756

Ding, Z., Fan, Z., Tam, V. W. Y., Bian, Y., Li, S., Illankoon, I. M. C. S., & Moon, S. (2018). Green building evaluation system implementation. *Building and Environment*, *133*, 32–40. DOI: 10.1016/j.buildenv.2018.02.012

Ditlev-Simonsen, C. D. (2022). Sustainability and finance: environment, social, and governance (ESG). *A Guide to Sustainable Corporate Responsibility: From Theory to Action*, 189-206. Cham. DOI: 10.1007/978-3-030-88203-7_9

Doan, D. T., Ghaffarianhoseini, A., Naismith, N., Zhang, T., Ghaffarianhoseini, A., & Tookey, J. (2017). A critical comparison of green building rating systems. In *Building and Environment* (Vol. 123, pp. 243–260). Elsevier Ltd., DOI: 10.1016/j.buildenv.2017.07.007

Donthu, N., Kumar, S., Mukherjee, D., Pandey, N., & Lim, W. M. (2021). How to conduct a bibliometric analysis: An overview and guidelines. *Journal of Business Research*, *133*, 285–296. DOI: 10.1016/j.jbusres.2021.04.070

Doukas, Y. E., Vardopoulos, I., & Petides, P. (2024). Challenging the status quo: a critical analysis of the common agricultural policy's shift towards sustainability. In Climate change in regional perspective. European Union and Latin American initiatives, challenges, and solutions. (pp. 115–134). Cham, Switzerland: United Nations University Series on Regionalism, Vol. 27, Springer Nature. DOI: 10.1007/978-3-031-49329-4_8

Doukas, Y. E., Salvati, L., & Vardopoulos, I. (2023). Unraveling the European agricultural policy sustainable development trajectory. *Land (Basel)*, *12*(9), 1749. DOI: 10.3390/land12091749

Douvas, L. (2007). Leisure parks: Their contribution to tourism development. Administration and Economy, 143–156.

Drempetic, S., Klein, C., & Zwergel, B. (2020). The influence of firm size on the ESG score: Corporate sustainability ratings under review. *Journal of Business Ethics*, *167*(2), 333–360. DOI: 10.1007/s10551-019-04164-1

Droege, P. (2011). *Urban Energy Transition: From Fossil Fuels to Renewable Power*. Elsevier.

Drosou, N., Soetanto, R., Hermawan, F., Chmutina, K., Bosher, L., & Hatmoko, J. U. D. (2019). Key factors influencing wider adoption of blue–green infrastructure in developing cities. *Water (Basel)*, *11*(6), 1234. DOI: 10.3390/w11061234

Ducharme, C. (2010). Technical and economic analysis of Plasma-assisted Waste-to-Energy processes M.Sc., Department of Earth and Environmental Engineering Fu Foundation of Engineering and Applied Science Columbia University September 2010 Research partially sponsored by the Earth Engineering Center, Columbia University.

Duchin, R., Gao, J., & Xu, Q. (2024). Sustainability or greenwashing: Evidence from the asset market for industrial pollution. *The Journal of Finance*.

Duran, R. E., & Tierney, P. (2023). Fintech data infrastructure for ESG disclosure compliance. *Journal of Risk and Financial Management*, *16*(8), 378. DOI: 10.3390/jrfm16080378

Durán-Romero, G., López, A. M., Beliaeva, T., Ferasso, M., Garonne, C., & Jones, P. (2020). Bridging the gap between circular economy and climate change mitigation policies through eco-innovations and Quintuple Helix Model. *Technological Forecasting and Social Change*, *160*, 120246. DOI: 10.1016/j.techfore.2020.120246

Durán-Romero, G., & Urraca-Ruiz, A. (2015). Climate change and eco-innovation. A patent data assessment of environmentally sound technologies. *Innovation (North Sydney, N.S.W.)*, *17*(1), 115–138. DOI: 10.1080/14479338.2015.1011062

Eames, M., Dixon, T., Lannon, S., Hunt, M., Laurentis, C. D., Marvin, S., Hodson, M., Guthrie, P., & Georgiadou, M. (2014). *Retrofit 2050: Critical challenges for urban transitions*. Cardiff University. https://centaur.reading.ac.uk/36187/1/critical%20challenges%20briefing-March%202014.pdf

East, M. (2018). Rising Cities: Continuity, Innovation and Deliberation of Vauban District of Freiburg, Germany. *Ecocycles*, *4*(2), 58–74. DOI: 10.19040/ecocycles.v4i2.106

EC. (2019). *The European Green Deal*. EU Publications.

EC. (2024, December 3). *Climate-neutral and smart cities*. https://research-and-innovation.ec.europa.eu/funding/funding-opportunities/funding-programmes-and-open-calls/horizon-europe/eu-missions-horizon-europe/climate-neutral-and-smart-cities_en

Eccles, R. G., Ioannou, I., & Serafeim, G. (2014). The impact of corporate sustainability on organizational processes and performance. *Management Science, 60*(11), 2835–2857. https://www.jstor.org/stable/24550546. DOI: 10.1287/mnsc.2014.1984

Egid, B. (2024). Mosquito-borne disease, waste pickers and health equity: a co-produced study in Andhra Pradesh, India. PhD, Liverpool School of Tropical Medicine.

Ejaz, U., Ramon, W., & Olaoye, G. (2025). The Role of Big Data and AI in Smart Cities and Urban Planning.

El Khoury, R., Nasrallah, N., & Alareeni, B. (2023). ESG and financial performance of banks in the MENAT region: Concavity–convexity patterns. *Journal of Sustainable Finance & Investment, 13*(1), 406–430. DOI: 10.1080/20430795.2021.1929807

Elahi, A., Mahanta, P., & Ahmed, N. (2024). Challenges and preservation efforts of biodiversity in Northeast India. In Journal of Emerging Technologies and Innovative Research (JETIR) [JETIR]. *Journal of Emerging Technologies and Innovative Research, 11*(4), •••.

Eliades, F., Doula, M. K., Papamichael, I., Vardopoulos, I., Voukkali, I., & Zorpas, A. A. (2022). Carving out a niche in the sustainability confluence for environmental education centers in Cyprus and Greece. *Sustainability (Basel), 14*(14), 8368. DOI: 10.3390/su14148368

Elias, O., Awotunde, O. J., Oladepo, O. I., Azuikpe, P. F., Adeleye, O. S., Oladele, O. R., & Ogunruku, O. O. (2023). The evolution of green fintech: Leveraging AI and IoT for sustainable financial services and smart contract implementation. *ResearchGate.* https://www.researchgate.net/publication/382785523_The_evolution_of_green_fintech_Leveraging_AI_and_IoT_for_sustainable_financial_services_and_smart_contract_implementation

Elkington, J. (1997). *Cannibals with Forks: The Triple Bottom Line of 21st Century Business.* Capstone.

Elliott, K., Price, R., Shaw, P., Spiliotopoulos, T., Ng, M., Coopamootoo, K., & van Moorsel, A. (2021). Towards an equitable digital society: Artificial intelligence (AI) and corporate digital responsibility (CDR). *Society, 58*(3), 179–188. DOI: 10.1007/s12115-021-00594-8 PMID: 34149122

Enemuo, M., & Ogunmodimu, O. (2025). Transitioning the mining sector: A review of renewable energy integration and carbon footprint reduction strategies. *Applied Energy, 384*, 125484. DOI: 10.1016/j.apenergy.2025.125484

Engels, F., Wentland, A., & Pfotenhauer, S. M. (2019). Testing future societies? Developing a framework for test beds and living labs as instruments of innovation governance. *Research Policy, 48*(9), 103826. DOI: 10.1016/j.respol.2019.103826

Epting, S. (2020). Participatory Budgeting for Environmental Justice. *Ethics, Policy & Environment, 23*(1), 22–36. DOI: 10.1080/21550085.2020.1746003

Erica Avrami, J. L. (2021). Energy and historic buildings: Toward evidence-based policy reform. *Journal of Cultural Heritage Management and Sustainable Development.*

Erin Seekamp, E. J. (2020). Resilience and transformation of heritage sites to accommodate for loss and learning in a changing climate. *Climatic Change, 162*(1), 41–55. DOI: 10.1007/s10584-020-02812-4

Esangbedo, C. O., Chang, Y., Ushie, A. M., & Hasan, M. M. (2024). Eco-Innovation and Firm Sustainability in Emerging Economies: The Role of Consumer Behavior and Green B2B Marketing. *Journal of Business-To-Business Marketing, 31*(4), 405–420. DOI: 10.1080/1051712X.2024.2369764

Escrivà Saneugenio, F., Sateriano, A., Salvati, L., & Vardopoulos, I. (2024). The extrinsic manifestation of sprawl: Complex dynamics of urbanization, land imperviousness, and sustainability in Mediterranean landscapes. In *Homage (and criticism) to the Mediterranean city. Regional sustainability and economic resilience* (pp. 65–84). River Publishers. DOI: 10.1201/9781003517306-3

Esfandi, S., Tayebi, S., Byrne, J., Taminiau, J., Giyahchi, G., & Alavi, S. A. (2024). Smart cities and urban energy planning: An advanced review of promises and challenges. *Smart Cities, 7*(1), 414–444. DOI: 10.3390/smartcities7010016

Etim, E. (2024). Leveraging public awareness and behavioural change for entrepreneurial waste management. *Heliyon, 10*(21), e40063. DOI: 10.1016/j.heliyon.2024.e40063 PMID: 39553649

European Commission. (2016). Flash Eurobarometer 419: Quality of life in European cities 2015. https://ec.europa.eu/regional_policy/en/information/publications/studies/2016/quality-of-life-in-european-cities-2015

European Commission. (2025). Green infrastructure. https://environment.ec.europa.eu/topics/nature-and-biodiversity/green-infrastructure_en

European Environment Agency. (2011). Green infrastructure and territorial cohesion: The concept of green infrastructure and its integration into policies using monitoring systems (Report No. 18/2011). https://www.eea.europa.eu/en/analysis/publications/green-infrastructure-and-territorial-cohesion

European Environment Agency. (2022). Urban adaptation in Europe: How cities and towns respond to climate change (EEA Report No. 12/2022). https://www.eea.europa.eu/publications/urban-adaptation-in-europe

European Parliament. (2020). The importance of urban and green infrastructure - European Year of Greener Cities 2022 (Texts adopted, P9_TA(2020)0241). Official Journal of the European Union. https://www.europarl.europa.eu/doceo/document/TA-9-2020-0241_EN.html

Everett, S., & Kittel, C. (2010). Sustainability and Australian Coastal Shipping: Some Issues. *Australian Journal of Maritime and Ocean Affairs*, *2*(3), 82–89. DOI: 10.1080/18366503.2010.10815661

EYDAP. (2018). Technical sheet: Construction of compact MBR unit in Antonis Tritsis Park [in Greek]. https://parkotritsis.gr/wp-content/uploads/Ereynhtiko-Yliko/EreynesKaiMeletes/Gnwm.%20ektheseis/Techniko%20Deltio_Compact%20Monada%20MBR_Eydap_2018.pdf

Faiz, F., Ninduwezuor-Ehiobu, N., Adanma, U. M., & Solomon, N. O. (2024). Blockchain for sustainable waste management: Enhancing transparency and accountability in waste disposal. *Comprehensive Research and Reviews in Science and Technology*, *2*(1), 45–69. DOI: 10.57219/crrst.2024.2.1.0032

Farooq, A., Haputta, P., & Gheewala, S. H. (2024). Economic feasibility assessment of waste to energy technologies for the development of a sustainable municipal solid waste management system in Thailand. *Renewable Energy*, *233*, 121155. DOI: 10.1016/j.renene.2024.121155

Fastenrath, S., & Braun, B. (2018). Ambivalent urban sustainability transitions: Insights from Brisbane's building sector. *Journal of Cleaner Production*, *176*, 581–589. DOI: 10.1016/j.jclepro.2017.12.134

Feng, W., Bilivogui, P., Wu, J., & Mu, X. (2023). Green finance: Current status, development, and future course of actions in China. *Environmental Research Communications*, *5*(3), 035005. DOI: 10.1088/2515-7620/acc1c7

Ferreira Gregorio, V., Pié, L., & Terceño, A. (2018). A systematic literature review of bio, green and circular economy trends in publications in the field of economics and business management. *Sustainability (Basel)*, *10*(11), 4232. DOI: 10.3390/su10114232

Ferreira, J., Santos, D., & Campos, L. (2024). Blue-green infrastructure in view of Integrated Urban Water Management: A novel assessment of an effectiveness index. *Water Research*, *257*, 121658. DOI: 10.1016/j.watres.2024.121658 PMID: 38696983

Foulds, C., Valkenburg, G., Ryghaug, M., Suboticki, I., Skjølsvold, T. M., Korsnes, M., & Heidenreich, S. (2023). Implementing Mission-oriented Experiments: Recommendations on Epistemic Inclusion for City Stakeholders Working in Climate Change Initiatives. *Journal of City Climate Policy and Economy*, *2*(1), 55–76. DOI: 10.3138/jccpe-2022-0014

Fraker, H. (2013). *The Hidden Potential of Sustainable Neighborhoods*. Island Press/ Center for Resource Economics., DOI: 10.5822/978-1-61091-409-3

Frasca, F. V., Verticchio, E., Bosco, E., Kuka, E., Lee, D. S.-H., Andersen, C. K., Bertolin, C., & Siani, A. M. (2024). Assessing microclimate thresholds for heritage preventive conservation to achieve sustainable and energy efficiency goals in a changing climate. *Scientific Reports*, *14*(1), 18707. Advance online publication. DOI: 10.1038/s41598-024-69395-3 PMID: 39134605

Free, C., Jones, S., & Tremblay, M. S. (2024). Greenwashing and sustainability assurance: A review and call for future research. *Journal of Accounting Literature*. Advance online publication. DOI: 10.1108/JAL-11-2023-0201

Furlow, N. E. (2010). 'Greenwashing in the new millennium', *Journal of Applied Business and Economics*, 10(6), pp. 22-25.

Fu, X., Liu, G., Wu, H., Zhuang, T., Huang, R., Yuan, F., & Zhang, Y. (2024). Dissecting behavioral inertia in shaping different resident participation behaviors in neighborhood regeneration: A quantitative behavioral experiment. *Environmental Impact Assessment Review*, *109*, 107632. DOI: 10.1016/j.eiar.2024.107632

Galeone, G., Ranaldo, S., & Fusco, A. (2024). ESG and FinTech: Are they connected? *Research in International Business and Finance*, *69*, 0275531924000175. DOI: 10.1016/j.ribaf.2024.102225

Galvão, N., Alves, I. R. F. S., & Bassin, J. P. (2023). Chapter 32 - Municipal solid waste management in Brazil: overview and trade-offs between different treatment technologies. Waste Management and Resource Recycling in the Developing World. P. Singh, P. Verma, R. Singh, A. Ahamad and A. C. S. Batalhão, Elsevier: 755-772.

Garrido, V., & Alberto, M. (2023). ESG-driven innovation in the fintech industry. *ResearchGate*. https://www.researchgate.net/publication/373238616_ESG-driven_innovation_in_the_fintech_industry

Gavanas, N., Pozoukidou, G., & Verani, E. (2016). Integration of LUTI models into sustainable urban mobility plans (SUMPs). *European Journal of Environmental Sciences*, *6*(1), 11–17. DOI: 10.14712/23361964.2016.3

Gazheli, A., Antal, M., & van den Bergh, J. (2015). The behavioral basis of policies fostering long-run transitions: Stakeholders, limited rationality and social context. *Futures*, *69*, 14–30. DOI: 10.1016/j.futures.2015.03.008

Geels, F. W. (2019). Socio-technical transitions to sustainability: A review of criticisms and elaborations of the Multi-Level Perspective. *Current Opinion in Environmental Sustainability*, *39*, 187–201. DOI: 10.1016/j.cosust.2019.06.009

George, G., Merrill, R. K., & Schillebeeckx, S. J. D. (2021). Digital sustainability and entrepreneurship: How digital innovations are helping tackle climate change and sustainable development. *Entrepreneurship Theory and Practice*, *45*(5), 999–1027. DOI: 10.1177/1042258719899425

Georgian Ionu PURCAREATA, M. B. (2015). A model of land suitability general analysis for new infrastructure projects in the Barsa country (Romania). Forum Geographic.

Georgiou, C., Maialetti, M., Zorpas, A. A., Salvati, L., & Vardopoulos, I. (2025). Policy and planning for "zero-net energy" real estate development in rural and ex-urban districts. In *Environmental Sustainability and Global Change* (pp. 291–303). Elsevier., DOI: 10.1016/B978-0-443-31596-1.00020-9

Geovetenskaper, G. U. F. (2023, June 22). Traditional ecological knowledge in agriculture for adaptation/ resilience to climate change: high mountain Asia. Greinke, L., & Mehnen, N. Urban Transformation through Public Participation. Möglichkeit und Macht Possibility and Power.

Getzner, M. (2024). Cultural infrastructure as part of social infrastructure: perspectives of cultural policy and economics. In *A.-T. R. Getzner, Handbook of Social Infrastructure* (pp. 232–248). Edward Elgar Publishing. DOI: 10.4337/9781800883130.00024

Geyer, R., Jambeck, J. R., & Law, K. L. (2017). Production, use, and fate of all plastics ever made. *Science Advances*, *3*(7), e1700782. DOI: 10.1126/sciadv.1700782 PMID: 28776036

Ghisellini, P., Cialani, C., & Ulgiati, S. (2016). A Review on Circular Economy: The Expected Transition to a Balanced Interplay of Environmental and Economic Systems. *Journal of Cleaner Production*, *114*, 11–32. DOI: 10.1016/j.jclepro.2015.09.007

Giakoumelou, A., Salvi, A., Bekiros, S., & Onorato, G. (2024). ESG and fintech funding in the EU. *Research in International Business and Finance*, *69*, 0275531924000254. DOI: 10.1016/j.ribaf.2024.102233

Giannelli, A., Giuffrida, S., & Trovato, M. (2018). Madrid Río Park: Symbolic values and contingent valuation. *Valori e Valutazioni*, *21*, 75–85.

Giannouli, I., Tourkolias, C., Zuidema, C., Tasopoulou, A., Blathra, S., Salemink, K., Gugerell, K., Georgiou, P., Chalatsis, T., Christidou, C., Bellis, V., Vasiloglou, N., & Koutsomarkos, N. (2018). A methodological approach for holistic energy planning using the living lab concept: The case of the prefecture of Karditsa. *European Journal of Environmental Sciences*, *8*(1), 14–22. DOI: 10.14712/23361964.2018.3

Gi-du Kang, G.-D., & Kim, Y.-D. (2009). An Analysis of the Measurement of the Shipping Service Quality. *The Asian Journal of Shipping and Logistics*, *25*(1), 41–55. DOI: 10.1016/S2092-5212(09)80011-4

Giglio, F., Frontera, P., Malara, A., & Armocida, F. (2024). Materials and Climate Change: A Set of Indices as the Benchmark for Climate Vulnerability and Risk Assessment for Tangible Cultural Heritage in Europe. *Sustainability (Basel)*, *16*(5), 2067. Advance online publication. DOI: 10.3390/su16052067

Gill, A., T, R., Upadhyay, R., & Gautam, A. A. (2024). Strategic Grid Integration of Renewable Using Electric Vehicles for Smart Cities Stability. *E3S Web of Conferences, 540*, 10024. DOI: 10.1051/e3sconf/202454010024

Giokas, K. (2019). The urban heat island phenomenon in western metropolitan Athens: Urban planning tools and natural methods to deal with it [Master's thesis, in Greek]. National Technical University of Athens. http://oldwww.arch.ntua.gr/sites/default/files/project/14369_/gkiokas_konstantinos.pdf

Giuseppe Forino, J. M. (2016). A proposed assessment index for climate change-related risk for cultural heritage protection in Newcastle (Australia). *International Journal of Disaster Risk Reduction*, *19*, 235–248. Advance online publication. DOI: 10.1016/j.ijdrr.2016.09.003

Gkotsis, P., Kougias, P., Mitrakas, M., & Zouboulis, A. (2023). Biogas upgrading technologies–Recent advances in membrane-based processes. *International Journal of Hydrogen Energy*, *48*(10), 3965–3993. DOI: 10.1016/j.ijhydene.2022.10.228

Glaeser, E. L., & Kahn, M. E. (2008). The Greenness of Cities: Carbon Dioxide Emissions and Urban Development. http://www.eia.doe.gov/oiaf/1605/ggrpt/carbon.html

Glantzi, A., & Ramfou, B. (2012). Urban green spaces [Lecture notes, in Greek]. National Technical University of Athens. https://dspace.lib.ntua.gr/xmlui/bitstream/handle/123456789/6594/Glantzi.pdf

Gobierno de Colombia. (2020). Bioeconomía para una Colombia Potencia viva y diversa: Hacia una sociedad impulsada por el conocimiento. Bogotá DC, Colombia.

Gong, X., Tao, X., Zhang, M., Xu, Y., Kwok, H. H. L., Dai, J., & Cheng, J. C. P. (2023). Secure environmental, social, and governance (ESG) data management for construction projects using blockchain. *OUCI*. https://ouci.dntb.gov.ua/en/works/loegEKz7/

Gorji, A. A., & Martek, I. (2023). The Role of Renewable Energy as a 'Green Growth' Strategy for the Built Environment. *Buildings (Basel, Switzerland)*, *13*(5), 1356. DOI: 10.3390/buildings13051356

Gospodini, A. (2004). Urban morphology and place identity in European cities: Built heritage and innovative design. *Journal of Urban Design*, *9*(2), 225–248. DOI: 10.1080/1357480042000227834

Government Gazette of Greece (2002). Establishment, organization and operation of a legal entity under private law named "Organization for the Administration and Management of the Antonis Tritsis Environmental Awareness Park" at the Vassilissis Tower site [FEK A' 172/2002, in Greek].

Government Gazette of Greece (2016). New support scheme for electricity generation stations from Renewable Energy Sources and High Efficiency Cogeneration of Heat and Power – Provisions for the legal and functional unbundling of supply and distribution sectors in the natural gas market and other provisions [FEK A' 149/2016, in Greek]

Government Gazette of Greece. (1985). The master plan and environmental protection programme of the wider Athens region [FEK A' 18/1985, in Greek].

Greene, M., Mora, R. I., Figueroa, C., & Waintrub, N., & de D. Ortúzar, J. (. (2017). Towards a sustainable city: Applying urban renewal incentives according to the social and urban characteristics of the area. *Habitat International*, *68*, 15–23. DOI: 10.1016/j.habitatint.2017.03.004

Grewal, J., Hauptmann, C., & Serafeim, G. (2021). Material sustainability information and stock price informativeness. *Journal of Business Ethics*, *171*(3), 513–544. DOI: 10.1007/s10551-020-04451-2

Griffith, R., & Nicholls, M. (2024). *Financing UK place-based climate action: From Westminster to Cumberland*. Grantham Research Institute on Climate Change and the Environment, London School of Economics and Political Science.

Guandalini, I. (2022). Sustainability through digital transformation: A systematic literature review for research guidance. *Journal of Business Research*, *148*, 456–471. DOI: 10.1016/j.jbusres.2022.05.003

Guberman, R. (2021). "What is Waste-to-Energy?" Retrieved 27/03/2024, 2024, from https://www.rts.com/blog/what-is-waste-to-energy/

Guinée, J. (2016). *Life Cycle Sustainability Assessment: What Is It and What Are Its Challenges? Taking Stock of Industrial Ecology. R. Clift and A. Druckman.* Springer International Publishing.

Gunawardena, K. R., Wells, M. J., & Kershaw, T. (2017). Utilising green and bluespace to mitigate urban heat island intensity. *The Science of the Total Environment, 584–585*, 1040–1055. DOI: 10.1016/j.scitotenv.2017.01.158 PMID: 28161043

Gustafsson, M., & Anderberg, S. (2022). Biogas policies and production development in Europe: A comparative analysis of eight countries. *Biofuels, 13*(8), 931–944. DOI: 10.1080/17597269.2022.2034380

Hajiyev, S. (2023). Blockchain technology in EU sustainability reporting: Legal analysis. *Lund University Publications*. https://lup.lub.lu.se/student-papers/search/publication/9157333

Hamiduddin, I., & Daseking, W. (2014). Community-based planning in Freiburg, Germany: The case of Vauban. In *Community Action and Planning* (pp. 237–258). Policy Press. https://bristoluniversitypressdigital.com/edcollchap/book/9781447315193/ch013.xml

Han, S., Cao, B., Fu, Y., & Luo, Z. (2018). A liner shipping competitive model with consideration of service quality management. *Annals of Operations Research, 270*(1-2), 155–177. DOI: 10.1007/s10479-016-2386-y

Haque, M. O., Mohammad, F., & Rahman, R. (2021). Shipping Container Housing for Architectural Resilience in Coastal Regions: *Addressing Construction and Material Sustainability. ZEMCH International Conference*, 163 - 177.

Hariram, N. P., Mekha, K. B., Suganthan, V., & Sudhakar, K. (2023). Sustainalism: An Integrated Socio-Economic-Environmental Model to Address Sustainable Development and Sustainability. *Sustainability (Basel), 15*(13), 10682. Advance online publication. DOI: 10.3390/su151310682

Hartl, R., Harms, P., & Egermann, M. (2024). Towards transformation-oriented planning: What can sustainable urban mobility planning (SUMP) learn from transition management (TM)? *Transport Reviews, 44*(1), 167–190. DOI: 10.1080/01441647.2023.2239497

Hasan, M. H., Hossain, M. Z., Hasan, L., & Dewan, M. A. (2023). FinTech and sustainable finance: How is FinTech shaping the future of sustainable finance? *ResearchGate*. https://www.researchgate.net/publication/386548986_FinTech_and_Sustainable_Finance_How_is_FinTech_Shaping_the_Future_of_Sustainable_Finance

Hassebo, A., & Tealab, M. (2023). Global models of smart cities and potential IoT applications: A review. *IoT*, *4*(3), 366–411. DOI: 10.3390/iot4030017

Haupt, W. (2018). European Municipalities Engaging in Climate Change Mitigation and Adaptation Networks: Examining the Case of the Covenant of Mayors. In Y. Yamagata & A. Sharifi (Eds.), *Resilience-Oriented Urban Planning: Theoretical and Empirical Insights* (pp. 93–110). Springer International Publishing. DOI: 10.1007/978-3-319-75798-8_5

Hellenic Statistical Authority. (2021). 2021 Population-Housing Census. https://www.statistics.gr/en/2021-census-pop-hous

Heravi, E. D. (2021). Linked Data and Cultural Heritage: A Systematic Review of Participation, Collaboration. [JOCCH]. *Journal on Computing and Cultural Heritage*, *14*(2), 18.

Herden, C. J., Alliu, E., Cakici, A., Cormier, T., Deguelle, C., Gambhir, S., Griffiths, C., Gupta, S., Kamani, S. R., Kiratli, Y.-S., Kispataki, M., Lange, G., Moles De Matos, L., Tripero Moreno, L., Betancourt Nunez, H. A., Pilla, V., Raj, B., Roe, J., Skoda, M., … Edinger-Schons, L. M. (2021). Corporate digital responsibility: New corporate responsibilities in the digital age. Sustainability Management Forum | NachhaltigkeitsManagementForum, 29(1), 13–29.

Hernández-Lamas, P., Gavilán, A., & Bernabeu-Larena, J. (2016). Parks and roads build the cities: The M-30 and Madrid-Río project, building landscape. In Territory, V. C. (Ed.), *Back to the sense of the city: International monograph book* (pp. 415–428). Centre de Política de Sòl i Valoracions. DOI: 10.5821/ctv.8121

Heshmati, A., & Rashidghalam, M. (2021). Assessment of the urban circular economy in Sweden. *Journal of Cleaner Production*, *310*, 127475. DOI: 10.1016/j.jclepro.2021.127475

Hess, D. J. (2014). Sustainability transitions: A political coalition perspective. *Research Policy*, *43*(2), 278–283. DOI: 10.1016/j.respol.2013.10.008

Hewitt, R. J., Bradley, N., Baggio Compagnucci, A., Barlagne, C., Ceglarz, A., Cremades, R., McKeen, M., Otto, I. M., & Slee, B. (2019). Social Innovation in Community Energy in Europe: A Review of the Evidence. *Frontiers in Energy Research*, *7*, 31. Advance online publication. DOI: 10.3389/fenrg.2019.00031

Heynen, N. (2017). Urban Political Ecology. In *International Encyclopedia of Geography* (pp. 1–9). John Wiley & Sons, Ltd., DOI: 10.1002/9781118786352.wbieg1110

Hiratsuka-Sasaki, A., & Kojima, M. (2020). Inter-municipal cooperation on solid waste management in Japan: Its challenges and implications for ASEAN countries. Regional Waste Management – Inter-municipal Cooperation and Public and Private Partnership. Jakarta, Indonesia. 61: 61-82.

Hiswara, A. A., Aziz, A. M., & Pujowati, Y. (2023). Cultural Preservation in a Globalized World: Strategies for Sustaining Heritage. *West Science Social and Humanities Studies*, *1*(03), 98–106. DOI: 10.58812/wsshs.v1i03.250

Hoefsloot, F. I., Martínez, J., & Pfeffer, K. (2024). An emerging knowledge system for future water governance: Sowing water for Lima. *Territory, Politics, Governance*, *12*(6), 825–845. DOI: 10.1080/21622671.2021.2023365

Hofstad, H., Sørensen, E., Torfing, J., & Vedeld, T. (2022). Designing and leading collaborative urban climate governance: Comparative experiences of co-creation from Copenhagen and Oslo. *Environmental Policy and Governance*, *32*(3), 203–216. DOI: 10.1002/eet.1984

Holling, C. S. (1973). Resilience and stability of ecological systems. *Annual Review of Ecology and Systematics*, *4*(1), 1–23. DOI: 10.1146/annurev.es.04.110173.000245

Hongo, T. (2016). Circular economy potential and public–private partnership models in Japan. Towards a Circular Economy: Corporate

Hoornweg, D., & Bhada-Tata, P. (2012). *What a waste: a global review of solid waste management*. DOI: 10.1596/17388

Hoppe, T., & Lulofs, K. (2008). The Impact of Multi-Level Governance on Energy Performance in the Current Dutch Housing Stock. *Energy & Environment*, *19*(6), 819–830. DOI: 10.1260/095830508785363587

Hospido, A., Moreira, T., Martín, M., Rigola, M., & Feijoo, G. (2005). Environmental evaluation of different treatment processes for sludge from urban wastewater treatments: Anaerobic digestion versus thermal processes. *The International Journal of Life Cycle Assessment*, *10*(5), 336–345. DOI: 10.1065/lca2005.05.210

Hossain, M. Z., & Ali, S. (2023). Exploring the role of artificial intelligence in ESG disclosure. *Journal of Business Ethics*, *168*(3), 1–17. DOI: 10.1007/s10551-023-04924-3

Hossain, S., Law, H. J., & Asfaw, A. (2022). *The waste crisis: roadmap for sustainable waste management in developing countries*. John Wiley & Sons. DOI: 10.1002/9781119811961

Hougaard, I.-M., & Christiansen, K. L. (2022, September 13). *Net zero: Copenhagen's failure to meet its 2025 target casts doubt on other major climate plans*. The Conversation. https://theconversation.com/net-zero-copenhagens-failure-to-meet-its-2025-target-casts-doubt-on-other-major-climate-plans-190033

Howard, E. (1965). Garden cities of tomorrow. MIT Press. (Original work published 1902) https://www.ipcc.ch/report/ar6/wg2/

Hsu, H.-W., Binyet, E., Nugroho, R. A. A., Wang, W.-C., Srinophakun, P., Chein, R.-Y., Demafelis, R., Chiarasumran, N., Saputro, H., Alhikami, A. F., Sakulshah, N., & Laemthong, T. (2024). Toward sustainability of Waste-to-Energy: An overview. *Energy Conversion and Management*, *321*, 119063. DOI: 10.1016/j.enconman.2024.119063

Huang, W.-H., & Chao, M.-C. (2014). The Impacts of the Mass Rapid Transit System on Household Car Ownership in Taipei. *[Journal of Sustainable Development of Energy, Water and Environment Systems]*, *[2]*([2]), [191]-[207].

Hubbart, J. (1994). *Reuniting man and nature: The need for natural landscapes in urban areas*. Macalester College.

Hufen, H., & de Bruijn, H. (2016). Getting the incentives right. Energy performance contracts as a tool for property management by local government. *Journal of Cleaner Production*, *112*, 2717–2729. DOI: 10.1016/j.jclepro.2015.10.036

Huovila, A., Siikavirta, H., Antuña Rozado, C., Rökman, J., Tuominen, P., Paiho, S., Hedman, Å., & Ylén, P. (2022). Carbon-neutral cities: Critical review of theory and practice. *Journal of Cleaner Production*, *341*, 130912. DOI: 10.1016/j.jclepro.2022.130912

Huston, S., & Lahbash, E. (2018). *Land Value Capture and Tax Increment Financing: Overview and Considerations for Sustainable Urban Investment* (SSRN Scholarly Paper 3103949). Social Science Research Network. DOI: 10.2139/ssrn.3103949

Hutton, J. G., Goodman, M. B., Alexander, J. B., & Genest, C. M. (2001). Reputation management: The new face of corporate public relations? *Public Relations Review*, *27*(3), 247–261. DOI: 10.1016/S0363-8111(01)00085-6

Iamonico, D. (2022). Biodiversity in urban areas: The extraordinary case of Appia Antica Regional Park (Rome, Italy). *Plants, 11*(16), 2122. DOI: 10.3390/plants11162122 PMID: 36015425

Ibanez, R., Andres, A., Viguri, J., Ortiz, I., & Irabien, J. (2000). Characterisation and management of incinerator wastes. *Journal of Hazardous Materials, 79*(3), 215–227. DOI: 10.1016/S0304-3894(00)00268-5 PMID: 11077160

Ignatov, K., & Rudolf, M. (2023). Sentimental sustainability: Does what companies say tell more than what companies do? *Financial Markets, Institutions and Instruments, 32*(4), 221–252. DOI: 10.1111/fmii.12181

Illankoon, I. M. C. S., Tam, V. W. Y., Le, K. N., Tran, C. N. N., & Ma, M. (2019). Review on green building rating tools worldwide: Recommendations for Australia. *Journal of Civil Engineering and Management, 25*(8), 831–847. DOI: 10.3846/jcem.2019.10928

Imbrenda, V., Rontos, K., Salvati, L., Maialetti, M., Vardopoulos, I., Alhussen, A., & Ciaschini, C. (2025). Traditional landscapes in Mediterranean areas: environmental protection vs economic growth. In Ciaschini, C., & Imbrenda, V. (Eds.), *Local development and socio-environmental systems. Experiences and practices toward sustainability in Mediterranean Europe* (pp. 19–38). River Publishers. DOI: 10.1201/9788770047616-2

International Energy Agency. (2020). *Outlook for biogas and biomethane: Prospects for organic growth*. IEA.

International Energy Agency. (2022). *World Energy Outlook*. International Energy Agency.

International Energy Agency. (2024). World Energy Outlook 2024. www.iea.org/terms

International Solid Waste Association. (2024). Global waste management outlook 2024: beyond an age of waste, turning rubbish into a resource. Nairobi, Kenya: 15.

IPCC. (2022). *Climate change 2022: Impacts, adaptation and vulnerability. Contribution of Working Group II to the Sixth Assessment Report of the Intergovernmental Panel on Climate Change*. Cambridge University Press.

Irimia-Vladu, M. (2014). “Green” electronics: Biodegradable and biocompatible materials and devices for sustainable future. *Chemical Society Reviews, 43*(2), 588–610. DOI: 10.1039/C3CS60235D PMID: 24121237

Islam, A., Islam, T., Mahmud, H., Raihan, O., Islam, M. S., Marwani, H. M., Rahman, M. M., Asiri, A. M., Hasan, M. M., Hasan, M. N., Salman, M. S., Kubra, K. T., Shenashen, M. A., Sheikh, M. C., & Awual, M. R. (2024). Accelerating the green hydrogen revolution: A comprehensive analysis of technological advancements and policy interventions. *International Journal of Hydrogen Energy*, *67*, 458–486. DOI: 10.1016/j.ijhydene.2024.04.142

Ismaeel, W. S. E., & Kassim, N. (2023). An environmental management plan for construction waste management. *Ain Shams Engineering Journal*, *14*(12), 102244. Advance online publication. DOI: 10.1016/j.asej.2023.102244

Ismagilova, Elvira, Laurie Hughes, Nripendra P. Rana, and Yogesh K. Dwivedi. 2020. "Security, Privacy, and Risks Within Smart Cities: Literature Review and Development of a Smart City Interaction Framework." *Information Systems Frontiers 2020 24:2* 24(2): 393–414. .DOI: 10.1007/s10796-020-10044-1

Ismail, C., Wiropranoto, F., Takama, T., Lieu, J., & Virla, L. D. (2021). Frugal Eco-innovation for Addressing Climate Change in Emerging Countries: Case of Biogas Digester in Indonesia. Handbook of Climate Change Management: Research, Leadership. *Transformation (Durban)*, *1*, 693–719. DOI: 10.1007/978-3-030-57281-5_177

Iwuanyanwu, O., Gil-Ozoudeh, I., Okwandu, A. C., & Ike, C. S. (2024). Cultural and social dimensions of green architecture: Designing for sustainability and community well-being. *International Journal of Applied Research in Social Sciences*, *6*(8), 1951–1968. DOI: 10.51594/ijarss.v6i8.1477

Jabareen, Y. R. (2006). Sustainable Urban Forms: Their Typologies, Models, and Concepts. *Journal of Planning Education and Research*, *26*(1), 38–52. DOI: 10.1177/0739456X05285119

Jain, P., & Singh, P. (2023). Lethargic approach in expediting financial reporting system: A case of construction firm. *International Journal of Teaching and Case Studies*, *14*(1), 1–10. DOI: 10.1504/IJTCS.2023.131663

Jain, P., Singh, P., & Kushwaha, R. (2024). Investigating change management as a novel digital wallet service innovation in banks: A case of Indian bank. *International Journal of Teaching and Case Studies*, *15*(1), 67–81. DOI: 10.1504/IJTCS.2025.143552

Jambeck, J. R., Geyer, R., Wilcox, C., Siegler, T. R., Perryman, M., Andrady, A., Narayan, R., & Law, K. L. J. (2015). Plastic waste inputs from land into the ocean. *Science*, *347*(6223), 768–771. DOI: 10.1126/science.1260352 PMID: 25678662

Jameel, M. K., Mustafa, M. A., Ahmed, H. S., Mohammed, A., Ghazy, H., Shakir, M. N., Lawas, A. M., Mohammed, S., Idan, A. H., Mahmoud, Z. H., Sayadi, H., & Kianfar, E. (2024). Biogas: Production, properties, applications, economic and challenges: A review. *Results in Chemistry*, *7*, 101549. DOI: 10.1016/j.rechem.2024.101549

Jansson, J., Marell, A., & Nordlund, A. (2010). Green consumer behavior: Determinants of curtailment and eco-innovation adoption. *Journal of Consumer Marketing*, *27*(4), 358–370. DOI: 10.1108/07363761011052396

Jayakumar Menon, V., & Palackal, A. (2022). Centralized and decentralized approaches to solid waste management–a case study. *The Holistic Approach to Environment*, *12*(4), 155–164. DOI: 10.33765/thate.12.4.3

Jen Heathcote, H. F. (2017). Predicting and Adapting to Climate Change: Challenges for the Historic Environment. *The Historic Environment*, *8*(2), 89–100. Advance online publication. DOI: 10.1080/17567505.2017.1317071

Jesic, J., Okanovic, A., & Panic, A. A. (2021). Net zero 2050 as an EU priroty: Modeling a system for efficient investments in eco innovation for climate change mitigation. *Energy, Sustainability and Society*, *11*(1), 50. DOI: 10.1186/s13705-021-00326-0

Jiang, P., Zhang, L., You, S., Van Fan, Y., Tan, R. R., Klemeš, J. J., & You, F. (2023). Blockchain technology applications in waste management: Overview, challenges and opportunities. *Journal of Cleaner Production*, *421*, 138466. DOI: 10.1016/j.jclepro.2023.138466

Jiang, X., Zhang, J., Yang, C., & Wan, R. (2023). Evaluating the service quality of insular and coastal recreational fisheries by integration of the SERVQUAL-fuzzy Kano model and importance-performance analysis. *Ocean and Coastal Management*, *243*, 106753. DOI: 10.1016/j.ocecoaman.2023.106753

Johansen, R. (2022). Oslo is Demonstrating Ambitious Leadership through its Climate Budget. *Journal of City Climate Policy and Economy*, *1*(1), 61–64. DOI: 10.3138/jccpe-2022.1.1.0004

Jordová, R., & Brůhová-Foltýnová, H. (2021). Rise of a New Sustainable Urban Mobility Planning Paradigm in Local Governance: Does the SUMP Make a Difference? *Sustainability (Basel)*, *13*(11), 11. Advance online publication. DOI: 10.3390/su13115950

Joynson, C. (2018, August). Corporate digital responsibility: Principles to guide progress. LinkedIn. Retrieved from https://www.linkedin.com/pulse/corporate-digital-responsibility-principles-guide-progress-joynson/

Julio Garrote, A. D.-H. (2020). A Framework Proposal for Regional-Scale Flood-Risk Assessment of Cultural Heritage Sites and Application to the Castile and León Region (Central Spain). *Water (Basel), 12*(2), 329. Advance online publication. DOI: 10.3390/w12020329

Kabir, Z., & Kabir, M. (2021). Solid waste management in developing countries: Towards a circular economy. Handbook of Solid Waste Management: Sustainability through Circular Economy, Springer: 1-34.

Kagias, P., Garefalakis, A., Passas, I., Kyriakogkonas, P., & Sariannidis, N. (2024). Whistleblowing Based on the Three Lines Model. *Administrative Sciences 2024, Vol. 14, Page 83, 14*(5), 83. DOI: 10.3390/admsci14050083

Kagias, P., Sariannidis, N., Garefalakis, A., Passas, I., & Kyriakogkonas, P. (2023). Validating the Whistleblowing Maturity Model Using the Delphi Method. Administrative Sciences 2023, Vol. 13, Page 120, 13(5), 120. https://doi.org/DOI: 10.3390/admsci13050120

Kaiser, Z. R. M. A. (2024). Smart Governance for Smart Cities and Nations. *Journal of Economy and Technology, 2*, 216–234. DOI: 10.1016/j.ject.2024.07.003

Kaklauskas, A., Lepkova, N., Raslanas, S., Vetloviene, I., Milevicius, V., & Sepliakov, J. (2021). COVID-19 and green housing: A review of relevant literature. In Energies (Vol. 14, Issue 8). MDPI AG. DOI: 10.3390/en14082072

Kanaujia, K. (2020). Green Buildings: Building a Greener City, a Greener Future—An Indian Perspective. Geoecology of Landscape Dynamics, 223-241.

Kaplan, A. M., & Haenlein, M. (2010). Users of the world, unite! The challenges and opportunities of Social Media. *Business Horizons, 53*(1), 59–68. Retrieved December 13, 2024, from. DOI: 10.1016/j.bushor.2009.09.003

Kaplan, R. (1985). The analysis of perception via preference: A strategy for studying how the environment is experienced. *Landscape Planning, 12*(2), 161–176. DOI: 10.1016/0304-3924(85)90058-9

Kapoor, A., Sindwani, R., & Goel, M. (2022a). Assessing mobile banking service quality dimensions using multi-criteria decision making. In *Future of Work and Business in Covid-19 Era: Proceedings of IMC-2021* (pp. 131-147). Singapore: Springer Nature Singapore. DOI: 10.1007/978-981-19-0357-1_12

Kapoor, A., Sindwani, R., & Goel, M. (2021a). Ranking mobile wallet service providers using fuzzy multi-criteria decision-making approach. [IJEBR]. *International Journal of E-Business Research, 17*(4), 1–21. DOI: 10.4018/IJEBR.2021100102

Kapoor, A., Sindwani, R., & Goel, M. (2021b). Prioritising the key factors influencing the adoption of mobile wallets: An Indian perspective in covid-19 era. *Journal of Information Technology Management, 13*(4), 161–182.

Kapoor, A., Sindwani, R., & Goel, M. (2023a). Exploring quick commerce service experience: A moderated mediated investigation using SEM and fsQCA. *Total Quality Management & Business Excellence, 34*(13-14), 1896–1919. DOI: 10.1080/14783363.2023.2213653

Kapoor, A., Sindwani, R., & Goel, M. (2023b). Evaluating mobile wallet acceptance factors using best worst method. *International Journal of Process Management and Benchmarking, 13*(4), 449–469. DOI: 10.1504/IJPMB.2023.129822

Kapoor, A., Sindwani, R., & Goel, M. (2024a). A novel framework for understanding the interplay between the mobile wallet service quality dimensions and loyalty intention. *The TQM Journal.*

Kapoor, A., Sindwani, R., & Goel, M. (2024b). Exploring the symmetric and asymmetric effects of mobile wallet service experience on brand loyalty in India: Evidence from SEM and fsQCA. *Journal of Financial Services Marketing, 29*(4), 1–18. DOI: 10.1057/s41264-024-00280-w

Kapoor, A., Sindwani, R., & Goel, M. (2024c). Mobile wallets: Theoretical and empirical analysis. *Global Business Review, 25*(2, suppl), S211–S228. DOI: 10.1177/0972150920961254

Kapoor, A., Sindwani, R., & Goel, M. (2025). Assessing mobile wallet service quality using analytic hierarchy process and fuzzy evaluation method. *International Journal of Services and Operations Management, 50*(2), 246–266. DOI: 10.1504/IJSOM.2025.144481

Kapoor, A., Sindwani, R., Goel, M., & Shankar, A. (2022b). Mobile wallet adoption intention amid COVID-19 pandemic outbreak: A novel conceptual framework. *Computers & Industrial Engineering, 172*, 108646. DOI: 10.1016/j.cie.2022.108646 PMID: 36118531

Kapoor, L. E. (2024). Climate Responsive Strategies in Vernacular Architecture: A Comparative Analysis at Various Latitude. *International Research Journal of Multidisciplinary Scope, 5*(4), 1047–1068. DOI: 10.47857/irjms.2024.v05i04.01637

Karytsas, S., Vardopoulos, I., & Theodoropoulou, E. (2019). Factors affecting sustainable market acceptance of residential microgeneration technologies. A two time period comparative analysis. *Energies, 12*(17), 3298. DOI: 10.3390/en12173298

Kashif, U., & Meo, M. S. (2024). Case Studies in Green Bond Portfolio Management: Best Practices and Lessons Learned. In Green Bonds and Sustainable Finance (pp. 50-68). Routledge.

Kassem, N., Hockey, J., Lopez, C., Lardon, L., Angenent, L. T., & Tester, J. (2020). Integrating anaerobic digestion, hydrothermal liquefaction, and biomethanation within a power-to-gas framework for dairy waste management and grid decarbonization: A techno-economic assessment. *Sustainable Energy & Fuels*, *4*(9), 4644–4661. DOI: 10.1039/D0SE00608D

Kassios, K., & Perperidou, D. (2005, March 18). The detergent contribution of urban green in the city's atmosphere [Conference presentation, in Greek]. Technical Chamber of Greece Conference "Air quality in urban areas: New data and perspectives", Athens, Greece.

Katsafadou, K. (2017). Listing report of hydrolytic sources in Antonis Tritsis Metropolitan Park [in Greek]. Antonis Tritsis Metropolitan Park. https://parkotritsis.gr/wp-content/uploads/Ereynhtiko-Yliko/EreynesKaiMeletes/Gnwm.%20ektheseis/Katsafadou_2017.pdf

Kaur, R., & Gupta, K. (2022). Blue-green infrastructure (BGI) network in urban areas for sustainable stormwater management: A geospatial approach. *City and Environment Interactions*, *16*, 100087. DOI: 10.1016/j.cacint.2022.100087

Kayan, B.A., H. I. (2017). Green Maintenance for Heritage Buildings: Low Carbon Repair Appraisal Approach on Laterite Stones. *Chemical Engineering Transactions*, *56*, 337–342. DOI: 10.3303/CET1756057

Kaza, S., Yao, L., Bhada-Tata, P., & Woerden, F. V. (2018). What a waste 2.0: A Global Snapshot of Solid Waste Management to 2050. Washington.

Kazachenok, O. P., Stankevich, G. V., Chubaeva, N. N., & Tyurina, Y. G. (2023). Economic and legal approaches to the humanization of FinTech in the economy of artificial intelligence through the integration of blockchain into ESG finance. *ResearchGate*. https://www.researchgate.net/publication/370027001_Economic_and_legal_approaches_to_the_humanization_of_FinTech_in_the_economy_of_artificial_intelligence_through_the_integration_of_blockchain_into_ESG_Finance

Kempton, Y., Salvati, L., & Vardopoulos, I. (2022). Long-term planning and development for urban and regional inclusion, safety, resilience, and sustainability. Insights from Singapore. *Region & Periphery*, *14*(10), 58–78. DOI: 10.12681/rp.32607

Kennedy, C., Pincetl, S., & Bunje, P. (2011). The study of urban metabolism and its applications to urban planning and design. *Environmental Pollution, 159*(8), 1965–1973. DOI: 10.1016/j.envpol.2010.10.022 PMID: 21084139

Keramitsoglou, I., Kiranoudis, C. T., Ceriola, G., Weng, Q., & Rajasekar, U. (2011). Identification and analysis of urban surface temperature patterns in Greater Athens, Greece, using MODIS imagery. *Remote Sensing of Environment, 115*(12), 3080–3090. DOI: 10.1016/j.rse.2011.06.014

Kern, S. J., Pfeifer, C., & Hofbauer, H. (2013). Cogasification of Polyethylene and Lignite in a Dual Fluidized Bed Gasifier. *Industrial & Engineering Chemistry Research, 52*(11), 4360–4371. DOI: 10.1021/ie303453e

Khakee, A. (2007). From Olympic village to middle-class waterfront housing project: Ethics in Stockholm's development planning. *Planning Practice and Research, 22*(2), 235–251. DOI: 10.1080/02697450701584402

Khan, M. A. S., Du, J., Malik, H. A., Anuar, M. M., Pradana, M., & Yaacob, M. R. B. (2022). Green innovation practices and consumer resistance to green innovation products: Moderating role of environmental knowledge and pro-environmental behavior. *Journal of Innovation and Knowledge, 7*(4), 100280. Advance online publication. DOI: 10.1016/j.jik.2022.100280

Khan, S., Ali, B., Alharbi, A. A., Alotaibi, S., & Alkhathami, M. (2024). Efficient IoT-assisted waste collection for urban smart cities. *Sensors (Basel), 24*(10), 3167. DOI: 10.3390/s24103167 PMID: 38794020

Khanum, F. (2021). The Northeast India in the midst of climate change.

Khawaja, M. K., Alkayyali, K., Almanasreh, M., & Alkhalidi, A. (2024). Waste-to-energy barriers and solutions for developing countries with limited water and energy resources. *The Science of the Total Environment, 926*, 172096. DOI: 10.1016/j.scitotenv.2024.172096 PMID: 38556009

Khemir, S. (2019). Perception of ESG criteria by mainstream investors: Evidence from Tunisia. *International Journal of Emerging Markets, 14*(5), 752–768. DOI: 10.1108/IJOEM-05-2017-0172

Khemir, S., Baccouche, C., & Ayadi, S. D. (2019). The influence of ESG information on investment allocation decisions: An experimental study in an emerging country. *Journal of Applied Accounting Research, 20*(4), 458–480. DOI: 10.1108/JAAR-12-2017-0141

Kim, A. M., & Eisenlohr, A. (2022). *Community Land Trusts for Sustainably Affordable Rental Housing Redevelopment: A Case Study of Rolland Curtis Gardens in Los Angeles. 24*(1). https://www.huduser.gov/portal/periodicals/cityscpe/vol23num4/community-land-trusts.html

Kimbrough, M. D., Wang, X., Wei, S., & Zhang, J. (2024). Does voluntary ESG reporting resolve disagreement among ESG rating agencies? *European Accounting Review, 33*(1), 15–47. DOI: 10.1080/09638180.2022.2088588

Kitchin, R. (2014). The Real-Time City? Big Data and Smart Urbanism. *GeoJournal, 79*(1), 1–14. DOI: 10.1007/s10708-013-9516-8

Kojima, M. (2020). *Regional Waste Management–Inter-municipal Cooperation and Public and Private Partnership*. Economic Research Institute for ASEAN and East Asia.

Kokatnur, T., Faris, F., Gunay, B., O'Brien, W., & Azar, E. (2025). The WELL Building Standard: A literature review and bibliometric analysis of a nascent field. In Journal of Building Engineering (Vol. 103). Elsevier Ltd. DOI: 10.1016/j.jobe.2025.112121

Koohestani, S. S., Mukheibir, P., Wakefield-Rann, R., & Santamouris, M. (2025). Adopting a socio-technical perspective on the challenges and barriers in transitioning to Blue-Green Infrastructure (BGI). *Blue-Green Systems, 7*(1), 79–94. DOI: 10.2166/bgs.2025.011

Koronis, G., Silva, A., & Fontul, M. (2013). Green composites: A review of adequate materials for automotive applications. *Composites. Part B, Engineering, 44*(1), 120–127. DOI: 10.1016/j.compositesb.2012.07.004

Kougias, P. G., & Angelidaki, I. (2018). Biogas and its opportunities—A review. *Frontiers of Environmental Science & Engineering, 12*(3), 1–12. DOI: 10.1007/s11783-018-1037-8

Kreitler, J., Papenfus, M., Byrd, K., & Labiosa, W. (2013). Interacting Coastal Based Ecosystem Services: Recreation and Water Quality in Puget Sound, WA. *PLoS One, 8*(2), 56670. DOI: 10.1371/journal.pone.0056670 PMID: 23451067

Krishnakumar, S. (2025). Blue-Green Infrastructure for Climate Resilience: Aiding Urban Development. In Integrating Blue-Green Infrastructure into Urban Development (pp. 99-128). IGI Global Scientific Publishing.

Kristia, K., & Rabbi, M. F. (2023). "Exploring the Synergy of Renewable Energy in the Circular Economy Framework: A Bibliometric Study." Sustainability (Switzerland) 15(17).

Krug, M., Di Nucci, M. R., Caldera, M., & De Luca, E. (2022). Mainstreaming Community Energy: Is the Renewable Energy Directive a Driver for Renewable Energy Communities in Germany and Italy? *Sustainability (Basel), 14*(12), 12. Advance online publication. DOI: 10.3390/su14127181

Küfeoğlu, S. (2024). Waste Management Emissions. Net Zero: Decarbonizing the Global Economies, Springer: 555-611.

Kulakovskaya, A., Knoeri, C., Radke, F., & Blum, N. U. (2023). Measuring the economic impacts of a circular economy: An evaluation of indicators. *Circular Economy and Sustainability, 3*(2), 657–692. DOI: 10.1007/s43615-022-00190-w PMID: 37408852

Kumar, A., Singh, P., Parihar, A.S. & Saraswat, S. (2025), "Analyzing the impact of psychological capital on the entrepreneurial intentions of students through mediating role of individual expectancy using PLS-SEM", *International Journal of Social Economics*, Vol. ahead-of-print No. ahead-of-print. DOI: 10.1108/IJSE-03-2024-0252

Kumar, M., Mehra, R., & Saini, M. (2024). Blockchain in Waste Tracking: Enhancing Transparency and Accountability. Municipal Solid Waste Management and Recycling Technologies, IGI Global: 57-70.

Kumar, A., Srivastava, V., Tabash, M. I., & Chawda, D. (2022). Profitability determinants of infrastructure public private partnerships (PPPs): Empirical evidence from Indian data. *Journal of Financial Management of Property and Construction, 27*(1), 91–111. DOI: 10.1108/JFMPC-09-2020-0062

Kumari, V., Ansari, A. I., & Singh, A. (2025). Empowering Urban Communities: The Role of Local Stakeholders in Solid Waste Management. Solid Waste Management for Rural Regions, IGI Global Scientific Publishing: 153-164.

Kuo, F. E., Bacaicoa, M., & Sullivan, W. C. (1998). Transforming inner-city landscapes: Trees, sense of safety, and preference. *Environment and Behavior, 30*(1), 28–59. DOI: 10.1177/0013916598301002

Kurdi, S. M. (2017). Heritage conservation and urban development: A supporting management model for the effective incorporation of archaeological sites in the planning process. *Journal of Cultural Heritage, 28*, 117–128. Advance online publication. DOI: 10.1016/j.culher.2017.05.007

Kushwaha, J., Sharma, A., & Singh, P. (2023). b. Exploration and prioritization of enablers to organization work–family balance planning for working sole Indian mothers integrating fuzzy Delphi and AHP. *International Journal of Social Economics, 50*(3), 398–418. DOI: 10.1108/IJSE-05-2022-0348

Kushwaha, J., Singh, P., & Kushwaha, R. (2024). Predicting working sole Indian mothers' satisfaction towards work–family balance integrating the Kano model and weighted average method. *International Journal of Social Economics*, *51*(6), 725–740. DOI: 10.1108/IJSE-02-2023-0074

Kushwaha, J., Singh, P., & Sharma, A. (2023). a, "Modelling the enablers of work-family balance among working single mothers integrating expert-mining and TISM-MICMAC". *Journal of Organizational Effectiveness: People and Performance*, *10*(4), 501–526. DOI: 10.1108/JOEPP-05-2022-0106

Kuzhin, M. F., Joshi, A., Mittal, V., Khatkar, M., & Guven, U. (2024). *Optimizing waste management through IoT and analytics: a case study using the waste management optimization test. BIO web of conferences*. EDP Sciences.

Kyramargiou, Z., & Vardopoulos, I. (2019). Use and sustainability of the ships' waste reception facilities. Evidence from the Port of Corinth, Greece. *Journal of Management and Regional Development*, *15*(3), 93–106.

Kyriakopoulos G.L. (2021). Environmental Legislation in European and International Contexts: Legal Practices and Social Planning toward the Circular Economy. Laws 10(1), 3, 15 pp. .DOI: 10.3390/laws10010003

Kyriakopoulos G.L., Sebos I. (2023). Enhancing Climate Neutrality and Resilience through Coordinated Climate Action: Review of the Synergies between Mitigation and Adaptation Actions. Climate 11(5), 105, 27 pp. DOI: DOI: 10.3390/cli11050105

Lakhouit, A. (2025). Revolutionizing Urban Solid Waste Management with AI and IoT: A review of smart solutions for waste collection, sorting, and recycling. *Results in Engineering*, *25*, 104018. DOI: 10.1016/j.rineng.2025.104018

Lamond, J., & Everett, G. (2019). Sustainable blue-green infrastructure: A social practice approach to understanding community preferences and stewardship. *Landscape and Urban Planning*, *191*, 103639. DOI: 10.1016/j.landurbplan.2019.103639

Latsoudis, P., Beriatos, G., & Probonas, N. (2017). Antonis Tritsis Environmental Information and Awareness Park's avifauna: Species-habitat management proposals [in Greek]. Hellenic Ornithological Society. https://parkotritsis.gr/wp-content/uploads/Ereynhtiko-Yliko/EreynesKaiMeletes/Gnwm.%20ektheseis/Latsoudis_Ornithopanida_2017.pdf

Latsoudis, P., Beriatos, G., & Probonas, N. (2023). List of bird species in the Antonis Tritsis Park [in Greek]. Hellenic Ornithological Society. https://old.ornithologiki.gr/page_list.php?lID=5&bc=1970

Laura Quesada-Ganuza, L. G. (2021). Do we know how urban heritage is being endangered by climate change? A systematic and critical review. *International Journal of Disaster Risk Reduction*, *65*, 65. DOI: 10.1016/j.ijdrr.2021.102551

Lee, S. U., Perera, H., Liu, Y., Xia, B., Lu, Q., Zhu, L., Cairns, J., & Nottage, M. (2024). Integrating ESG and AI: A comprehensive responsible AI assessment framework. *arXiv*. https://doi.org//arXiv.2408.00965DOI: 10.48550

Lee, J., Lin, K.-Y. A., Jung, S., & Kwon, E. E. (2023). Hybrid renewable energy systems involving thermochemical conversion process for waste-to-energy strategy. *Chemical Engineering Journal*, *452*, 139218. DOI: 10.1016/j.cej.2022.139218

Lee, J., Serafin, A. M., & Courteau, C. (2023). Corporate disclosure, ESG and green fintech in the energy industry. *SSRN*. https://papers.ssrn.com/sol3/papers.cfm?abstract_id=4090792 DOI: 10.1093/jwelb/jwad018

Leme, M. M. V., Rocha, M. H., Lora, E. E. S., Venturini, O. J., Lopes, B. M., & Ferreira, C. H. (2014). Techno-economic analysis and environmental impact assessment of energy recovery from Municipal Solid Waste (MSW) in Brazil. *Resources, Conservation and Recycling*, *87*, 8–20. DOI: 10.1016/j.resconrec.2014.03.003

Leonidou, L. C., Katsikeas, C. S., & Samiee, S. (2013). Marketing strategy determinants of export performance: A meta-analysis. *Journal of International Business Studies*, *44*(5), 493–510. Retrieved December 13, 2024, from. DOI: 10.1057/jibs.2013.13

Levine, D., Sussman, S., Ayalon, S. Y., & Aharon-Gutman, M. (2022). Rethinking Gentrification and Displacement: Modeling the Demographic Impact of Urban Regeneration. *Planning Theory & Practice*, *23*(4), 578–597. DOI: 10.1080/14649357.2022.2117399

Lind, A., & Espegren, K. (2018). Mitigation of Greenhouse Gas Emissions in Urban Areas: The Case of Oslo. In Giannakidis, G., Karlsson, K., Labriet, M., & Gallachóir, B. Ó. (Eds.), *Limiting Global Warming to Well Below 2 °C: Energy System Modelling and Policy Development* (pp. 391–405). Springer International Publishing., DOI: 10.1007/978-3-319-74424-7_23

Li, Q., Zou, G., Zeng, W., Gao, J., He, F., & Zhang, Y. (2024). ESG guidance and artificial intelligence support for power systems analytics in the energy industry. *Scientific Reports*, *14*(1), 11347. DOI: 10.1038/s41598-024-61491-8 PMID: 38762582

Liu, T., Chen, L., Yang, M., Sandanayake, M., Miao, P., Shi, Y., & Yap, P. S. (2022). Sustainability Considerations of Green Buildings: A Detailed Overview on Current Advancements and Future Considerations. In Sustainability (Switzerland) (Vol. 14, Issue 21). MDPI. DOI: 10.3390/su142114393

Liu, X., Wu, H., Wu, W., Fu, Y., & Huang, G. Q. (2019). Blockchain-enabled ESG reporting framework for sustainable supply chain. *ResearchGate*. https://www.researchgate.net/publication/344274654_Blockchain-Enabled_ESG_Reporting_Framework_for_Sustainable_Supply_Chain

Lobo, A. (2010). Assessing the service quality of container shipping lines in the international supply chain network - shippers' perspective. *International Journal of Value Chain Management*, *4*(3), 256–266. DOI: 10.1504/IJVCM.2010.033615

Lobschat, L., Mueller, B., Eggers, F., Brandimarte, L., Diefenbach, S., Kroschke, M., & Wirtz, J. (2021). Corporate digital responsibility. *Journal of Business Research*, *122*, 875–888. DOI: 10.1016/j.jbusres.2019.10.006

Lockie, Sean, Pasqual Capizzi, Tim Hawley, Luke Thompson, Max Russels, Ralf-Martin Soe, Semira Osmanovic Design, et al. 2024. "Acknowledgements Smarter and Inclusive Cities Module 1 Inclusivity for Smarter City Development Smarter and Inclusive Cities Smarter Governance Planning for Smarter and Inclusive Cities Smarter and Inclusive Cities Projects and Pilots Introduction to the Course."

Loizia, P., Voukkali, I., Zorpas, A. A., Navarro Pedreño, J., Chatziparaskeva, G., Inglezakis, V. J., Vardopoulos, I., & Doula, M. (2021). Measuring the level of environmental performance in insular areas, through key performed indicators, in the framework of waste strategy development. *The Science of the Total Environment*, *753*, 141974. DOI: 10.1016/j.scitotenv.2020.141974 PMID: 32906046

Lombardini, G., Pilogallo, A., & Tucci, G. The Provision of Ecosystem Services Along the Italian Coastal Areas: A Correlation Analysis Between Environmental Quality and Urbanization (2022). Lecture Notes in Computer Science 2022, 13380 LNCS, 298 - 314. DOI: DOI: 10.1007/978-3-031-10542-5_21

Looijen Bader, E. M., & Oostra, M. A. R. (2022). The Dutch energy transition in practice. *IOP Conference Series. Earth and Environmental Science*, *1085*(1), 012019. DOI: 10.1088/1755-1315/1085/1/012019

Loonen, L. C. M. (2020). Timing the Level of Citizen Participation in Spatial Planning Processes (Master's thesis).

Lopes, J. M., Gomes, S., & Trancoso, T. (2023). The dark side of green marketing: How greenwashing affects circular consumption? *Sustainability (Basel), 15*(15), 11649. DOI: 10.3390/su151511649

Lotha, T. N., Ritse, V., Nakro, V., Rudithongru, L., Hazarika, N., & Jamir, L. (2024). Climate Change Impact and Traditional Adaptation Practices in Northeast India: A Review. *Current World Environment, 19*(2), 558–575. DOI: 10.12944/CWE.19.2.5

Loures, L., Santos, R., & Panagopoulos, T. (2007). Urban parks and sustainable development: The case study of Portimão city [Conference paper]. Proceedings of the 3rd IASME/WSEAS International Conference on Energy, Environment, Ecosystems and Sustainable Development, Agios Nikolaos, Greece. University of Algarve.

Lu, H., Fu, Y., Xia, C., Lu, C., Wang, B., Yang, Q., & Wang, D. (2023). Low-carbon urban experiments from vision to reality: A systematic review of the literature from 2005 to 2020. *Climate Policy, 23*(8), 1058–1077. DOI: 10.1080/14693062.2023.2205371

Lyon, T. P., & Maxwell, J. W. (2011). Greenwash: Corporate environmental disclosure under threat of audit. *Journal of Economics & Management Strategy, 20*(1), 3–41. Retrieved December 13, 2024, from. DOI: 10.1111/j.1530-9134.2010.00282.x

Lyon, T. P., & Montgomery, A. W. (2015). 'The means and ends of greenwashing', *Organization & Environment*, 28(2), pp. 223-249. Retrieved December 13, 2024, from. DOI: 10.1108/SAMPJ-06-2017-0181

Mabrouk, M., Han, H., Fan, C., Abdrabo, K. I., Shen, G., Saber, M., Kantoush, S. A., & Sumi, T. (2023). Assessing the effectiveness of nature-based solutions-strengthened urban planning mechanisms in forming flood-resilient cities. *Journal of Environmental Management, 344*, 118260. DOI: 10.1016/j.jenvman.2023.118260 PMID: 37354590

Macchiavello, E., & Siri, M. (2020). Sustainable finance and fintech: Can technology contribute to achieving environmental goals? A preliminary assessment of 'green FinTech'. *SSRN*. https://papers.ssrn.com/sol3/papers.cfm?abstract_id=3672989 DOI: 10.2139/ssrn.3672989

Machiba, T. (2010). Eco-innovation for enabling resource efficiency and green growth: Development of an analytical framework and preliminary analysis of industry and policy practices. *International Economics and Economic Policy*, 7(2), 357–370. DOI: 10.1007/s10368-010-0171-y

Madrid City Council. (2025). Madrid Río. https://madrides.es/en/madrid-rio/

Madrid City Council. (2025, March 17). Parque Madrid Río. https://www.madrid.es/portales/munimadrid/es/Inicio/El-Ayuntamiento/Arganzuela/Direcciones-y-telefonos/Parque-Madrid-Rio/

Mahanta, N. R., Samuel, A., & Rajput, B. (2020). Community participation and bio rights possibilities in the conservation of urban wetlands: A study of Deepor Beel (Lake), Guwahati, India. 2022 10th International Conference on Reliability, Infocom Technologies and Optimization (Trends and Future Directions) (ICRITO), 1367–1372.

Mahmoudi, S., Baeyens, J., & Seville, J. P. K. (2010). NOx formation and selective non-catalytic reduction (SNCR) in a fluidized bed combustor of biomass. *Biomass and Bioenergy*, *34*(9), 1393–1409. DOI: 10.1016/j.biombioe.2010.04.013

Mahzouni, A. (2015). The 'Policy Mix' for Sustainable Urban Transition: The city district of Hammarby Sjöstad in Stockholm. *Environmental Policy and Governance*, *25*(4), 288–302. DOI: 10.1002/eet.1688

Maialetti, M., Halbac-Cotoara-Zamfir, R., Vardopoulos, I., & Salvati, L. (2024). Early desertification risk in advanced economies: Summarizing past, present and future trends in Italy. *Earth (Basel, Switzerland)*, *5*(4), 690–706. DOI: 10.3390/earth5040036

Maier, G., & Gluszak, M. (2024). The Diffusion of Green Building Certifications in Europe-A preliminary descriptive analysis. https://www.researchgate.net/publication/386336541

Manaktala, S. (2020). Green bonds in sustainable finance: exploring the case of India. Available at *SSRN* 3644116. DOI: 10.2139/ssrn.3644116

Mandal, D. (2018, June 1). *How Portland's "Right to Return" is Indeed Right to Return Housing to the Underrepresented—Berkeley Political Review*. https://bpr.studentorg.berkeley.edu/2018/06/01/how-portlands-right-to-return-is-indeed-right-to-return-housing-to-the-underrepresented/

Mandal, R., Nishant, N., Chutia, D., Aggarwal, S. P., & Sharma, B. (2024, February). LoRa Enabled IoT Sensor Framework for Monitoring Urban Flood in Guwahati City. In NIELIT's International Conference on Communication, Electronics and Digital Technologies (pp. 55-76). Singapore: Springer Nature Singapore. DOI: 10.1007/978-981-97-3604-1_5

Mandal, R. K., Alam, M. A., Prasad, R. R., & Kundra, S. (2022). Urbanization Trend In North East India: An Empirical Study. *Journal of Positive School Psychology*, *6*(9), 4641–4653.

Maranghi, S., Parisi, M. L., Facchini, A., Rubino, A., Kordas, O., & Basosi, R. (2020). Integrating urban metabolism and life cycle assessment to analyse urban sustainability. *Ecological Indicators*, *112*, 106074. DOI: 10.1016/j.ecolind.2020.106074

Marchesi, M., & Tweed, C. (2021). Social innovation for a circular economy in social housing. *Sustainable Cities and Society*, *71*, 102925. DOI: 10.1016/j.scs.2021.102925

Marique, A.-F., & Reiter, S. (2011). Towards more sustainable neighbourhoods: Are good practices reproducible and extensible? A review of a few existing "sustainable neighbourhoods." *Proceedings of PLEA 2011 - 27th Conference on Passive and Low Energy Architecture*, 27–32. https://orbi.uliege.be/bitstream/2268/88755/1/PLEA2011_afm.pdf

Marius Mosoarca, A. I. (2017). Failure analysis of historical buildings due to climate change. *Engineering Failure Analysis*, *82*, 666–680. DOI: 10.1016/j.engfailanal.2017.06.013

Markopoulos, E., Kirane, I. S., Gann, E. L., & Vanharanta, H. (2020). A democratic, green ocean management framework for environmental, social and governance (ESG) compliance. In *Human Interaction, Emerging Technologies and Future Applications II: Proceedings of the 2nd International Conference on Human Interaction and Emerging Technologies: Future Applications (IHIET–AI 2020), April 23-25, 2020, Lausanne, Switzerland* (pp. 21-33). Springer International Publishing. DOI: DOI: 10.1007/978-3-030-44267-5_4

Marlianingrum, P. R., Kusumastanto, T., Adrianto, L., & Fahrudin, A. (2021). Valuing habitat quality for managing mangrove ecosystem services in coastal Tangerang District, Indonesia. *Marine Policy*, *133*, 104747. DOI: 10.1016/j.marpol.2021.104747

Marquet, O., Anguelovski, I., Nello-Deakin, S., & Honey-Rosés, J. (2025). Decoding the 15-Minute City Debate: Conspiracies, Backlash, and Dissent in Planning for Proximity. *Journal of the American Planning Association*, *91*(1), 117–125. DOI: 10.1080/01944363.2024.2346596

Martin, L., & March, L. (1972). *Urban Space and Structures* (Martin, L., & March, L., Eds.). Cambridge University Press.

Marvin Ravan, M. J. (2023). A Vulnerability Assessment Framework for Cultural Heritage Sites: The Case of the Roman Ruins of Tróia. *International Journal of Disaster Risk Science*, *14*(1), 26–40. Advance online publication. DOI: 10.1007/s13753-023-00463-4

Matzarakis, A., Rammelberg, J., & Junk, J. (2013). Assessment of thermal bioclimate and tourism climate potential for central Europe-the example of Luxembourg. *Theoretical and Applied Climatology*, *114*(1-2), 193–202. DOI: 10.1007/s00704-013-0835-y

Mba, E. J., Okeke, F. O., Igwe, A. E., Ozigbo, C. A., Oforji, P. I., & Ozigbo, I. W. (2024). Evolving trends and challenges in sustainable architectural design; a practice perspective. *Heliyon*, *10*(20), e39400. Advance online publication. DOI: 10.1016/j.heliyon.2024.e39400 PMID: 39502211

McPherson, G., Simpson, J. R., Peper, P. J., Maco, S. E., Gardner, S. L., Cozad, S. K., & Xiao, Q. (2005). Municipal tree resource analysis. U.S. Department of Agriculture, Forest Service, Pacific Southwest Research Station.

McTarnaghan, S., Morales-Burnett, J., & Marx, R. (2022). *Urban resilience: From global vision to local practice*. Urban Institute.

Medda, F. (2012). Land value capture finance for transport accessibility: A review. *Journal of Transport Geography*, *25*, 154–161. DOI: 10.1016/j.jtrangeo.2012.07.013

Mediterranean Experts on Climate and Environmental Change (MedECC). (2020). Climate and environmental change in the Mediterranean Basin – Current situation and risks for the future (First Mediterranean Assessment Report). https://www.medecc.org/first-mediterranean-assessment-report-mar1/

Mehta, M., Mehta, D., Nair, N., & Mohanty, A. (2023). *ESG assessments for cities in India*. ResearchGate.

Meister, T., Schmid, B., Seidl, I., & Klagge, B. (2020). How municipalities support energy cooperatives: Survey results from Germany and Switzerland. *Energy, Sustainability and Society*, *10*(1), 18. DOI: 10.1186/s13705-020-00248-3 PMID: 32215236

Meitei, K. M., & Das, B. (2024). *Trends, Patterns and impacts of urbanization in North East India*. ResearchGate.

Melissas, D. (2009). *Urban greenery and its management by local government* [in Greek]. Institute of Local Government.

Merrill, N. H., Mulvaney, K. K., Martin, D. M., Chintala, M. M., Berry, W., Gleason, T. R., Balogh, S., & Humphries, A. T. (2018). A Resilience Framework for Chronic Exposures: Water Quality and Ecosystem Services in Coastal Social-Ecological Systems. *Coastal Management*, *46*(4), 242–258. DOI: 10.1080/08920753.2018.1474066 PMID: 31178625

Mhlanga, D. (2023). FinTech, financial inclusion, and sustainable development: Disruption, innovation, and growth. *Routledge*. https://www.routledge.com/FinTech-Financial-Inclusion-and-Sustainable-Development-Disruption-Innovation-and-Growth/Mhlanga/p/book/9781032657912?srsltid=AfmBOoo5NDLStJnH0bUpvjOzreX0FwibhTUcJ3boSiBRJBLMO1Td0zqc

Michael, C., & Hall, T. B. (2015). Climate change and cultural heritage: Conservation and heritage tourism in the Anthropocene. *Journal of Heritage Tourism*.

Ministry of Environment and Energy. (2015). Energy Efficiency Plan for Buildings of Regions and Municipalities. *Υπουργείο Περιβάλλοντος και Ενέργειας*. https://ypen.gov.gr/energeia/energeiaki-exoikonomisi/ktiria/schedio-energeiakis-apodosis-ktirion-perifereion-kai-dimon/

Mir, A. H., Sarma, K., & Upadhaya, K. (2021). Assessing the effectiveness of community managed forests for plant diversity conservation in Meghalaya, Northeast India. *Plant Diversity*, *44*(3), 243–254. DOI: 10.1016/j.pld.2021.11.010 PMID: 35769587

Mitra, S., Madhuvanthi, S., & Sabumon, P. C. (2024). Integrating Green Infrastructure. Nature-Based Solutions in Achieving Sustainable Development Goals, 167.

Mitra, A., & Paul, S. (2025). Empowering Indigenous Communities Through Regenerative Tourism: Cultural Preservation and Socio-Economic Resilience in Northeast India. In Sousa, B. B. (Eds.), *Regenerative Tourism for Social Development* (pp. 281–310). IGI Global. DOI: 10.4018/979-8-3373-0235-5.ch012

Mlindelwa, S., Chowdhury, S. D., & Lencwe, M. J. (2024). Overview of the challenges, developments, and solutions for electric vehicle charging infrastructure. *2024 32nd Southern African Universities Power Engineering Conference (SAUPEC)*, 1–6. DOI: 10.1109/SAUPEC60914.2024.10445031

Mohammed ElAdl, F. F. (2022). Managing microclimate challenges for museum buildings in Egypt. *Ain Shams Engineering Journal*.

Mohanty, A. K., Misra, M., & Drzal, L. T. (2002). Sustainable bio-composites from renewable resources: Opportunities and challenges in the green materials world. *Journal of Polymers and the Environment*, *10*(1-2), 19–26. DOI: 10.1023/A:1021013921916

MoHUA. (2021). *Smart Cities Mission: Government of India*. Ministry of Housing and Urban Affairs.

Montuori, M. (2024). Bridging past and present: Cutting-edge technologies for predictive conservation of built cultural heritage. *The International Archives of the Photogrammetry, Remote Sensing and Spatial Information Sciences*, *XLVIII-4/W10-2024*, 139–145. DOI: 10.5194/isprs-archives-XLVIII-4-W10-2024-139-2024

Moshood, T. D., Nawanir, G., Mahmud, F., Mohamad, F., Ahmad, M. H., & Abdul Ghani, A. (2022). Biodegradable plastic applications towards sustainability: A recent innovations in the green product. *Cleaner Engineering and Technology, 100404*, 100404. Advance online publication. DOI: 10.1016/j.clet.2022.100404

Mourtzis, D., Angelopoulos, J., & Panopoulos, N. (2022). A literature review of the challenges and opportunities of the transition from Industry 4.0 to Society 5.0. *Energies, 15*(17), 6276. DOI: 10.3390/en15176276

Moussa, A. S., & Elmarzouky, M. (2024). Beyond compliance: How ESG reporting influences the cost of capital in UK firms. *Journal of Risk and Financial Management, 17*(8), 1–21. DOI: 10.3390/jrfm17080326

Moustairas, I., Vardopoulos, I., Kavouras, S., Salvati, L., & Zorpas, A. A. (2022). Exploring factors that affect public acceptance of establishing an urban environmental education and recycling center. *Sustainable Chemistry and Pharmacy, 25*, 100605. DOI: 10.1016/j.scp.2022.100605

Mukherjee, M. (2022). *India's progress on its climate action plan–an update in early 2022*. Oxford Institute for Energy Studies.

Mukwevho, P., Radzuma, P., & Roos, C. (2024). Exploring Barriers to the Effective Implementation of Integrated Waste Management Plans in Developing Economies: Lessons Learned from South African Municipalities. *Sustainability (Basel), 16*(22), 9643. DOI: 10.3390/su16229643

Mullinger, P., & Jenkins, B. (2022). *Industrial and process furnaces: principles, design and operation*. Butterworth-Heinemann.

Munoz-Torres, M. J., Fernández-Izquierdo, M. Á., Rivera-Lirio, J. M., & Escrig-Olmedo, E. (2019). Can environmental, social, and governance rating agencies favor business models that promote a more sustainable development? *Corporate Social Responsibility and Environmental Management, 26*(2), 439–452. DOI: 10.1002/csr.1695

Münster, M., & Meibom, P. (2011). Optimization of use of waste in the future energy system. *Energy, 36*(3), 1612–1622. DOI: 10.1016/j.energy.2010.12.070

Murray, C., & Landry, C. (2020). Urban Regeneration 3.0: Realising the potential of an urban psychology. *Journal of Urban Regeneration and Renewal, 13*(3), 231–240. DOI: 10.69554/FYIO9002

Muthukumar, J. (2017). CONSERVATION OF HERITAGE MONUMENTS FROM ENVIRONMENTAL CHANGE: A CASE STUDY. Research Directions.

Náthia-Neves, G., Berni, M., Dragone, G., Mussatto, S. I., & Forster-Carneiro, T. (2018). Anaerobic digestion process: Technological aspects and recent developments. *International Journal of Environmental Science and Technology, 15*(9), 2033–2046. DOI: 10.1007/s13762-018-1682-2

National Research Council, Commission on Life Sciences, Board on Environmental Studies, Committee on Health Effects of Waste Incineration Council (2000). Waste incineration and public health.

Navaratnam, S., Nguyen, K., Selvaranjan, K., Zhang, G., Mendis, P., & Aye, L. (2022). Designing Post COVID-19 Buildings: Approaches for Achieving Healthy Buildings. *Buildings (Basel, Switzerland), 12*(1), 74. Advance online publication. DOI: 10.3390/buildings12010074

Neehaul, N., Jeetah, P., & Deenapanray, P. (2020). Energy recovery from municipal solid waste in Mauritius: Opportunities and challenges. *Environmental Development, 33*, 100489. DOI: 10.1016/j.envdev.2019.100489

Nemes, N., Scanlan, S. J., Smith, P., Smith, T., Aronczyk, M., Hill, S., Lewis, S. L., Montgomery, A. W., Tubiello, F. N., & Stabinsky, D. (2022). An integrated framework to assess greenwashing. *Sustainability (Basel), 14*(8), 4431. DOI: 10.3390/su14084431

Neuwahl, F., Cusano, G., Gómez, B. J., Holbrook, S., & Roudier, S. (2019). *Best available techniques (BAT) reference document for waste incineration: industrial emissions directive 2010/75/EU (Integrated Pollution Prevention and Control).* European Commission.

Nevens, F., Frantzeskaki, N., Gorissen, L., & Loorbach, D. (2013). Urban Transition Labs: Co-creating transformative action for sustainable cities. *Journal of Cleaner Production, 50*, 111–122. DOI: 10.1016/j.jclepro.2012.12.001

Neves, R. A. F., Naveira, C., Miyahira, I. C., Portugal, S. G. M., Krepsky, N., & Santos, L. N. (2020). Are invasive species always negative to aquatic ecosystem services? The role of dark false mussel for water quality improvement in a multi-impacted urban coastal lagoon. *Water Research, 184*, 116108. DOI: 10.1016/j.watres.2020.116108 PMID: 32688149

Nguyen, K. H. (2023). Urban Adaptation and Resilience Planning for a Changing Climate: A Study of New York City (Doctoral dissertation, Rutgers The State University of New Jersey, School of Graduate Studies).

Nicholls, S. (2006). Climate change, tourism and outdoor recreation in Europe. *Managing Leisure, 11*(3), 151–163. DOI: 10.1080/13606710600715226

Nicholls, S., & Amelung, B. (2008). Climate change and tourism in northwestern Europe: Impacts and adaptation. *Tourism Analysis*, *13*(1), 21–31. DOI: 10.3727/108354208784548724

Nickayin, S. S., Jahelka, A., Ye, S., Perrone, F., & Salvati, L. (2023). Planning for just cities with nature-based solutions: Sustainability and socio-environmental inequalities in San José de Chamanga, Ecuador. *Land (Basel)*, *12*(3), 604. DOI: 10.3390/land12030604

Nicolini, E. (2024). Climate change adaptation and mitigation and historic centers preservation. Underway and repeatable technological design solutions. *Cities (London, England)*, *152*, 152. DOI: 10.1016/j.cities.2024.105174

Nieuwenhuijsen, M. J., & Khreis, H. (2016). Car free cities: Pathway to healthy urban living. *Environment International*, *94*, 251–262. DOI: 10.1016/j.envint.2016.05.032 PMID: 27276440

Nobre, E. A. C. (2023). Implementing land value capture in a Global South city: Evaluation of the experience in the City of São Paulo, Brazil. *Revista Brasileira de Estudos Urbanos e Regionais*, *25*(1). Advance online publication. DOI: 10.22296/2317-1529.rbeur.202327en

Nordestgaard, P. M., & Arndt, C. H. (2019). AMAGER BAKKE: A steel building with the design challenge of creating a world famous recreational roof. *ce/papers*, *3*(3–4), 151–156. DOI: 10.1002/cepa.1156

Norton, R. K. (2014). Agenda 21 and Its Discontents: Is Sustainable Development a Global Imperative or Globalizing Conspiracy. *The Urban Lawyer*, *46*(2), 325–360.

Nouri, B. A., & Emkani, P. (2020). Effect of consumer innovation process on green consumption behaviour: An application of planned behaviour theory. *International Journal of Business Environment*, *11*(4), 347–368. DOI: 10.1504/IJBE.2020.111387

Nucci, C. (2023). *Città intelligenti e neutralità climatica: Verso la transizione energetica*. Istituto Lombardo - Accademia Di Scienze e Lettere - Rendiconti Di Scienze., DOI: 10.4081/scienze.2022.825

Nutkani, I., & Khan, W. (2023). Electric Vehicles Impact on Distribution Network Demand and Power Quality. *2023 IEEE International Conference on Energy Technologies for Future Grids (ETFG)*, 1–6. DOI: 10.1109/ETFG55873.2023.10407442

Nwogugu, M. C. (2023). ESG, inequality and international political economy problems inherent in regulation of DeFi, Web5/Web3, cryptocurrencies and fractional-NFTs. *SSRN*. https://papers.ssrn.com/sol3/papers.cfm?abstract_id=4164776

Nygaard, A., & Silkoset, R. (2023). Sustainable development and greenwashing: How blockchain technology information can empower green consumers. *Business Strategy and the Environment*, *32*(6), 3801–3813. DOI: 10.1002/bse.3338

O'Brien, J. K. (2006). Comparison of air emissions from waste-to-energy facilities to fossil fuel power plants. *North American Waste-to-Energy Conference*. DOI: 10.1115/NAWTEC14-3187

O'Donnell, E. C., Lamond, J. E., & Thorne, C. R. (2017). Recognising barriers to implementation of Blue-Green Infrastructure: A Newcastle case study. *Urban Water Journal*, *14*(9), 964–971. DOI: 10.1080/1573062X.2017.1279190

O'Donnell, E. C., Netusil, N. R., Chan, F. K. S., Dolman, N. J., & Gosling, S. N. (2021). International perceptions of urban blue-green infrastructure: A comparison across four cities. *Water (Basel)*, *13*(4), 544. DOI: 10.3390/w13040544

Ojuri, O., Ayodele, F., & Oluwatuyi, O. E. (2018). Risk assessment and rehabilitation potential of a millennium city dumpsite in Sub-Saharan Africa. *Waste Management (New York, N.Y.)*, *76*, 621–628. DOI: 10.1016/j.wasman.2018.03.002 PMID: 29548830

Okedara, O., & Odubiyi, M. A. (2024). The Role of Smart Buildings and Automated Energy Management Systems in Enhancing Building Energy Efficiency. https://www.researchgate.net/publication/384107270

Oke, T. R. (1987). *Boundary layer climates* (2nd ed.). Routledge.

Olabi, A. G., Shehata, N., Issa, U. H., Mohamed, O. A., Mahmoud, M., Abdelkareem, M. A., & Abdelzaher, M. A. (2025). The role of green buildings in achieving the sustainable development goals. *International Journal of Thermofluids*, *25*, 101002. Advance online publication. DOI: 10.1016/j.ijft.2024.101002

Oladeji, S. O., Grace, O., & Ayodeji, A. A. (2022). Community Participation in Conservation and Management of Cultural Heritage Resources in Yoruba Ethnic Group of South Western Nigeria. *SAGE Open*, *12*(4), 21582440221130987. Advance online publication. DOI: 10.1177/21582440221130987

Ongpeng, J. M. C., Dungca, J. R., Aviso, K. B., & Tan, R. R. (2019). Minimizing the carbon footprint of urban reconstruction projects. *Journal of Cleaner Production*, *240*, 118222. DOI: 10.1016/j.jclepro.2019.118222

Opferkuch, K., Caeiro, S., Salomone, R., & Ramos, T. B. (2022). Circular economy disclosure in corporate sustainability reports: The case of European companies in sustainability rankings. *Sustainable Production and Consumption*, *32*, 436–456. DOI: 10.1016/j.spc.2022.05.003

Orsi, N. (2013). The engineering behind the Via Appia. Engineering Rome. https://engineeringrome.org/the-engineering-behind-the-via-appia/

Outapa, P., & Roi-et, V. N. (2018). Evaluation of Greenhouse Gas Emissions from Municipal Solid Waste (MSW) Management: Case Study of Lampang Municipality, Thailand. *Applied Environmental Research*, *40*(1), 46–56. DOI: 10.35762/AER.2018.40.1.5

Ovaere, M., & Proost, S. (2022). Cost-effective reduction of fossil energy use in the European transport sector: An assessment of the Fit for 55 Package. *Energy Policy*, *168*, 113085. DOI: 10.1016/j.enpol.2022.113085

Owens, S. (1986). *Energy, planning, and urban form*. Pion.

Ozbay, G., Jones, M., Gadde, M., Isah, S., & Attarwala, T. (2021). Design and operation of effective landfills with minimal effects on the environment and human health. *Journal of Environmental and Public Health*, *2021*(1), 13. DOI: 10.1155/2021/6921607 PMID: 34531916

Ozturk, M., & Dincer, I. (2020). "An efficient waste management system with municipal solid waste incineration plant." Greenhouse Gases. *Science, Technology & Society*, *10*(4), 855–864.

Padilla-Rivera, A., Russo-Garrido, S., & Merveille, N. (2020). Addressing the social aspects of a circular economy: A systematic literature review. *Sustainability (Basel)*, *12*(19), 7912. DOI: 10.3390/su12197912

Pajula, T., Behm, K., Vatanen, S., & Saarivuori, E. (2017). Managing the life cycle to reduce environmental impacts. *Dynamics of long-life assets: From technology adaptation to upgrading the business model*, 93-113. https://doi.org/.DOI: 10.1007/978-3-319-45438-2_6

Pal, P. (2023). Present status and future outlooks of renewable energy in India for sustainable development. A Basic Overview of Environment and Sustainable Development [Volume 2], International Academic Publishing House (IAPH), 408-433.

Panagiotopoulos, P., Vardopoulos, I., Maialetti, M., Ciaschini, C., Koundouri, P., & Salvati, L. (2024). Reimagining sustainable development and economic performance indicators: a human-centric Maslow–Bossel blueprint. *Economies*, *12*(12), 338. DOI: 10.3390/economies12120338

Pandey, R. U., Mitra, T., Wadwekar, M., Nigam, J., & Trivedi, K. (2021). Green Infrastructure as a Tool for Improving Livability of Area Based Development Projects Under Smart City Mission. Geospatial Technology and Smart Cities: ICT, Geoscience Modeling, GIS and Remote Sensing, 447-468.

Pandis Iveroth, S., Vernay, A.-L., Mulder, K. F., & Brandt, N. (2013). Implications of systems integration at the urban level: The case of Hammarby Sjöstad, Stockholm. *Journal of Cleaner Production*, *48*, 220–231. DOI: 10.1016/j.jclepro.2012.09.012

Pandis Iverot, S., & Brandt, N. (2011). The development of a sustainable urban district in Hammarby Sjöstad, Stockholm, Sweden? *Environment, Development and Sustainability*, *13*(6), 1043–1064. DOI: 10.1007/s10668-011-9304-x

Pandit, N. S.SILPIREKHA PANDIT. (2016). Transformation of urban wetlands as an effect of urban development: An analysis of Deepor Beel in Guwahati, Assam. *CREATIVE SPACE*, *4*(1), 103–114. DOI: 10.15415/cs.2016.41006

Pandiyan, P., Saravanan, S., Usha, K., Kannadasan, R., Alsharif, M. H., & Kim, M. K. (2023). Technological Advancements toward Smart Energy Management in Smart Cities. *Energy Reports*, *10*, 648–677. DOI: 10.1016/j.egyr.2023.07.021

Pan, S. L., & Zhang, S. (2020). From fighting COVID-19 pandemic to tackling sustainable development goals: An opportunity for responsible information systems research. *International Journal of Information Management*, *55*, 102196. DOI: 10.1016/j.ijinfomgt.2020.102196 PMID: 32836647

Pan, S.-Y., Du, M. A., Huang, I.-T., Liu, I.-H., Chang, E., & Chiang, P.-C. (2015). Strategies on implementation of waste-to-energy (WTE) supply chain for circular economy system: A review. *Journal of Cleaner Production*, *108*, 409–421. DOI: 10.1016/j.jclepro.2015.06.124

Paola, D. R., & Rosenthal-Sabroux, C. 2014. "Smart City How to Create Public and Economic Value with High Technology in Urban Space" eds. Renata Paola Dameri and Camille Rosenthal-Sabroux. *Smart City How to Create Public and Economic Value with High Technology in Urban Space*: VIII, 238. https://www.springer.com/series/10440 (November 14, 2024).

Papamichael, I., Voukkali, I., Economou, F., Loizia, P., Demetriou, G., Esposito, M., Naddeo, V., Liscio, M. C., Sospiro, P., & Zorpas, A. A. (2023). Mobilisation of textile waste to recover high added value products and energy for the transition to circular economy. *Environmental Research*, 117716. PMID: 37995999

Papamichael, I., Voukkali, I., Loizia, P., Stylianou, M., Economou, F., Vardopoulos, I., Klontza, E. E., Lekkas, D. F., & Zorpas, A. A. (2023). Measuring Circularity: Tools for monitoring a smooth transition to Circular Economy. *Sustainable Chemistry and Pharmacy*, *36*, 101330. DOI: 10.1016/j.scp.2023.101330

Park, S. R., & Jang, J. Y. (2021). The impact of ESG management on investment decision: Institutional investors' perceptions of country-specific ESG criteria. *International Journal of Financial Studies*, *9*(3), 1–27. DOI: 10.3390/ijfs9030048

Passas, I. (2023). *Accounting for integrity: ESG and financial disclosures: the challenge of internal fraud in management decision-making* [Hellenic Mediterranean University]. DOI: 10.12681/eadd/55359

Passas, I. (2024). Bibliometric Analysis: The Main Steps. Encyclopedia 2024, Vol. 4, Pages 1014-1025, 4(2), 1014–1025. https://doi.org/DOI: 10.3390/encyclopedia4020065

Passas, I. Accounting for Integrity: ESG and Financial Disclosures: The Challenge of Internal Fraud in Management Decision-Making. Ph.D. Thesis, Hellenic Mediterranean University, Heraklion, Greece, 2023

Passas, I., Ragazou, K., Zafeiriou, E., Garefalakis, A., & Zopounidis, C. (2022). ESG Controversies: A Quantitative and Qualitative Analysis for the Sociopolitical Determinants in EU Firms. Sustainability 2022, Vol. 14, Page 12879, 14(19), 12879. https://doi.org/DOI: 10.3390/su141912879

Passas, I. (2024). The Evolution of ESG: From CSR to ESG 2.0. *Encyclopedia*, *4*(4), 1711–1720. DOI: 10.3390/encyclopedia4040112

Patil, R. A., Ghisellini, P., & Ramakrishna, S. (2021). Towards sustainable business strategies for a circular economy: environmental, social and governance (ESG) performance and evaluation. *An introduction to circular economy*, 527-554. DOI: DOI: 10.1007/978-981-15-8510-4_26

Pavate, V., Deshmukh, V. D., Kolekar, A., Mendapara, A., Patil, S., & Amrutatti, S. (2024). Green Building and Energy-Efficient Design. *Journal of Environmental Engineering and Studies*, *9*(2), 33–52. DOI: 10.46610/JoEES.2024.v09i02.003

Pavlides, T., Vardopoulos, I., Papamichael, I., Voukkali, I., Stylianos, M., & Zorpas, A. A. (2023). Environmental sustainability assessment of excavation, construction, and demolition waste conditions and practices across Greece and Cyprus. *IOP Conference Series. Earth and Environmental Science*, *1196*(1), 012037. DOI: 10.1088/1755-1315/1196/1/012037

Peattie, K., & Crane, A. (2005). Green marketing: Legend, myth, farce or prophesy? *Qualitative Market Research*, *8*(4), 357–370. Retrieved December 13, 2024, from. DOI: 10.1108/13522750510619733

Peloza, J., Loock, M., Cerruti, J., & Muyot, M. (2012). Sustainability: How stakeholder perceptions differ from corporate reality. *California Management Review, 55*(1), 74–97. DOI: 10.1525/cmr.2012.55.1.74

Peña, R., Meek, C., & Davis, D. (2017). The Bullitt Center: A Comparative Analysis Between Simulated and Operational Performance. *Technology Architecture + Design, 1*(2), 163–173. DOI: 10.1080/24751448.2017.1354611

Perch-Nielsen, S. L., Amelung, B., & Knutti, R. (2010). Future climate resources for tourism in Europe based on the daily Tourism Climatic Index. *Climatic Change, 103*(3), 363–381. DOI: 10.1007/s10584-009-9772-2

Perini, K. (2017). Madrid Río, Spain: Opportunities and policies. In K. Perini & P. Sabbion (Eds.), Urban sustainability and river restoration: Green and blue infrastructure (pp. 227–234). Wiley.

Perini, K. (2017). Madrid Río, Spain: Strategies and techniques. In K. Perini & P. Sabbion (Eds.), Urban sustainability and river restoration: Green and blue infrastructure (pp. 117–126). Wiley.

Peštová, V. (2021, November). Sustainable Urban Development: Case Study Copenhagen. In *Proceedings of the EDAMBA 2021 Conference*, Bratislava, Slovakia (Vol. 390, pp. 390-399).

Petrea, M. I., & Ursache, I. M. (2023, June). Inside the World's most sustainable smart city: lessons from Copenhagen. In Proceedings of the International Conference on "Sustainable Development of European Smart Cities"—SmartEU.

Petropoulou, A., Angelaki, E., Rompogiannakis, I., Passas, I., Garefalakis, A., & Thanasas, G. (2024). Digital Transformation in SMEs: Pre- and Post-COVID-19 Era: A Comparative Bibliometric Analysis. Sustainability 2024, Vol. 16, Page 10536, 16(23), 10536. DOI: 10.3390/su162310536

Pham, T. Y., & Yeo, G.-T. (2019). Evaluation of Transshipment container terminals' service quality in Vietnam: From the shipping companies' perspective. *Sustainability (Basel), 11*(5), 1503. DOI: 10.3390/su11051503

Piechota, G., Generowicz, N., Abd, A. A., Othman, M. R., Kowalczyk-Juśko, A., Kumar, G., & Veeremuthu, A. (2024). An introduction to biogas and biomethane. Biogas to Biomethane, Elsevier: 3-40. DOI: 10.1016/B978-0-443-18479-6.00008-9

Pimonenko, T., Bilan, Y., Horák, J., Starchenko, L., & Gajda, W. (2020). Green brand of companies and greenwashing under sustainable development goals. *Sustainability (Basel), 12*(4), 1679. DOI: 10.3390/su12041679

Pincetl, S., Bunje, P., & Holmes, T. (2012). An expanded urban metabolism method: Toward a systems approach for assessing urban energy processes and causes. *Landscape and Urban Planning*, *107*(3), 193–202. DOI: 10.1016/j.landurbplan.2012.06.006

Pioppi, B. I. P. (2019). Cultural heritage microclimate change: Human-centric approach to experimentally investigate intra-urban overheating and numerically assess foreseen future scenarios impact. *The Science of the Total Environment*. Advance online publication. DOI: 10.1016/j.scitotenv PMID: 31757533

Polyzos, I. (2016, October 7). Sustainable urban development policies in metropolitan Athens: Urban avenues and metropolitan rings [Conference presentation, in Greek]. Metropolitan Interventions in West Athens Conference: Antonis Tritsis Park & Mount Egaleo-Poikilo, Peristeri, Greece.

Polyzos, I., Balabanidis, D., & Triantis, L. (2014). Sustainable urban development policies in metropolitan Athens: The strategy of integrated urban interventions - urban avenues/green routes/cultural routes [in Greek]. Region of Attica - National Technical University of Athens.

Polzin, F., von Flotow, P., & Nolden, C. (2016). What encourages local authorities to engage with energy performance contracting for retrofitting? Evidence from German municipalities. *Energy Policy*, *94*, 317–330. DOI: 10.1016/j.enpol.2016.03.049

Popescu, C., Hysa, E., Kruja, A., & Mansi, E. (2022). Social innovation, circularity and energy transition for environmental, social and governance (ESG) practices—A comprehensive review. *Energies*, *15*(23), 9028. DOI: 10.3390/en15239028

Potting, J., Hekkert, M., Worrell, E., & Hanemaaijer, A. (2017). *Circular economy: measuring innovation in the product chain*. PBL Publishers.

Pougkakioti, V. M., Lazoglou, M., & Maistrou, E. (2023). Climate Change vs Cultural Heritage: An Adaptation Strategy for the Archaeological Site of Ancient Messene. *American Journal of Climate Change*, *12*(3), 456–488. Advance online publication. DOI: 10.4236/ajcc.2023.123021

Poyyamozhi, M., Murugesan, B., Rajamanickam, N., Shorfuzzaman, M., & Aboelmagd, Y. (2024a). IoT—A Promising Solution to Energy Management in Smart Buildings: A Systematic Review, Applications, Barriers, and Future Scope. In Buildings (Vol. 14, Issue 11). Multidisciplinary Digital Publishing Institute (MDPI). DOI: 10.3390/buildings14113446

Pozoukidou, G., Gavanas, N., & Verani, E. (2017). Land Use Models And Sustainable Urban Mobility Plans: An Integrative Approach For Strategic Planning. *International Journal of Sustainable Development and Planning*, *12*(5), 867–882. DOI: 10.2495/SDP-V12-N5-867-882

Price, R., Dörr, S., Frick, T., Joynson, C., Merx, O., & Wade, M. (2022). The international CDR manifesto. Retrieved from https://corporatedigitalresponsibility.net/cdr-manifesto-english

Pullen, T. (2015). *Anaerobic digestion-making biogas-making Energy: the earthscan expert guide*. Routledge. DOI: 10.4324/9781315770772

Pune, A. (2018). *Coastal Ecology and Fishing Community in Mumbai CRZ Policy*. Sustainability and Livelihoods.

Puppim de Oliveira, J. A., Bellezoni, R. A., Shih, W., & Bayulken, B. (2022). Innovations in Urban Green and Blue Infrastructure: Tackling local and global challenges in cities. *Journal of Cleaner Production*, *362*, 132355. DOI: 10.1016/j.jclepro.2022.132355

Purchase, C. K., Al Zulayq, D. M., O'Brien, B. T., Kowalewski, M. J., Berenjian, A., Tarighaleslami, A. H., & Seifan, M. (2022). Circular economy of construction and demolition waste: A literature review on lessons, challenges, and benefits. *Materials (Basel)*, *15*(1), 76. DOI: 10.3390/ma15010076 PMID: 35009222

Pushkar, S. (2024). Leadership in Energy and Environmental Design Commercial Interior Version 4 (LEED-CI v4) Gold-Certified Office Space Projects: A Pairwise Comparative Analysis between Three Mediterranean Countries. *Buildings (Basel, Switzerland)*, *14*(3), 815. Advance online publication. DOI: 10.3390/buildings14030815

Qayyum, A., Jamil, R. A., & Sehar, A. (2023). Impact of green marketing, greenwashing and green confusion on green brand equity. *Spanish Journal of Marketing - ESIC*, *27*(3), 286–305. DOI: 10.1108/SJME-03-2022-0032

Rabbani, M. R., Kiran, M., Bhuiyan, A. B., & Al-Hiyari, A. (2024). Does diversity in top management and boards affect ESG performance? Evidence from Islamic and conventional banks in the MENA region. *International Journal of Islamic and Middle Eastern Finance and Management*, *17*(6), 1244–1266. DOI: 10.1108/IMEFM-06-2024-0281

Rabinovitch, J. (1996). Innovative land use and public transport policy. *Land Use Policy*, *13*(1), 51–67. DOI: 10.1016/0264-8377(95)00023-2

Rafey, A., Prabhat, K., & Samar, M. (2020). Comparison of technologies to serve waste to energy conversion. *International Journal of Waste Resources, 10*, 372.

Ragazou, K., Garefalakis, A., Zafeiriou, E., & Passas, I. (2022). Agriculture 5.0: A New Strategic Management Mode for a Cut Cost and an Energy Efficient Agriculture Sector. Energies 2022, Vol. 15, Page 3113, 15(9), 3113. DOI: 10.3390/en15093113

Ragazou, K., Lemonakis, C., Passas, I., Zopounidis, C., & Garefalakis, A. (2024), "ESG-driven ecopreneur selection in European financial institutions: entropy and TOPSIS analysis", *Management Decision*, Vol. ahead-of-print No. ahead-of-print. https://doi.org/DOI: 10.1108/MD-12-2023-2425

Ragazou, K., Lemonakis, C., Passas, I., Zopounidis, C., & Garefalakis, A. (2024). ESG-driven ecopreneur selection in European financial institutions: entropy and TOPSIS analysis. Management Decision, ahead-of-print(ahead-of-print). https://doi.org/DOI: 10.1108/MD-12-2023-2425/FULL/PDF

Ragazou, K., Passas, I., & Garefalakis, A. (2022). It Is Time for Anti-Bribery: Financial Institutions Set the New Strategic "Roadmap" to Mitigate Illicit Practices and Corruption in the Market. *Administrative Sciences 2022, Vol. 12, Page 166, 12*(4), 166. DOI: 10.3390/admsci12040166

Ragazou, K., Garefalakis, A., Papademetriou, C., & Passas, I. (2023). Well-being Human Resource Model In The Collaborative Economy: The Keystone of ESG Strategy In *The Tourism Sector. International Conference on Tourism Research*, 6(1), 251–260. DOI: 10.34190/ictr.6.1.1066

Ragazou, K., Passas, I., Garefalakis, A., & Dimou, I. (2022). Investigating the Research Trends on Strategic Ambidexterity, Agility, and Open Innovation in SMEs: Perceptions from Bibliometric Analysis. *Journal of Open Innovation*, *8*(3), 118. DOI: 10.3390/joitmc8030118

Ragazou, K., Passas, I., Garefalakis, A., Galariotis, E., & Zopounidis, C. (2023). Big Data Analytics Applications in Information Management Driving Operational Efficiencies and Decision-Making: Mapping the Field of Knowledge with Bibliometric Analysis Using R. *Big Data and Cognitive Computing*, *7*(1), 13. DOI: 10.3390/bdcc7010013

Ragazou, K., Passas, I., Garefalakis, A., Zafeiriou, E., & Kyriakopoulos, G. (2022). The Determinants of the Environmental Performance of EU Financial Institutions: An Empirical Study with a GLM Model. *Energies*, *15*(15), 5325. DOI: 10.3390/en15155325

Rahman, S. U., & Nguyen-Viet, B. (2023). Towards sustainable development: Coupling green marketing strategies and consumer perceptions in addressing greenwashing. *Business Strategy and the Environment, 32*(4), 2420–2433. DOI: 10.1002/bse.3256

Ram, M., & Bracci, E. (2024). Waste Management, Waste Indicators and the Relationship with Sustainable Development Goals (SDGs): A Systematic Literature Review. *Sustainability (Basel), 16*(19), 8486. DOI: 10.3390/su16198486

Ramos, A., R. Tavares and A. Rouboa (2018). Microplastics Co-gasification with Biomass: Modelling Syngas Characteristics at Low Temperatures. Technologies and Materials for Renewable Energy, Environment and Sustainability. C. T. Salame, M. Aillerie, P. Papageorgas et al. Melville, Amer Inst Physics. 1968.

Ramos, A. (2025). Integrating waste thermal conversion and lifecycle analysis for sustainable energy production: Reflecting upon environmental and economic impacts. *Sustainable Energy Technologies and Assessments, 78*, 104342. DOI: 10.1016/j.seta.2025.104342

Ramos, A. T., Teixeira, C. A., & Rouboa, A. (2019). Environmental Assessment of Municipal Solid Waste by Two-Stage Plasma Gasification. *Energies, 12*(1), 137. DOI: 10.3390/en12010137

Ramos, A., Afonso Teixeira, C., & Rouboa, A. (2018). Environmental Analysis of Waste-to-Energy—A Portuguese Case Study. *Energies, 11*(3), 548. DOI: 10.3390/en11030548

Ramos, A., Berzosa, J., Clarens, F., Marin, M., & Rouboa, A. (2020b). Environmental and socio-economic assessment of cork waste gasification: Life cycle and cost analysis. *Journal of Cleaner Production, 249*, 119316. DOI: 10.1016/j.jclepro.2019.119316

Ramos, A., Berzosa, J., Espí, J., Clarens, F., & Rouboa, A. (2020a). Life cycle costing for plasma gasification of municipal solid waste: A socio-economic approach. *Energy Conversion and Management, 209*, 112508. DOI: 10.1016/j.enconman.2020.112508

Ramos, A., Monteiro, E., & Rouboa, A. (2019). Numerical approaches and comprehensive models for gasification process: A review. *Renewable & Sustainable Energy Reviews, 110*, 188–206. DOI: 10.1016/j.rser.2019.04.048

Ramos, A., Monteiro, E., & Rouboa, A. (2022). Biomass pre-treatment techniques for the production of biofuels using thermal conversion methods – A review. *Energy Conversion and Management, 270*, 116271. DOI: 10.1016/j.enconman.2022.116271

Ramos, A., Monteiro, E., Silva, V., & Rouboa, A. (2018). Co-gasification and recent developments on waste-to-energy conversion: A review. *Renewable & Sustainable Energy Reviews, 81*, 380–398. DOI: 10.1016/j.rser.2017.07.025

Ramos, A., & Rouboa, A. (2020). Renewable energy from solid waste: Life cycle analysis and social welfare. *Environmental Impact Assessment Review, 85*, 106469. DOI: 10.1016/j.eiar.2020.106469 PMID: 32952252

Ramos, A., & Rouboa, A. (2020). Syngas production strategies from biomass gasification: Numerical studies for operational conditions and quality indexes. *Renewable Energy, 155*, 1211–1221. DOI: 10.1016/j.renene.2020.03.158

Ramos, A., & Rouboa, A. (2022). Life cycle thinking of plasma gasification as a waste-to-energy tool: Review on environmental, economic and social aspects. *Renewable & Sustainable Energy Reviews, 153*, 111762. DOI: 10.1016/j.rser.2021.111762

Randazzo, L., Collina, M., Ricca, M., Barbieri, L., Bruno, F., Arcudi, A., & La Russa, M. (2020). *Damage Indices and Photogrammetry for Decay Assessment of Stone-Built Cultural Heritage: The Case Study of the San Domenico Church Main Entrance Portal*. Sustainability.

Rane, N. L., Choudhary, S. P., & Rane, J. (2023). Blockchain and artificial intelligence (AI) integration for revolutionizing security and transparency in finance.

Rane, N., Choudhary, S., & Rane, J. (2024). Artificial intelligence driven approaches to strengthening Environmental, Social, and Governance (ESG) criteria in sustainable business practices: A review. *SSRN*. https://doi.org/DOI: 10.2139/ssrn.4843215

Region of Attica. (2025). Places of refuge for citizens in the event of an earthquake [in Greek]. http://www.patt.gov.gr/site/index.php?option=com_content&view=article&id=6562&Itemid=36

Regionale dell'Appia Antica, P. (2025). Parco Regionale dell'Appia Antica [in Italian]. https://www.parcoappiaantica.it

Rejeb, A., Rejeb, K., Simske, S., Treiblmaier, H., & Zailani, S. (2022). The Big Picture on the Internet of Things and the Smart City: A Review of What We Know and What We Need to Know. *Internet of Things : Engineering Cyber Physical Human Systems, 19*, 100565. DOI: 10.1016/j.iot.2022.100565

Ren, Z., & Zuo, G. (2024). Challenges of Implementing Municipal Solid Waste Separation Policy in China. *Sustainability (Basel), 16*(18), 8081. DOI: 10.3390/su16188081

Reuter, J., Hellmund, T., Moßgraber, J., & Hertweck, P. (2023). The KERES Ontology: Protecting Cultural Heritage from Extreme Climate Events. Heritage., 10.3390.

Revez, M. J. (2019). Analysing the Cost-Effectiveness of Heritage Conservation Interventions: A Methodological Proposal within Project STORM. *International Journal of Architectural Heritage*, •••, 985–999.

Rezania, S., Oryani, B., Nasrollahi, V. R., Darajeh, N., Lotfi Ghahroud, M., & Mehranzamir, K. (2023). Review on Waste-to-Energy Approaches toward a Circular Economy in Developed and Developing Countries. *Processes (Basel, Switzerland)*, *11*(9), 2566. DOI: 10.3390/pr11092566

Ribeiro Hoffmann, A., Sandrin, P., & Doukas, Y. E. (Eds.). (2024). Climate change in regional perspective. European Union and Latin American initiatives, challenges, and solutions. Cham, Switzerland: United Nations University Series on Regionalism, Vol. 27, Springer Nature. https://doi.org/DOI: 10.1007/978-3-031-49329-4

Riccardo Cacciotti, A. K. (2021). Climate change-induced disasters and cultural heritage: Optimizing management strategies in Central Europe. *Climate Risk Management*.

Ritch, E. L. (2023). Sustainable fashion marketing: green or greenwash? In *Marketing fashion: Critical perspectives on the power of fashion in contemporary culture*. Routledge. DOI: 10.4324/9781003263753-16

Riyath, M. I. M., & Jariya, A. M. I. (2024). The role of ESG reporting, artificial intelligence, stakeholders and innovation performance in fostering sustainability culture and climate resilience. *Journal of Financial Reporting and Accounting*. https://doi.org/DOI: 10.1108/JFRA-10-2023-0621

Roessler, J. G., Olivera, F. D., Wasman, S. J., Townsend, T. G., McVay, M. C., Ferraro, C. C., & Blaisi, N. I. (2016). Construction material properties of slag from the high temperature arc gasification of municipal solid waste. *Waste Management (New York, N.Y.)*, *52*, 169–179. DOI: 10.1016/j.wasman.2016.03.031 PMID: 27020344

Ross, A., Alstine, J. M. V., Cotton, M., & Middlemiss, L. (2021). Deliberative democracy and environmental justice: Evaluating the role of citizens' juries in urban climate governance. *Local Environment*, *26*(12), 1512–1531. DOI: 10.1080/13549839.2021.1990235

Rouf, M. A., Rahman, M. M., Rahman, S. M., & Ahsan, M. N. (2019). Coming stakes in the ocean: Food Production, Shipping and Trade, Tourism, Ecosystem-biodiversity, New Technologies and Climate Change Challenges in Bangladesh. *Journal of Ocean and Coastal Economics*, *6*(2), 5. DOI: 10.15351/2373-8456.1102

Roversi, R., Boeri, A., Pagliula, S., & Turci, G. (2022). Energy Community in Action—Energy Citizenship Contract as Tool for Climate Neutrality. *Smart Cities*, *5*(1), 1. Advance online publication. DOI: 10.3390/smartcities5010018

Roy, S. B., & Basak, M. (2013) Journal of Documentation: a bibliometric study. Library Philosophy and Practice (e-Journal). Paper.;945.: https://digitalcommons.unl.edu/libphilpracPart of the Library and Information Science Commons.

Rozkrut, D., Tarczynska-Luniewska, M., Singh, G. A., & Piwowarski, M. (2024). The use of big data in ESG monitoring and evaluation. In Grima, S., Maditinos, D., Noja, G. G., Stankevičienė, J., Tarczynska-Luniewska, M., & Thalassinos, E. (Eds.), *Exploring ESG challenges and opportunities: Navigating towards a better future* (Vol. 116, pp. 25–40). Emerald Publishing Limited., DOI: 10.1108/S1569-375920240000116003

Rudden, P. J., O'Neill, K., McEvoy, B., & Treanor, A. (2015). Environmental sustainability of European cities. *Proceedings of the Institution of Civil Engineers. Civil Engineering*, *168*(2), 75–80. DOI: 10.1680/cien.14.00037

Ruiz, J. A., Juarez, M. C., Morales, M. P., Munoz, P., & Mendivil, M. A. (2013). Biomass gasification for electricity generation: Review of current technology barriers. *Renewable & Sustainable Energy Reviews*, *18*, 174–183. DOI: 10.1016/j.rser.2012.10.021

Saadabadi, S. A., Thattai, A. T., Fan, L., Lindeboom, R. E., Spanjers, H., & Aravind, P. (2019). Solid Oxide Fuel Cells fuelled with biogas: Potential and constraints. *Renewable Energy*, *134*, 194–214. DOI: 10.1016/j.renene.2018.11.028

Saberi, S., Kouhizadeh, M., Sarkis, J., & Shen, L. (2019). Blockchain technology and its relationships to sustainable supply chain management. *International Journal of Production Research*, *57*(7), 2117–2135. DOI: 10.1080/00207543.2018.1533261

Sadeq, A. M., Homod, R. Z., Hussein, A. K., Togun, H., Mahmoodi, A., Isleem, H. F., Patil, A. R., & Moghaddam, A. H. (2024). Hydrogen energy systems: Technologies, trends, and future prospects. *The Science of the Total Environment*, *939*, 173622. DOI: 10.1016/j.scitotenv.2024.173622 PMID: 38821273

Sadiq, M., Ngo, T. Q., Pantamee, A. A., Khudoykulov, K., Ngan, T. T., & Tan, L. P. (2023). The role of environmental social and governance in achieving sustainable development goals: Evidence from ASEAN countries. *Economic research-. Ekonomska Istrazivanja*, *36*(1), 170–190. DOI: 10.1080/1331677X.2022.2072357

Sakhamuri, H., & Sanagani, V. (2024). SDG Localization and Lessons from India: Role of NITI-Aayog. In *SDGs in the Asia and Pacific Region* (pp. 447–461). Springer International Publishing. DOI: 10.1007/978-3-031-17463-6_58

Salvati, L. (Ed.). (2025). *Environmental sustainability and global change*. Elsevier., DOI: 10.1016/C2023-0-52570-4

Salvati, L., & Vardopoulos, I. (2024). Intricacies of Mediterranean urban and regional dynamics. *Euro-Mediterranean Journal for Environmental Integration*, *9*(1), 277–292. DOI: 10.1007/s41207-023-00434-7

Salvati, L., & Vardopoulos, I. (2025). Into the kaleidoscopic urbanization of Southern Europe: Modelling 'form-function' relationships through partial least squares regression. *Cities (London, England)*, *162*, 105933. DOI: 10.1016/j.cities.2025.105933

Samant Raja, Deepti, and World Bank Group. "Digital Dividends Bridging the Disability Divide through Digital Technologies Bridging the Disability Divide through Digital Technologies." https://www.worldbank.org/en/publication/wdr2016. (October 18, 2024).

Sandra Fatoric, E. S. (2017). Securing the Future of Cultural Heritage by Identifying Barriers to and Strategizing Solutions for Preservation under Changing Climate Conditions. *Sustainability (Basel)*, *9*(11), 2143. Advance online publication. DOI: 10.3390/su9112143

Sandra Fatoric, L. E. (2020). Realising the potential of cultural heritage to achieve climate change actions in the Netherlands. *Journal of Environmental Management*. PMID: 32791326

Santosuosso, P. (2022). A satisfaction-based model for risk indexing in cultural heritage conservation. *Journal of Cultural Heritage*, *57*, 173–183. DOI: 10.1016/j.culher.2022.08.007

Santo-Tomás Muro, R., & Rodríguez Romero, E. J. (2020). City thresholds: The role of urban green infrastructures in Madrid. *ZARCH*, *14*(14), 176–187. DOI: 10.26754/ojs_zarch/zarch.2020144289

Sapountzaki, K., & Dandoulaki, M. (2015). *Hazards and disasters: Concepts and tools for evaluation, protection, and management* [in Greek]. Association of Greek Academic Libraries.

Saraswat, S. & Singh, P. (2025). "The moderating role of perceived trust in predicting the adoption intention of consumers toward online pharmacies using the application of UTAUT model", *International Journal of Pharmaceutical and Healthcare Marketing*, Vol. ahead-of-print No. ahead-of-print. DOI: 10.1108/IJPHM-05-2024-0042

Sarfraz, M., Raza, M., Khalid, R., Liu, T., Li, Z., & Niyomdecha, L. (2022). Consumer Purchasing Behavior Toward Green Environment in the Healthcare Industry: Mediating Role of Entrepreneurial Innovation and Moderating Effect of Absorptive Capacity. *Frontiers in Public Health*, *9*, 823307. Advance online publication. DOI: 10.3389/fpubh.2021.823307 PMID: 35186878

Sateriano, A., D'Agata, A., & Vardopoulos, I. (Eds.). (2024). *Socio-environmental change: the path toward sustainability*. Nova Science Publishers., DOI: 10.52305/NZKT5653

Satiada, M. A., & Calderon, A. (2021). Comparative analysis of existing waste-to-energy reference plants for municipal solid waste. *Cleaner Environmental Systems*, *3*, 100063. DOI: 10.1016/j.cesys.2021.100063

Saxena, A., Singh, R., Gehlot, A., Akram, S. V., Twala, B., Singh, A., Montero, E. C., & Priyadarshi, N. (2022). Technologies empowered environmental, social, and governance (ESG): An Industry 4.0 landscape. *Sustainability (Basel)*, *15*(1), 309. DOI: 10.3390/su15010309

Scarpa, G. M., Zaggia, L., Manfè, G., Lorenzetti, G., Parnell, K., Soomere, T., Rapaglia, J., & Molinaroli, E. (2019). The effects of ship wakes in the Venice Lagoon and implications for the sustainability of shipping in coastal waters. *Scientific Reports*, *9*(1), 19014. DOI: 10.1038/s41598-019-55238-z PMID: 31831769

Schepelmann, P. (2018). *Governance of Low-Carbon Energy System Transitions: A Case Study from North-Rhine Westphalia, Germany. 32*. https://www.adb.org/publications/low-carbon-energy-system-transitions

Schmid, B., Fricke, C., & Zengerling, C. (2024). Towards a" Freiburg Model" of Housing for the Common Good? Fostering Collaborative Housing in Urban Development. *Urban Planning*, *9*, 9. DOI: 10.17645/up.8191

Schoenefeld, J. J., Stein, B., & Renz, I. (2023). Renovating Europe: How to start and steer a wave? In *Making the European Green Deal Work*. Routledge. DOI: 10.4324/9781003246985-7

Schroeder, H., & Bulkeley, H. (2009). Global Cities and the Governance of Climate Change: What is the Role of Law in Cities. *The Fordham Urban Law Journal*, *36*(2), 313–360.

Schulz, K. A., Gstrein, O. J., & Zwitter, A. J. (2020). Exploring the governance and implementation of sustainable development initiatives through blockchain technology. *Futures*. https://papers.ssrn.com/sol3/papers.cfm?abstract_id=3663149

Schulze, G. (2016). Growth Within: A Circular Economy Vision for a Competitive Europe, Ellen MacArthur Foundation, the McKinsey Center for Business Environment: 1-22.

Scott, D., Rutty, M., Amelung, B., & Tang, M. (2016). An inter-comparison of the Holiday Climate Index (HCI) and the Tourism Climate Index (TCI) in Europe. *Atmosphere (Basel)*, *7*(6), 80. DOI: 10.3390/atmos7060080

Sellers, B. C., & Fiore, S. M. (2013). Sustainable and user-centered: Applying human factors solutions towards improving the effectiveness of "green" buildings. *Proceedings of the Human Factors and Ergonomics Society*, 550–554. DOI: 10.1177/1541931213571118

Senadheera, S. S., Gregory, R., Rinklebe, J., Farrukh, M., Rhee, J. H., & Ok, Y. S. (2022). The development of research on environmental, social, and governance (ESG): A bibliometric analysis. *Sustainable Environment*, *8*(1), 2125869. DOI: 10.1080/27658511.2022.2125869

Sevieri, G. G., Galasso, C., D'Ayala, D., De Jesus, R., Oreta, A., Grio, M. E. D. A., & Ibabao, R. (2020). A multi-hazard risk prioritisation framework for cultural heritage assets. *Natural Hazards and Earth System Sciences*, *20*(5), 1391–1414. DOI: 10.5194/nhess-20-1391-2020

Shafik, W. (2024). Incorporating Artificial Intelligence for Urban and Smart Cities' Sustainability. In Maintaining a Sustainable World in the Nexus of Environmental Science and AI (pp. 23-58). IGI Global.

Shah, H. H., Amin, M., Iqbal, A., Nadeem, I., Kalin, M., Soomar, A. M., & Galal, A. M. (2023). A review on gasification and pyrolysis of waste plastics. *Frontiers in Chemistry*, *10*, 38. DOI: 10.3389/fchem.2022.960894 PMID: 36819712

Shahzad, U., Ghaemi Asl, M., & Tedeschi, M. (2023). Is there any market state-dependent contribution from blockchain-enabled solutions to ESG investments? Evidence from conventional and Islamic ESG stocks. *Research in Economics*, *86*, 139–154. https://ideas.repec.org/a/eee/reveco/v86y2023icp139-154.html

Shanwen Zheng, B. H. (2018). Ecological Wisdom and Inspiration Underlying the Planning and Construction of Ancient Human Settlements: Case Study of Hongcun UNESCO World Heritage Site in China. *Sustainability*.

Sharma, G. D., Reppas, D., Muschert, G., & Pereira, V. (2021). Investigating digital sustainability: A retrospective bibliometric analysis of literature leading to future research directions. *Clinical Hemorheology and Microcirculation*.

Sharma, S. B., Singh, A. K., & Rajawat, A. S. (2022). Flood Vulnerability and Risk Assessment with Parsimonious Hydrodynamic Modeling and GIS. In *Geospatial Modeling for Environmental Management* (pp. 3–18). CRC Press. DOI: 10.1201/9781003147107-2

Shchipunov, Y. (2012). Bionanocomposites: Green sustainable materials for the near future. *Pure and Applied Chemistry, 84*(12), 2579–2607. DOI: 10.1351/PAC-CON-12-05-04

Sheth, K. N. (n.d.). WATER EFFICIENT TECHNOLOGIES FOR GREEN BUILDINGS. https://www.researchgate.net/publication/316582554

Shi Y., Zhang D. (2017). International regulatory responses to global challenges in marine pollution and climate change. Marine Pollution and Climate Change, 279 - 322. DOI: DOI: 10.1201/9781315119243

Shonder, J., Morofsky, E., Schmidt, F., Morck, O., & Himanen, M. (2010). *Best Practice Guidelines for Using Energy Performance Contracts To Improve Government Buildings* (IEA ECBCS Annex 46).

Sikdar, S., Siddaiah, A., & Menezes, P. L. (2020). Conversion of waste plastic to oils for tribological applications. *Lubricants (Basel, Switzerland), 8*(8), 78. DOI: 10.3390/lubricants8080078

Sinclair, S., & Rockwell, G. (2020). "Voyant-tools". available at: http://voyant-tools .org (accessed 2 June 2022). DOI: DOI: 10.1002/9781118680605.ch19

Singh, A. K., Meshram, S. A., Khandelwal, V., Tiwari, P., & Singh, P. (2024)c. Implementation of Industry-Based Perspective in Technical Education: A Qualitative and Quantitative Analysis. *Industrial and Manufacturing Designs: Quantitative and Qualitative Analysis*, 321-346. https://doi.org/.DOI: 10.1002/9781394212668.ch11

Singh, P. & Kushwaha, R. (2024). "Envisaging Indian farmers' desires from agricultural index insurance integrating rank sum weighting method and MoSCoW technique: an approach to requirements prioritization", *International Journal of Social Economics*, Vol. ahead-of-print No. ahead-of-print. DOI: 10.1108/IJSE-03-2024-0233

Singh, P. (2014). Unorganized sector and India's informal economy: challenges and prospects. *New Man International Journal of Multidisciplinary Studies, 1*(2), 209-218. https://www.newmanpublication.com/dash/issueworkfiles/71.pdf?1721465584

Singh, P., & Kushwaha, R. (2025), "Conceptualizing the corporate-social accountability and sustainability assessment of weather-risk management policy towards stakeholders using qualitative insights: a case of weather index insurance", *Qualitative Research in Financial Markets*, Vol. ahead-of-print No. ahead-of-print. DOI: 10.1108/QRFM-05-2024-0113

Singh, P., Agrawal, G., & Kushwaha, R. (2021, November) a. Evaluating Customer Satisfaction of Weather Index Insurance Service Quality Using Kano Model. In *2021 IEEE Bombay Section Signature Conference (IBSSC)* (pp. 1-6). IEEE. **DOI:** DOI: 10.1109/IBSSC53889.2021.9673370

Singh, P., Kushwaha, R., & Kushwaha, J. (2021) b. Assessing visitors' satisfaction about museum service quality using Kano model. In *2021 IEEE 9th Region 10 Humanitarian Technology Conference (R10-HTC)* (pp. 1-6). IEEE. **DOI:** DOI: 10.1109/R10-HTC53172.2021.9641707

Singh, P., Kushwaha, R., & Kushwaha, J. (2023)a. Analyzing Accountability of Weather Index Insurance Service in Attainment of Sustainable Development Goals: A Sustainable Accounting Perspective Review. In Pratik Arte et al. (Eds): *Sustainable International Business, 2023*. Springer Nature, 2023. doi: , 531340. DOI: 978-3-031-43784-7

Singh, P., Kushwaha, R., & Kushwaha, J. (2023)b. Evaluating Farmer Participation in Weather Index Insurance in a Socio-Demographic Perspective: Evidence From the Crop Insurance Data of India. In A. Popoola, H. Magidimisha-Chipungu, & L. Chipungu (Eds.), *Handbook of Research on Managing the Urban-Rural Divide Through an Inclusive Framework* (pp. 69-88). IGI Global. DOI: 10.4018/978-1-6684-6258-4.ch005

Singh, P., Kushwaha, R., & Kushwaha, J. (2024)a. Foresight the future of Industry 5.0 in perspective of sustainable services in developing countries. *Modern Technologies and Tools Supporting the Development of Industry 5.0*, 381. DOI: .DOI: 10.1201/9781003489269-11

Singh, P., Kushwaha, R., & Kushwaha, J. (2025), "A Qualitative Study on Analysing the Role of Mimicking Technology in Electronic Customer Relationship Management: A Case of Service Sector", Kumar, A., Ciddikie, M.D., Kashyap, A.K. and Akram, H.W. (Ed.) *Marketing 5.0*, Emerald Publishing Limited, Leeds, pp. 103-119. DOI: 10.1108/978-1-83797-815-120251008

Singh, P., Kushwaha, R., Kushwaha, J., & Singh, A. K. (2024)b. A Qualitative Study to Rank Non-Conventional Energy Sources for Industrial Sustainability and Energy Management Decisions Using MoSCoW Prioritization Method. *Industrial and Manufacturing Designs: Quantitative and Qualitative Analysis*, 105-137. https://doi.org/.DOI: 10.1002/9781394212668.ch4

Singh, A. K., Wadhwa, D., Tripathi, D., Pal, S., Singh, P. K., & Arora, G. (2025). Waste To Energy Technologies In Urban Setting: Comparative Analysis Of Environmental And Economic Performance. *African Journal of Biomedical Research, 28*, 1181–1189.

Singhania, M., & Saini, N. (2021). Systems approach to environment, social and governance (ESG): Case of Reliance industries. *Sustainable Operations and Computers, 3*, 103–117. DOI: 10.1016/j.susoc.2021.11.003

Singh, M. A., & Singha, K. (Eds.). (2020). *Understanding Urbanisation in Northeast India: issues and challenges*. Routledge. DOI: 10.4324/9781003032625

Singh, P. (2024). Analyzing Responsible Leadership Effectiveness for Attainment of Sustainable Development Goals in Developing Countries Using Environmental Scanning Techniques. In Bhattacharya, S., Venkatesh, V. G., & Chatterjee, S. R. (Eds.), *Responsible Corporate Leadership Towards Attainment of Sustainable Development Goals. Responsible Leadership and Sustainable Management*. Springer., DOI: 10.1007/978-981-97-1386-8_13

Singh, P. (2024). Weather index insurance viability in mitigation of climate change impact risk: A systematic review and future agenda. *Journal of Science and Technology Policy Management, 15*(1), 142–163. DOI: 10.1108/JSTPM-07-2021-0102

Singh, P., & Agrawal, G. (2019). Efficacy of weather index insurance for mitigation of weather risks in agriculture: An integrative review. *International Journal of Ethics and Systems, 35*(4), 584–616. DOI: 10.1108/IJOES-09-2018-0132

Singh, P., & Agrawal, G. (2020). Development, present status and performance analysis of agriculture insurance schemes in India: Review of evidence. *International Journal of Social Economics, 47*(4), 461–481. DOI: 10.1108/IJSE-02-2019-0119

Singh, P., & Agrawal, G. (2022). Mapping the customer centric weather index insurance service design using quality function deployment. *The TQM Journal, 34*(6), 1800–1822. DOI: 10.1108/TQM-08-2021-0236

Sinha, A., Gupta, R., & Kutnar, A. (2013). Održivi razvoj i zelena gradnja. In Drvna Industrija (Vol. 64, Issue 1, pp. 45–53). DOI: 10.5552/drind.2013.1205

Şirin, C., Goggins, J., & Hajdukiewicz, M. (2023). A review on building-integrated photovoltaic/thermal systems for green buildings. In *Applied Thermal Engineering* (Vol. 229). Elsevier Ltd., DOI: 10.1016/j.applthermaleng.2023.120607

Solano Meza, J. K., Orjuela Yepes, D., Rodrigo-Ilarri, J., & Rodrigo-Clavero, M.-E. (2023). Comparative analysis of the implementation of support vector machines and long short-term memory artificial neural networks in municipal solid waste management models in megacities. *International Journal of Environmental Research and Public Health, 20*(5), 4256. DOI: 10.3390/ijerph20054256 PMID: 36901265

Sombat Prajonsant, V. F. (2018). *THE STUDY OF THERMAL COMFORT OF I-SAN VERNACULAR BUDDHIST HOLY TEMPLES (SIMS) IN BURIRAM PROVINCE. Humanities*. Arts and Social Sciences Studies.

Song, Y., Kalacska, M., Gašparović, M., Yao, J., & Najibi, N. (2023). Advances in geocomputation and geospatial artificial intelligence (GeoAI) for mapping. *International Journal of Applied Earth Observation and Geoinformation, 120*, 103300. DOI: 10.1016/j.jag.2023.103300

Soni, N., & Soni, N. (2016). Benefits of pedestrianization and warrants to pedestrianize an area. *Land Use Policy, 57*, 139–150. DOI: 10.1016/j.landusepol.2016.05.009

Sonnleithner, M. (2021). New Opportunities for Increasing the Renovation Rate of Buildings. *Architecture Papers of the Faculty of Architecture and Design STU, 26*(1), 2–9. DOI: 10.2478/alfa-2021-0002

Spronken-Smith, R. A., Oke, T. R., & Lowry, W. P. (2000). Advection and the surface energy balance across an irrigated urban park. *International Journal of Climatology, 20*(9), 1033–1047. DOI: 10.1002/1097-0088(200007)20:9<1033::AID-JOC508>3.0.CO;2-U

Squires, G., & Lord, A. D. (2012). The transfer of Tax Increment Financing (TIF) as an urban policy for spatially targeted economic development. *Land Use Policy, 29*(4), 817–826. DOI: 10.1016/j.landusepol.2011.12.007

Stahl, B. C. (2024). From corporate digital responsibility to responsible digital ecosystems. *Sustainability (Basel), 16*(12), 4972. DOI: 10.3390/su16124972

Stamoulidis, K., & Stafylas, P. (2017). The area of the Queen's Tower–Tritsis Park in Athens: Strategic management for the formation of a recognizable identity with metropolitan–supra-local reach [Conference presentation]. 2nd Panhellenic Marketing Conference & Site Branding, University of Thessaly, Larissa, Greece.

Statista. (2025). Largest urban agglomerations in Europe in 2023. https://www.statista.com/statistics/1101883/largest-european-cities/

Stavropoulou, E., Spinthiropoulos, K., Ragazou, K., Papademetriou, C., & Passas, I. (2023). Green Balanced Scorecard: A Tool of Sustainable Information Systems for an Energy Efficient Business. *Energies, 16*(18), 6432. DOI: 10.3390/en16186432

Stefani, F., & Tsilimigas, G. (2023). *Landscape in spatial planning: a critical evaluation of the treatment of the degraded landscapes of the Greek island area in regional spatial planning*. Aichoros.

Stopka, O., Šimková, I., & Konečný, V. (2015). The quality of service in the public transport and shipping industry. *Nase More (Dubrovnik), 62*(3), 126–130. DOI: 10.17818/NM/2015/SI7

Streimikiene D., Lekavičius V., Baležentis T., Kyriakopoulos G.L., Abrhám J. (2020). Climate Change Mitigation Policies Targeting Households and Addressing Energy Poverty in European Union. Energies 13(13), 3389; 24 pp. .DOI: 10.3390/en13133389

Streimikiene, D., Kyriakopoulos G.L. (2024). Climate Change Mitigation Performance in the EU Tourism Destination Sector. Journal of Tourism and Services 15(28), 268–284. . https://doi.org/DOI: 10.29036/jots.v15i28.758

Study, C., & Verticale, B. (2015). Tall buildings: design, construction, and operation | 2015 Issue I. www.iuav.it

Sulkowski, A. J. (2024). AI, ESG, and law: Potential, limitations, and strategies concerning artificial intelligence in sustainability reporting. *SSRN*. https://doi.org/DOI: 10.2139/ssrn.4972787

Sun, J., Meng, X., & Qiao, J. (2023). Data-driven optimal control for municipal solid waste incineration process. *IEEE Transactions on Industrial Informatics, 19*(12), 11444–11454. DOI: 10.1109/TII.2023.3246467

Surrey County Council. (2023). Green and blue infrastructure: Best practice and case studies [Policy report]. https://www.surreycc.gov.uk/community/climate-change/what-are-we-doing/green-and-blue-infrastructure

Sustainalytics, (2021). available at: https://www.sustainalytics.com/esg-ratings

Syed, Abbas Shah, Daniel Sierra-Sosa, Anup Kumar, and Adel Elmaghraby. 2021. "IoT in Smart Cities: A Survey of Technologies, Practices and Challenges." *Smart Cities 2021, Vol. 4, Pages 429-475* 4(2): 429–75. .DOI: 10.3390/smartcities4020024

Szabo, S., & Webster, J. (2021). Perceived greenwashing: The effects of green marketing on environmental and product perceptions. *Journal of Business Ethics, 171*(4), 719–739. DOI: 10.1007/s10551-020-04461-0

T K, G. K.T K. (2024). Valorization of intangible cultural heritage through documentation: An Indian scenario. *Library Hi Tech News, 41*(8), 10–14. Advance online publication. DOI: 10.1108/LHTN-02-2024-0031

Tabassum, R. R., & Park, J. (2024). Development of a Building Evaluation Framework for Biophilic Design in Architecture. *Buildings (Basel, Switzerland), 14*(10), 3254. Advance online publication. DOI: 10.3390/buildings14103254

Tallini, M., Di Leo, A., Rossetti, C., & Berardi, F. (2013). The Sacred Almone River of the Appian Way Regional Park in Rome: Ancient myths, a ritual link between Rome and Asia Minor and water uses in the modern age. *Water Science and Technology: Water Supply, 13*(3), 727–734. DOI: 10.2166/ws.2013.098

Tambach, M., & Visscher, H. (2012). Towards Energy-neutral New Housing Developments. Municipal Climate Governance in The Netherlands. *European Planning Studies, 20*(1), 111–130. DOI: 10.1080/09654313.2011.638492

Tanchangya, T., Raihan, A., Mia, M. R., Tafsirun, U., Naher, K., Islam, N., Rashid, F., & Sarker, S. R. (2023). Greenwashing within the context of financial technology and sustainable development: Conceptual frameworks and theoretical perspectives.

Tang, Y., Dong, J., Li, G., Zheng, Y., Chi, Y., Nzihou, A., Weiss-Hortala, E., & Ye, C. (2020). Environmental and exergetic life cycle assessment of incineration- and gasification-based waste to energy systems in China. *Energy, 205*, 118002. DOI: 10.1016/j.energy.2020.118002

Tavares, G., Zsigraiová, Z., Semiao, V., & da Graca Carvalho, M. (2011). Monitoring, fault detection and operation prediction of MSW incinerators using multivariate statistical methods. *Waste Management (New York, N.Y.), 31*(7), 1635–1644. DOI: 10.1016/j.wasman.2011.02.005 PMID: 21376557

Tavares, R., Ramos, A., & Rouboa, A. (2018). Microplastics thermal treatment by polyethylene terephthalate-biomass gasification. *Energy Conversion and Management, 162*, 118–131. DOI: 10.1016/j.enconman.2018.02.001

Teixeira Dias, F., de Aguiar Dutra, A. R., Vieira Cubas, A. L., Ferreira Henckmaier, M. F., Courval, M., & de Andrade Guerra, J. B. S. O. (2023). Sustainable development with environmental, social and governance: Strategies for urban sustainability. *Sustainable Development (Bradford), 31*(1), 528–539. DOI: 10.1002/sd.2407

TerraChoice. (2009) 'The Seven Sins of Greenwashing: Environmental Claims in Consumer Markets', *TerraChoice Group*. Available at: https://sinsofgreenwashing.com/ (Accessed: 13 December 2024).

Thai, V. V., Tay, W. J., Tan, R., & Lai, A. (2014). Defining service quality in tramp shipping: Conceptual model and empirical Evidence. *Asian Journal of Shipping and Logistics, 30*(1), 1–29. DOI: 10.1016/j.ajsl.2014.04.001

Thilakshan, T., Saranjan, P., Fernando, G. V. K., & Bandara, J. M. S. J. Analysis of Passenger Perception on Quality of Service of Railway Transportation: Coastal Railway Line, *Sri Lanka. Moratuwa Engineering Research Conference* 2023, MERCon, 420 - 425. DOI: DOI: 10.1109/MERCon60487.2023.10355413

Thorne, C. R., Lawson, E. C., Ozawa, C., Hamlin, S. L., & Smith, L. A. (2015). Overcoming uncertainty and barriers to adoption of Blue-Green Infrastructure for urban flood risk management. *Journal of Flood Risk Management, 11*(S2), S960–S972. DOI: 10.1111/jfr3.12218

Tianchen Dai, X. Z. (2022). A systematic review of studies at the intersection of urban climate and historical urban landscape. *Environmental Impact Assessment Review, 97*, 106894. Advance online publication. DOI: 10.1016/j.eiar.2022.106894

Tiwari, C., & Sunny, A. (2024). To the Ones Who Listen to the Mountains: Evaluating the Role of Traditional Ecological Knowledge in Conservation Efforts in the Himalayas. In *Learning 'from'and 'with'the Locals: Traditional Knowledge Systems for Environmental Sustainability in the Himalayas* (pp. 3–22). Springer Nature Switzerland. DOI: 10.1007/978-3-031-51696-2_1

Toli, A. M., & Murtagh, N. (2020a). The Concept of Sustainability in Smart City Definitions. *Frontiers in Built Environment, 6*, 496662. DOI: 10.3389/fbuil.2020.00077

Tollefson, J. (2025). Earth breaches 1.5 °C climate limit for the first time: What does it mean? *Nature, 637*(8047), 769–770. DOI: 10.1038/d41586-025-00010-9 PMID: 39794429

Torres, L., Ripa, D., Jain, A., Herrero, J., & Leka, S. (2023). The potential of responsible business to promote sustainable work–An analysis of CSR/ESG instruments. *Safety Science, 164*, 106151. DOI: 10.1016/j.ssci.2023.106151

Tousi, E., & Mela, A. (2024). Supralocal Role of Medium to Large Scale Urban Parks, in Greater Athens Region in Greece. Issues of Meso Car Dependence During the Covid-19 Pandemic. *Journal of Sustainable Architecture and Civil Engineering, 35*(2), 201–215. DOI: 10.5755/j01.sace.35.2.34661

Tozer, L. (2018). *Urban Decarbonization: Politics and Practices of Carbon Neutrality*. https://utoronto.scholaris.ca/server/api/core/bitstreams/a94786de-7403-469a-b2d0-c9c2313c81dd/content

Tranos, E., & Davoudi, S. (2014). The Regional Impact of Climate Change on Winter Tourism in Europe. *Tourism Planning & Development, 11*(2), 163–178. DOI: 10.1080/21568316.2013.864992

Trotta, A., Rania, F., & Strano, E. (2023). Exploring the linkages between FinTech and ESG: A bibliometric perspective.

Tsigdinos, S. (2017). Aspects of the urban landscape and public space: City, nature and new technologies - Tracing accessibility in the Antonis Tritsis Metropolitan Park [Master's thesis, in Greek] National Technical University of Athens. http://oldwww.arch.ntua.gr/sites/default/files/project/11940_/opseis_tsigdinos_2017.pdf

Tsilika, E., & Vardopoulos, I. (2022). The FIX-up mix-up; undue façadism or adaptive reuse? Examining the former FIX brewery transformation into the National Museum of Contemporary Art in Athens. *Archnet-IJAR: International Journal of Architectural Research, 16*(3), 688–709. DOI: 10.1108/ARCH-09-2021-0255

Tsimnadis K., Kyriakopoulos G.L. (2024). Investigating the Role of Municipal Waste Treatment within the European Union through a Novel Created Common Sustainability Point System. Recycling 9(3), 42, 25 pp. DOI: .DOI: 10.3390/recycling9030042

Tunis Stock Exchange. (2024). available at: https://www.bvmt.com.tn

Tun, M. M., Palacky, P., Juchelkova, D., & Síťař, V. (2020). Renewable Waste-to-Energy in Southeast Asia: Status, Challenges, Opportunities, and Selection of Waste-to-Energy Technologies. *Applied Sciences (Basel, Switzerland), 10*(20), 7312. DOI: 10.3390/app10207312

Turner, J. S., & Soar, R. C. (2008). Beyond biomimicry: What termites can tell us about realizing the living building. www.archpaper.com/features/2007_14_imitation.htm

Uddin, S. M. N., Gutberlet, J., Ramezani, A., & Nasiruddin, S. M. (2020). Experiencing the everyday of waste pickers: A sustainable livelihoods and health assessment in Dhaka city, Bangladesh. *Journal of International Development, 32*(6), 833–853. DOI: 10.1002/jid.3479

Udiyana, I. B. G., Suastama, I. B. R., Astini, N. N. S., Mahanavami, G. A., Karwini, N. K., & Maretta, Y. A. (2018). Innovation strategy the development of competitifeness of eco-based coastal tourism destination, management organization and quality of services. *Journal of Environmental Management and Tourism, 9*(4), 851–860. DOI: 10.14505//jemt.9.4(28).19

Ulloa-Murillo, L. M., Villegas, L. M., Rodríguez-Ortiz, A. R., Duque-Acevedo, M., & Cortés-García, F. J. (2022). Management of the Organic Fraction of Municipal Solid Waste in the Context of a Sustainable and Circular Model: Analysis of Trends in Latin America and the Caribbean. *International Journal of Environmental Research and Public Health, 19*(10), 6041. DOI: 10.3390/ijerph19106041 PMID: 35627577

Ulrich, R. S. (1981). Natural versus urban scenes: Some psychophysiological effects. *Environment and Behavior, 13*(5), 523–556. DOI: 10.1177/0013916581135001

UN. (2019, September 18). *Cities: A "cause of and solution to" climate change | UN News*. https://news.un.org/en/story/2019/09/1046662

UNEP. (2022). Annual Report 2022.

United Nations (2016). Transforming our world: The 2030 agenda for sustainable development. A/RES/70/1.

United Nations Habitat. (2010). WTE industry in Latin America. Rio de Janeiro, Brazil, Habitat, UN: 73.

Urban Sustainability Exchange. (2025). The Madrid Río project. https://use.metropolis.org/case-studies/the-madrid-rio-project

van den Dobbelsteen, A., Roggema, R., Tillie, N., Broersma, S., Fremouw, M., & Martin, C. L. (2018). 4.6—Urban Energy Masterplanning—Approaches, Strategies, and Methods for the Energy Transition in Cities. In Droege, P. (Ed.), *Urban Energy Transition* (2nd ed., pp. 635–660). Elsevier., DOI: 10.1016/B978-0-08-102074-6.00045-0

Van Eck N & Waltman L. (2010). Software survey: VOSviewer, a computer program for bibliometric mapping. scientometrics. 84 (2):523-38. .DOI: 10.1007/s11192-009-0146-3

Vardopoulos, I., & Karytsas, S. (2019). An exploratory path analysis of climate change effects on tourism. *Sustainable Development, Culture, Traditions Journal, Special Volume in Honor of Professor George I. Theodoropoulos,* 132–152. DOI: 10.26341/issn.2241-4002-2019-sv-13

Vardopoulos, I., & Konstantinou, Z. (2017). Study of the possible links between CO2 emissions and employment status. Sustainable Development, Culture, Traditions Journal, 1(B), 100–112. https://doi.org/DOI: 10.26341/issn.2241-4002-2017-1b-8

Vardopoulos, I., Chroni, C., Synani, K., Abeliotis, K., & Lasaridi, K. (2023). More than meets the eye: Uncovering the social impacts of a household hazardous waste management system in Athens. In D. F. Lekkas (Ed.), *Book of Abstracts of the 18th International Conference on Environmental Science and Technology (CEST2023), 30 August - 2 September, 2023. Athens, Greece* (p. 142). Marousi, Greece: Cosmos. ISBN: 9786188629219.

Vardopoulos, I., Georgiou, C., Voukkali, I., Vassiliades, C., & Salvati, L. (2024). Case-based exploration of the nexus between transformative urban megadevelopments and sustainability. *Theoretical and Empirical Researches in Urban Management, 19*(3), 5–30. https://www.jstor.org/stable/27321372

Vardopoulos, I. (2022). Industrial building adaptive reuse for museum. Factors affecting visitors' perceptions of the sustainable urban development potential. *Building and Environment, 222*, 109391. DOI: 10.1016/j.buildenv.2022.109391

Vardopoulos, I. (2023). Adaptive reuse for sustainable development and land use: A multivariate linear regression analysis estimating key determinants of public perceptions. *Heritage, 6*(2), 809–828. DOI: 10.3390/heritage6020045

Vardopoulos, I. (2024). Urban planning policies and tools for sustainable development: insights into adaptive reuse, smart cities, and urban metabolism. In *Socio-enviromental change: The path toward sustainability* (pp. 121–140). Nova Science Publishers.

Vardopoulos, I. (2024). Urban planning policies and tools for sustainable development: insights into adaptive reuse, smart cities, and urban metabolism. In *Socio-environmental change: The path toward sustainability* (pp. 121–140). Nova Science Publishers.

Vardopoulos, I. (2025). Cities in flux: Waste, the shadow of circularity, and the (elusive) promise of sustainability. *Sustainability (Basel), 17*(2), 705. DOI: 10.3390/su17020705

Vardopoulos, I., Abeliotis, K., & Lasaridi, K. (2025). A systematic informetric analysis and literature review of food waste quantification studies in the food service industry. *Sustainability (Basel), 17*(1), 103. DOI: 10.3390/su17010103

Vardopoulos, I., D'Agata, A., Escrivà Saneugenio, F., & Salvati, L. (2024). *Sprawl and the city: realizing a sustainable Mediterranean urbanization.* Nova Science Publishers., DOI: 10.52305/GTVV1244

Vardopoulos, I., Falireas, S., Konstantopoulos, I., Kaliora, E., & Theodoropoulou, E. (2018). Sustainability assessment of the agri-environmental practices in Greece. Indicators' comparative study. *International Journal of Agricultural Resources, Governance and Ecology, 14*(4), 368–399. DOI: 10.1504/IJARGE.2018.098024

Vardopoulos, I., Giannopoulos, K., Kalli, S., & Salvati, L. (2024). From tech-centric to human-centric smart tourist destinations: innovations in real estate, urban planning, and sustainability. In *Socio-enviromental change: the path toward sustainability* (pp. 65–96). Nova Science Publishers.

Vardopoulos, I., Ioannides, S., Georgiou, M., Voukkali, I., Salvati, L., & Doukas, Y. E. (2023). Shaping sustainable cities: A long-term GIS-emanated spatial analysis of settlement growth and planning in a coastal Mediterranean European city. *Sustainability (Basel), 15*(14), 11202. DOI: 10.3390/su151411202

Vardopoulos, I., Konstantopoulos, I., Zorpas, A. A., Limousi, L., Bennici, S., Inglezakis, V., & Voukkali, I. (2021). Sustainable metropolitan areas perspectives through assessment of the existing waste management strategies. *Environmental Science and Pollution Research International, 28*(19), 24305–24320. DOI: 10.1007/s11356-020-07930-1 PMID: 32072414

Vardopoulos, I., Maialetti, M., Scarpitta, D., & Salvati, L. (2024). Spatially explicit analysis of landscape structures, urban growth, and economic dynamics in Metropolitan regions. *Urban Science (Basel, Switzerland), 8*(4), 150. DOI: 10.3390/urbansci8040150

Vardopoulos, I., Papoui-Evangelou, M., Nosovo, B., & Salvati, L. (2023). Smart 'tourist cities' revisited: Culture-led urban sustainability and the global real estate market. *Sustainability (Basel), 15*(5), 4313. DOI: 10.3390/su15054313

Vardopoulos, I., & Poli, M. (2025). Decoding Europe's "Renovation Wave" energy strategy: real estate shifts for sustainable cities. In Mitoula, R. (Ed.), *2nd Open-Air Cities International Conference: Local and Regional Sustainable Development and Urban Reconstruction, 14-16 February, 2025. Book of Abstracts* (p. 173). Athens, Greece: Open Air Cities Institute. ISBN: 9786188707030.

Vardopoulos, I., Santamouris, M., Zorpas, A. A., Barone, G., Italos, C., & Vassiliades, C. (2024). A comparative study on discrepancies in residential building energy performance certification in a Mediterranean context. *Buildings (Basel, Switzerland), 14*(4), 1009. DOI: 10.3390/buildings14041009

Vardopoulos, I., Stamopoulos, C., Chatzithanasis, G., Michalakelis, C., Giannouli, P., & Pastrapa, E. (2020). Considering urban development paths and processes on account of adaptive reuse projects. *Buildings (Basel, Switzerland), 10*(4), 73. DOI: 10.3390/buildings10040073

Vardopoulos, I., Tsilika, E., Sarantakou, E., Zorpas, A. A., Salvati, L., & Tsartas, P. (2021). An integrated SWOT-PESTLE-AHP model assessing sustainability in adaptive reuse projects. *Applied Sciences (Basel, Switzerland), 11*(15), 7134. DOI: 10.3390/app11157134

Vardopoulos, I., Vannas, I., Xydis, G., & Vassiliades, C. (2023). Homeowners' perceptions of renewable energy and market value of sustainable buildings. *Energies, 16*(10), 4178. DOI: 10.3390/en16104178

Varma, C. R. S., & Palaniappan, S. (2019). Comparision of green building rating schemes used in North America, Europe and Asia. *Habitat International, 89*, 101989. DOI: 10.1016/j.habitatint.2019.05.008

Vartholomaios, A. (2015). The Residential Solar Block envelope: A method for enabling the development of compact urban blocks with high passive solar potential. *Energy and Building, 99*, 303–312. DOI: 10.1016/j.enbuild.2015.04.046

Vasilopoulou, C. (2017). Water-sensitive urban design and natural water infrastructure: The case of Antonis Tritsis Park [Master's thesis, in Greek]. National Technical University of Athens. https://dspace.lib.ntua.gr/xmlui/handle/123456789/46433

Vassilev, S. V., Baxter, D., Andersen, L. K., & Vassileva, C. G. (2013). An overview of the composition and application of biomass ash. Part 2. Potential utilisation, technological and ecological advantages and challenges. *Fuel, 105*, 19–39. DOI: 10.1016/j.fuel.2012.10.001

Vassiliades, C., Kyriakides, N., Pissourios, I., & Vardopoulos, I. (2024). Sustainable architectural innovations for energy-efficient multi-storey buildings through a novel taxonomy and evaluation approach. *Strategic Planning for Energy and the Environment, 43*(4), 1195–1224. DOI: 10.13052/spee1048-5236.43415

Vassiliades, C., Minterides, C., Astara, O. E., Barone, G., & Vardopoulos, I. (2023). Socio-economic barriers to adopting energy-saving bioclimatic strategies in a Mediterranean sustainable real estate setting: A quantitative analysis of resident perspectives. *Energies, 16*(24), 7952. DOI: 10.3390/en16247952

Vassiliades, C., Vardopoulos, I., Barone, G., Santamouris, M., & Kalogirou, S. (2025). Exposing the limits in the residential building energy performance certification model. *Energy Reports, 14*, 1008–1019. DOI: 10.1016/j.egyr.2025.07.011

Velenturf, A. P., & Purnell, P. (2021). Principles for a sustainable circular economy. *Sustainable Production and Consumption, 27*, 1437–1457. DOI: 10.1016/j.spc.2021.02.018

Verdouw, C., Tekinerdogan, B., Beulens, A., & Wolfert, S. (2021). Digital twins in smart farming. *Agricultural Systems, 189*, 103046. DOI: 10.1016/j.agsy.2020.103046

Vidović, T., Šimunović, J., Radica, G., & Penga, Ž. (2023). Systematic Overview of Newly Available Technologies in the Green Maritime Sector. *Energies, 16*(2), 641. DOI: 10.3390/en16020641

Vinci, S., Vardopoulos, I., & Salvati, L. (2023). A tale of a shrinking City? Exploring the complex interplay of socio-demographic dynamics in the recent development of Attica, Greece. *Cities (London, England), 132*, 104089. DOI: 10.1016/j.cities.2022.104089

Vodák, J., Šulyová, D., & Kubina, M. (2021). Advanced technologies and their use in smart city management. *Sustainability (Basel), 13*(10), 5746. DOI: 10.3390/su13105746

Vogiatzi, C., Gemenetzi, G., Massou, L., Poulopoulos, S., Papaefthimiou, S., & Zervas, E. (2018). Energy use and saving in residential sector and occupant behavior: A case study in Athens. *Energy and Building, 181*, 1–9. DOI: 10.1016/j.enbuild.2018.09.039

Vohra, K., & Franklin, M. L. (2021). Water Resources Management—An Indian Perspective. In Springer transactions in civil and environmental engineering (pp. 237–252).

Vortelinos, D., Menegaki, A., Passas, I., Garefalakis, A., & Viskadouros, G. (2024). Heterogeneous Responses of Energy and Non-Energy Assets to Crises in Commodity Markets. Energies 2024, Vol. 17, Page 5438, 17(21), 5438. DOI: 10.3390/en17215438

Vortelinos, D., Passas, I., Floros, C., & Garefalakis, A. (2024). Corporate Social Responsibility and Country Governance: An International Comparative Study Amid the COVID-19 Pandemic. International Journal of Financial Studies 2024, Vol. 12, Page 110, 12(4), 110. DOI: 10.3390/ijfs12040110

Võsa, K.-V., Kurnitski, J., Felsmann, C., Meinsenbach, A., Carli, M. D., Tonon, M., & Iivonen, M. (2023). National methods fail to calculate standardized deep renovation concepts for dwellings: Benchmarking in three EU climates. *E3S Web of Conferences, 396*, 04014. DOI: 10.1051/e3sconf/202339604014

Voukkali, I., Vardopoulos, I., Papamichael, I., & Zorpas, A. A. (2024). Urban metabolism considering the "15-minute-city" concept. In *1st Open-Air Cities International Conference: Local and Regional Sustainable Development and Urban Reconstruction, 16-18 February, 2024, Kallithea, Greece. Book of Abstracts* (p. 99). Athens, Greece: Open-Air Cities Institute. ISBN: 9786188707009.

Voytenko, Y., McCormick, K., Evans, J., & Schliwa, G. (2016). Urban living labs for sustainability and low carbon cities in Europe: Towards a research agenda. *Journal of Cleaner Production, 123*, 45–54. DOI: 10.1016/j.jclepro.2015.08.053

Vreugdenhil, B., Zwart, R., & Neeft, J. P. A. (2009). *Tar formation in pyrolysis and gasification*. ECN.

Vukovic, N., & Makogon, E. (2022). Waste-to-energy generation: Complex efficiency analysis of modern technologies. *Sustainability (Basel), 14*(21), 13814. DOI: 10.3390/su142113814

Waheed, A., Zhang, Q., Rashid, Y., Tahir, M. S., & Zafar, M. W. (2020). Impact of green manufacturing on consumer ecological behavior: Stakeholder engagement through green production and innovation. *Sustainable Development (Bradford), 28*(5), 1395–1403. DOI: 10.1002/sd.2093

Walker, B., & Salt, D. (2006). *Resilience Thinking: Sustaining Ecosystems and People in a Changing World*. Island Press.

Walker, S., Coleman, N., Hodgson, P., Collins, N., & Brimacombe, L. (2018). Evaluating the environmental dimension of material efficiency strategies relating to the circular economy. *Sustainability (Basel), 10*(3), 666. DOI: 10.3390/su10030666

Wamsler, C. (2015). Mainstreaming ecosystem-based adaptation: Transformation toward sustainability in urban governance and planning. *Ecology and Society, 20*(2), art30. https://www.jstor.org/stable/26270196. DOI: 10.5751/ES-07489-200230

Wang, C., Jiang, Q., Dittrich, Y., & Østerlund, C. (2023). ESG in fintech: An overview. *IT University of Copenhagen*. https://pure.itu.dk/ws/portalfiles/portal/101319385/TR-2023-209.pdf

Wang, Y., Song, Q., He, J., & Qi, Y. (2015). Developing low-carbon cities through pilots. *Climate Policy, 15*(sup1), S81–S103. DOI: 10.1080/14693062.2015.1050347

Wang, D., Peng, K., Tang, K., & Wu, Y. (2023). Does fintech development enhance corporate ESG performance? Evidence from an emerging market. *SSRN*.

Wang, D., Xu, P. Y., An, B. W., & Guo, Q. P. (2024). Urban green infrastructure: Bridging biodiversity conservation and sustainable urban development through adaptive management approach. *Frontiers in Ecology and Evolution*, *12*, 1440477. DOI: 10.3389/fevo.2024.1440477

Wang, J., Liu, J., Wang, H., & Mei, C. (2020). Approaches to multi-objective optimization and assessment of green infrastructure and their multi-functional effectiveness: A review. *Water (Basel)*, *12*(10), 2714. DOI: 10.3390/w12102714

Wang, K., & Fouseki, K. (2025). Sustaining the Fabric of Time: Urban Heritage, Time Rupture, and Sustainable Development. *Land (Basel)*, *14*(1), 193. Advance online publication. DOI: 10.3390/land14010193

Wang, L., Weller, C. L., Jones, D. D., & Hanna, M. A. (2008). Contemporary issues in thermal gasification of biomass and its application to electricity and fuel production. *Biomass and Bioenergy*, *32*(7), 573–581. DOI: 10.1016/j.biombioe.2007.12.007

Wassenhoven, L. (2018). *Hellinikon passion: The development of the former Hellinikon Airport* [in Greek]. Sakkoulas.

Webb, A. L. (2017). Energy retrofits in historic and traditional buildings: A review of problems and methods. *Renewable & Sustainable Energy Reviews*, *77*, 748–759. Advance online publication. DOI: 10.1016/j.rser.2017.01.145

Wee, J.-H., Bae, Y., Ahn, H., Choi, Y. O., Jeong, E., & Yeo, S. Y. (2022). Fibrous and granular activated carbon mixed media for effective gas removal as a cabin air filter. *Carbon Letters*, *32*(4), 1111–1118. DOI: 10.1007/s42823-022-00345-7

Wekesa, C., Ndalilo, L., & Swiderska, K. (2021). Case study for the project 'Indigenous biocultural heritage for sustainable development'.

Wellinger, A., Murphy, J. P., & Baxter, D. (2013). *The biogas handbook: science, production and applications*. Elsevier.

WEST8. (2025, March 18). Vision for the riverfront. https://www.west8.com/projects/madrid-rio/

Weston, P., & Nnadi, M. (2023). Evaluation of strategic and financial variables of corporate sustainability and ESG policies on corporate finance performance. *SSRN*. DOI: 10.1080/20430795.2021.1883984

Williams, J. (2024). Greenwashing: Appearance, illusion and the future of 'green' capitalism. *Geography Compass*, *18*(1), e12736. DOI: 10.1111/gec3.12736

Wilson, D. C. (2007). "Development drivers for waste management." 25(3): 198-207.

Wilson, D. C., Rodic, L., Modak, P., Soos, R., Carpintero, A., Velis, K., Iyer, M., & Simonett, O. (2015). *Global waste management outlook*. Unep.

Winfield, A. F. T., & Jirotka, M. (2019). *Ethical governance is essential to building trust in robotics and artificial intelligence systems*. ResearchGate.

Winkler, M. S., Jackson, D., Sutherland, D., Lim, J. M. U., Srikantaiah, V., Fuhrimann, S., & Medlicott, K. (2017). Sanitation safety planning as a tool for achieving safely managed sanitation systems and safe use of wastewater. *WHO South-East Asia Journal of Public Health*, *6*(2), 34–40. DOI: 10.4103/2224-3151.213790 PMID: 28857061

Woltering, L., Fehlenberg, K., Gerard, B., Ubels, J., & Cooley, L. (2019). Scaling – from "reaching many" to sustainable systems change at scale: A critical shift in mindset. *Agricultural Systems*, *176*, 102652. DOI: 10.1016/j.agsy.2019.102652

World Bank Group. (2017). Converting Biomass to Energy - A Guide for Developers and Investors. Washington D.C., USA, International Finance Corporation - Energy & Water Advisory: 216.

World Green Building Council. (2013). THE BUSINESS CASE FOR GREEN BUILDING A Review of the Costs and Benefits for Developers, Investors and Occupants Project Partner Principal Sponsors Sponsors. www.worldgbc.org

World Health Organization. (2016). Urban green spaces and health: A review of evidence. WHO Regional Office for Europe. https://iris.who.int/handle/10665/345751

Wut, T. M., Lee, D., Ip, W. M., & Lee, S. W. (2021). Digital sustainability in the organization: Scale development and validation. *Sustainability (Basel)*, *13*(6), 3530. DOI: 10.3390/su13063530

Xaxa, A. (2019). *Tribes and urbanisation in North East India - issues and challenges*. Iitg.

Xiao Xiao, P. L. (2024). Sustainable Adaptation Planning for Cultural Heritage in Coastal Tourism Destinations Under Climate Change: A Mixed-Paradigm of Preservation and Conservation Optimization. *Journal of Travel Research*, *63*(1), 215–233. DOI: 10.1177/00472875221143479

Xiao, G., Wang, Y., Wu, R., Li, J., & Cai, Z. (2024). Sustainable Maritime Transport: A Review of Intelligent Shipping Technology and Green Port Construction Applications. *Journal of Marine Science and Engineering*, *12*(10), 1728. DOI: 10.3390/jmse12101728

Xu, J. (2024). AI in ESG for financial institutions: An industrial survey. *arXiv*. https://ideas.repec.org/p/arx/papers/2403.05541.html

Yadav, A., Anwer, N., Mahapatra, K., Shrivastava, M. K., & Khatiwada, D. (2024). Analyzing the Role of Polycentric Governance in Institutional Innovations: Insights from Urban Climate Governance in India. *Sustainability (Basel), 16*(23), 10736. DOI: 10.3390/su162310736

Yang, D., Geng, H., & Wang, F. (2020). Marine regulations' impact on environmental sustainability in maritime shipping and coastal economic activities. *Journal of Coastal Research, 109*(sp1), 79–84. DOI: 10.2112/JCR-SI109-013.1

Yang, J., Gu, L., Liu, K., & Deng, C. (2025). Unabsorbed slack, capability reconfiguration and green innovation behavior: The moderating role of consumer green pressure. *Journal of Business and Industrial Marketing, 40*(2), 391–414. DOI: 10.1108/JBIM-02-2024-0068

Yang, Q. (2021). Fiscal Transparency and Public Service Quality Association: Evidence from 12 Coastal Provinces and Cities of China. *Journal of Risk and Financial Management, 14*(1), 13. DOI: 10.3390/jrfm14010013

Yang, T. H., Hu, K. M., Li, R. D., Sun, Y., & Kai, X. P. (2015). Cogasification of typical plastics and rice straw with carbon dioxide. *Environmental Progress & Sustainable Energy, 34*(3), 789–794. DOI: 10.1002/ep.12041

Yang, Z., Nguyen, T. T. H., Nguyen, H. N., Nguyen, T. T. N., & Cao, T. T. (2020). Greenwashing behaviours: Causes, taxonomy and consequences based on a systematic literature review. *Journal of Business Economics and Management, 21*(5), 1486–1507. DOI: 10.3846/jbem.2020.13225

Yiannakou, A., & Salata, K. (2018). Green infrastructure: An overview of its contribution to urban areas and an investigation of the provisions in the Greek planning system [in Greek]. Aichoros: Texts on Spatial Planning, Urban Planning and Development, 1(1), 43–75. DOI: 10.26253/heal.uth.ojs.aei.2018.435

Yiannakou, A. (2020). Urban regeneration as a perpetual planning process: Understanding the role of stakeholders in property-led regeneration projects in Greek cities. *Local Economy, 35*(2), 83–104. DOI: 10.1177/0269094220914470

Yiannoulis, Y., Vortelinos, D., & Passas, I. (2025). Exploring Audit Opinions: A Deep Dive into Ratios and Fraud Variables in the Athens Exchange. Accounting and Auditing 2025, Vol. 1, Page 3, 1(1), 3. DOI: 10.3390/accountaudit1010003

Yildirim, S. (2023). Greenwashing: A rapid escape from sustainability or a slow transition? *LBS Journal of Management & Research*, *21*(1), 53–63. DOI: 10.1108/LBSJMR-11-2022-0077

You, S. M., Wang, W., Dai, Y. J., Tong, Y. W., & Wang, C. H. (2016). Comparison of the co-gasification of sewage sludge and food wastes and cost-benefit analysis of gasification- and incineration-based waste treatment schemes. *Bioresource Technology*, *218*, 595–605. DOI: 10.1016/j.biortech.2016.07.017 PMID: 27416510

Yu, J., Safarov, B., Wang, C., Buzrukova, M., & Janzakov, B. (2023). The Effect of Transportation Networks on Heritage Tourism and New Urbanization—Empirical Research Based on Rich Heritage Sites in a Chinese Province. Heritage, 7293–7315. Retrieved from DOI: 10.3390/heritage6120383

Yuen, K. F., & Thai, V. V. (2015). Service quality and customer satisfaction in liner shipping. *International Journal of Quality and Service Sciences*, *7*(2-3), 170–183. DOI: 10.1108/IJQSS-02-2015-0024

Yuen, K. F., & Thai, V. V. (2017). Corporate social responsibility and service quality provision in shipping firms: Financial synergies or trade-offs? *Maritime Policy & Management*, *44*(1), 131–146. DOI: 10.1080/03088839.2016.1237782

Yu, H., Wei, W., Li, J., & Li, Y. (2022). The impact of green digital finance on energy resources and climate change mitigation in carbon neutrality: Case of 60 economies. *Resources Policy*, *79*, 103116. DOI: 10.1016/j.resourpol.2022.103116

Yukalang, N., Clarke, B., & Ross, K. (2017). Barriers to Effective Municipal Solid Waste Management in a Rapidly Urbanizing Area in Thailand. *International Journal of Environmental Research and Public Health*, *14*(9), 1013. DOI: 10.3390/ijerph14091013 PMID: 28869572

Zabed, H. M., Akter, S., Yun, J. H., Zhang, G. Y., Zhang, Y. F., & Qi, X. H. (2020). Biogas from microalgae: Technologies, challenges and opportunities. *Renewable & Sustainable Energy Reviews*, *117*, 21. DOI: 10.1016/j.rser.2019.109503

Zafeiriou, E., Garefalakis, A., Passas, I., & Ragazou, K. (2023). Illicit and Corruption Mitigation Strategy in the Financial Sector: A Study with a Hybrid Methodological Approach. Sustainability 2023, Vol. 15, Page 1366, 15(2), 1366. DOI: 10.3390/su15021366

Zanon, B., & Verones, S. (2013). Climate change, urban energy and planning practices: Italian experiences of innovation in land management tools. *Land Use Policy*, *32*, 343–355. DOI: 10.1016/j.landusepol.2012.11.009

Zeng, J., Yu, S., & Zhan, Y. (2023). A Novel Approach to Green Emissions Reduction and Reliability Improvement in the Maritime Industry: Multi-Table Fusion-Based Full-Life Weight Control Technology. *7th IEEE International Conference on Transportation Information and Safety, ICTIS 2023*, 1655 - 1661. DOI: DOI: 10.1109/ICTIS60134.2023.10243738

Zhang, C., Kazanci, O. B., Levinson, R., Heiselberg, P., Olesen, B. W., Chiesa, G., Sodagar, B., Ai, Z., Selkowitz, S., Zinzi, M., Mahdavi, A., Teufl, H., Kolokotroni, M., Salvati, A., Bozonnet, E., Chtioui, F., Salagnac, P., Rahif, R., Attia, S., & Zhang, G. (2021). Resilient cooling strategies – A critical review and qualitative assessment. *Energy and Building*, *251*, 111312. Advance online publication. DOI: 10.1016/j.enbuild.2021.111312

Zhang, K., Troitzsch, S., Zhang, S. Y., Teh, E. S. P., Subramanian, L., & Massier, T. (2021). Transactive Energy in an Urban Environment: A Case Study of Local Generation and Flexibility Potentials in a Singaporean Urban District. *Frontiers in Energy Research*, *9*, 632992. Advance online publication. DOI: 10.3389/fenrg.2021.632992

Zhang, Y. (2024). Role of green finance, green bonds, public private partnership, and technology innovation in carbon neutrality and sustainable development. *Heliyon*, *10*(18), e37189. Advance online publication. DOI: 10.1016/j.heliyon.2024.e37189 PMID: 39328522

Zhang, Y., Niu, Y., & Zhang, T. (2022). *Introductory chapter: The overview of recent advances of sustainable waste management*. The Circular Economy-Recent Advances in Sustainable Waste Management.

Zhu, E., Gao, H., Chen, L., Yao, J., Liu, T., & Sha, M. (2024). Interactions between coastal protection forest ecosystems and human activities: Quality, service and resilience. *Ocean and Coastal Management*, *254*, 107190. DOI: 10.1016/j.ocecoaman.2024.107190

Zopounidis, C., Garefalakis, A., Lemonakis, C., & Passas, I. (2020). Environmental, social and corporate governance framework for corporate disclosure: A multicriteria dimension analysis approach. *Management Decision*, *58*(11), 2473–2496. DOI: 10.1108/MD-10-2019-1341

Zorpas, A. A. (2020). Strategy development in the framework of waste management. *The Science of the Total Environment*, *716*, 137088. DOI: 10.1016/j.scitotenv.2020.137088 PMID: 32059326

About the Contributors

Ioannis Vardopoulos, studied civil engineering, did his postgraduate studies first in environmental management and conservation and later in sustainable development with honors distinction. Holding a doctorate in applied urban sustainability, he conducts high-quality research towards addressing key concerns of the contemporary build environment through principles of applied sustainability. His research interests include a wide range of urban sustainability aspects.

Ioannis Passas is an Assistant Professor with a Ph.D. in financial accounting and auditing from the Department of Business Administration & Tourism of the Hellenic Mediterranean University, where he teaches accounting courses at the undergraduate and postgraduate levels. He also teaches at the Department of Accounting and Finance at Neapolis University of Paphos in the Forensic Accounting postgraduate program. He is also a Certified Internal Control Auditor (CICA), a Certified Control Specialist (CCS) by the Institute of Internal Auditors (IIC), a Certified Management Accountant (CMA) by the Institute of Chartered Accountants, and a Certified Internal Auditor by the State of Greece. He has participated in international conferences, contributed, and published numerous papers in peer-reviewed journals and volumes in foreign-language books. He has also authored six academic books, five in his discipline (Accounting and Management Accounting) and the other in Research Methodology. His main research interests focus on the scientific fields: Financial Accounting; Auditing; ESG/CSR; Financial integrity; Governance. He is a member of the Hellenic Association of Economists and Accountants, the research laboratories OramaLab and LAFIM, the Institute of Public Accountants ACPAI, the Hellenic Economic Chamber of Greece, and a founding member of ELINESEE.

***

Tarek Ben Noamene is an Assistant Professor of Management Dr. Tarek Ben Noamene is an Assistant Professor of Management at Jendouba University, specializing in Corporate Social Responsibility (CSR) and Diversity Management. His research explores the strategic integration of CSR practices within organizations, focusing on how businesses can achieve sustainable growth while addressing social and environmental concerns. Additionally, his work in diversity management examines the role of inclusive policies in enhancing organizational performance, leadership effectiveness, and workplace equity. Dr. Ben Noamene has published and presented his research in peer-reviewed journals and international conferences. His expertise extends to business ethics, stakeholder engagement, and sustainable development, particularly in emerging markets. Beyond academia, he collaborates with industry partners and policymakers to promote responsible business practices. He is committed to advancing knowledge on CSR's impact on corporate governance, employee well-being, and investor relations.

Kiranbhai Ramabhai Dodiya is an experienced Cyber Forensic Expert and Assistant Professor specializing in cyber security and digital forensics. He has taught at prestigious institutions and is currently pursuing a PhD in Biochemistry and Forensic Science. With a strong background in ethical hacking and malware analysis, Kiranbhai is passionate about educating students and contributing to the field of cyber forensics. His expertise includes malware forensic, computer forensics, and data recovery software. Kiranbhai's commitment to professional growth is evident through his participation in workshops, seminars, and research publications. He is dedicated to shaping the next generation of professionals in the field and combatting cybercrimes.

Alexandros Garefalakis is Chartered Accountant (Fellow of CPA), Management Accountant (CMA Diploma) and Business Consultant in numerous of companies and organizations. Also, he is a Assistant Professor at the Dept. of Business Administration and Tourism, Hellenic Mediterranean University (HMU). He holds a degree in Financial Applications, a Master degree in Accounting and Finance from University of Southampton and PhD in Accounting from Hellenic Open University. He has co-authored 2 books on Audit Accounting and Management Accounting issues. He has more than 70 publications in peer reviewed journals and participation in more than 21 national and international conferences. His areas of research interest include the quality of Narrative information of Financial Reports and business performance assessment as well as management efficiency and productivity.

Pooja Jain is an Associate Professor in the area of Finance and Accounting at Amity University, Gwalior. She has over 12 years of work experience in academics. Her research interests include Financial Management and Financial Market and she published several Scopus indexed articles.

Vishal Jain is presently working as a Professor (CSE) at the Department of Computer Science & Engineering, School of Engineering & Technology, Vivekananda Institute of Professional Studies - Technical Campus, New Delhi. Before that, he worked as Professor at Sharda University, Greater Noida and Associate Professor at Bharati Vidyapeeth's Institute of Computer Applications and Management (BVICAM), New Delhi.

Ashwarya Kapoor received her bachelors and masters degree from University of Delhi. She holds PhD from J.C. Bose University of Science & Technology, YMCA, Faridabad, India. Her research interest is service quality management. She has presented many papers in international conferences. Currently, she is working as assistant professor at School of Commerce, Manav Rachna International Institute of Research and Studies, Faridabad.

Konstantinos Kavouras is a PhD Candidate at the Department of Geography, Harokopio University of Athens. His research focuses on climate governance, with an emphasis on the challenges of renewable energy development in Greece in the context of climate change. He holds MSc degrees in Applied Geography and Spatial Planning and in European Social Policy, as well as a BSc in Economics. He has contributed to academic conferences and research initiatives in the fields of sustainable development, climate risk and energy transition. He was a member of the Local Organizing Committee of the SRA-Europe 2024 conference "Risk in Time and Space".

Ruchi Kushwaha is a Ph.D. scholar in Management, Jiwaji University, Gwalior, India. She holds a Master's and M. Phil. in Economics and her particular research interest in the area of Digital Financial Service and Entrepreneurship Intention. She published several Scopus indexed journal articles, book chapters and IEEE conference paper.

Jyoti Kushwaha completed her Ph.D. in home-science domain from Jiwaji University, Gwalior, India. She holds a Master's and M. Phil. in Home Science and her particular research in the area of Women empowerment, Work-family balance and Women entrepreneurship. She published several Scopus indexed journal articles, book chapters and IEEE conference proceedings.

Grigorios L. Kyriakopoulos (2 PhDs) is Teaching and Research Associate at the School of Electrical and Computer Engineering in NTUA (Greece). He worked more than 25-years in NTUA, having also a total of 10-years teaching service in Hellenic Naval Academy (2 years), Merchant Marine Academy of Aspropyrgos, Greece (2 years), and School of Pedagogical and Technological Education of Athens (ASPAITE) (4 years). He holds 2 PhDs, in Low Carbon Economy (DUTH, Greece) and in Chemical Engineering (NTUA, Greece), Chemical Engineering (MEng, MSc in Technical-Economic Systems, PhD, PostDoc), Environment (BSc, 2MSc in Environmental Design of: a) large infrastructure works and b) cities/buildings, PhD), Business Administration (BA and MA), Hellenic Civilization (BA and MA), Energy (MSc), Education (PGCE), Psychology (CertHE, HND, PGCert), Human Resource Management (CertHE). He co-authored more than 140 papers in 70 journals, 20 in book chapters and 30 in conferences, all obtaining more than 6300 citations. He is reviewer of more than 7000 manuscripts in 525 journals. He is recorded in the top 2% of scientists, Stanford University list, 5 years: 2020-2024 and career periods 1996-2022-2023-2024.

Archan Mitra is an Assistant Professor at School of Media Studies (SOMS) at Presidency University, Bangalore. He is the author of two book "Cases for Classroom Media and Entertainment Business" and "Multiverse and Media", he also has other several edited books to his credit. He has done his doctorate from Visva-Bharati Santiniketan, West Bengal in the field of "environmental informatics and communication for sustainability". In addition to that he is a certified Science Communicator and Journalism from Indian Science Communication Society (ISCOS), certified Corporate Trainer with Amity Institute of Training and Development, Certified Social Media Network Analyst. He has a strong interest in environmental communication. He was awarded certificate of merit by PRSI, Kolkata Chapter and Medal of Honor by Journalistic Club of Kolkata. He was working as a research assistant with the World Bank's "Environmental Capacity Building in Southeast Asia" project at IIM Kashipur. He was instrumental in launching the World Bank's Green MBA MOOC, he has also assisted in the research project on Uttarakhand disaster mitigation by ICSSR, the leading research on Uttarakhand disaster.

Sanchita Paul is a dedicated and accomplished content strategist with extensive experience in managing diverse content formats across different digital platforms. With a strong background in content strategy, editing, and UX writing, she has honed her skills in branding, social media management, and technical writing. Holding an MA in Communication and Journalism from Visva-Bharati University, Ms. Paul is proficient in tools like Google Analytics, Canva, and MS Office Suite, making her a versatile professional committed to producing high-quality, impactful content to improve brand visibility.

Aikaterini Petropoulou holds a Master's Degree in Hospitality and Tourism Management and a Bachelor's Degree in Business Administration and Tourism from the Hellenic Mediterranean University, both with distinction (MSc 9.58, BSc 8.69). She is currently Education Officer and Career Consultant at the Technical Schools of the Heraklion Chamber of Commerce (ISAEC). Her research and publications focus on ESG, Artificial Intelligence, Blockchain, and Digital Transformation in SMEs. She has co-authored book chapters and papers presented at international conferences and journals such as Sustainability and IGI Global. She is a member of the Economic Chamber of Greece and holds a Class B Accountant-Tax Consultant License.

Sneha Rajput was born in India. She is a full-time Associate Professor and Head of the Marketing Department at Prestige Institute of Management and Research, Gwalior (PIMR-G), India. Dr. Sneha received her doctorate from Jiwaji University in 2015 and holds a postgraduate degree in Management from the School of Economics, Devi Ahilya University, Indore (2005). She has qualified UGC NET and has 18 years of experience in teaching and research.

Ana Ramos, Cum Laude PhD in Environmental Engineering (University of Porto, 2020) and Stanford/Elsevier World's Top 2% Scientist (2024). Researcher in national and international R&D projects in academy, technology centers and international SMEs, with expertise in fundamental and applied research in chemistry and environment, mapping and supporting the development of circular economy strategies, and carrying out sustainability assessments (life cycle thinking, focusing on aspects of environmental and socioeconomic impacts) in the fields of renewable energy, urban solid waste and biomass management. Author and editor of scientific publications in environment, sustainability, waste-to-energy, and life cycle assessment topics. Guest speaker at scientific events, and lecturer on Circular Economy, Life Cycle Assessment and Social Impact Assessment in national and international scientific and technological system and higher education entities.

Kalliopi Sapountzaki, Professor Emerita (Spatial Planning and Disaster/Risk Management) at Harokopio University of Athens, Department of Geography. She has been Dean of the Faculty of "Environment, Geography and Applied Economics" (2017-2024). She holds PhD on Risk Management and Spatial Planning from the National Technical University of Athens (1990), MA diploma in Urban Design from the University of Manchester and Diploma in Architecture from the National Technical University of Athens. She has been Board member of the Society for Risk Analysis - Europe and she has offered lectures as visiting professor at several universities across Europe (University of Vienna, Technical University of Dortmund, University of Valencia etc). She has contributed to the UNISDR Global Assessment Report (GAR) 2019 and in several chapters of the report of the European Commission “Science for Disaster Risk Management 2020: Acting today, protecting tomorrow”. She has been Guest Editor of International Scientific Journals for open access Special Issues featuring disasters, climate change and spatial planning. She has been the principal author of a co-authored textbook on Risks and Disasters (in Greek) and the scientific leader of European Commission funded and national research projects focusing on risk and vulnerability analysis, resilience building and risk governance, mitigation and adaptation to climate change, sustainable development, urban and regional planning. She has published her work in international books and scientific journals and she has chaired the organizing and scientific committee of the 2024 international conference “Risk in Time and Space” (under the auspices of the Society for Risk Analysis – Europe) that was held in Athens (2-5 June 2024).

Satyaki Sarkar is an Architect and Urban Planner who is in the field of academics and research for the last two decades. He has published numerous research papers, and books and has been involved with numerous research projects throughout his career.

Pankaj Singh is an Associate Professor of Marketing Science in School of Business at Galgotias University, Greater Noida India. He completed his doctoral research in marketing domain at IIIT, Gwalior and his particular research interest in the area of Service Marketing and Public Policy. Pankaj Singh is the corresponding author and can be contacted at: pankajabviiitm@gmail.com

Aris Vartholomaios is Assistant Professor of Urban Planning at the Department of Planning and Regional Development of the University of Thessaly (UTh), Greece. His research interests lie in sustainable urban forms, urban decarbonization and climate resilience, energy and microclimate simulations and GIS. He has extensive experience in teaching, research and consultancy and is co-author of 'Sustainable (?) Urban Design' student handbook (in Greek). ORCID: 0000-0001-6177-1130.

Index

A

B

C

D

E

S

T

U

W

Z

www.ingramcontent.com/pod-product-compliance
Lightning Source LLC
LaVergne TN
LVHW081318300926
849161LV00008B/1805
* 9 7 9 8 3 3 7 3 2 3 2 7 5 *